Practical Cost-Saving Techniques for Housing Construction

Bart Jahn

McGraw-Hill, Inc.

New York San Francisco Washington, D.C. Auckland Bogotá
Caracas Lisbon London Madrid Mexico City Milan
Montreal New Delhi San Juan Singapore
Sydney Tokyo Toronto

Library of Congress Cataloging-in-Publication Data

Jahn, Bart.
Practical cost-saving techniques for housing construction / by Bart Jahn.
p. cm.
Includes index.
ISBN 0-07-005208-5
1. House construction—Cost control. I. Title.
TH4812.J35 1994
690'.8—dc20 94-30226
CIP

1 2 3 4 5 6 7 8 9 0 DOH/DOH 9 9 8 7 6 5 4

ISBN 0-07-005208-5

The editor of this book was Annette M. Testa and the production supervisor was Katherine G. Brown. This book was set in ITC Century Light. It was composed by TAB Books.

Printed and bound by R.R. Donnelly & Sons, Harrisonburg.

Contents

Acknowledgments

My sincerest thanks to the following friends, coworkers, tradespeople, contractors, and family who contributed topics, proofreading, criticism, patience, and encouragement toward the writing of this book: Paul Dettenmaier, Matt Ford, Jeff Ford, Robert Jahn, Brian Jahn, Barry Jahn, Michael King, Pete Lopez, John Montoya, Aaron Salvatico, Paul Shadle, Tim Snyder, John Vander Lans, and my mother, Barbara Shadle, for typing several earlier versions of the manuscript over the last 14 years.

Introduction

One of the problems affecting cost and quality control in housing construction today is that two similar projects going up side-by-side, built by different companies can each be making the same costly construction mistakes without either one benefiting from the other's experience. On project after project, people encounter many of the same construction problems and mistakes, totally unaware that these problems were solved on another project last week, last month, or last year. The unfortunate result is that hundreds of thousands of developers, contractors, superintendents, foremen, tradespeople, architects, engineers, inspectors, and do-it-yourself remodelers, all find themselves on the uphill slope of the learning curve, repeating many, if not most, of the same hard-earned lessons. This is an enormous waste of time, energy, and money.

Until houses can be built 10,000 or 20,000 at a time on mass-production assembly lines, the solution to this breakdown in communication is for the housing construction industry as a whole, and each construction company individually, to begin compiling checklists of what specifically goes wrong during construction. This would reduce the amount of debugging that must be done on each new project. The time, money, and aggravation that could be saved simply by knowing in advance the problems that went wrong on a project of similar design and construction could then be redirected toward improving quality and lowering costs.

This book contains more than 600 housing construction problems and mistakes that I've observed and recorded throughout a 20 year period while working as a carpenter, customer service repairman, superintendent, and project manager, and through interviews with tradespeople and contractors. These problems and mistakes actually occurred on real housing projects built by large, well-known development companies that were staffed with competent supervisors, contractors, and tradespeople. All of these problems resulted in either costly repairs, delays to the construction schedule, or unhappy homebuyers.

The value of this book lies in forewarning the reader of construction problems and mistakes *before* having to discover them the hard way. Because construction problems are difficult to foresee and anticipate, the many years I've invested in first-hand research observing and recording these problems cannot be duplicated without going through a similar experience. *Practical Cost-Saving Techniques for Housing Construction* can spare you from having to master that portion of the learning curve that covers the construction problems and mistakes discussed in the book.

The importance of avoiding numerous small mistakes during housing construction cannot be overemphasized. Although everyone working in building construction has heard of major construction mistakes, most projects get into trouble by the accumulation of small mistakes that grow and gain momentum. Most building construction projects in trouble don't die from one large gaping wound, but slowly bleed to death from hundreds of small cuts.

If houses were built in large numbers on a mass production assembly line, the problems and mistakes discussed in this book are among the minor assembly line-type bugs that would be identified and removed during construction of the first several houses. The objective on the assembly line would be to debug all the problems on the first few houses so that the remaining houses could be assembled smoothly and efficiently. The same idea holds true on real-life housing construction projects, except that you don't have the benefit or luxury of a trial run. By the time the last of the construction problems are discovered, the project is complete and we move on to a new location with a new product design, new contractors, new tradespeople, and a whole new set of problems to solve.

Making housing construction mistakes is a lot like making mistakes in sports. Basketball players practice lay-ups and free throws, baseball infielders practice double plays, and football running backs practice holding on to the football because all these skills are important to winning the game. The advantage in a close, evenly matched ball game often goes to the team that makes the fewest errors.

When housing construction projects run behind schedule and over budget, the people in charge can often look back and point to a series of small problems and mistakes as the cause. Like missing easy lay-ups and free throws early in the fourth quarter of a close basketball game, walking the first batter in the ninth inning of a tie baseball game, or dropping a third-down pass during the crucial stretch drive in a football game, the losing team always looks back and wonders what might have been.

This book may well be the most cost-effective book you will ever buy on housing construction. It takes only a few of the problems and mistakes covered in this book to add up to a cost-savings equal to or greater than the cost of this book. In fact, considering the time it takes to look at a problem, go and find the contractor or tradesperson on the job site, discuss the problem, check the repair after it is made, and approve the billing invoice, the cost of avoiding even one of these mistakes will probably exceed the price of this book.

Whether you are building a $200,000 single-family house or a $20 million tract housing project, the cost of this book is cheap money compared to the potential cost of construction problems. Like the sports scenarios, your chances of winning are increased when you can minimize mistakes and errors.

One of the reasons a book of this type has not been written until now is the difficulty in researching the subject. If you want to discover building construction problems, you literally have to put yourself in the middle of the action. There is a limit, therefore, to how much one person can personally observe and record and how much can be gained through interviews with contractors and tradespeople. If you would like to contribute to the next volume, please send examples of housing construction problems and mistakes you've observed to:

5595 East 7th Street, #351
Long Beach, CA 90804

Include with your comments and contributions your name, address, and telephone number. If I can include these housing construction problems and mistakes in future editions, I will be happy to include your name in the acknowledgments.

Chapter

1

Masonry Block

Working Space behind Block Walls

Ample space between a block retaining wall and a dirt berm is necessary for maneuverability and the safety of the workers. Figure 1.1 shows a block retaining wall and an adjacent dirt berm. In this example, a group of condominium buildings were built side-by-side in the high-density project with narrow corridors of landscaped walkways between each building. The block retaining walls separated the ground level garages from the landscaped walkways, one floor level above. It was decided to rough grade the entire project leaving the dirt berms for the landscaped walkways already in place because there is no practical way of getting the quantity of dirt between the buildings after the retaining walls are built.

To ensure that ample space is left for work purposes, check the proximity of the soon-to-be-built block wall at the time surveyor's stakes are placed. If the rough grading of the dirt berm is slightly off, the vertical slope of the berm will be too close to the block wall. The time to find this out is before the block footing trenches are dug. In areas where the dirt berm is too fat, the trenching backhoe can scrape away the sides of the berm and remove the dirt before work is started. As the building of the block wall progresses, the discovery and solution of inadequate space becomes increasingly more difficult. For example, it becomes harder to remove the fat part of the berm without damaging the block wall. Furthermore, when shaving off the fat part, a mess is created behind the wall that must be cleaned up before drainage pipes can be installed and waterproofing applied.

The builder does not want to be placed in the position of having to decide whether to shave the side of the dirt berm after the retaining wall is already up or to ask people to illegally work within a narrow space between a block wall and an unstable ver-

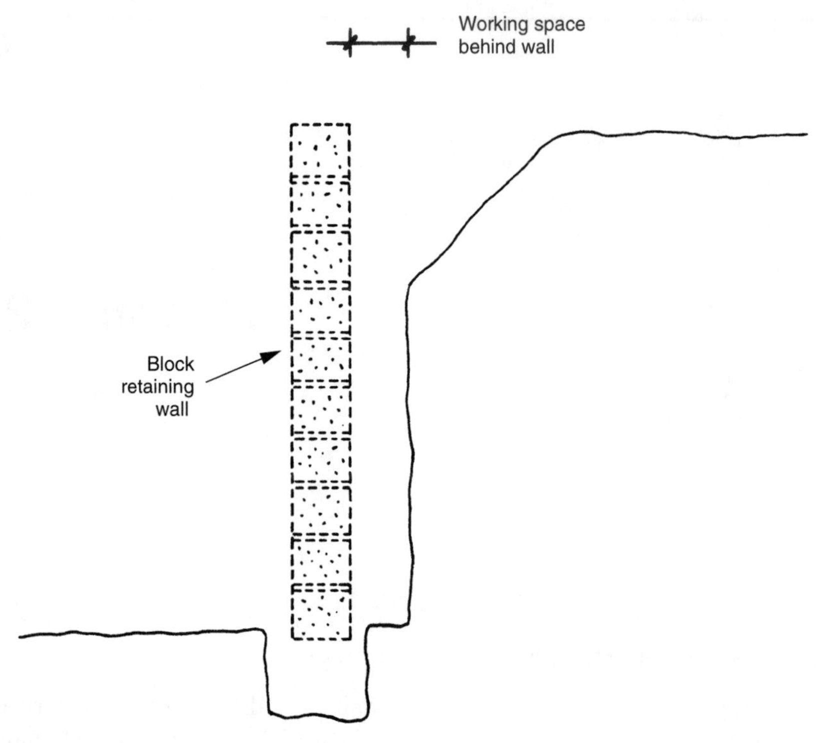

Figure 1.1

tical earth slope. The ample-space scenario can also occur at the block retaining wall that separates the lower half of a split-level house from the upper level.

Angle Split-Level House Slope Cuts

For safety reasons, the top section of the slope adjacent a retaining wall for a split-level house should be cut back at a 45-degree angle (Figure 1.2). The unsupported height of a vertical slope next to work areas should be no higher than about 4 feet. If the 8-foot slope is left as is, the earth could cave in on the workers as they work at the base of the wall setting in drainpipes, placing gravel backfill, or applying waterproofing to the block.

The builder should include the angling back of vertical slopes in the rough grading contract. The angling cuts can be made by the tractor during rough grading by setting the tractor blades at a 45-degree angle and scraping off the tops of the vertical slopes. If vertical slopes are not cut back during the rough grading, the problem becomes much more difficult to correct. For example, any scraping back of the vertical slope at the top after the wall is in place will result in fallen debris at the base of the wall, and shoring makes it more difficult to move and work within the space between the wall and the slope. The only other alternative to cutting off the top of the slope is to ask people to work in an unsafe condition—which is not a good alternative.

Block Footings and Building Footings

The joint between a retaining wall footing and the building footing is pictured in Figure 1.3. This block retaining wall footing is 3-feet-6-inches deep and encroaches on the building footing at one location where the block indents 6 inches. The cross-section of the building footing had to be shortened by 6 inches, thus reducing the cross-section of the reinforcing bar cage in the footing. This in turn made it more difficult to get plumbing pipes through the footing.

To alleviate the 6-inch discrepancy in future buildings, the block retaining wall footing was formed 6 inches less in depth at this location so as to allow for the full-size width of the building footing.

Block Footing and the Grade Beam

Figure 1.4 illustrates a conflict between a block retaining wall footing at the side of a three-story condominium building and a grade beam at the front of a garage concrete slab.

In this example, the block wall footing extends into the building 3½ feet, but was correctly designed to be below the elevation of the building concrete slab. However, the grade beam, at the front of the garage, being the same depth as a foundation footing was at the same elevation as the block retaining wall footing. On the next building, the block wall footing was formed off and held back at the front of the garage to allow ample space for the garage grade beam, thus alleviating the conflict.

Plumbing Allowance Near the Block Footing

In order to provide clearance for the plumbing pipes for a washing machine in the garage of this three-story condominium, the block wall footing (Figure 1.5) needs to be lowered one full course of block or 8 inches.

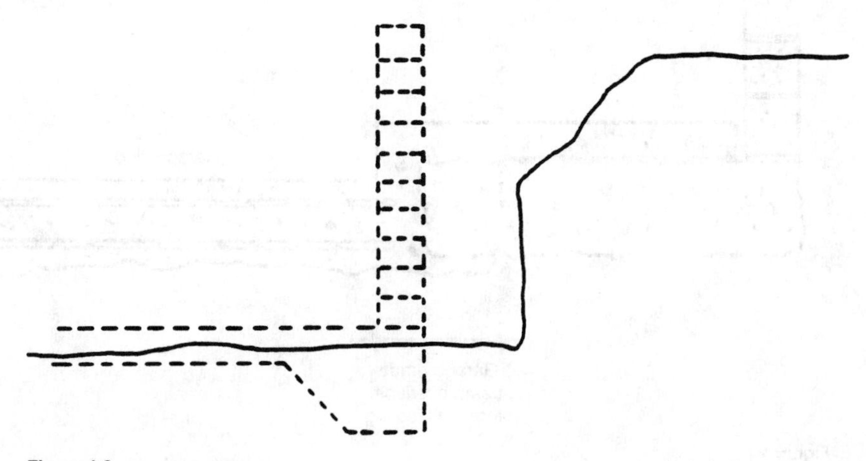

Figure 1.2

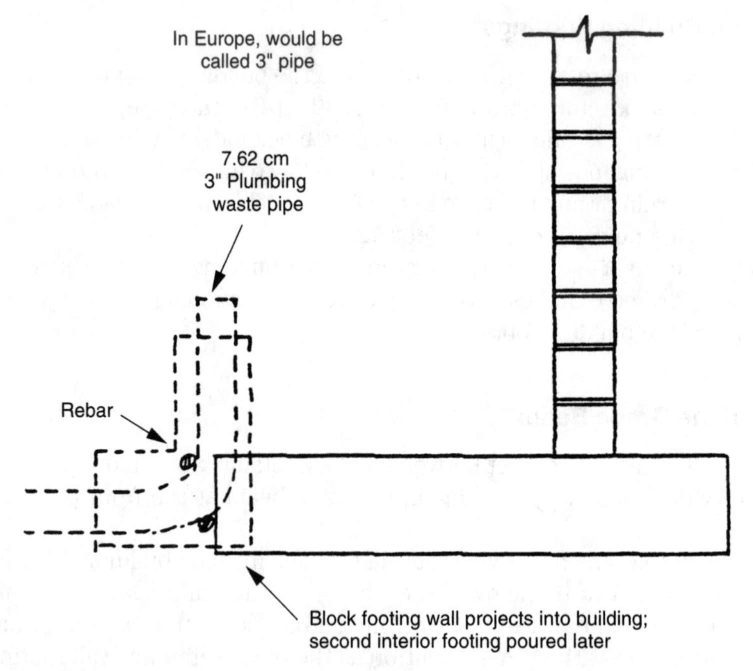

Figure 1.3

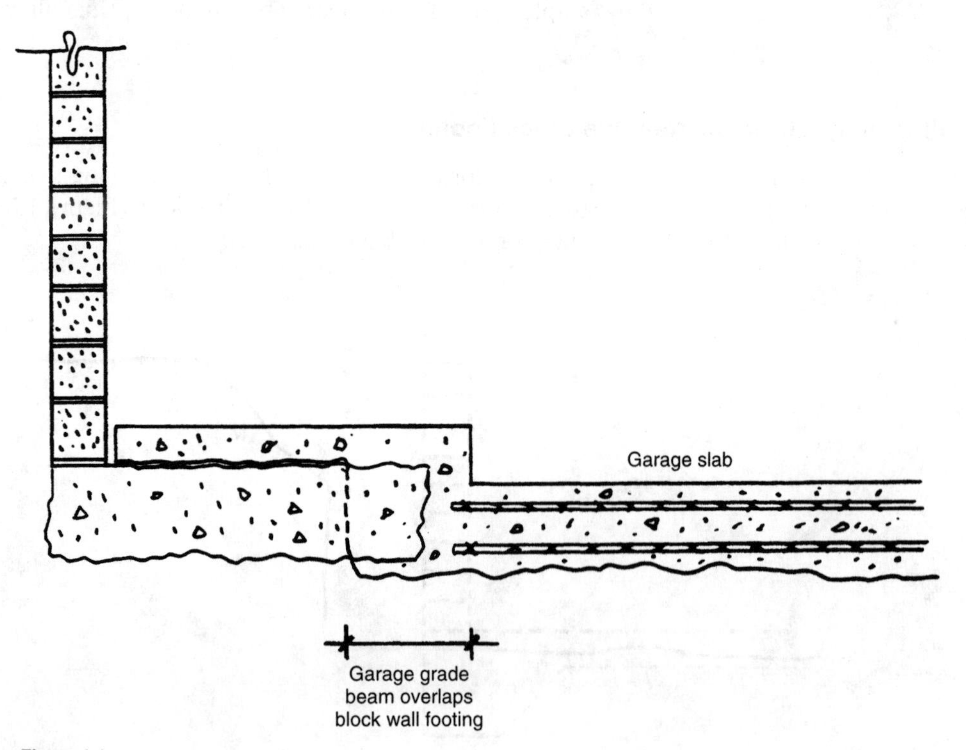

Figure 1.4

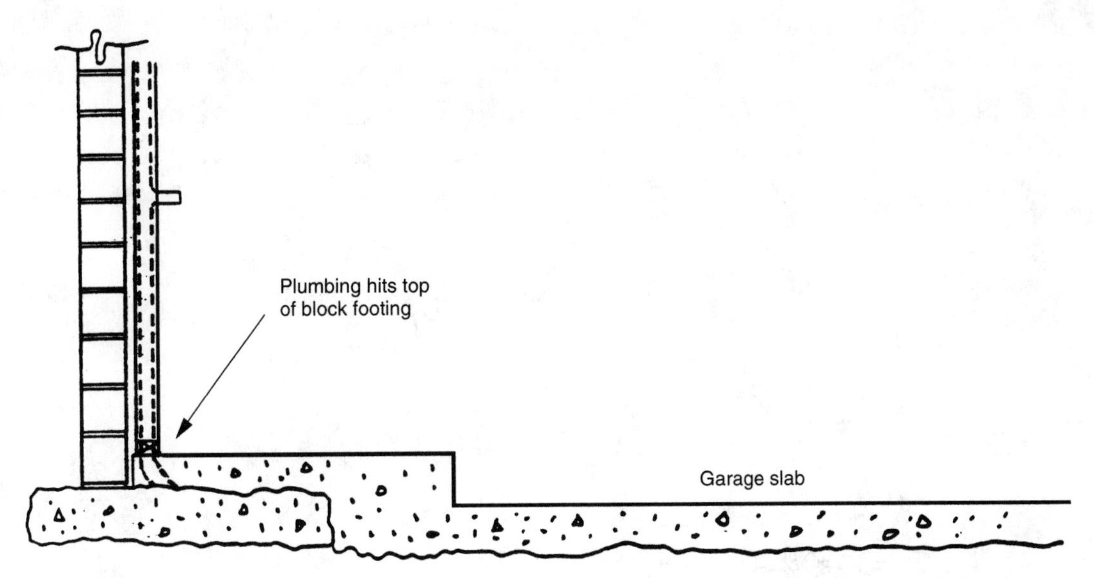

Figure 1.5

In this example, it was discovered that the top of the block wall footing was too high for the plumbing pipes to rest in a downward-sloping position, miss the concrete step, and still connect up with the house sewer that was running out the garage floor. By lowering the block footing 8 inches, the plumber had enough vertical allowance to place the pipes at the correct elevation to slope downward into the house sewer pipe without interfering with the block footing.

Ample Root Space at Block Footings

A decorative planting area can often be formed next to a block retaining wall by adding a shorter block wall a few feet from the retaining wall (Figure 1.6). The footing for the retaining wall should be turned outward, away from the planter area, to allow the root structures of the trees or large shrubs to expand. If the footing is mistakenly turned inward toward the planter, there may not be enough soil depth in the planter box for large landscaping members.

Because the civil engineer designing the block retaining wall does not always think about landscaping, the orientation of the block wall footing is something the builder should check.

Block Footing and Conduit

A block wall and footing for an apartment/condominium building electrical meter room is illustrated in Figure 1.7. To avoid spacing problems with the electrical components, the block wall footing should be low enough to provide clearance for the plastic sleeve pipe to enter and sweep up into the meter room and the sleeve pipe should be placed inside the block footing before it is poured.

Figure 1.6

The easiest method of ensuring that ample space is allotted is to lower the block footing at the meter room 24 inches—a space equal to three courses of block. The sleeve conduit can then be placed on top of the footing, through the wet-set bottom course block, and up the block wall, in accurate alignment, to the meter box location.

Lower Block Wall Footings at Gas Meters

The gas meter area for the condominium/apartment building that is pictured in Figure 1.8 is enclosed with a three-foot high block garden wall. To ensure adequate space for underground gas piping to the meters, the block wall footings on both walls should be formed off. If the ends of the block wall footings are not formed off, the concrete is allowed to run to the edges of the excavated footings and there may not be enough space left over between the footings for the number of gas pipes coming through this opening.

One simple method to avoid these types of conflicts is to simply lower the block footing at this area a few courses. Three extra courses of block at the opening to the gas meter area gives you a top of footing that is now 24 inches lower than the previous top of footing elevation. This means that the plumber can now run the gas pipes on top of the block wall footing and still be at a level below grade depth without any interference from block wall footing concrete

Why would the block masonry contractor be willing to lower the footing and add a few courses of block in this area? Because it is easier and cheaper to lower the footing and run a few more courses of block to avoid a problem up front than it is to come

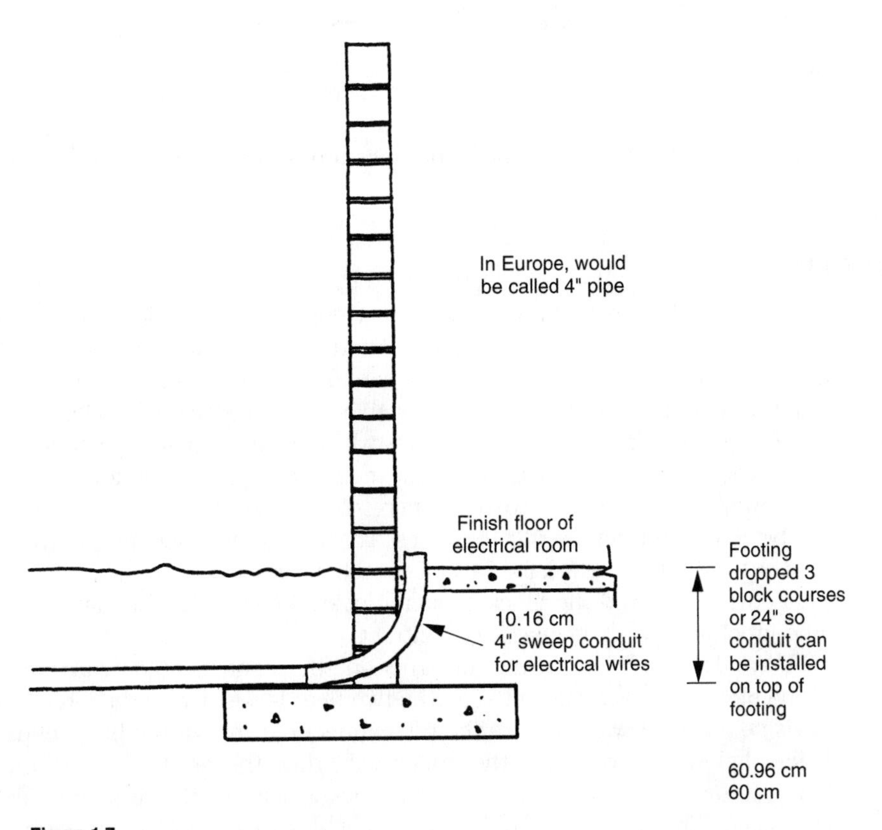

Figure 1.7

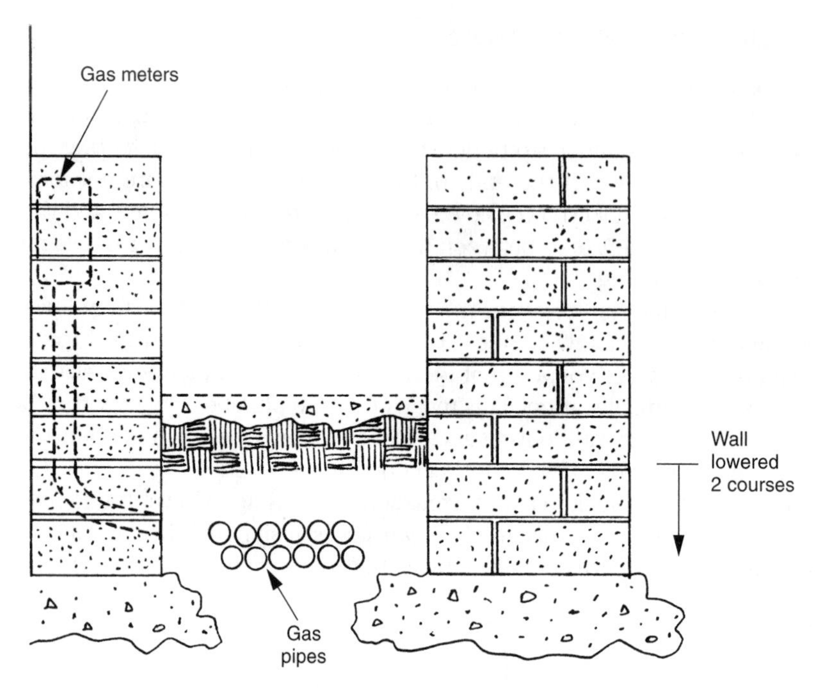

Figure 1.8

back later and jack-hammer out footing concrete that is in the way of another subcontractor's work.

Split-Level Wall Height

Illustrated in Figure 1.9 is the top of a masonry block retaining wall that separates the lower and upper floors of a split-level house. In this example, the lower floor ceiling height was 9 feet, the ceiling joists were 9½-inch truss joists, and the upper floor slab was a 15-inch thick concrete mat slab over expansive adobe soil.

Because the block wall height did not work out to an even number of blocks with the above conditions, the structural engineer had to approve extending the top of the block wall by 1½ inches into the concrete, leaving 13½ inches of concrete above the block rather than the full 15 inches. This saved the block mason from having to cut the top course of block for 14 split-level houses, each approximately 43 feet long. This obviously saved the block mason a lot of time and the builder a considerable amount of money on the masonry contract.

To avoid this problem during construction, the builder should figure out the height of the block retaining wall between the two floor levels before the job is started and devise the easiest way to make the top connection work. If the block ends up a few inches below the bottom of the upper slab, then thicken the concrete above the block. If the number of courses puts the block wall inside the concrete slab, depending upon its thickness, see if the engineer can design a detail for projecting the block into the concrete.

Accelerate Schedule for Special Inspector

When the plans call for a special inspection of masonry block retaining walls for split-level houses or earth slopes, the builder should try to accelerate the masonry block work. A special inspection is a continuous, on-site inspection of the block work by a certified inspector, usually hired through a testing laboratory. Because these inspectors are not cheap, the planning and coordinating of the work should be done prior to the day of the inspection. Similarly, the block work should be accelerated in an effort to reduce the length of time the special inspector is required on the job site. It makes little sense to have a high-priced special inspector watching one or two block masons and a few mason's helpers lay block, for weeks or months when the work could be speeded up by using more workers. This is something the builder must think of and work out ahead of time with the masonry block contractor.

Grout Block in Morning

When masonry block construction requires special inspection by a licensed inspector, the grouting of the block, whenever possible, should be scheduled for mornings. Because the inspector is required to be on-site throughout the entire grouting process, if the job is scheduled to begin in the late afternoon and it runs late, the inspector will have to be paid time-and-one-half for overtime. By scheduling grouting in the morning, the process will seldom run past quitting time and the final hours of the work day can be spent laying block—a job that can be stopped at any time. Once an afternoon grouting begins, however, it must continue until complete, even if it runs into an overtime expense for the builder.

Block Anchor Bolts and Beams

Figure 1.10 shows the top of a block retaining wall that is holding up a mudsill plate with floor joists and structural floor beams resting on it. The anchor bolts securing the mudsill to the top of the block wall should be laid-out with consideration for the

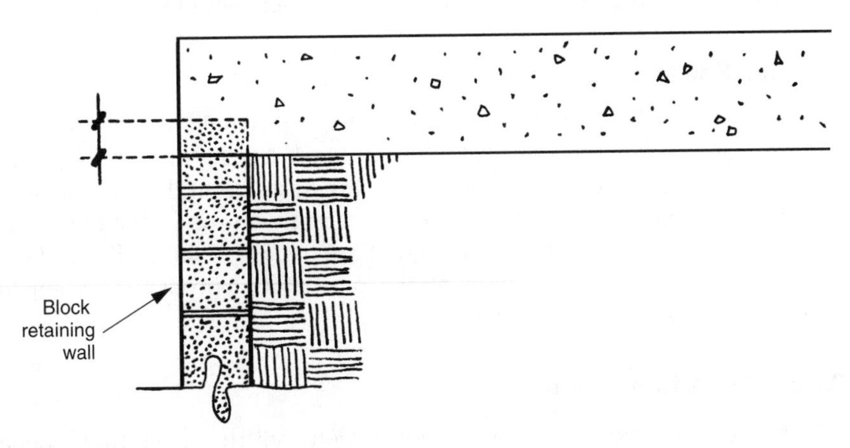

Figure 1.9

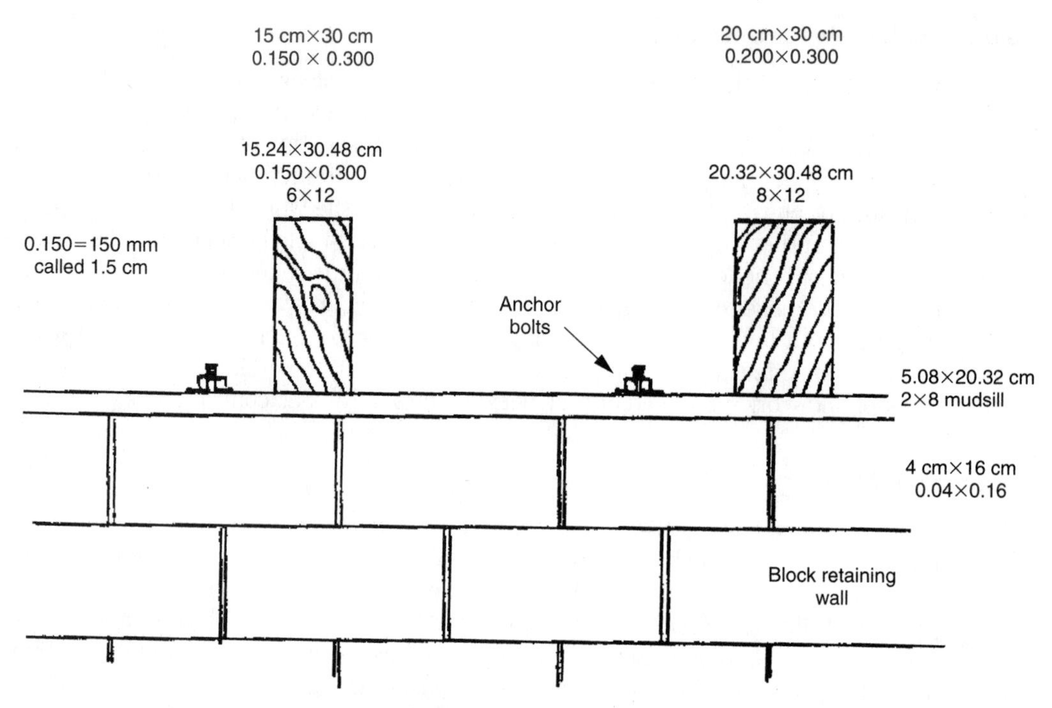

Figure 1.10

location of the structural floor beams. Floor joists can be easily notched out on the bottom to fit over anchor bolts, but it is a nuisance to have to notch out the bottoom of larger sized beams. A better alternative is to set the anchor bolts away from these beams during the grouting of the top course of block.

The builder should meet with the framer and the block mason prior to the grouting of the top lift of block to determine the layout of the structural floor beams. These locations can then be spray-painted or otherwise marked out on the top of the block wall so that anchor bolts are not placed within the grout cells of the block at these areas.

Block Wall Grout Spillage

When block retaining walls have back drains, care should be taken to reduce excess grout spillage over the wall and onto the block footing during the grouting of the block cells (Figure 1.11). Not only does this waste grout, but the slope of the back drain sometimes requires it to be placed directly on top of the block footing at its lowest point. When 3 to 6 inches of grout are deposited on the backside footing due to carelessness, the placement of the pipe at the proper elevation for correct drainage becomes more difficult.

Anchor Garden Walls to Building

Figure 1.12 shows a typical stuccoed garden wall that joins to a standard-height stuccoed exterior building wall. The three-foot garden wall in this example retains a

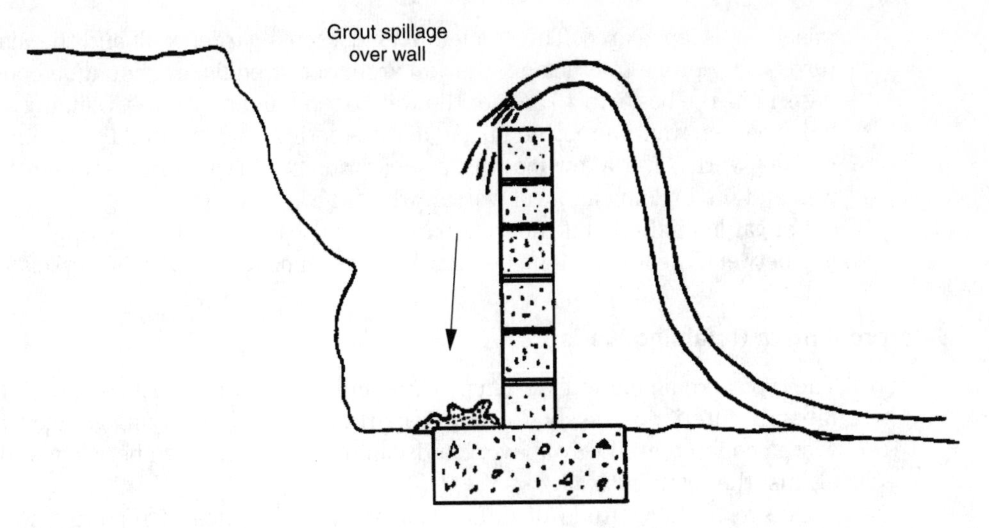

Figure 1.11

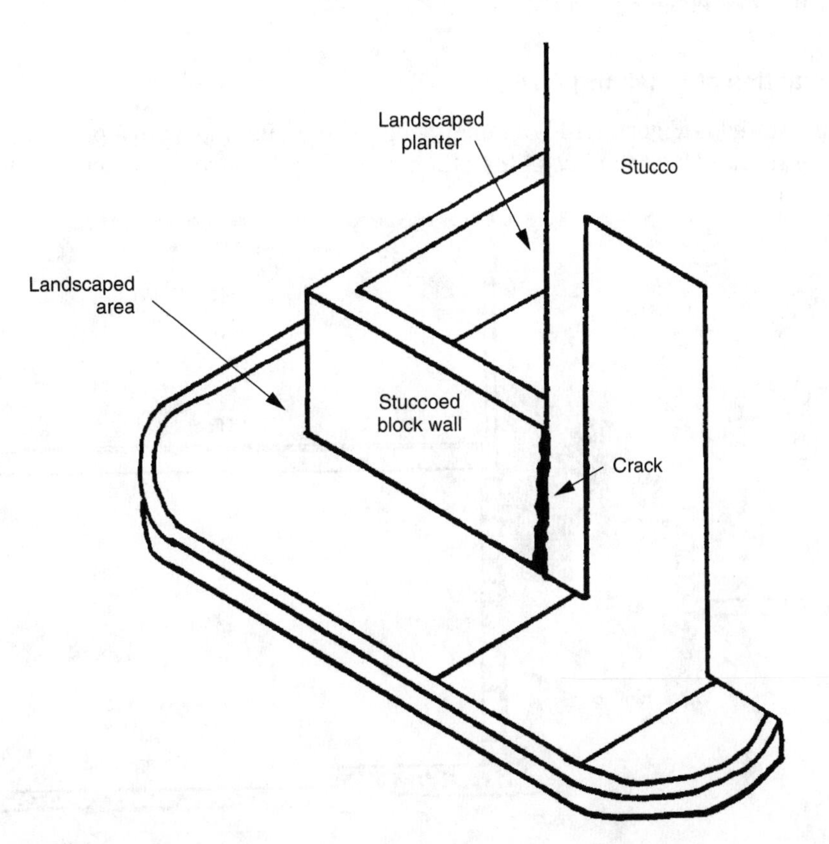

Figure 1.12

raised landscaped area. The connection between a garden wall and the main structure is very prone to cracking, thus an architect or engineer should be consulted to determine the best way to anchor the wall to the building. Merely building a masonry block garden wall, supported on a concrete footing, and grouted to the building wall will not work. After a few months, a separation crack will form between the garden wall and the building and will worsen with time.

For garden walls that do not retain raised landscaped areas, by leaving a 1- or 2-inch gap between the wall and the building you will eliminate the risk of a crack forming.

Waterproofing at Retaining Walls

The typical retaining wall design pictured in Figure 1.13 consists of a wall that separates the interior of the condominium/apartment building parking garage from an elevated concrete walkway above. Landscaping is often planted between and directly against the retaining wall.

Due to the magnitude of difficulty and inconvenience of repairing water leaks through a block wall in a garage or through the landscaping from the outside of the wall, the builder should spend time analyzing the design and monitoring the waterproofing application at the block wall. This is one area where an ounce of prevention really does equal a pound of cure.

Check Compaction at Retaining Walls

Illustrated in Figure 1.14 is a building block retaining wall with a parking garage on one side and landscaping on the other side. With this layout, the builder should pay special attention to the quality and compaction of the elevated fill side of the block

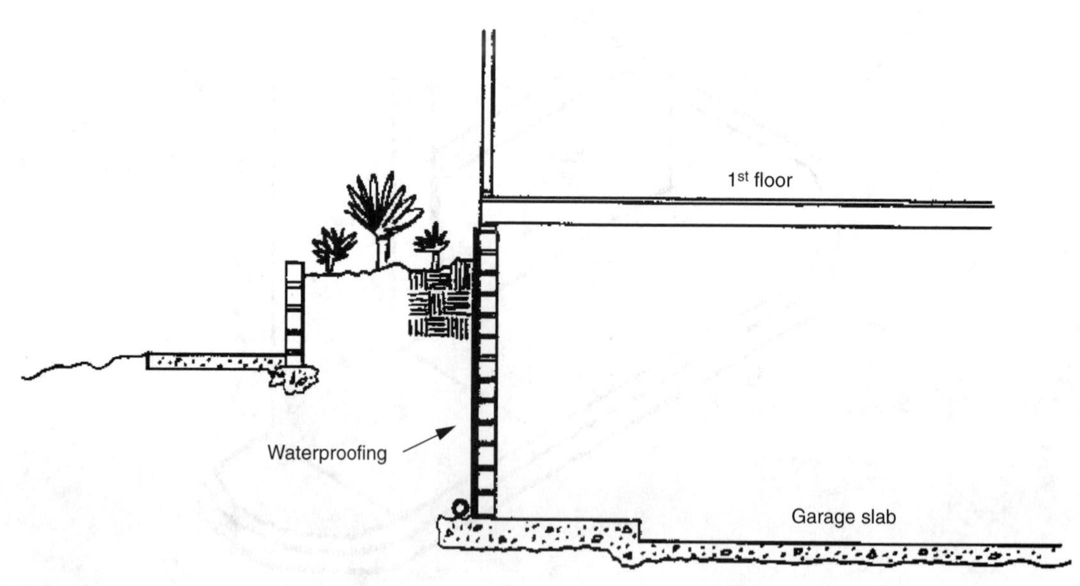

Figure 1.13

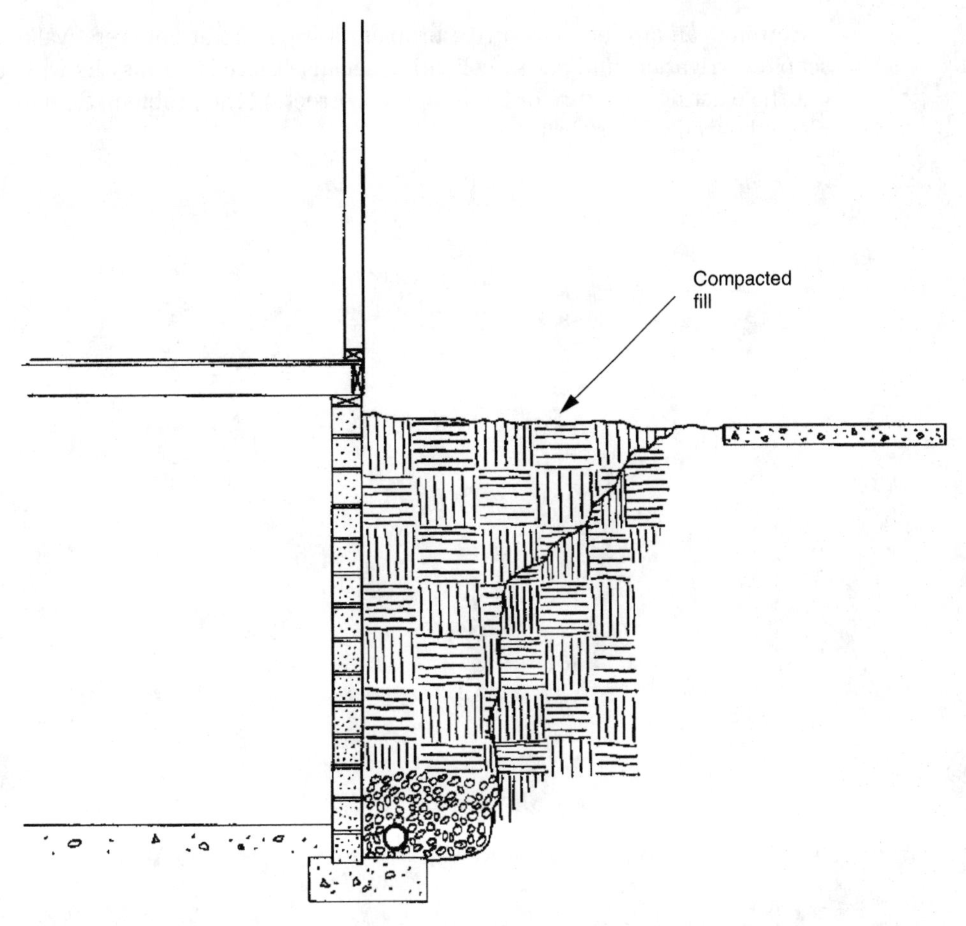

Figure 1.14

retaining wall. Typical for this situation is a drainpipe at the base of the wall, covered by gravel, then dirt placed and compacted in 1- or 2-foot lifts. The gravel placement and the dirt backfill is sometimes required to be inspected by a soils engineer.

Soil tech inspections and tests are crucial. For example, the dirt backfill around a block retaining wall generally settles at the edge of the block wall creating a downward sloping grade toward the building. To further compound this problem, the water intended for the landscaping then runs toward the building instead of draining away from it as designed. This water forms puddles at the building, runs down through the backfill at the wall, creates further settling at the wall, and finds any waterproofing problems, resulting in leaks. Another common problem occurs at the air-conditioning condensers, which are placed on concrete or plastic pads around the exterior perimeter of a condominium or apartment building. As the dirt backfill settles, the condensers can tilt out of level toward the building causing problems.

Settling can also become a real problem if it occurs at concrete walkways, landscaping, irrigation, and pathway lighting because there is no easy fix without tearing out the existing structure or landscaping, correcting the problem, then having to replace the walkway or shrubs.

Chapter 2

Concrete

Don't Stake on Friday

Whenever possible, the surveyor's wood staking of property lines and grade elevations for the layout, trenching, and forming of building foundations should be scheduled for a day at the beginning or middle of the week—not on Friday. Three-foot-high wood stakes with pointed ends and colored ribbons tied to their tops often attract the attention of neighborhood kids over the weekend. The stakes become swords, spears, javelins, and rockets, and by Monday morning much of the expensive surveying work must be redone. If surveyed and staked on Monday or Tuesday, at least the layout and trenching work starts the following day—with night time darkness in between.

Staking for Split-Level Houses

For split-level houses, the builder should have the surveyor setting the property line and grade stakes for the layout, forming, and trenching of building footings as if each level were a separate house (Figure 2.1). For example, stakes should be placed at four corners of the lower level pad, and at four corners of the upper level pad, as a minimum (Figure 2.2). This simplifies the layout process by having complete measuring points at each split-level, without having to pull a string-line or sight from upper pad stakes to lower pad stakes to get a building line.

Clearance for Trenching

When a house is placed within 4 or 5 feet of an adjacent earth slope (Figure 2.3), the builder should be sure there is adequate clearance for trenching equipment before concrete footings are dug.

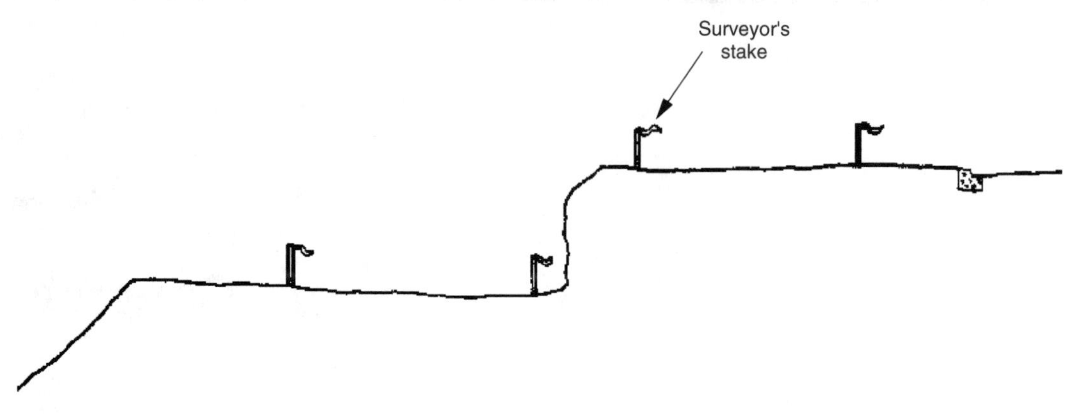

Figure 2.1

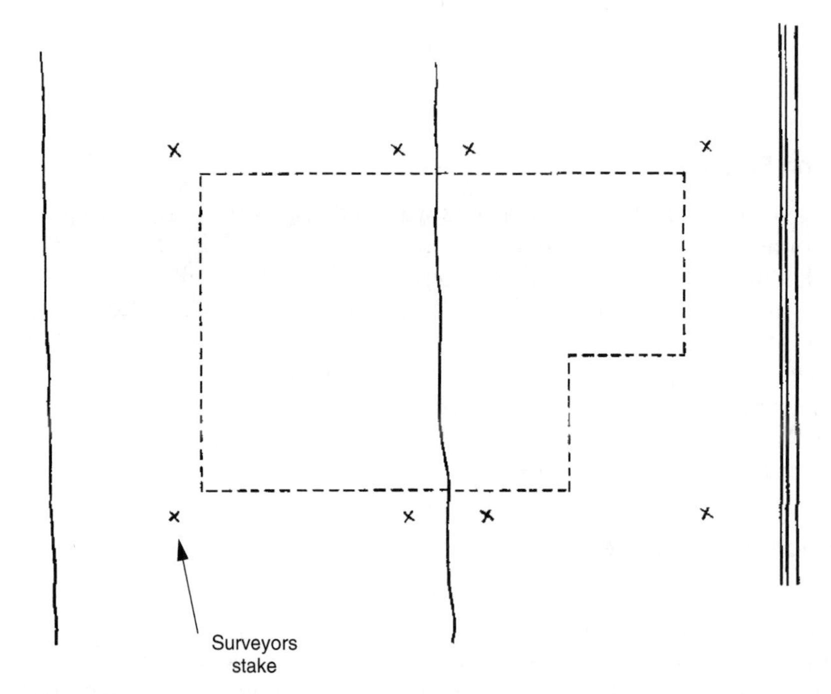

Figure 2.2

In this example, the concrete contractor wanted to dig the footings with a wheel trencher, which can dig footings faster than a backhoe but requires more working space. Because of the space constraints, however, the footings adjacent to the slope had to be specially trenched with a backhoe, while the remainder of the trenches could be dug with the wheel trencher.

Because these slopes were merely temporary until small retaining garden walls were built later in the construction to retain the differences in pad elevations, they could have been cut at a steeper vertical during the rough grading to provide more working space for trench digging.

Get Plumbing Trench Layout

When foundation footing and under slab plumbing trenching are done at the same time and the cost is shared between the concrete and plumbing contractors, the plumbing trench layout should be given to the builder before trenching.

Because it is usually not necessary for the plumber to be at the job site for the entire time during trenching, the supervision of the plumbing trench locations and depths is the responsibility of the builder, to some degree. Detailed sketches of the plumbing trench locations should be drawn by the plumber, with copies given to the builder and the trencher. This at least gives the builder the chance to check the trench layout and prevent possible mistakes such as the reversal of floor plans or layout errors due to the absence of the plumber on the job site.

Close Off Ends of Footing Formboards

Illustrated in Figure 2.4 is the edge of a concrete slab as it meets the footing for a sideyard block retaining wall. In this example, rows of split-level houses shared common adjoining side-yard retaining walls. The problem that occurred in this project was concrete boil-out from the slab footings interfering with the placement of the rebar in the block wall footing. The solution to the problem was to block off the sides of the concrete slab formboards at the footing elevation to prevent concrete from flowing over into the future block wall footing area.

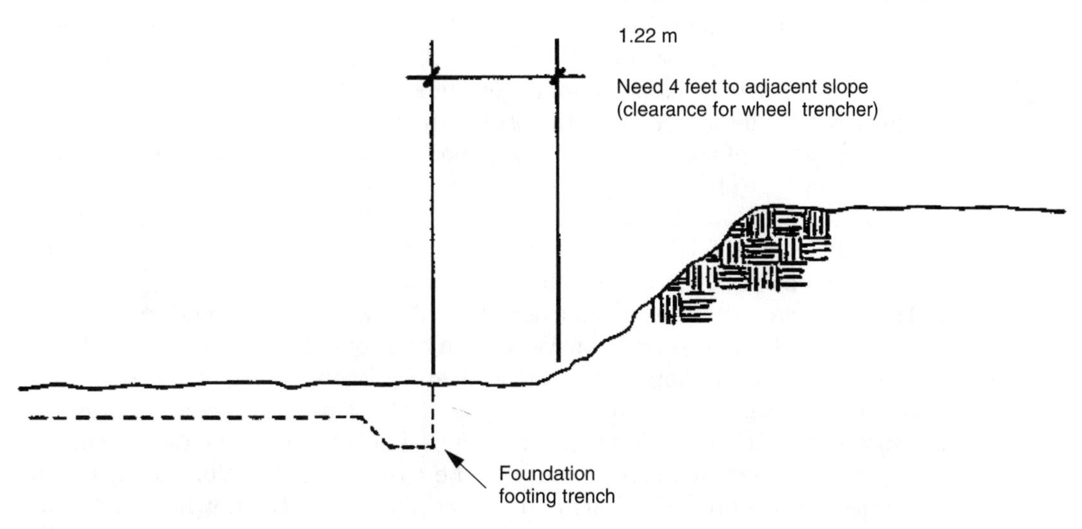

Figure 2.3

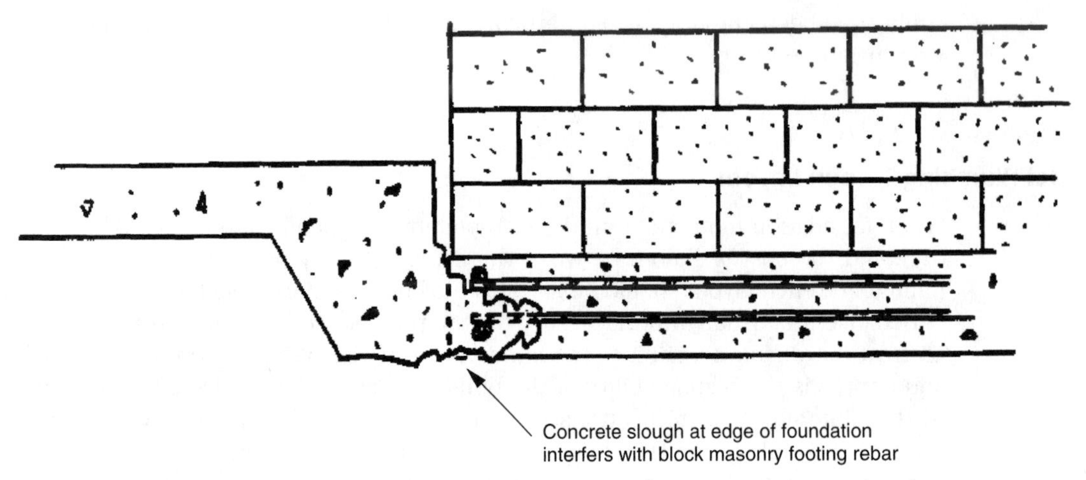

Figure 2.4

Design Uniform Structural Members

It simplifies the construction if structural members are designed to be the same or vary between two or three different types at most. For example, instead of having four different sizes of post hold-downs, four different nailing schedules for shear panel, two thicknesses of shear panel plywood, and four or five different foundation anchor bolt spacings, it helps simplify materials management and reduce mistakes if the options are fewer.

For instance, one brand name has hold-down product number types varying from 2, 5, 7, and 20, to 2a, 5a, 7a, and 20a, all requiring different bolt sizes, bolt heights above the concrete, and dimensions from the center of each bolt to a post. Instead of specifying four or five different varieties of hold-downs, the structural engineer should pick two medium sizes only and use them throughout the design. Instead of specifying anchor bolt spacings of 8, 16, 24, 32, 48, and 72 inches on center spread throughout the structure, the structural engineer should try to use 32 inches on center for most of the structure, with some 16-inch and 72-inch on center spacings mixed in. Instead of specifying ⅜-inch plywood shear panel one place and ½-inch or ⅝-inch somewhere else, the engineer should try to use ⅜-inch exclusively with as few areas of ½-inch as possible. Instead of using edge nail spacing on shear panel plywood of 2, 3, 4, and 6 inches on center, the engineer should try to design all of the spacings at 4 inches only throughout the structure.

There is a variety of products available that have been designed to meet specific design criteria. However, I have worked on projects where it seemed the engineer was using the houses for a test of every known structural member and shear panel design in the catalog. Instead of choosing each piece of structural hardware to match the minimum size required according to the design calculations, the engineer should also consider the ease of construction. By upgrading the design in one area of the structure to accommodate the downgrading of design in another area, the variety of members used can be standardized to a smaller number, greatly simplifying the ordering, handling, placement, and inspection of structural members.

Offset Grade Beam Rebar

Illustrated in Figure 2.5 is the top corner reinforcing bar for a grade beam along the perimeter of a foundation concrete footing. Sometimes the structural engineer inadvertently details this grade beam to be positioned right in the way of the anchor and hold-down bolts at the exterior edge of the concrete. In this instance, the builder should try to get the engineer to approve offsetting this one top bar 2 or 3 inches to provide room for bolt placement.

Allow Space for Pipes and Rebar

A typical problem in condominiums is the 12-inch-wide stem wall garage footing that is shared by the adjoining units (Figure 2.6). In this wall, two, 3-inch plumbing waste pipes service back-to-back plumbing fixtures on second floors above, leaving only 6 inches of remaining space in the concrete footing for the reinforcing steel bars.

The problem is that the structural detail shows the rebar at the outside edges of the footing next to the formboards and not in the center 6 inches of the footing.

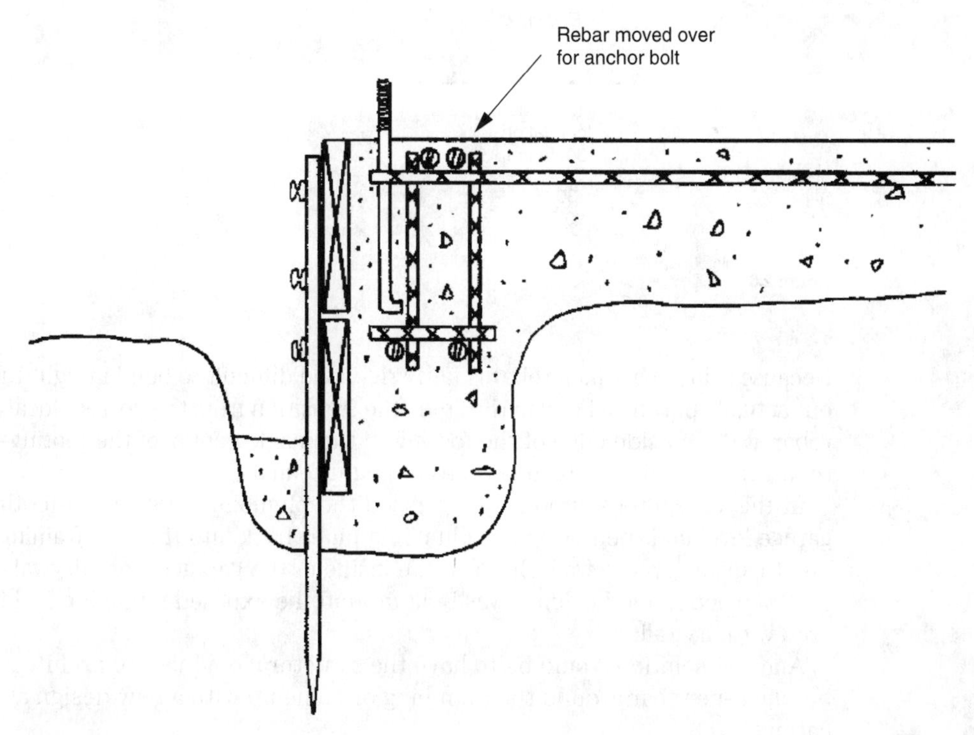

Figure 2.5

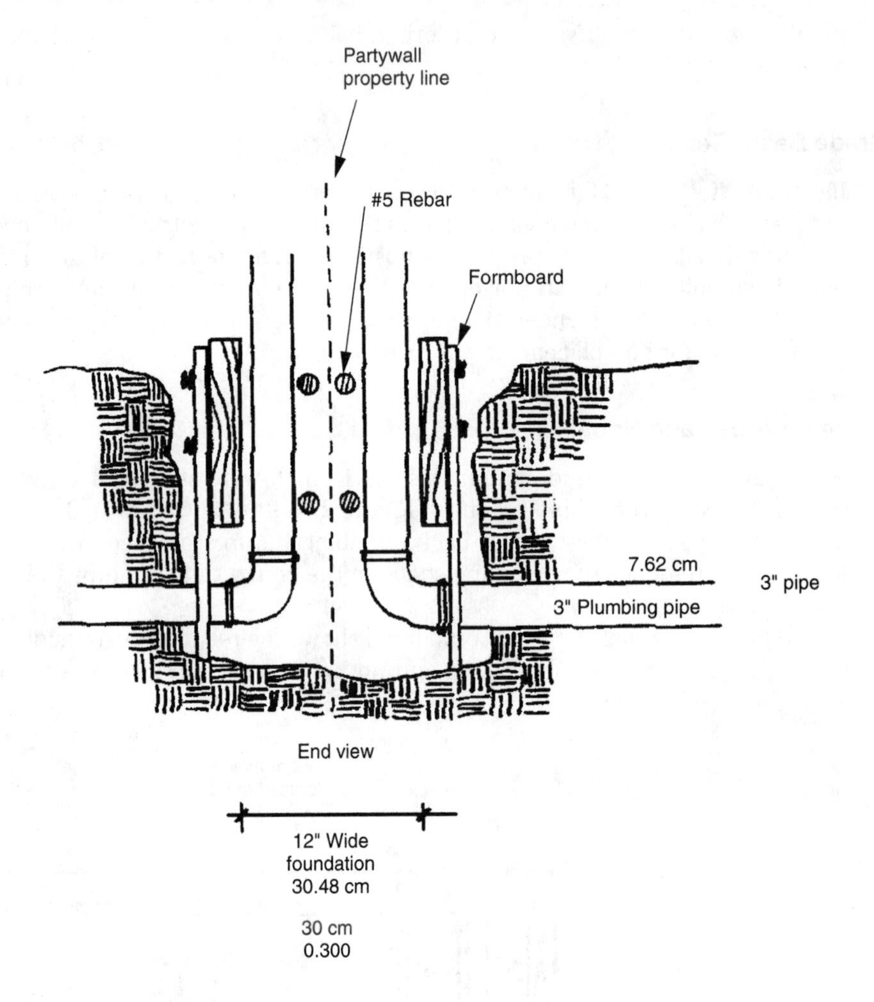

Figure 2.6

Because ½-inch or ⅝-inch rebar is fairly rigid and difficult to bend in tight turns without actually putting a bend in the steel, the transition from the correct location of the rebar at the outside edge of the footing to the inside center of the footing—needed to miss the plumbing—is sometimes too tight to make.

In this case, one solution was to place the plumbing outside the footing at the garage level and then bring the plumbing pipes back into the wall framing at some point halfway up the wall. Because the inside of the garages were drywalled in this case, a wood framed column was built around the exposed plumbing and then covered with drywall.

Another solution would be to have the structural engineer or architect approve bending the rebar around the plumbing or come up with a new design at these locations.

Anchor Bolt Layout

Structural posts and wall framing stud layout should be considered when setting foundation anchor bolts. For example, garage front walls are sometimes plywood shear paneled with two 4×4s or larger posts and one or two studs between them as shown in Figure 2.7. The anchor bolts should be laid-out and set during the concrete pour so they do not interfere with installation of the posts or studs during the framing.

Because the framer is not on the job site during the start of the concrete work, the builder must represent the framer in getting the anchor bolt placement correct. Allowing the concrete finishers to throw in wet-set anchor bolts anywhere during the pouring of the concrete usually results in conflicts between concrete hardware and framing members.

Hold-Down Bolt Heads

Illustrated in Figure 2.8 is a shear panel wall section. The end posts and the hold-down hardware of the wall were laid-out in the concrete so that the projecting bolt heads, nuts, and threaded ends would remain within the wall after it was framed. This planning and orientation of the hardware eliminates having to fur out the wall or countersink the bolt heads (see Figures. 2.9 and 2.10). If not done, the bolt heads would project past the wood posts.

Depth for Steel Base Plate Bolts

Figure 2.11 shows the side view of a concrete slab poured on top of a wide footing for a block retaining wall. In this example, the block footing extends into the interior of the building 7 feet and interferes with the anchor bolts for the base plate of a struc-

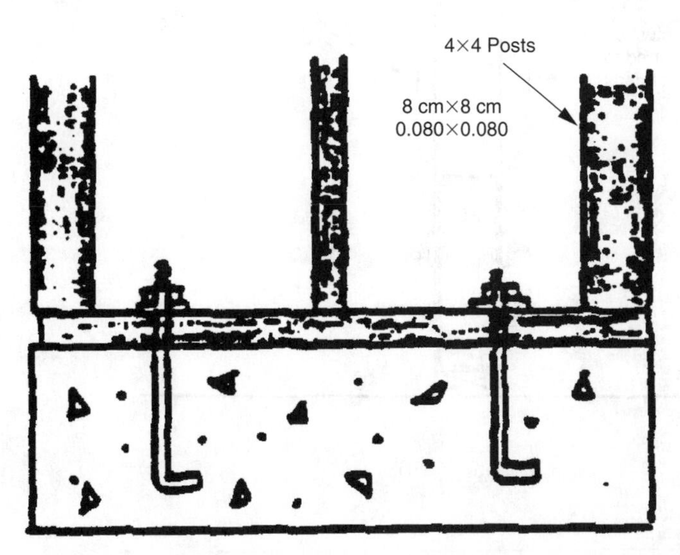

Figure 2.7

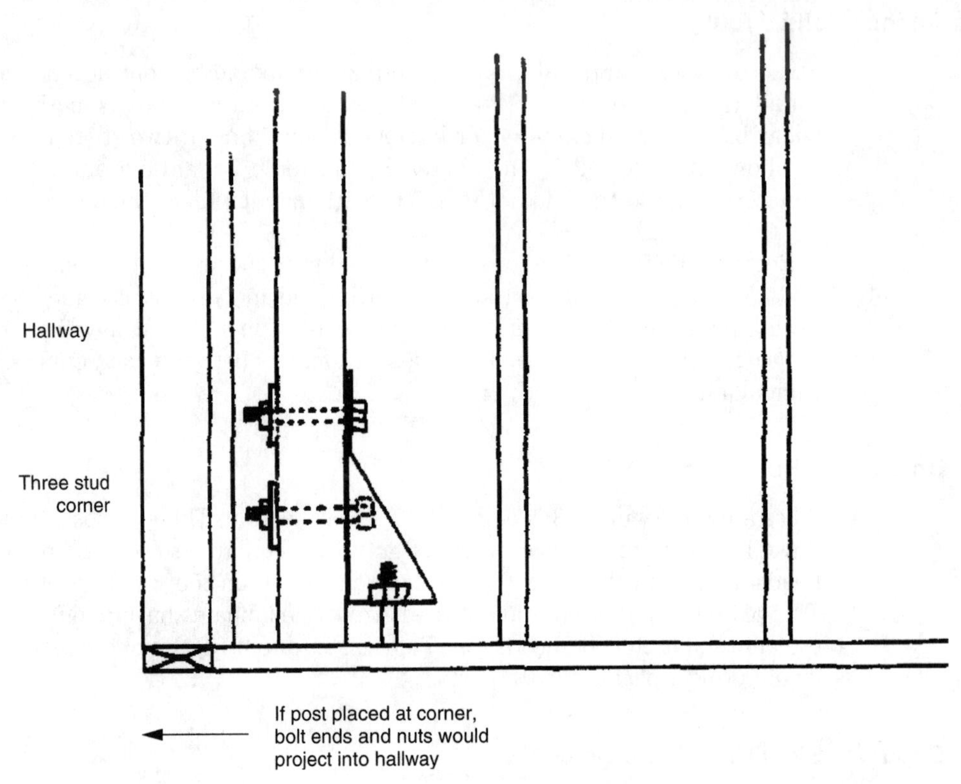

Figure 2.8

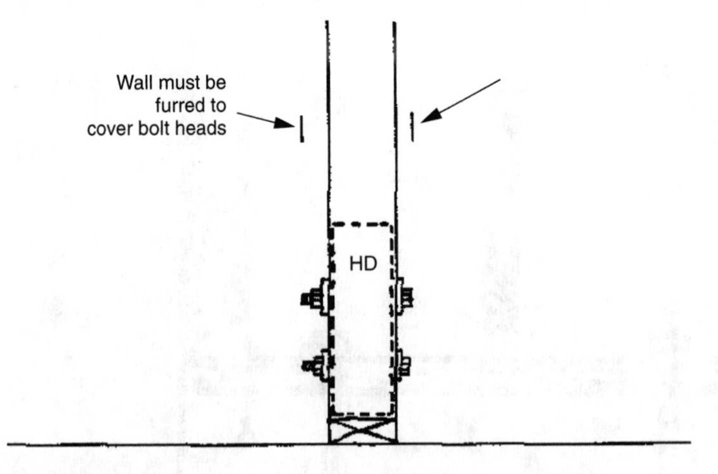

Figure 2.9

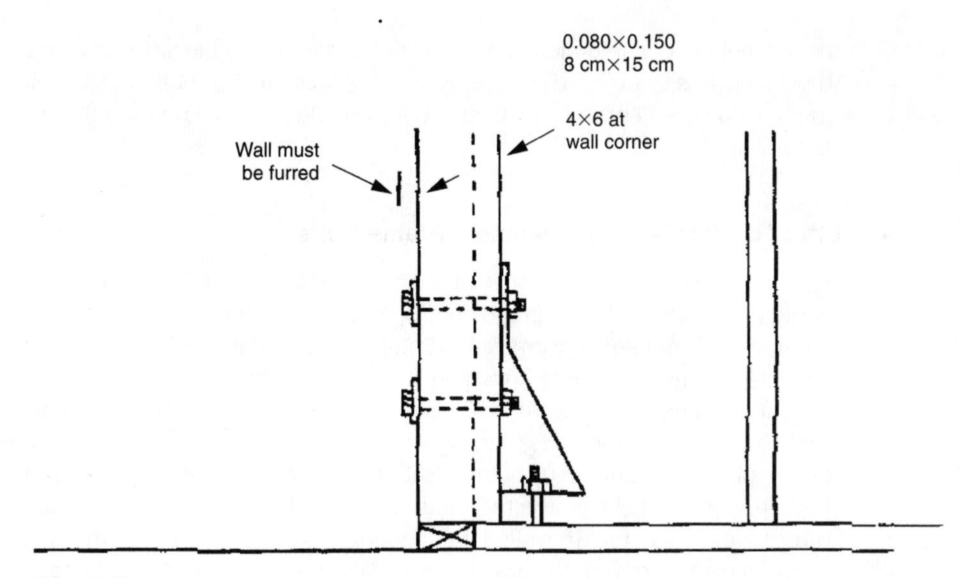

Figure 2.10

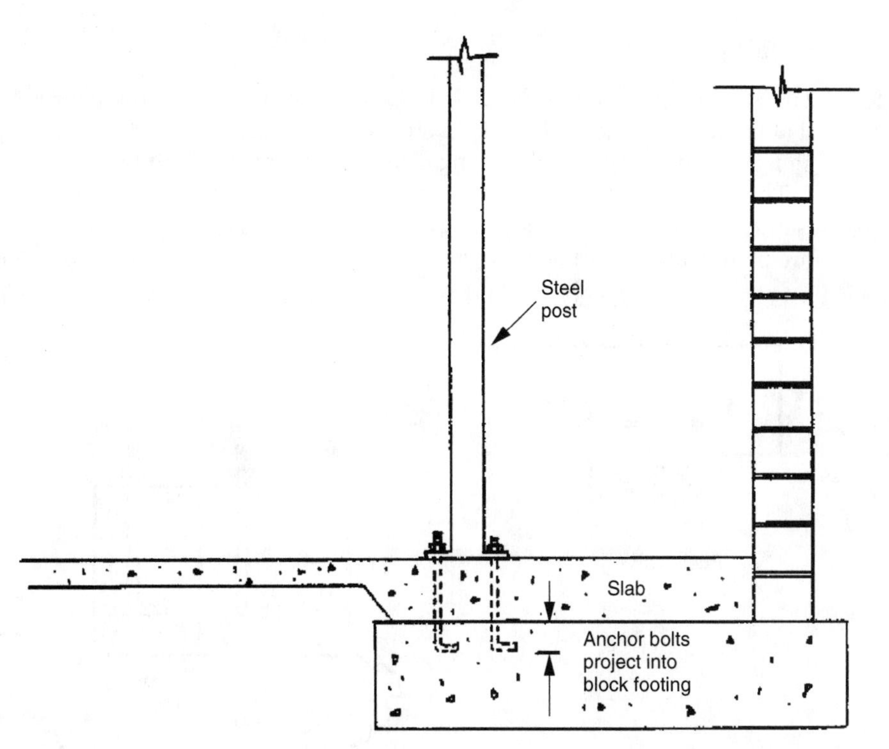

Figure 2.11

tural steel post. The dimension between the top of the block footing and the top of the slab is less than the 10-inch length of the steel anchor bolts. A cavity low enough to allow for the depth of the anchor bolts should have been left in the block footing at the location of the steel post.

Check Width of Concrete around Moment Frame Bolts

Figure 2.12 shows the anchor bolts for a steel moment frame at the front corner of a garage foundation at the garage door opening. This front foundation should be wide enough to completely cover the bottom base-plate flange of the moment frame and provide adequate concrete coverage around the moment frame anchor bolts.

In this example, it was discovered during the construction of the model building that the foundation at the garage door opening of one of the units, as dimensioned on the plans, was not deep enough. Not only did the edge of the moment frame base-plate project past the concrete foundation, and therefore be exposed after the interior of the garage was drywalled, but adequate concrete coverage around the anchor bolts was not provided. By making the foundation a few inches thicker at the front of the garage, these two problems were solved.

Moment Frame Jigs

The steel jig pictured in Figure 2.13 is used for positioning the plywood templates that contain the anchor bolts for a steel moment frame at the front of a garage. This jig is made by the steel fabricating company that makes the moment frames.

This method of setting the anchor bolts prior to pouring the concrete foundations is much more accurate than using a tape measure. The jig positions the bolts at the precise distance apart so that when the moment frames are delivered and installed,

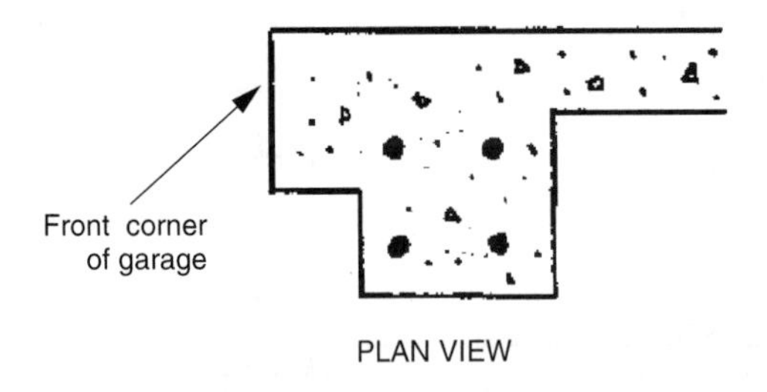

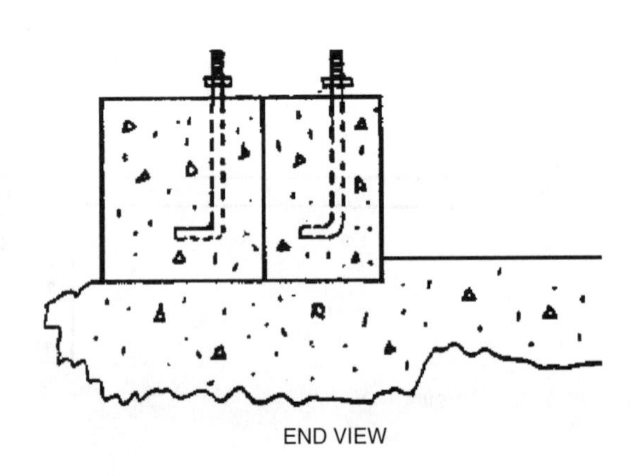

Figure 2.12

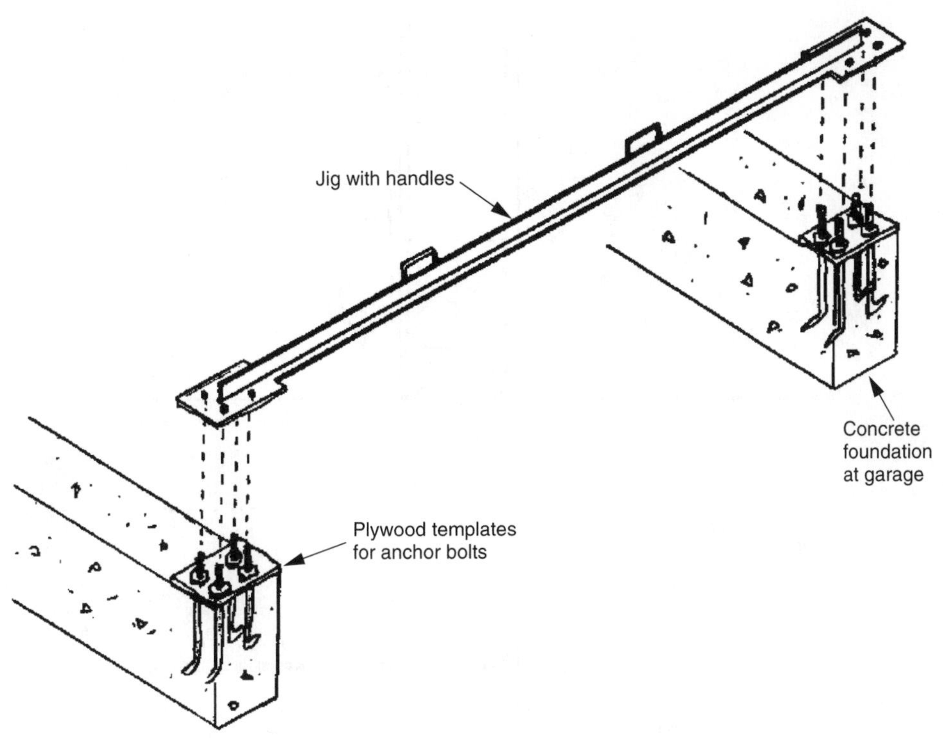

Figure 2.13

the holes drilled in the base plate of the moment frames match the holes in the jig, and the moment frame fits easily over the bolts.

Plumbing Wall Bulges

The layout of plumbing pipes should be checked before concrete is poured so that the pipes will fall within the center of future interior and exterior walls. Not only must the pipes be within the walls, but the elevation of elbow fittings must be low enough below the slab floor surface to not interfere later with the finish flooring (Figure 2.14).

Another problem to watch for is that a 3-inch drainpipe will not fit within a 2×4 wall because the hub connection required to join the pipe coming up through the slab to the riser pipe is 3¾ inches in diameter. The hub will cause a wall bulge even though the pipe itself is perfectly centered within the wall. A 2×6 inch wall should be framed for 3-inch plumbing pipes. Most homebuyers will notice and complain about bowed walls, especially if the bulge is along the floor baseboard.

Place Pipes Outside of Shear Wall

Figure 2.15 illustrates a situation where a shear wall containing plumbing pipes is adjacent to the underside of a stair plenum. Instead of having the plumbing pipes come

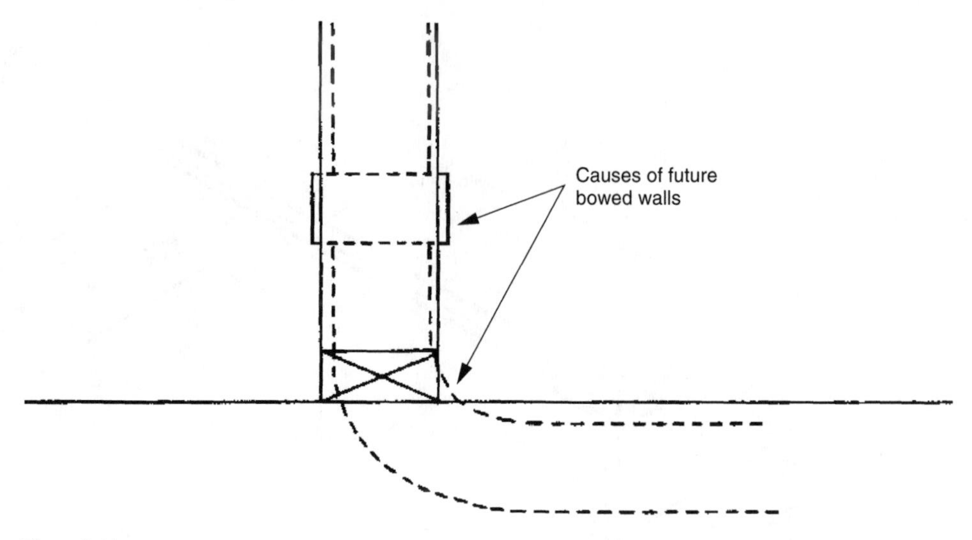

Figure 2.14

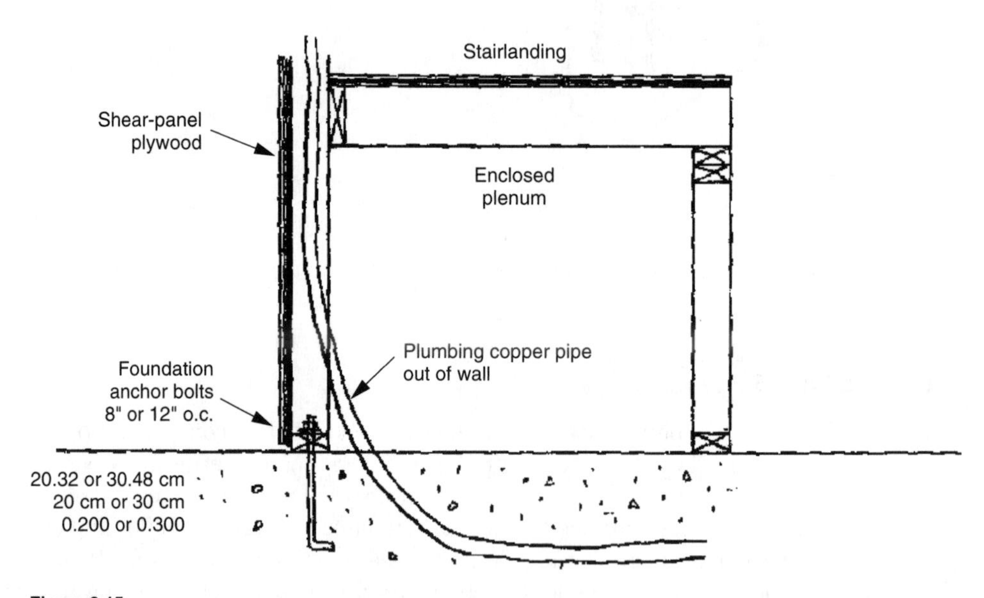

Figure 2.15

up through the shear wall, thus breaking up the mudsill into small pieces, the pipes can be placed outside the wall in the stair plenum concrete, and then bent into the wall a few feet above the floor.

In this instance, there is no reason why the plumbing cannot come up through the plenum. By side-stepping the shear wall, the mudsill can be continuous instead of in chopped-up pieces.

Dryer Vents

Dryer vents are sometimes placed within the floor slab and run from an interior laundry room or closet to an exterior wall at the foundation. When this is done, be sure that the end of the dryer vent at the exterior wall actually touches the outside formboard and is wrapped with plastic tape.

If the dryer vent is accidentally moved during the pouring of concrete, the end of the vent may be surrounded by concrete and its location concealed after the formboards are removed. The plastic tape on the end of the dryer vent prevents concrete from entering the vent during pouring, thereby later restricting the air flow (Figure 2.16).

Utility Sleeve Placement

Plastic conduit sleeves are placed in the garage footing for the future connection of underground electrical, telephone, and television cable to the breaker panel and utility cans on the garage wall.

The sleeves must be placed far enough back from the garage front corner so as not to interfere with brick, stone, or wood siding returning down the garage side walls. If the plans show brick on the garage front and 4-foot returns down each garage side, then the first utility sleeve must be laid-out and placed in the first framing bay beyond the 4-foot brick return (Figure 2.17) in order to miss this brickwork.

Use Straight Pipe for Utility Sleeves

Figure 2.18 shows utility sleeve plastic pipe through a concrete foundation. These sleeve pipes should be straight all the way through rather than have a ¼ bend at the bottom. This frees each underground electrical, telephone, and television cable to turn up into the building at the levels they are found in the trench instead of having to be forced into fixed levels required by using ¼ bend sweeps. Otherwise, too much

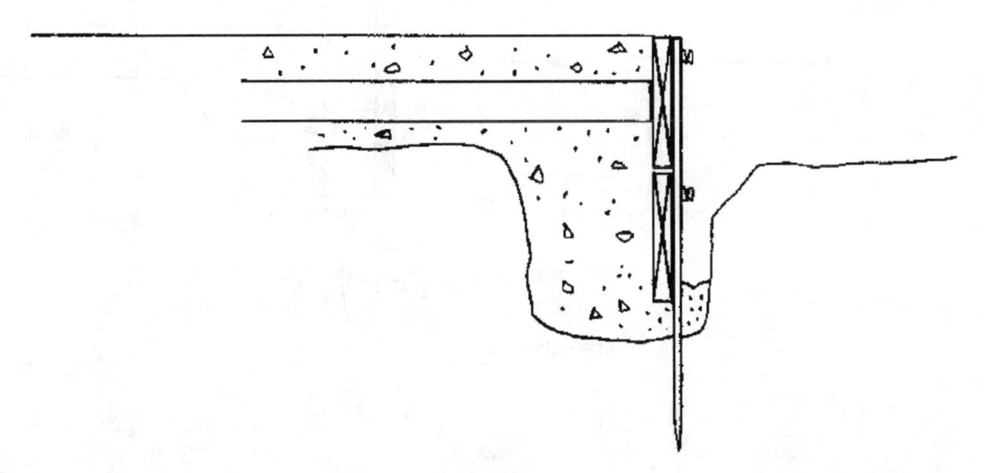

Figure 2.16

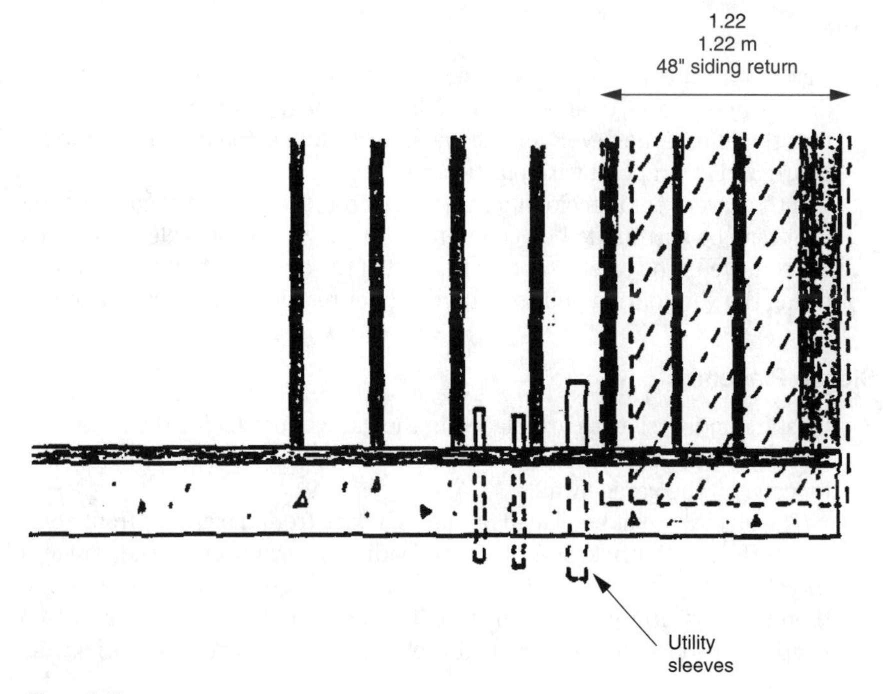

Figure 2.17

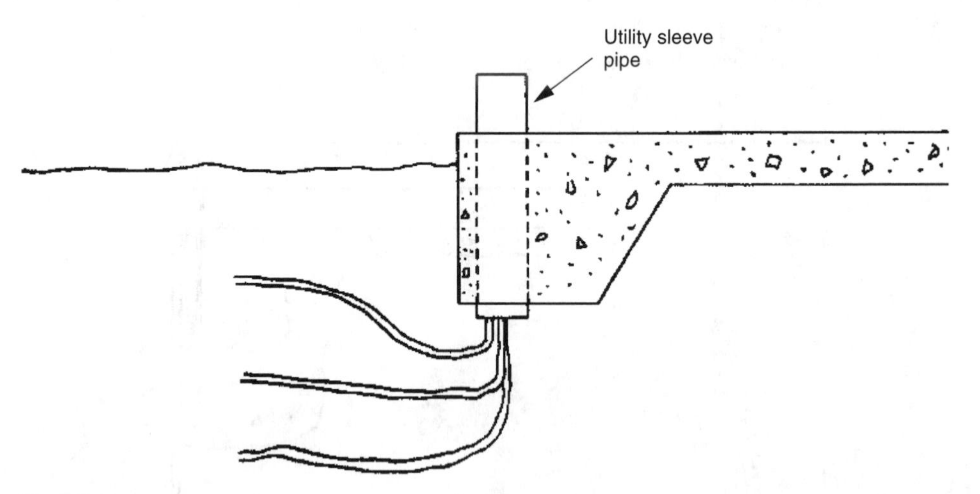

Figure 2.18

time and effort is spent trying to get each ¼ bend sweep at the correct elevation relative to each utility cable in the trench.

Coupling Ends of Sleeve Pipes

Plastic pipe is placed as sleeves underneath or within concrete for many future uses and situations such as to house utility cables or conduit that can come up from the ground, through the concrete foundation, and into the house interior. Drainage pipe sleeves are placed below or within block retaining wall footings so that area drains can later be installed behind the blocked off areas and yet still be connected later to the drainage system. Plastic pipe is placed beneath concrete driveways and sidewalks so irrigation lines can later be connected from the front yard hose bib to other areas of the yard irrigation system.

When a plastic sleeve pipe will have another piece of pipe connected to it later, the end of the sleeve pipe should be a coupling. This allows the future pipe to be slipped into the coupling, without having to chip away any concrete around the end of the sleeve pipe (Figure 2.19).

Entry Post Footing Height

Shown in Figure 2.20 is a foundation and footing for a solitary post that supports a roof overhang at an exterior entry. In this example, the elevation of the top of the post foundation should be the same as the house slab finish floor height. This makes framing simplified and keeps the wood post the required minimum clearance above finish grade.

The top of footing, however, should be formed slightly lower to allow for the grading slope away from the house. If grade is 6 inches below finish floor, with a 2% fall away from the house, the top of footing concrete may be exposed by the time the grading reaches the solitary post. By making the post footing 4 or 6 inches deeper and lowering the top of footing, the chance that grading will expose the footing is eliminated.

Trenching Next to Post Footing

Figure 2.21 illustrates a situation where the trenching for the house water service is adjacent the concrete footing for a solitary post at the exterior entry. When the

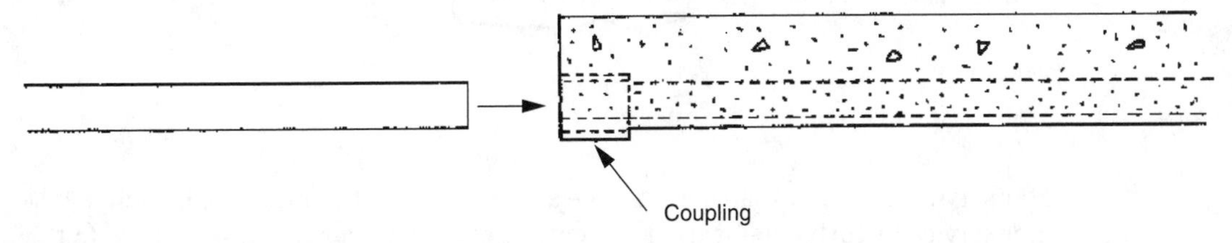

Figure 2.19

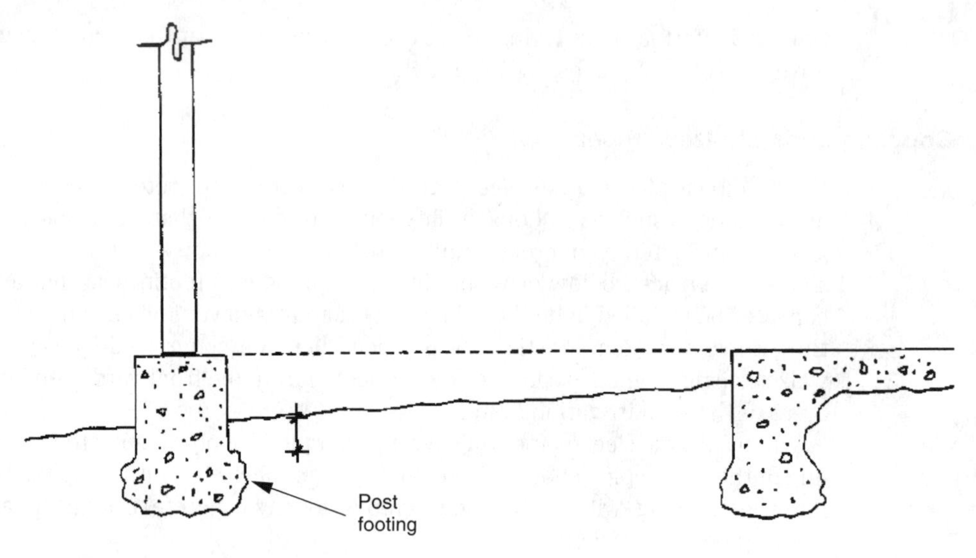

Figure 2.20

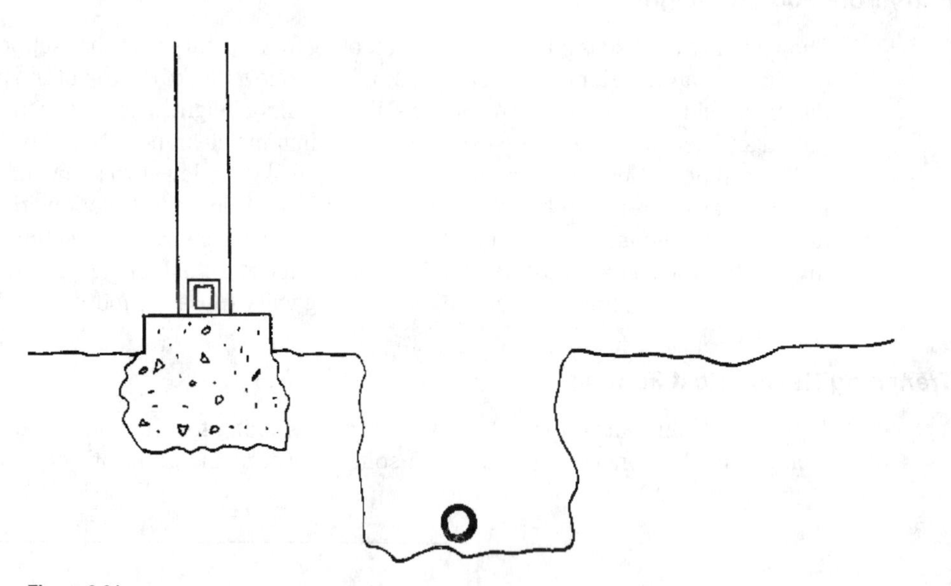

Figure 2.21

water service is deeper than a 45-degree angle to the post footing, the trench for the water service disturbs the bearing capacity of the soil under the post footing. In this situation, the builder should lower the depth of the post footing equal to or below the depth of the water service trench.

Raised Interior Slab

A typical mistake made when a raised entry concrete slab steps down into a sunken living room is cited in Figure 2.22. In this example, the concrete is formed straight across so that the footing for the lower level living room is in line with the higher entry slab. When drywall is applied to this living room wall, the lower corner of the drywall at the concrete step has nowhere to end. The drywall cannot continue across the vertical face of the step, and cannot turn the corner 90 degrees and butt into the step because there is no corner. The raw edge of the drywall must then be trimmed-out with metal mill-core or cornerbead at the stepdown.

To prevent this problem, the stem wall footing and the entry raised slab should be offset at least a few inches. This can be done easily by forming the raised slab 1½ inches in toward the entry, providing a 1½-inch corner at the living room stem wall (Figure 2.23).

Low Spots at Front of Garage Slabs

A typical recurring problem in tract housing is the low spots at the front of garage concrete slabs. These low spots collect puddles of water noticed by the homebuyer after the first rain or after the garage is washed out for the first time. Surprisingly,

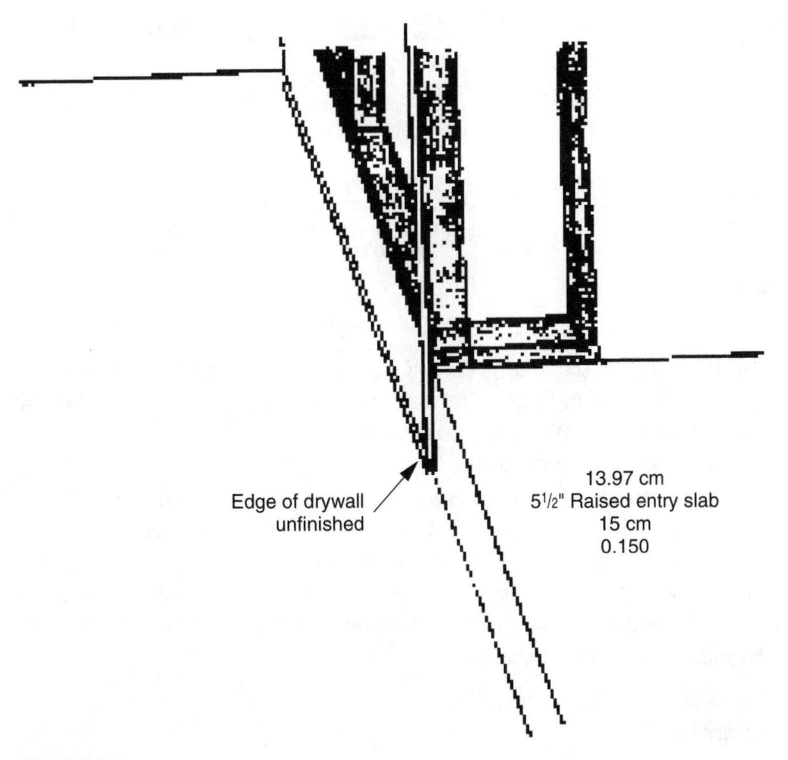

Figure 2.22

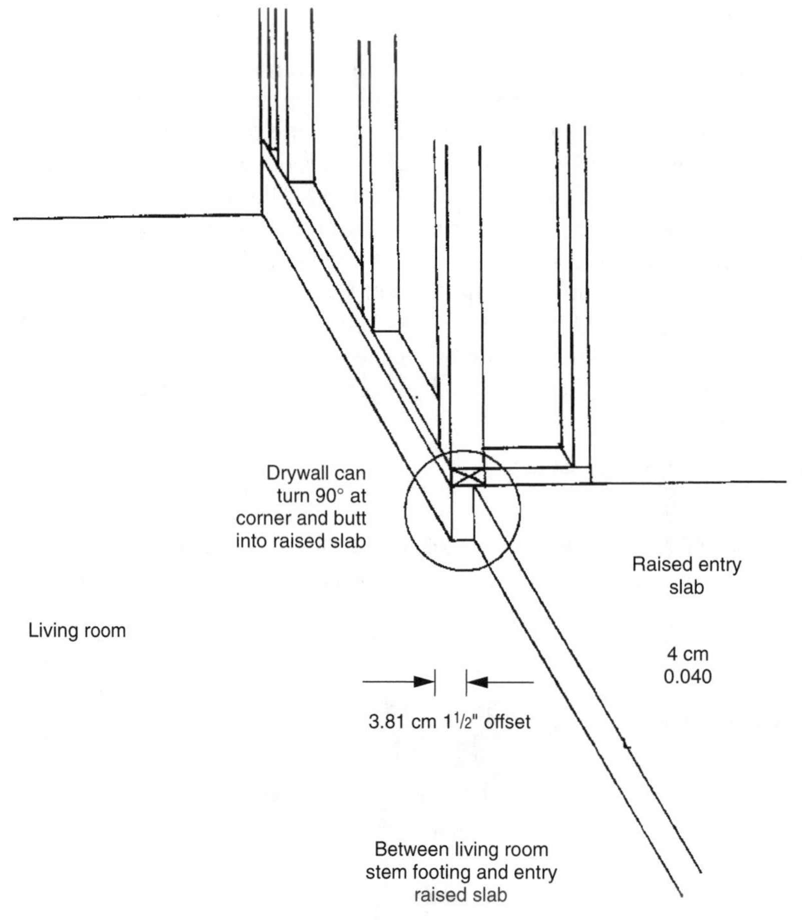

Figure 2.23

water puddles at the front of garage slabs are sometimes treated with the same degree of hysteria by homebuyers as a roof or plumbing leak—as if the water will not evaporate or be swept off with a push broom.

However, to avoid having to tear out and repair large sections of the garage concrete slabs, the builder should check during the pouring of the concrete that the rodding off and finishing of the garage fronts is done with the same high degree of care as with the house interior slab. The builder should also check during the pouring of the concrete driveway that there is at least 2 percent downward slope away from the front edge of the garage slab, for at least the first 5 feet. This will prevent water from collecting in front of the garage.

Straightedge Concrete

For moderately low-priced to expensive houses, it is good practice to check concrete slabs a day or two after it is finished for humps or valleys in areas that may be later finished with floor materials such as marble, tile, slate, or wood plank.

For low-priced houses, the homebuyers seldom have the money for large areas of marble, tile, or wood flooring in entries, kitchens, bathrooms, and family rooms. Because carpet and padding smooth out and soften minor variations on the concrete slab, the occasional need for slab repairs for this level of housing can best be handled as they arise late in the construction.

For moderately low-priced to expensive houses, large portions of the concrete slab may have expensive hard surface floor coverings that require a flatter slab surface. The general standard to aim for is no more than a ¼-inch variation over a 6-foot area.

The mistake to avoid here is that if problem areas on slabs are to be discovered, bush-hammered out, and then patched, the best time to do it is right after the concrete is poured, not late in the construction when dust will get all over cabinets, ceramic tile countertops, and freshly painted walls. By checking and repairing humps and valleys in kitchens, entries, bathrooms, and family rooms right after pouring, the concrete is easier to chip or bush-hammer because it is still "green." More importantly, the mess and dust generated from these repairs are confined to a time period early in the construction when it will not damage anything else.

Concrete Pickup before Framing

After the concrete slabs have been poured and the formboards removed, the concrete subcontractor should straighten the anchor bolts and clean off any concrete that may have landed on the anchor bolts with a wire-brush. Wire-brush cleaning of the anchor bolts removes the concrete from the bolt threads and allows the framer to install anchor bolt nuts faster and more easily.

The concrete subcontractor should also grind down any perimeter slab edges that are not even and flush with the slab. An irregular ridge along the perimeter of a concrete slab (Figure 2.24) will cause the framing mudsill bottom plate to slant upward, raising the wall framing. This can cause uneven framing of the top plate elevations over and above the slight normal variations in the slab and cause framing problems for the entire house. By grinding these perimeter edges smooth and flat, the mudsill can rest flat upon the concrete.

Camber Wood Sleepers

When wood sleepers are placed on the edge of concrete foundations to be used as knockouts to form channels for the future installation of water and gas services, they should be cambered as shown in Figure 2.25. This allows the wood sleepers to be hammered loose from the concrete easily. If square pieces of wood are used as inserts, they may be jammed tightly in place by the surrounding concrete and have to be split and broken out in order to be removed from the concrete.

Clean Out Concrete Slough

One way to get underground electrical and gas utility services up into condominiums and apartments is to place wood sleepers in the concrete forms (see Figure 2.25). The sleepers are then knocked out during the form-stripping to create a channel for the pipes.

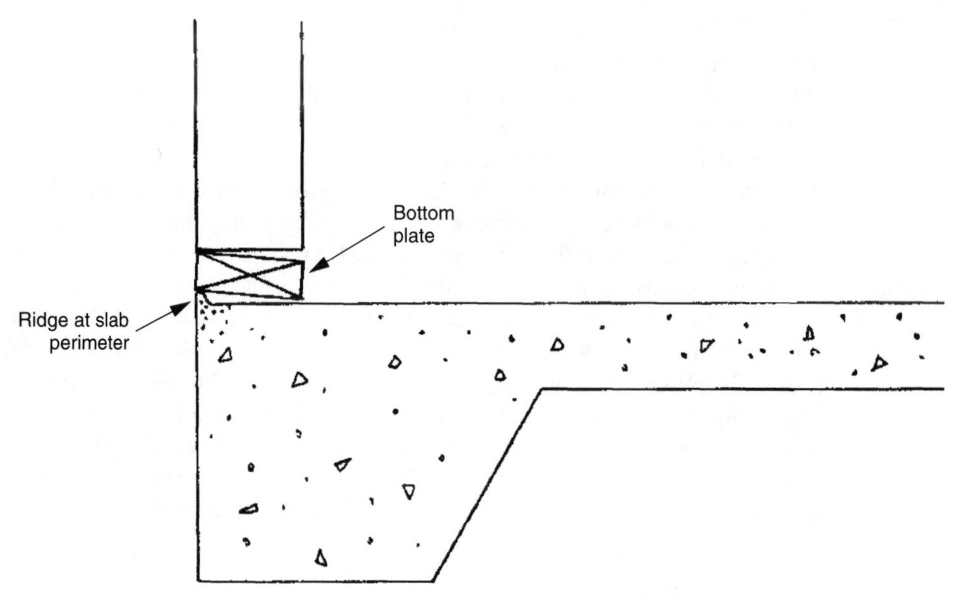

Figure 2.24

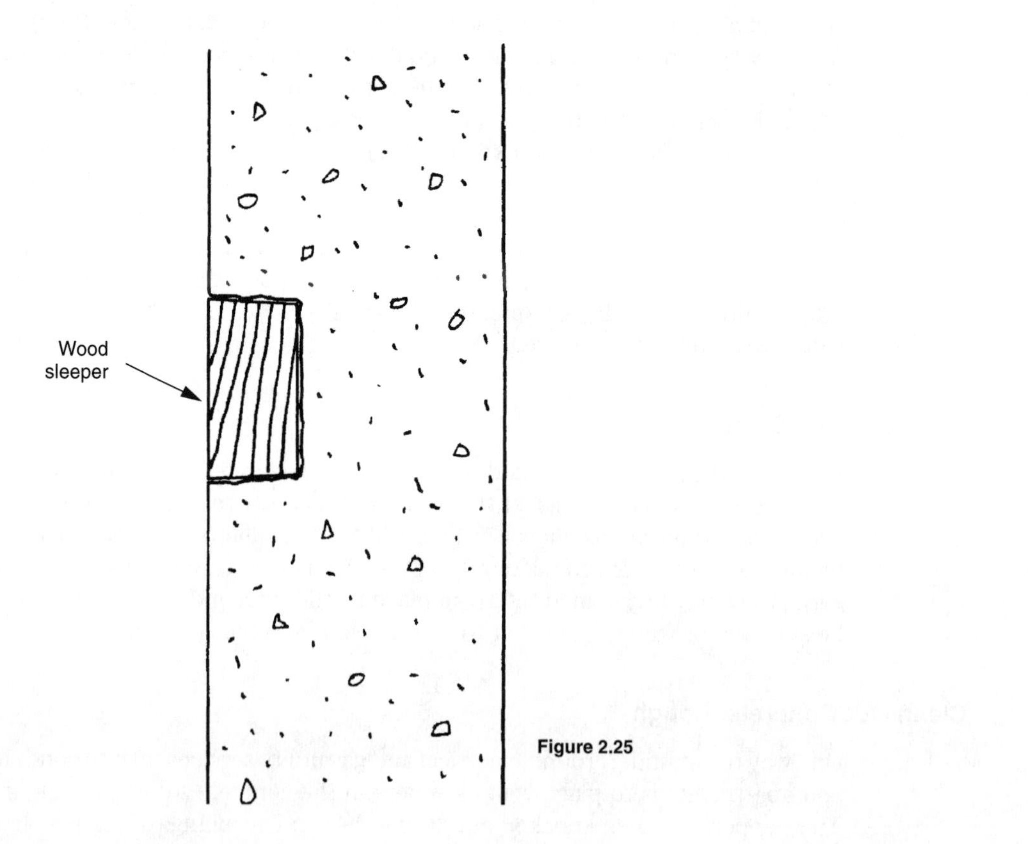

Figure 2.25

One of two things must be done, however, to make these knock-outs effective. The first is to have the plumber or electrician install the sleepers from the top of the formboard all the way down to the bottom of the footing (Figure 2.26). The idea of the sleeper is to displace the concrete, leaving an open channel in the side of the foundation for pipes and sleeves (Figure 2.27). By extending the sleeper all the way down to the bottom of the footing, concrete is prevented from pooling underneath the sleeper, and thus plugging up the bottom of the channel.

The second thing that must be done is for the formboard strippers to check that no concrete slough has worked its way around the front of the wood sleeper. If this is the case, they should chip out this extra concrete along with any other boil-out or slough that is chipped away during the formboard stripping phase.

Surveyors Stake Irregular Driveways

Figure 2.28 illustrates a cul-de-sac street in which some driveways to individual houses are joined on property lines. When this occurs, it is a good idea to have the engineer's surveyors establish the corners where these driveways turn along the property lines. This helps place driveways exactly where they belong and eliminates any question that could arise about accuracy if the driveways were laid-out by the concrete formsetters alone.

The builder should remember to include this activity on the construction schedule and the project budget.

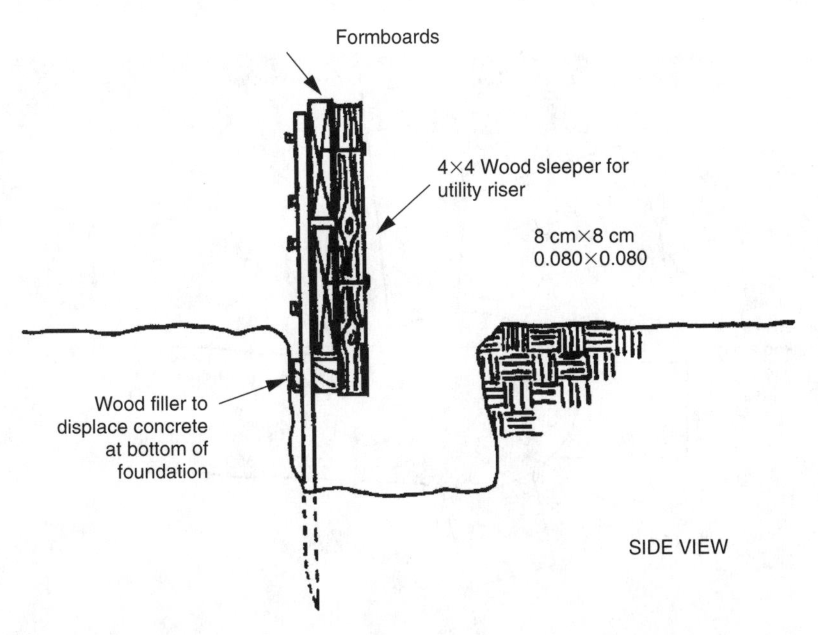

Figure 2.26

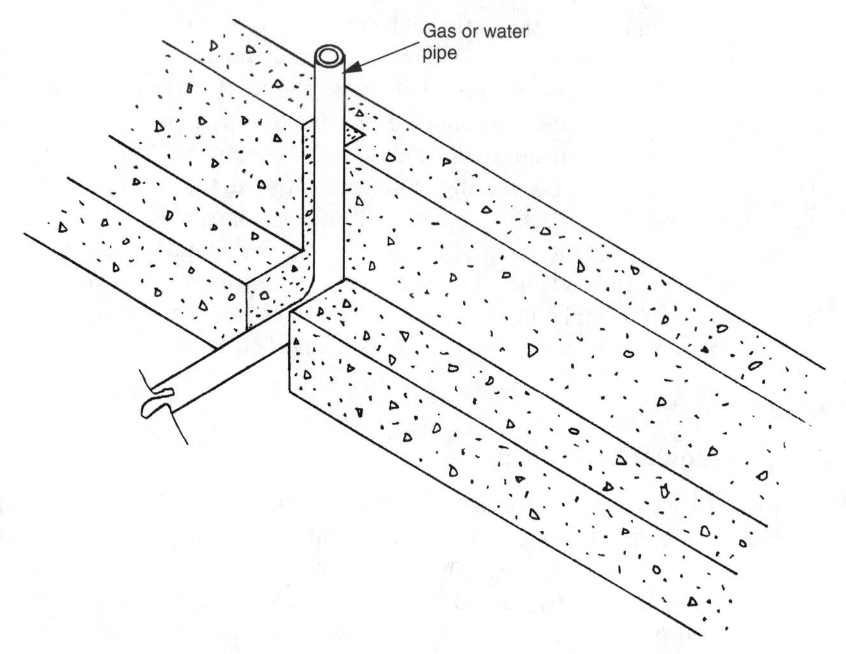

Figure 2.27

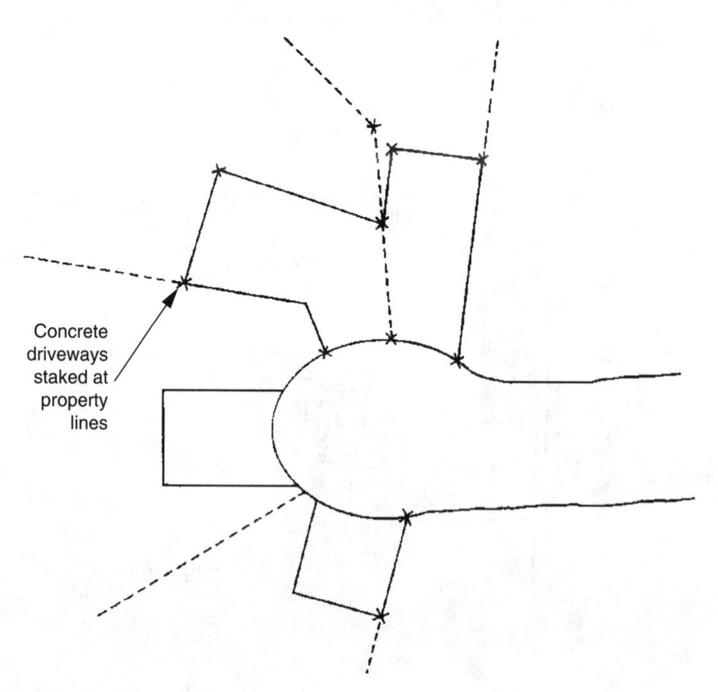

Figure 2.28

Sloped Driveways

For garage concrete slabs that are below street level with the driveway sloping downhill toward the house, the control joint material placed between the driveway and garage slabs should be slightly above grade (Figure 2.29). This helps prevent the water from the driveway from entering the garage.

When the driveway slopes downward away from the house, the control joint should be placed slightly below grade between the two slabs. This allows the home-buyer to wash out the garage without the water being dammed up and puddled at the control joint.

Driveway Setback Mix

Illustrated in Figure 2.30 is a common situation in tract housing. In this example, the setback easement to the city or county right-of-way is 12 feet from the curb edge. This 12 feet includes 6 feet of the private driveway. The concrete sidewalk, curb and gutter, approach, and the 6 feet of the driveway belong to the city or county and must comply with their standards.

The builder must be careful here. The concrete mix for the private driveway may require less number of bags of cement per cubic yard than required for the public right-of-way. If, for example, the driveway can have 4½ sacks of cement, whereas public areas require 6½ sacks, then the builder must pour the driveways with 6½ sacks of concrete throughout, or make a joint at the setback line and have two different concrete mixes in the driveway. What must be avoided is pouring the weaker mix in the public area of the driveway and then having to tear it out later because it did not meet the public standards.

Some form of evidence such as a ticket or receipt from the ready-mix concrete truck showing the mix design for the public concrete work should be kept for the building inspector.

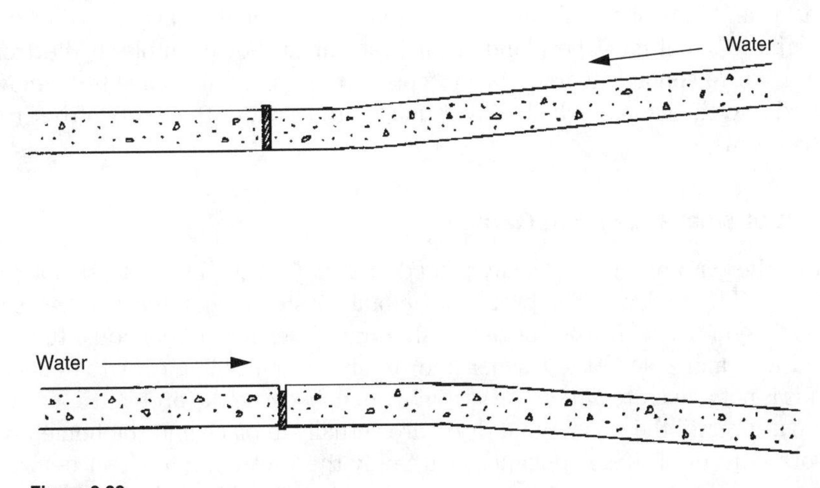

Figure 2.29

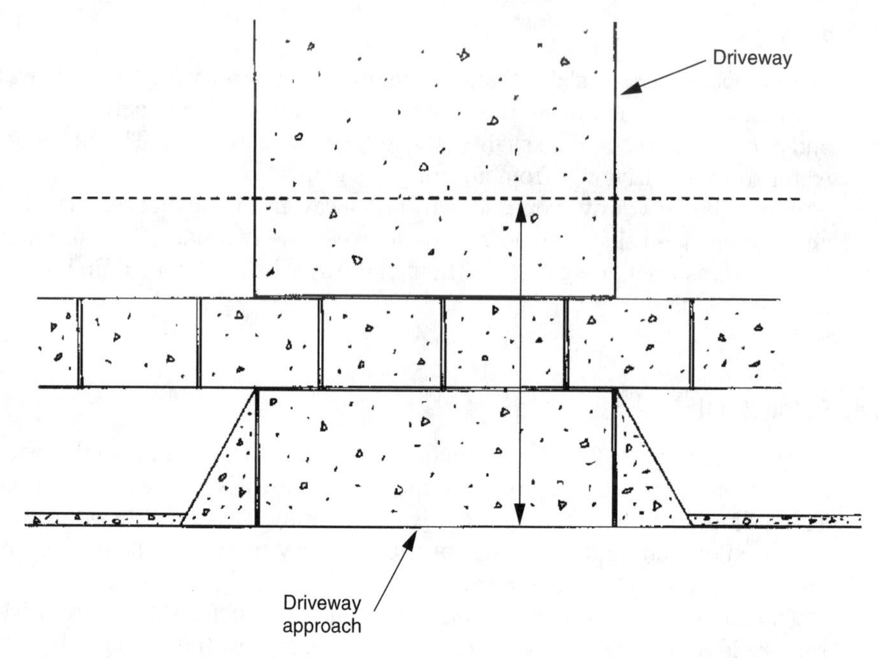

Figure 2.30

Colored Concrete at Garage

The side view of a colored, stamped concrete driveway as it meets the gray garage slab concrete under the garage door is illustrated in Figure 2.31. If it is desired, for aesthetic purposes, that only the colored concrete be seen after the garage door is down, with the gray garage slab concrete hidden behind the interior side of the garage door, then the garage slab must be held back several inches from the front edge of the garage. This means that if the garage slab is poured along with the foundation and the house slab, and therefore far in advance of the colored driveway, this aesthetic detail must be planned for in advance. For example, if the front of the garage slab contains a grade beam footing, this would also have to be moved backward into the garage, which may require a redesign of the garage front reinforcing steel bars.

Space in Front of Washer/Dryer in Garage

A washer and dryer is typically placed on a flat, raised concrete slab area inside a garage (Figure 2.32). The problem the builder should look for with this situation is insufficient space in front of the washer and dryer for an individual to safely stand and load and unload the washer and dryer without falling over the concrete step. This is not as easy to anticipate as it may seem because the problem does not become apparent until the washer and dryer are actually in place and the homebuyer has to work in front of these appliances with only the width of a few feet before stepping over the edge, falling, and spraining or breaking an ankle, as happened in one case.

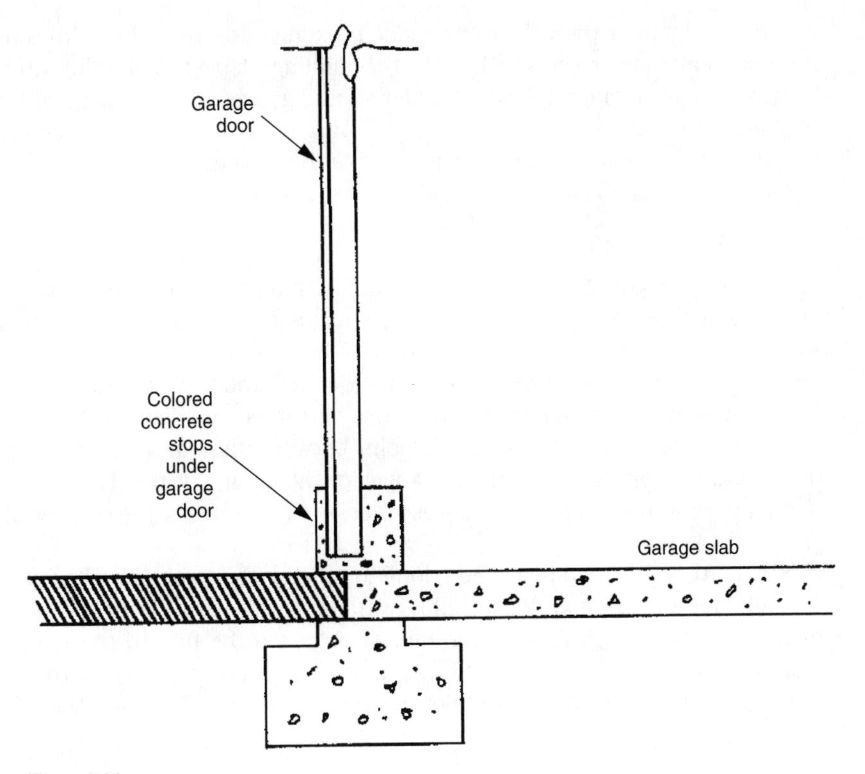

Figure 2.31

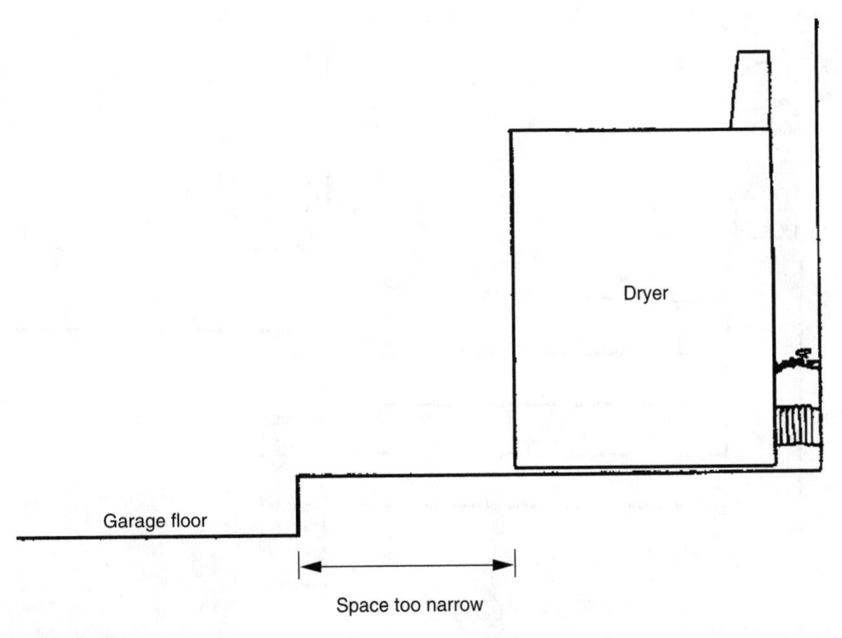

Figure 2.32

I do not know if there is a particular building code regarding this situation, although I have never seen or heard of a building inspector bringing up this issue. However, to avoid a lawsuit, the builder should go over this problem with the architect during the drawing of the plans or during the construction, and come up with the correct design and clearances in front of the washer and dryer.

Slope at Exterior Entries

Illustrated in Figure 2.33 is an entry porch with a precast concrete handrail placed on top of a short stuccoed pony wall. The mistake here was that not enough downward slope was built into the entry porch to channel the water away from the entry door and no weep holes were placed at the outside wall to provide adequate drainage for the porch. Whenever it rained, water remained puddled on the entry porch, which eventually leaked through the joint between the porch and the building and down into the top of a masonry block wall of the garage below. This in turn created damp areas, water stains, and efflorescence on the surface of the block wall at the inside of the garage.

Several things should have been done to prevent this problem. The first is to have enough slope in the porch to enable water to drain away from the building. The caulking at the porch-to-building joint appeared to be fine to the naked eye, but standing water will find any small hole in caulking and leak downward. Next, if any flooring material is added to the porch, such as stone, as it was in this case, the in-

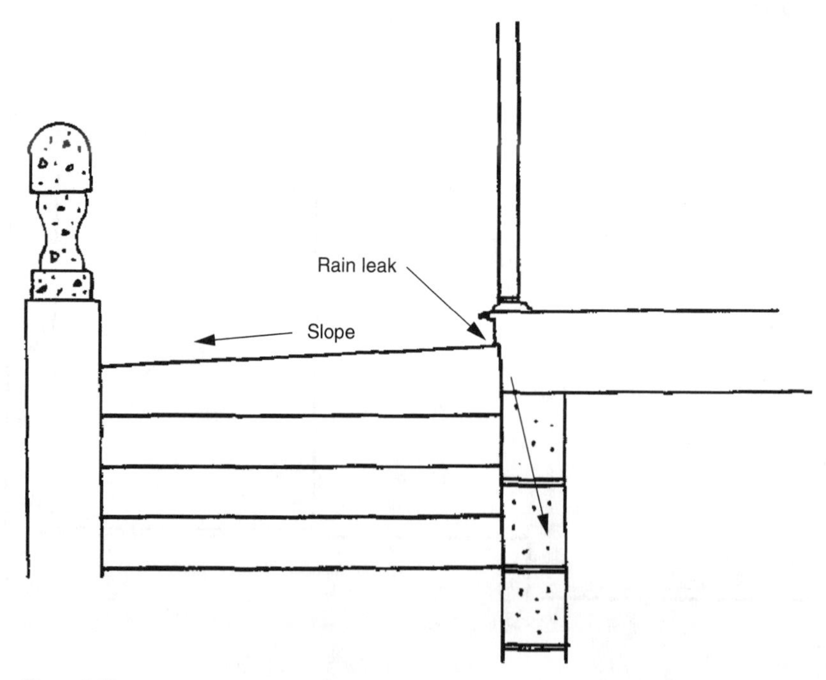

Figure 2.33

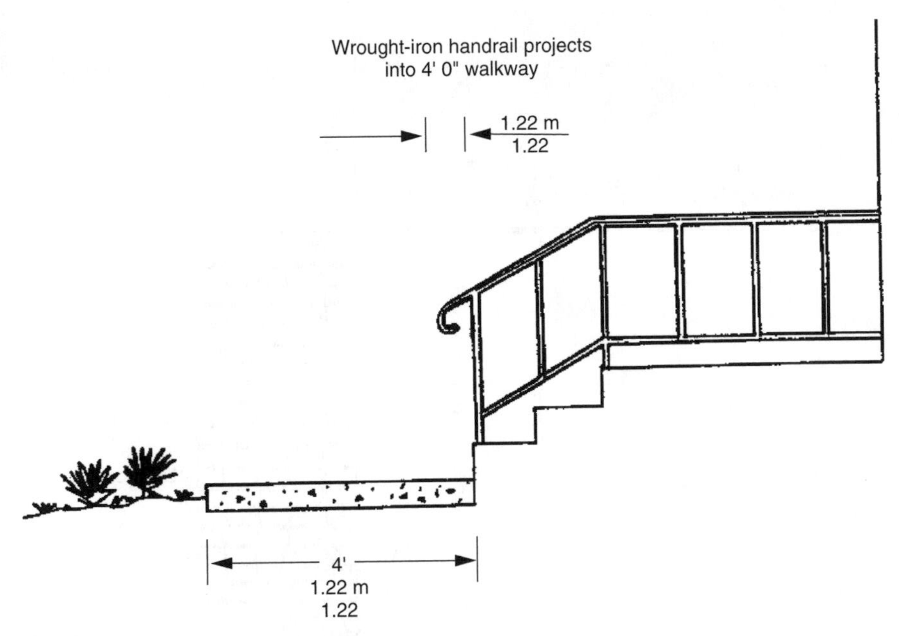

Figure 2.34

stallers must follow the slope of the porch. The installation of stone or clay paver tiles should not diminish the slope of the porch that has been built in by the concrete finishers. Any additional measures, such as weep holes at the base of the pony wall or extra flashing at the porch-to-building joint to help drain the porch or divert water away from the building should be looked at by the builder.

Common Area Sidewalk

A concrete walkway in a condominium or townhouse project is required to have a minimum 4-foot width. To ensure that this space requirement is met, the metal handrail from the entry steps should be set accordingly (Figure 2.34). The builder should consider the design and dimensions of the private handrails to each unit when laying out the location of common area concrete walkways.

Water Puddles at Exterior Entry

Figure 2.35 illustrates a problem that occurred at the exterior entry porch to a million dollar spec house. In this case, the walkway leading to the entry door had no step-up. The colored concrete panel and the red brick porch were at the same elevation. The circled area shows a low spot in the brick porch that collected water to form a standing pool. This pool never drained off. The water just sat until it evaporated.

This problem came about because the builder never checked for low spots in the brick porch, even though it did correctly have some downward slope away from the entry door.

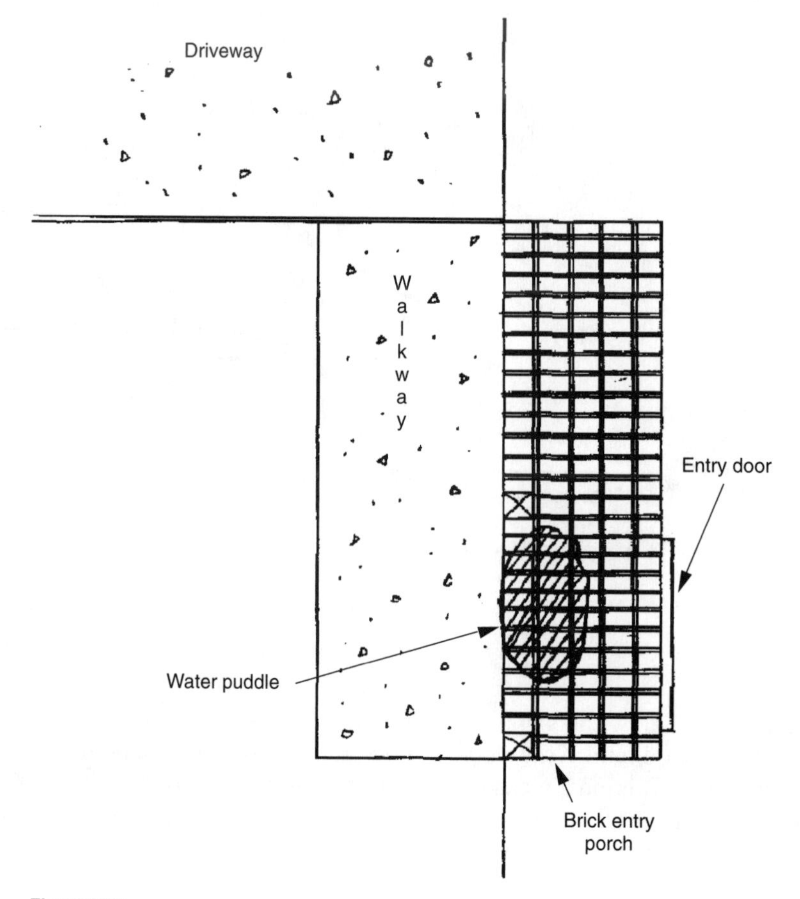

Figure 2.35

Check Width of Leftover Space for A/C Condensers

The builder should always check that adequate space is left over between these walkways and the building for the A/C condensers (Figure 2.36). If the space is too tight for the normal size condenser, the builder can go to a "slim-line" model, which is longer and narrower than the standard condenser. However, the prefabricated plastic or concrete pads that the condensers sit on top of may not fit in the space between the walkways and the building. When this occurs, the builder must pour custom-sized concrete pads for the condensers, filling in solidly the spaces between the concrete walkways and the building at the location of the condensers.

This is something the builder should check and anticipate prior to the pouring of the concrete walkways so that any custom A/C pads can be included with this concrete work. It is better to find out about this problem before the A/C condensers are to be installed.

No Floating Entry Porches on Compacted Fingers

The above ground level entry porch pictured in Figure 2.37 is supported by columns placed on top of compacted dirt fingers between the condominium/apartment buildings. In this example, parking garages were located at ground level in a courtyard with concrete steps leading up to the entry door level. This made the exteriors of the buildings look two-story high rather than three stories high.

The problem here is that these entry porches should have been placed on top of a grade beam coming out from the block perimeter building wall or carried all the way down to ground level. Even though in this case the dirt in these elevated fingers was compacted and tested, there still was some settling over time, which resulted in stucco cracks and the entry concrete pads dropping away from the building a half-inch or more in some places.

This is the type of problem where an ounce of prevention is definitely worth a pound of cure. The difficulty and expense of tearing out concrete porches and digging out and beefing up footings can be great when settling of this type occurs. Not only will the porches have to be redone, but the builder may be faced with having to fix surrounding concrete walkways, wrought-iron handrail, and landscaping. Once such settling cracks start, there is no way of predicting at what point they will stop.

From the builder's standpoint, it is better to overdesign the foundation support for an entry porch built over compacted fill and spend a little more money in an effort to eliminate the possibility of settling. Cost-cutting in these types of areas can backfire in an expensive way.

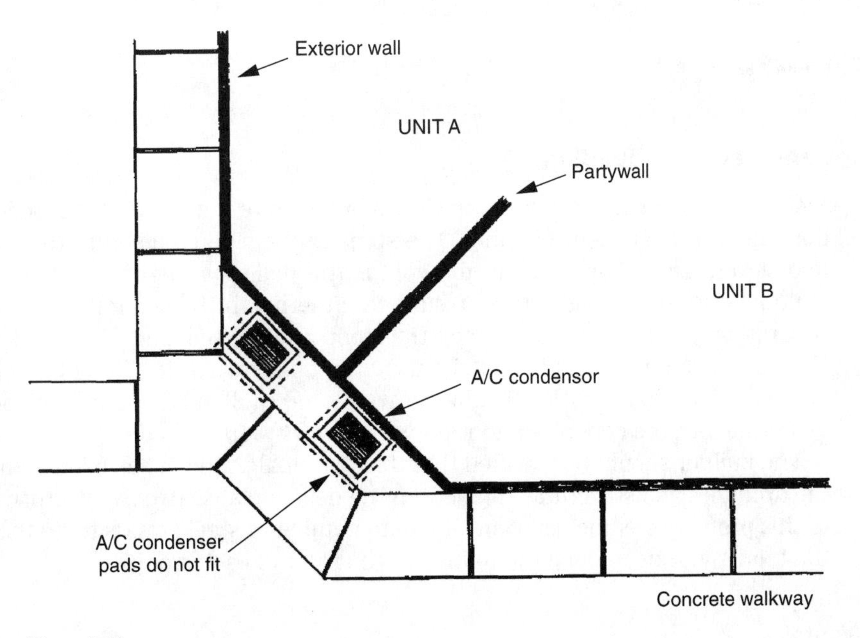

Figure 2.36

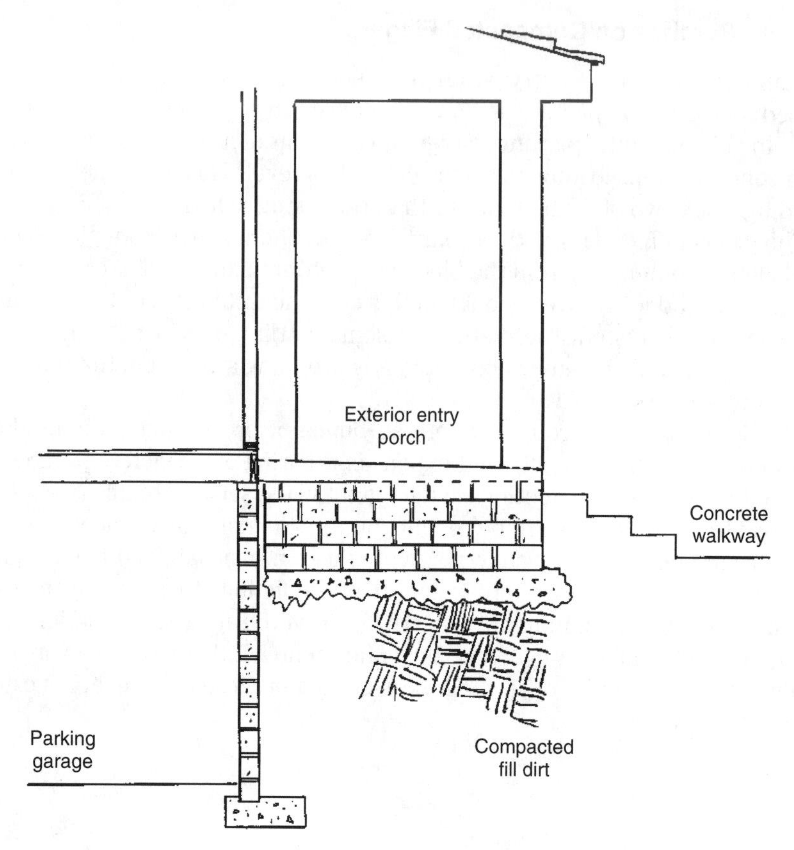

Figure 2.37

Tie Stoops and Patios to Building

Pictured in Figure 2.38 is the side view of a concrete entry stoop as it meets a condominium or apartment building. The steps to the side go down to the street, one floor level below. The problem to avoid is the pulling away of the stoop from the building due to settling of the subsurface underneath the concrete. This creates a separation gap at the joint between the stoop and the building. The crack is almost always noticed and complained about by the homeowner. It can be fixed either by filling the joint crack with a flexible, gray-colored caulking to match the concrete or by tearing out the concrete and repouring a new stoop.

The builder should be satisfied that this potential problem has been addressed by the architect and structural engineer in the design plans. Two satisfactory solutions to this problem include using an adequate number of steel dowels to tie the stoop to the building or deepening the footing at the front of each stoop.

Garage Vents

The builder must be careful that garage vents are installed above the level of concrete flatwork that steps up toward the front door. This is more of a problem for duplexes and townhouses where vents that would normally be placed on the garage wall on the opposite side to the entry cannot be placed there because these walls are now garage partywalls for attached garage, therefore not able to be vented to the outside. Figure 2.39 shows a garage vent inaccurately placed during the rough framing. The vent ends up partially buried by the entry concrete step.

Landscape Lighting Sleeves

The side view of a concrete walkway pictured in Figure 2.40 shows a 1-inch plastic pipe placed underneath the concrete to provide routes for electrical wiring for land-

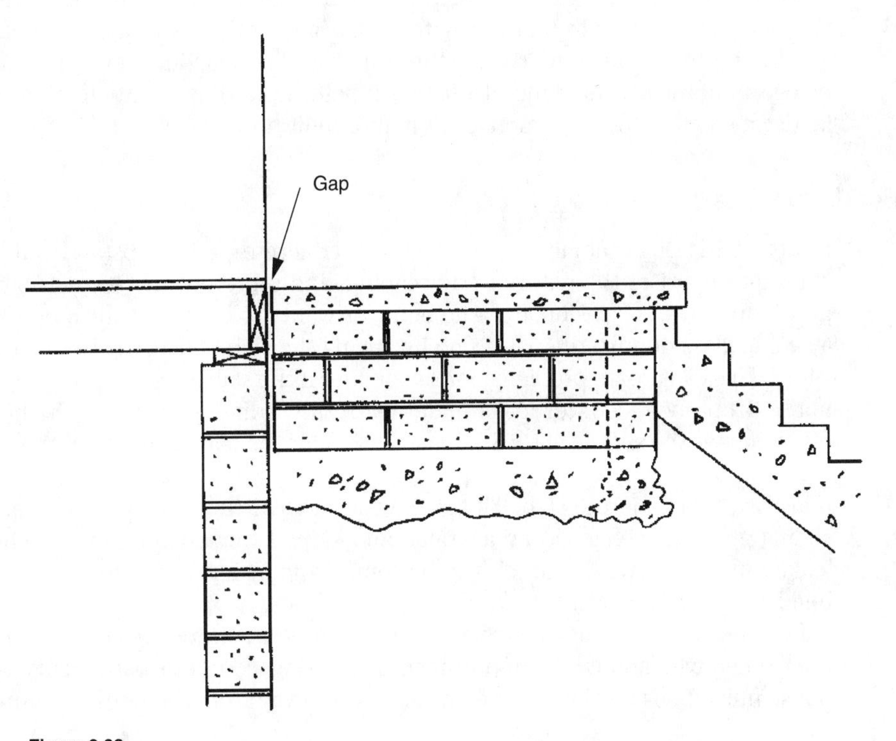

Figure 2.38

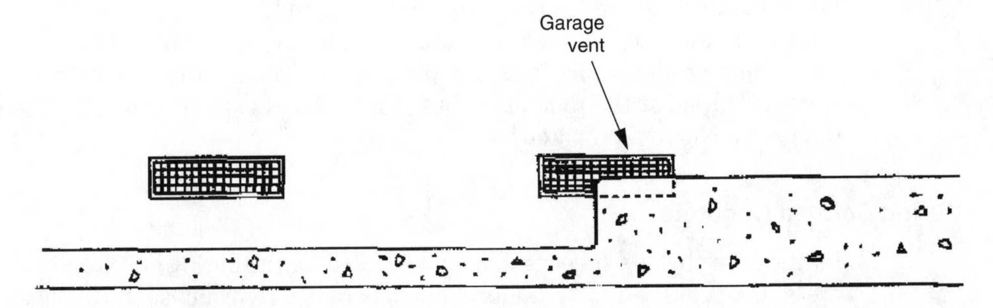

Figure 2.39

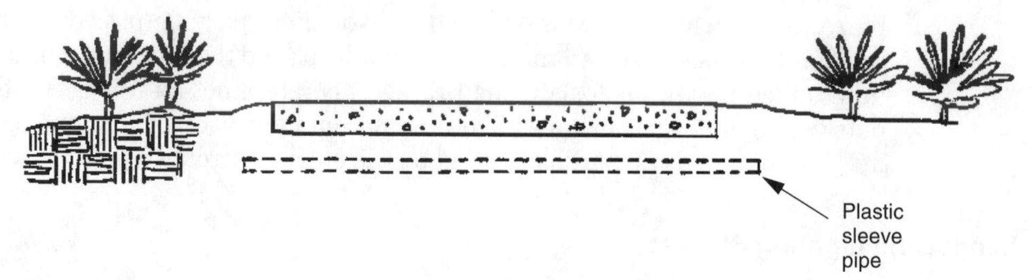

Figure 2.40

scape lighting, typically added later. By adding this plastic pipe, the landscaper or the electrician is able to pull wiring from one side of the walkway to the other without having to dig out underneath the walkway. The plastic sleeve pipes should be placed underneath the concrete flatwork before pouring in anticipation for future landscape lighting for common areas in multiunit projects.

Exterior Patio Slabs

In detached, single family houses, exterior concrete patio slabs and walkways are considered part of the exterior landscaping, and therefore the responsibility of the new homebuyer. Occasionally, a homebuyer wants to eliminate the step down from a sliding glass door to the patio slab by having the slab poured up to the level of the slider. This can create problems for the builder if the slab is not sloped away from the house properly and water leaks underneath the sliding glass door. The homebuyer often blames the sliding glass door installation for the leak rather than the patio slab.

In one particular case, the homebuyer had his landscaper pour almost the entire rear yard as a concrete slab. The slab, which was at the same elevation as the rear sliding glass door, had no area drains and sloped toward the house. When water leaked into the house through the bottom of the slider, the homebuyer called the builder not the landscaper.

For the builder's sanity, instructional brochures and literature should be offered to the homebuyer about the importance of pouring concrete patio slabs below the house finish floor level with a downward slope away from the house (Figure 2.41).

Stamped Concrete in Streets

When stamped, colored concrete is used on the streets, it is crucial that street manholes and other steel cover plates are raised to the finish street grade before pouring. Unlike paving, which is easily patched around steel covers or manholes that were not raised at the time of paving, stamped and colored concrete is much more difficult to cut out and patch.

Curing Colored Concrete

Plastic polyethylene sheets should not be used for the curing of colored or plain concrete that is desired to be uniform in appearance. By using such plastic as a means of curing, shades of light and dark concrete may result from where the plastic sheets either do or do not come in contact with the concrete due to wrinkles and folds. Concrete that contains calcium chloride and is steel-trowel smooth finished is particularly affected.

Colored Concrete

There are two methods of introducing color into concrete flatwork. One is to add the color to the concrete as it is mixed—called integral color. The second is to sprinkle coloring powder on the surface of the concrete and to work it in during the finishing—called broadcast color.

Integral coloring is the better of the two methods for several reasons. If some small chips should break off the concrete surface, integral concrete will show color beneath every broken chip, whereas the broadcast color may show gray, uncolored concrete. If some ridges or high spots must be ground down, the integral color for the same reason will remain colored below the repair. If the broadcast powder does not work around the edges of the formboards, the grading and landscaping will have to cover the exposed edges of the concrete slab so that the gray concrete will not show. The integral colored concrete will be colored around every exposed edge.

Patch Curbs before Colored Concrete

When a colored concrete driveway meets the city street curb, the edges of the curb should be patched before the driveway is poured. This provides a smooth and straight edge for the colored concrete to join rather than a crooked line with colored concrete lapping over into the gray colored curb.

Enough time should also be provided between the patching of the street curb and the pouring of the driveway. If poured the same day, the coloring agent in the wet driveway concrete may leach into the gray-colored, still-wet concrete curb patch.

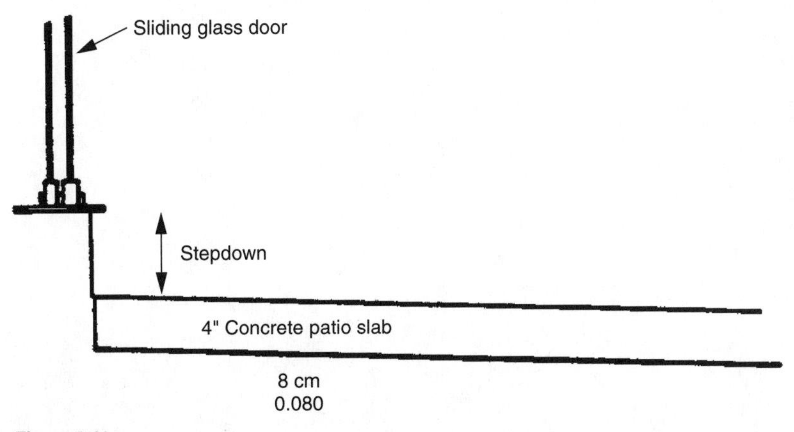

Figure 2.41

Stamped Concrete

Concrete flatwork with stamped patterns should have the same person doing the stamping for each house or separate area. If two people do the job, the pressure applied to the stamping may vary, resulting in slightly different patterns. For example, a different force may be applied to the stamping tool by a 200-pound person as compared to a 150-pound person. Furthermore, everyone has their own slightly different individual technique. If uniformity is important, then only one person should do the concrete stamping.

Masonry Grout Splatter on Concrete

Masonry block garden walls are sometimes installed inside condominium courtyards after the concrete walkways are poured. When this is the case, the mason should cover the concrete walkways with building paper or plastic sheets so that grout from the masonry work does not stain the concrete.

On one particular project, masonry grout splatter stained light-red colored concrete walkways in a condominium courtyard. Water-blasting could not remove the gray stains from the light-red adobe colored concrete.

Chapter

3

Layout and Wall Framing

Check Layout Snap Lines

During the layout and snapping of lines in preparation for wall plating and framing, the measurements are taped off and usually marked on the floor with a crayon keel using a V-shaped bird's mouth to locate the spot where the bottom plate wall framing chalk lines are to be snapped. The problem is that the crayon mark can be as thick as a ¼ inch or more, depending on the dullness of the crayon head. If the framer snapping the chalk lines doesn't measure to find the center of the V-shaped marks, the lines can be snapped 3¾ to 4 inches apart for a wall that is only supposed to be 3½ inches wide. With this level of sloppiness, the wall framers don't know which lines to use for positioning the walls and the framing can end up out of square and out of plumb.

Figure 3.1 shows an example of wall framing layout with chalked lines varying from 3¾ to 4 inches apart, all because the crayon marks were too thick and the line snapper used both V-shaped bird's mouths and eye-balled the chalking of the lines instead of taking the time to measure the second line off of the first chalked line. Walls that are off the true layout an ⅛ inch here, a ¼ inch there, and another ⅛ inch somewhere else make it difficult for the plumb and line crew to later get all of the walls square and plumb.

The builder should check the measurements between chalked lines during the framing layout and snapping phase prior to the walls being framed and stood up.

Check Plates

One thing the builder can check during the framing layout phase is that the plates are saw-cut on the pencil lines. This can help the plumb and line operation later and

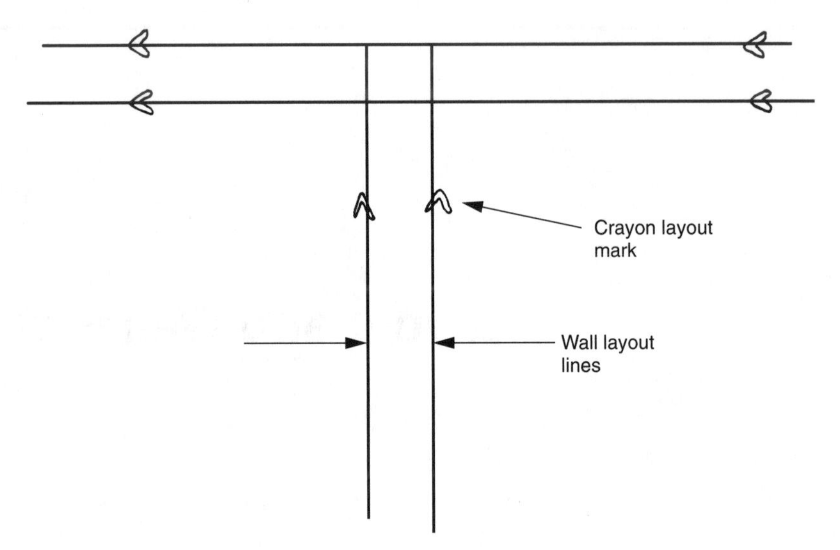

Figure 3.1

thereby improve the squareness of the interior walls. A section of wall that has its plates cut ⅛-inch too long past the pencil line, joined to a second wall with its plates cut ¼-inch too long, joined to a third wall with its plates cut ⅛-inch long, can quickly deteriorate into walls with inside and outside corners that are out of square and that are slightly out of parallel.

The builder should go through the framing as it is being laid-out, prior to the framing and standing up of the walls, and check the lengths of the plates with a tape measure. If a certain wall is supposed to be 6 feet, 4½ inches long and tapes out at 6 feet, 4⅝ inches long, then the layout man can trim off the extra ⅛ inch at that time. This saves a lot of work for the plumb and line crew later and helps catch small mistakes at the time they are made. This also reduces the possibility that the mistakes will remain throughout the construction and become part of the final product.

Cripples

The builder should insist that the framing contractor cut all window sill, header cripples, and trimmers on a cut-off saw rather than hand-cutting them with a skil-saw during the wall framing. Cripples and trimmers cut on a cut-off saw will be uniform in length. The result is framed openings around windows that have a better chance of being square and having equal reveals around all sides of the window frame.

Hand-cutting cripples and trimmers with a skil-saw is not as accurate. Variations between cripples can be undercut or overcut as much as an ⅛ of an inch, adding up to a ¼-inch difference overall. Headers and lower headers (sills) end up being slightly out of square when the framing is nailed all together. This can result in openings that are out of square and reveals around window frames that are not equal.

The hand-cutting of cripples is sometimes added to the piece-work in production tract housing because the framing contractor can save money by getting the layout crew or the wall framers to cut cripples as part of their work without raising the piece-work rate. This saves having to pay the cut-off saw man to cut the cripples.

Straighten Hold-Down Bolts

One of the things the builder should check during the plate and layout phase of the rough framing is that hold-down bolts are straight and fall within the walls (Figure 3.2). It is much easier for the framer to bend crooked HD bolts straight before all of the surrounding walls are framed and stood up. An HD bolt that is slightly off the center of the wall framing plates can also be bent over so that the top of the bolt is centered. This allows the HD to remain within the wall space that would otherwise end up outside the wall.

Shift Walls for Plumbing

During the snapping of chalk lines for wall framing layout, it may be discovered that some pipes, vents, or other items coming up through the slab within a wall are slightly off and fall just outside the wall by ⅛ to ¼ inch. When this is the case, the builder should give the layout carpenter the freedom to move the wall slightly to accommodate the off-layout plumbing. Many minor wall bulges at the floor line can be softened or eliminated during the framing layout by simply shifting a wall over an ⅛ or ¼ inch.

Laundry Closet Dimensions

Figure 3.3 shows in plan view a typical laundry closet for a side-by-side washer and dryer arrangement. On one particular project, the plans gave a dimension of 36 inches for the laundry closet depth. Subtracting the thickness of a 2×4 wall and one layer of drywall on the inside face of the closet, plus the one layer of drywall on the

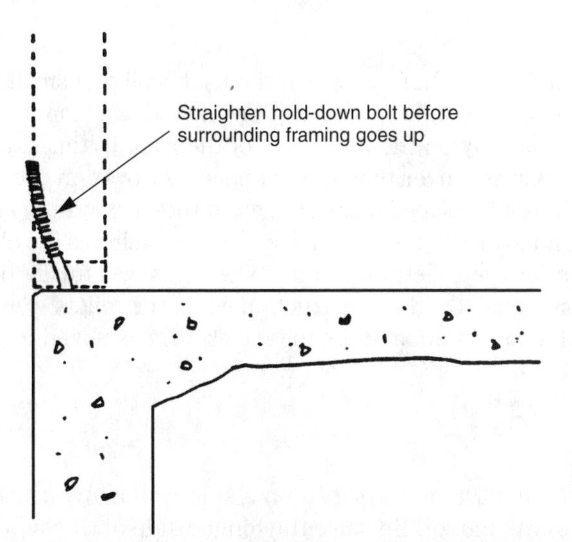

Figure 3.2

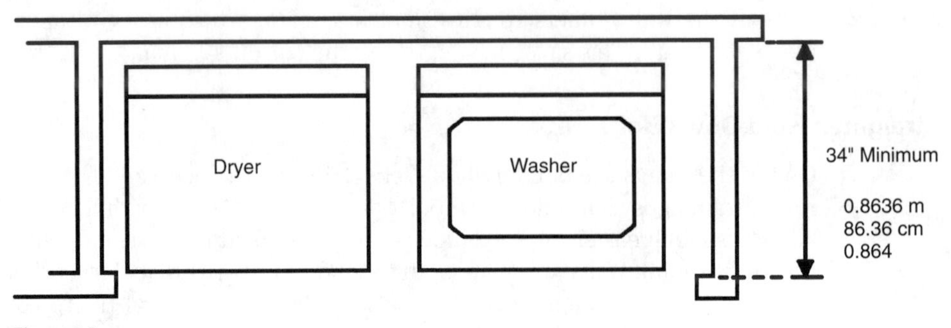

Figure 3.3

closet rear wall, the inside clearance for the washer and dryer turned out to be only 31¼ inches. This was not enough clearance for the dryer without squashing its vent hose and restricting its ability to dry the clothing.

We found that a minimum of 34 inches clearance inside the laundry closet was needed for everything to fit. The builder should check this laundry closet dimension during the plan review phase and deepen the closet if necessary.

Clearance for F.A.U. Closets

The builder should check before the start of the construction that the architect has allowed enough depth in F.A.U. closets for the required clearances (Figure 3.4). The furnace manufacturer will have recommended clearances behind and in front of the furnace that will be measured by the building inspector as part of the final inspection.

The time to find out that the required 2 inches and 6 inches minimum clearance behind and in front of the furnace is not available is not at the end of the construction when the house is complete.

Bath Walls and Toilets

Figure 3.5 illustrates a peculiar mistake in which the back wall of a small bathroom had its two corners designed and built at a 45-degree angle to accommodate the architectural design of the hallway on the other side of the wall. In this example, the lid to the toilet water tank was longer than the flat space leftover on the back wall, therefore the toilet could not be placed close enough to the back wall to engage the floor closet ring. The standard 15 inches away from the back wall was too close to the two angled wall corners for the toilet tank to fit. When this was finally discovered during the finish plumbing phase of the construction when the toilets were to be installed, the subfloor had to be cut out and the toilet closet rings moved further away from the back walls.

Medicine Cabinets

Bathroom medicine cabinets are often placed on a side wall above the bathroom cabinet and in front of a wall mirror. Because the hinged side of the medicine cabinet moves toward the mirror when opened, adequate clearance must be provided during the framing for the medicine cabinet (Figure 3.6). The rough framed opening should be at least 3 inches from the wall corner. This allows clearance for drywall, the thickness of the mirror, and the swinging action of the medicine cabinet door.

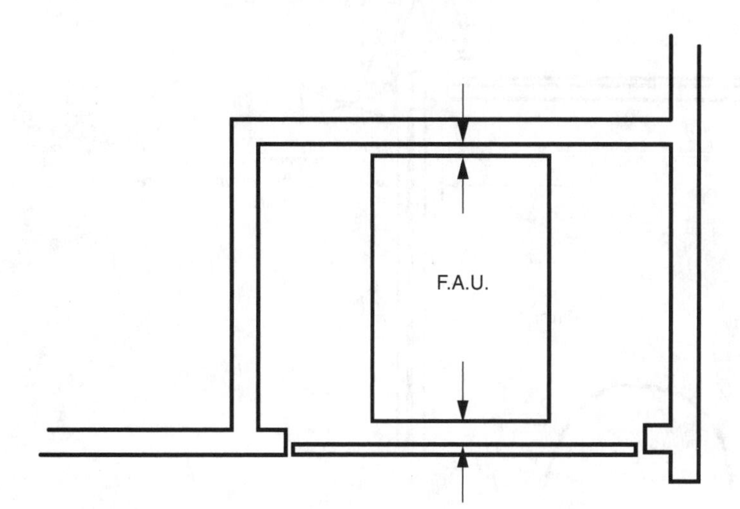

Figure 3.4

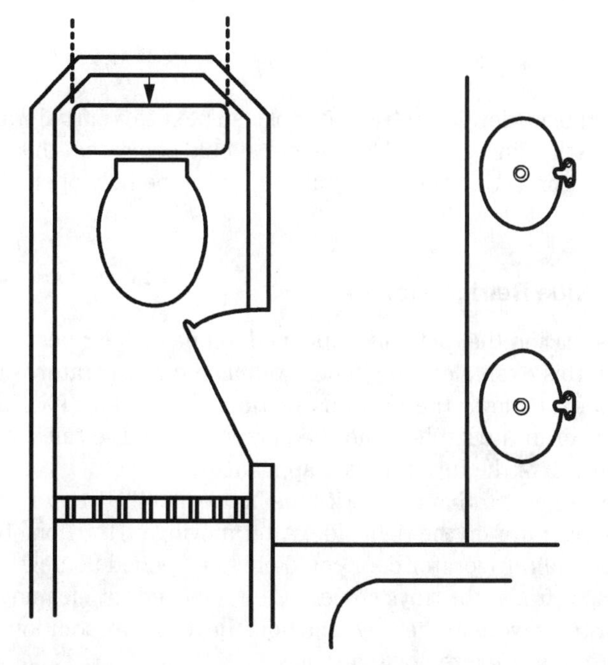

Figure 3.5

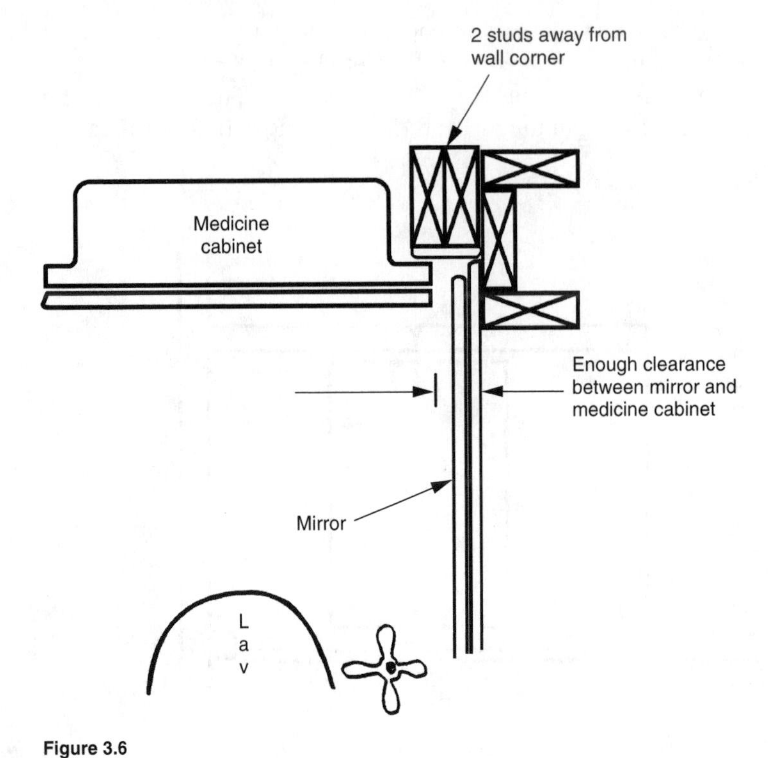

Figure 3.6

Wall Next to Refrigerator

Shown in Figure 3.7 in plan view is a refrigerator placed next to a partial wall that divides the kitchen from the dining room. The builder should check that this wall is not deeper than a refrigerator so that the refrigerator door can be fully opened.

Kitchen Wall Next to Double-Wide Refrigerator

Figure 3.8 shows a situation that actually happened in one of the sales models in a particular project. In this example, the kitchen wall at the refrigerator was deeper than the partial wall example in the previous section. The builder had installed a double-wide refrigerator in the kitchen and the person buying the sales model purchased the house with all of the furniture and appliances.

The double-wide refrigerator that came with the house had the freezer on the left-hand side and the refrigerator on the right side. After moving in, the homebuyer discovered that because the refrigerator door could only be opened 90 degrees due to the adjacent wall, the refrigerator trays could not be removed for cleaning because the refrigerator door trays were in the way. The homebuyer's only solution would be to pull the refrigerator out, then remove the trays.

Chase Dimensions in Closets

The plan view of the wardrobe closet pictured in Figure 3.9 shows a mechanical chase in one corner of the closet. The builder should check, during the preconstruction plan analysis period, that the dimensions of this chase are large enough to accommodate the closet shelf and pole. For example, a standard shelf and pole needs a wall space of at least 15 to 16 inches deep. This allows for the depth of an 11½-inch-wide shelf plus another 2 or 3 inches for the pole, which is placed just outside the

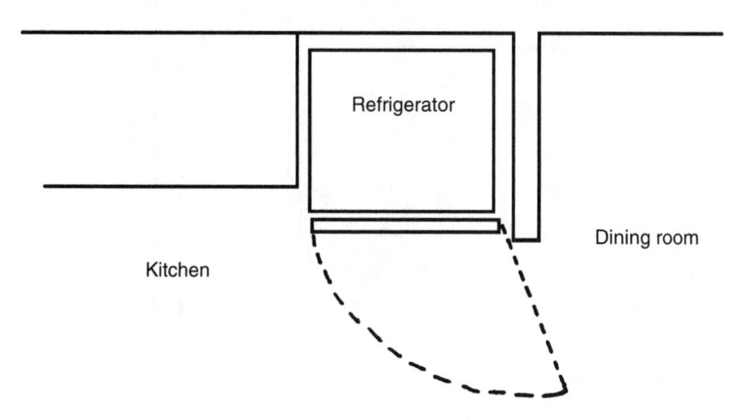

Figure 3.7

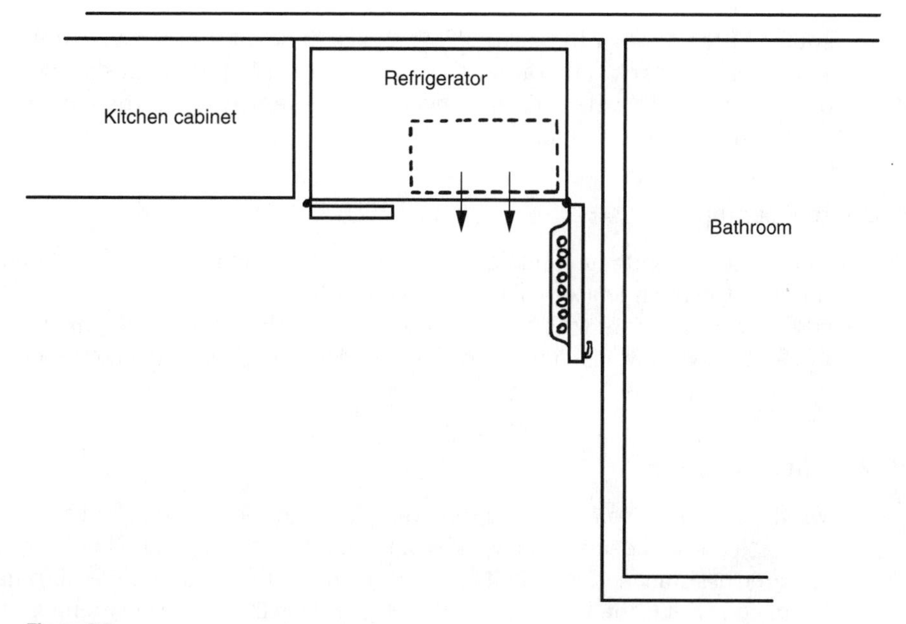

Figure 3.8

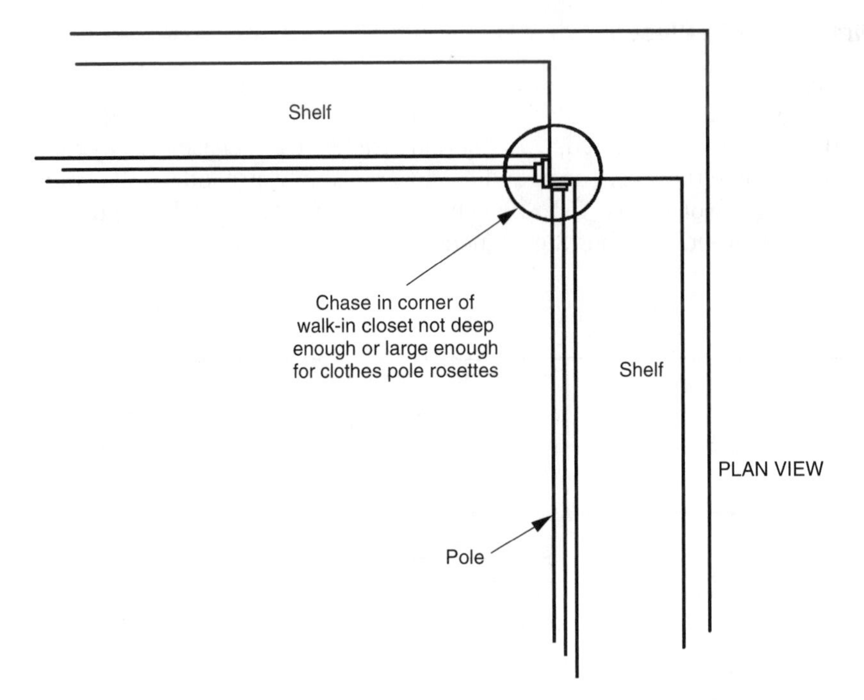

Figure 3.9

front edge of the shelf. If the chase is mistakenly designed on the plans to be 12 inches deep or is built that way as a later addition, the clothes pole must be installed at the edge of the chase. This makes the center of the pole only 11 inches away from the back wall, rather than the normal 14 or 15 inches away. The homeowner must fight with this narrower space every time clothes are hung on or taken off the clothes pole.

Trash Chute Openings

In multistory condominiums and apartments, trash chute openings should be centered on the trash chute wall or be placed with enough distance from each side wall so as not to interfere with the door opening into the trash room (Figure 3.10). Figure 3.11 shows how placing the trash chute door too close to the corner can interfere with the opening of the trash room door.

Wardrobe Shelf Supports

Wardrobe closet shelf-and-pole require metal or wood supports, spaced apart every 4 feet, to assist in bearing the weight of the clothing and storage. In order to lend the required assistance, these shelf supports must be attached to the wall framing studs contained within the back wall of the closet. This limits the positioning of the supports within the closet dependent upon the wall framing layout. If the wall framing is laid out without regard for shelf-and-pole supports, as shown in Figure 3.12, the supports will be off-center within the closet.

By starting the framing layout within the back wall of the closet from the center and working outward, the shelf supports can be placed either in the center or equally spaced apart from the center in the closet.

Pantry Closet Shelving

Kitchen pantry or hallway linen closet shelving that is cut and installed on the job site by the finish carpenters (Figure 3.13) differs from wardrobe closet shelving. Unlike wardrobe shelving, shelf supports or cleats for these closets are all precut to a

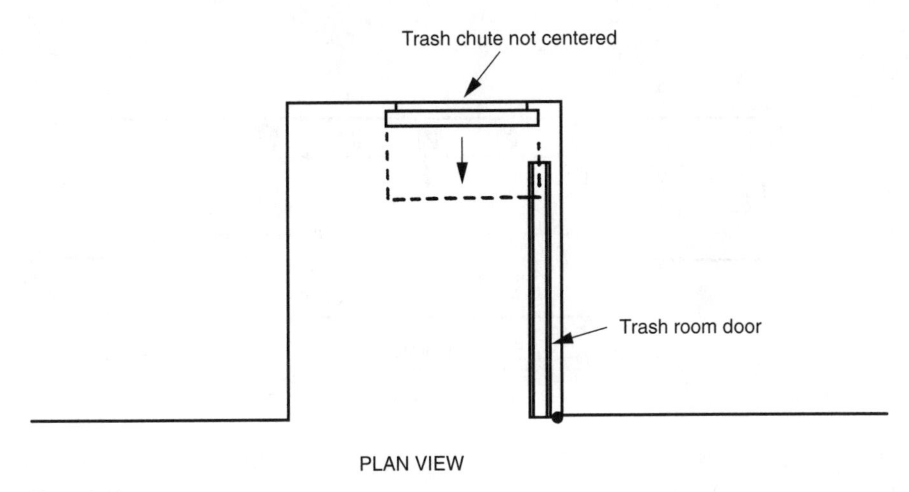

Figure 3.10

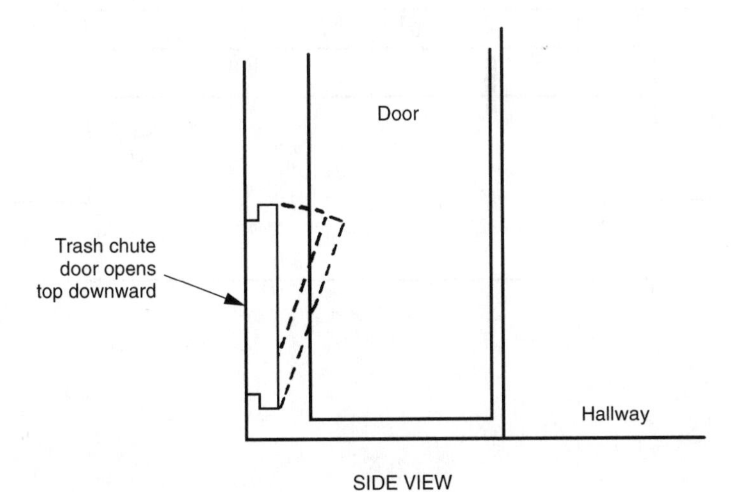

Figure 3.11

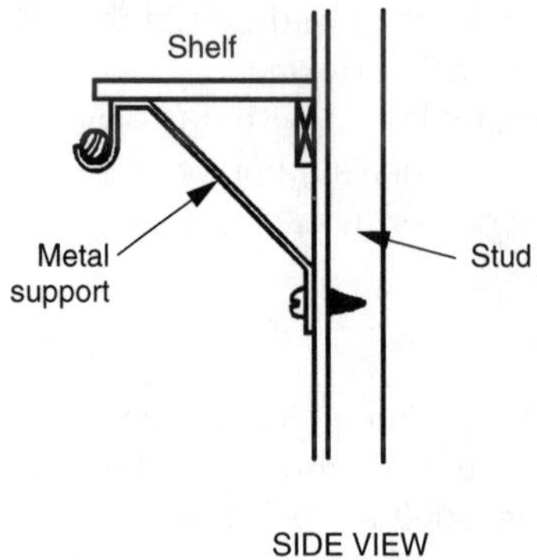

SIDE VIEW

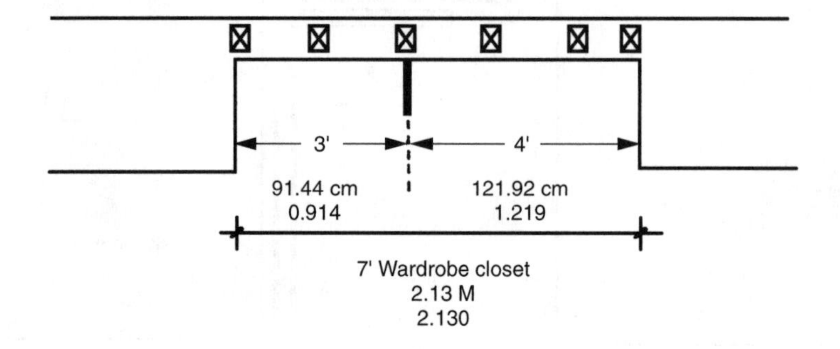

PLAN VIEW

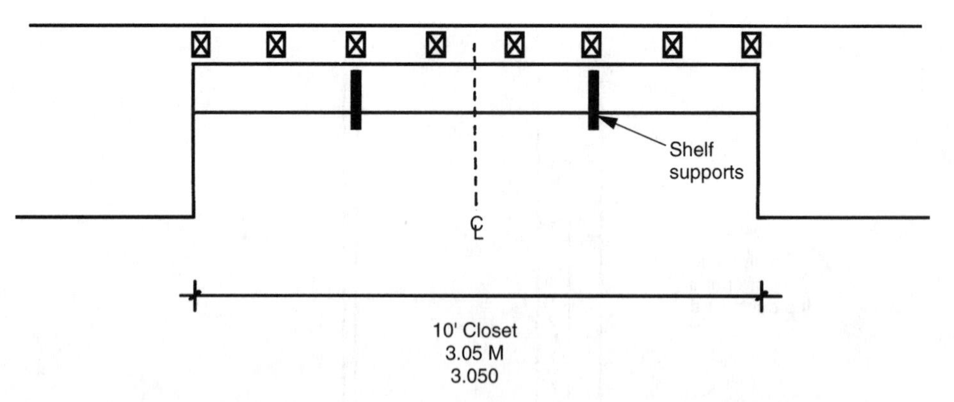

Figure 3.12

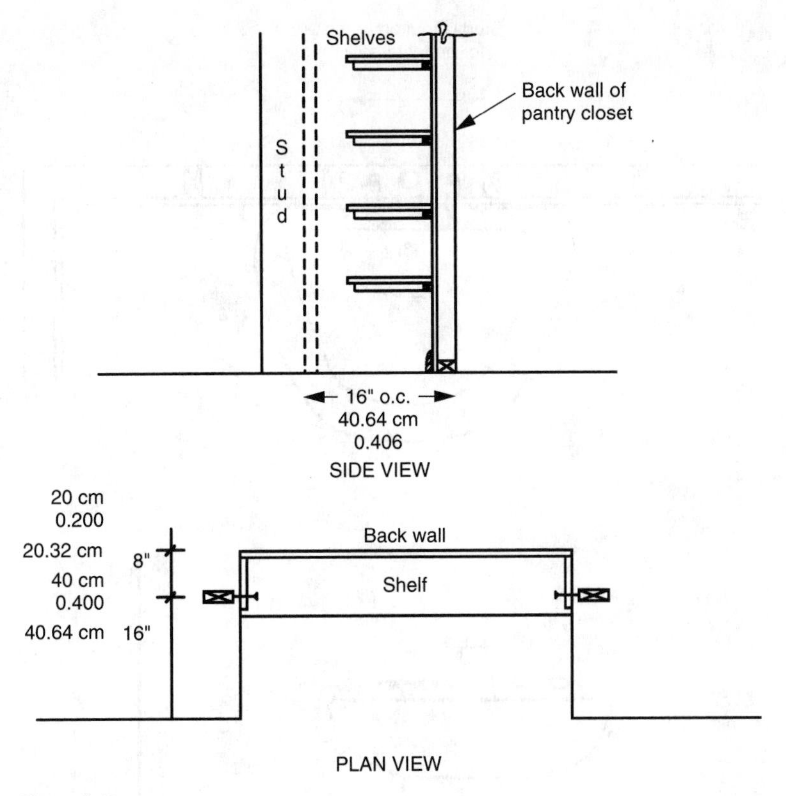

Figure 3.13

length slightly less than the depth of the shelving. This simplifies and speeds up the installation of a large number of shelves.

The problem is that the wall framing studs on the sides of the closet may be laid out 16 inches from the back wall and the shelf support cleats may not reach that far. To ensure that you hit solid wood for nailing for the shelf supports, the side wall studs should be laid out 16 inches from the front of the closet or the cleats should be precut to at least 16 inches long.

Layout for Pedestal Sinks

Figure 3.14 shows in plan view the framing stud layout for a bathroom pedestal sink. The studs are placed to provide a full open bay centered on the pedestal sink. This allows the plumber to bunch together the drainpipe and copper water lines so they are hidden underneath the sink bowl and behind the narrow pedestal base. Pedestal sinks look best when the water control valves and flex lines are hidden behind the sink (Figure 3.15). This requires an open framing bay to position everything correctly.

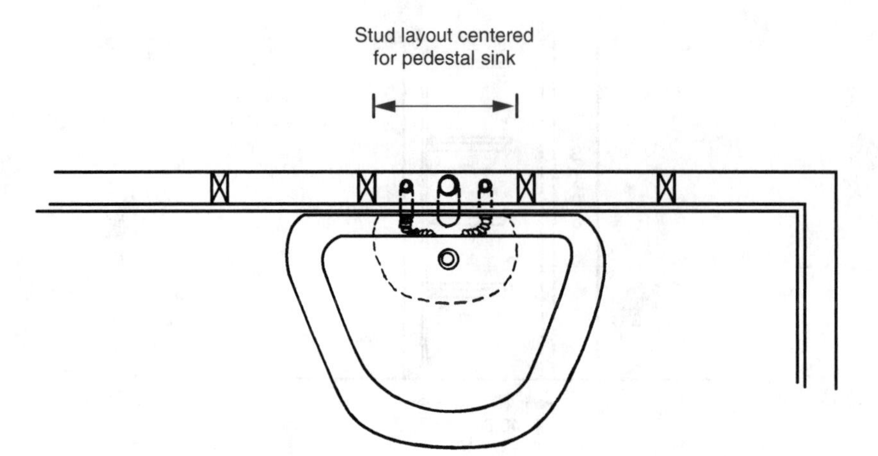

Figure 3.14

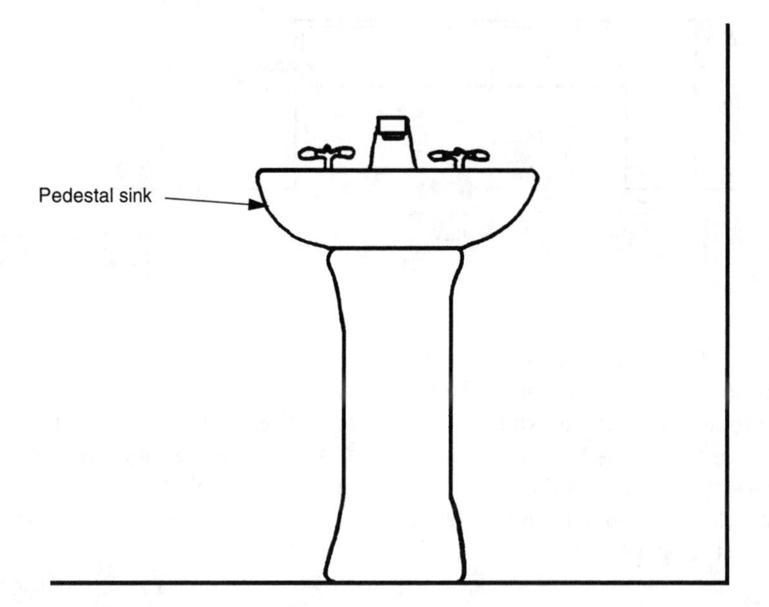

Figure 3.15

Kitchen Countertop Pony Wall

The end view of a kitchen cabinet attached to a half height pony wall is shown in Figure 3.16. The construction on this pass-through countertop can be simplified by deleting the pony wall and building the cabinet with a solid back of face frame and cabinet doors or a veneer plywood to match the cabinet fronts. This eliminates having to frame the pony wall at the correct height and location and adds 4 more inches of leg room underneath the countertop.

Space for Painting

Figure 3.17 illustrates a second floor hallway with linen cabinets and a partition/stairwell wall that is topped with a wood cap. The height of the partition/stairwell wall must be such that the space between the top of the cabinet and the wood cap is wide enough to paint.

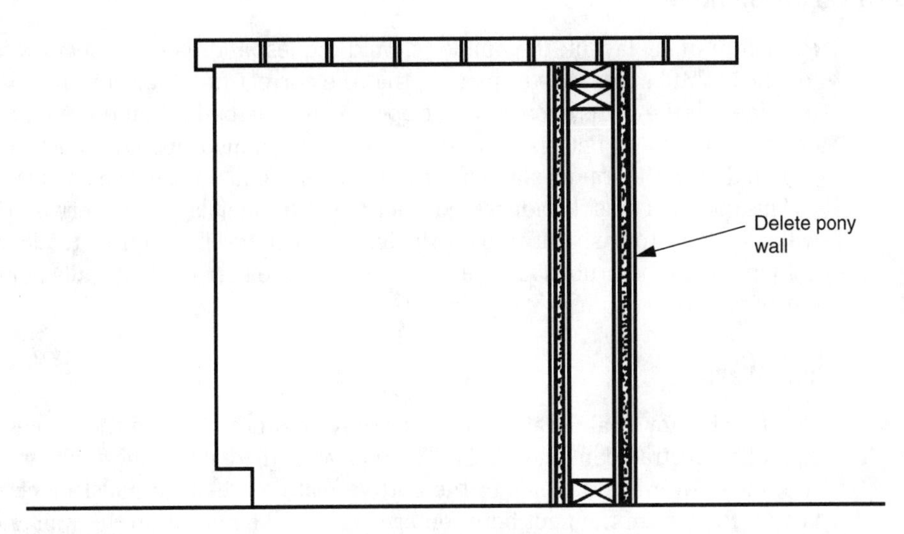

Figure 3.16

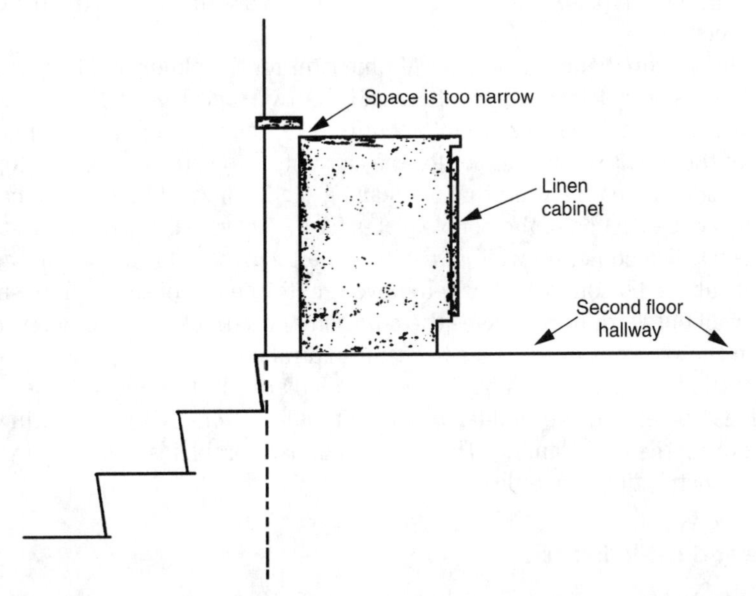

Figure 3.17

If the partition wall is casually built at 36 inches and the top of the cabinet is 35½ inches (standard rough-top height above floor), then the remaining gap between the cabinet and the cap leaves only a narrow ½ inch of exposed drywall to be painted.

If the wall is framed at 38 or 39 inches high, a 2- or 3-inch drywall band is exposed for easier painting and better appearance.

Allowance for Bathtubs

Prior to framing layout, the builder should request and receive specification sheets for the bathtubs and shower pans so that the correct framing allowances can be determined. If the plans show a 5-foot opening for the bathtub and the bathtub specs call for a 60½-inch net opening, then the rough-framed opening must be enlarged ½ inch during the framing layout for the bathtub to fit later in the construction.

This question must be addressed prior to the framing layout, otherwise the framer will later have to bash out walls with a sledge hammer in an effort to clear an extra ½ inch for the bathtub. This will then result in the surrounding walls being thrown out of plumb and square.

Framing Rake Walls

The framing for a rake wall while it is still lying on the floor in the process of being framed is illustrated in Figure 3.18. The typical method of framing a rake wall is to nail the uncut, over-length studs to the bottom plate on their normal 16-inch on-center layout. Blocks are then cut between each stud to be moved up the wall and used as spacers to get 16-inch on-center spacing at the top of the wall. The top plate is then laid on top of the studs at the correct rake angle. The top of the studs are marked with a pencil at the top plate line. The tops of each stud are then cut at the pencil mark at the correct angle.

One helpful tip that can save a lot of time later for the plumb and line crew is to cut the two outside blocks a ¼-inch short and the two top plates a ½-inch short overall (¼-inch short on each end) so that the entire rake wall is either parallel or slightly narrower at the top. It is much easier for the carpenters to hammer over the tops of walls on each side of the rake walls that are leaning slightly inward than it is to have to take apart and cut each end of the double plates for a rake wall that is too wide at the top.

Rake walls become too wide at the top due to spacer blocks that are cut slightly too long at the bottom, which when moved up to the top of the wall, push the ends of the wall outward. It only takes the addition of the thicknesses of a few saw-blade cuts on the fat sides of the lines to result in a rake wall that is ¼-inch or ½-inch too wide at the top. This throws off the perpendicular walls that meet it at each end.

This is something the builder can quality-control check by measuring the rake walls during the wall framing. This helps eliminate problems and thereby save time on the construction schedule.

Three-Inch Pipe and Let-In Brace

The end of a framed wall section in which a 3-inch waste pipe comes up through the concrete floor slab is pictured in Figure 3.19. A nail-on metal brace should be used

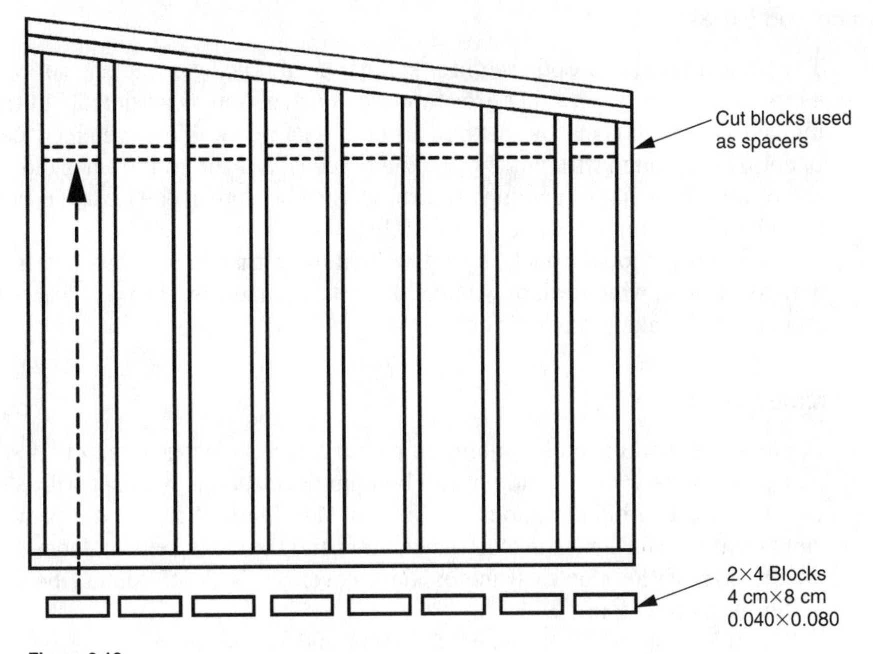

Figure 3.18

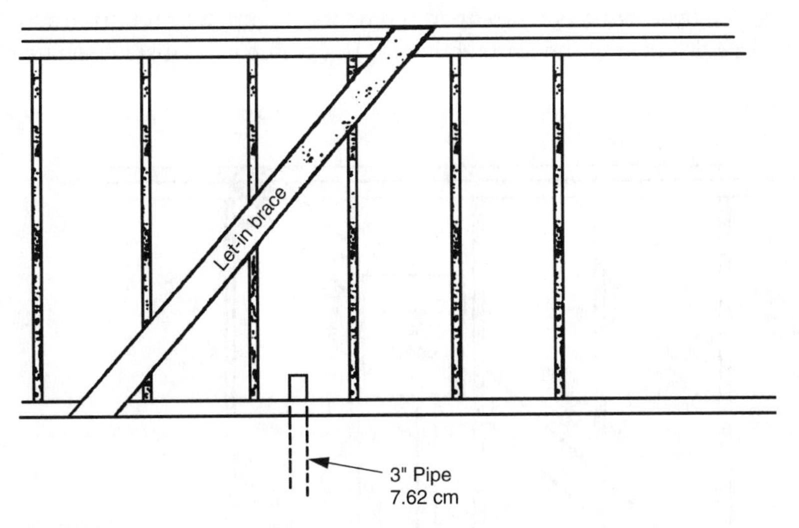

Figure 3.19

instead of the standard 1×6 let-in brace because the 3-inch pipe and the ¾-inch thick wood brace will not fit within a 3½-wide wall. Although no interference may be noticed at the time the wall is framed and stood up, the plumber will have to knock out the wall brace when it comes time to install the waste pipe.

Wall Brace and Breaker Panel

The Uniform Building Code requires some type of bracing at the ends of walls and at every 25 feet for houses that are standard construction. Occasionally in tract housing, a let-in brace is cut out at the corner of a garage wall for the electrical breaker panel. The problem with cutting out the brace is that the wall framer did not notice the plastic sleeve pipes coming up through the concrete at the garage corner, meaning a breaker panel would be installed there later.

The builder should check during wall framing that braces are not framed into garage corners, which will only have to be cut out later by the electrician who is installing the breaker panels (Figure 3.20).

Electrical Meter Closets

For apartments and condominiums, a row or bank of electrical meters is sometimes placed at the rear or one side of the building, enclosed in a closet with sliding plywood doors. The builder should check that the depth of this closet provides sufficient clearance for both the electrical meters and the sliding closet doors. Depending on the size of the electrical meter glass covers, this depth should be at least 16 inches or more (Figure 3.21).

Depending upon the number of meters and the space available for the meter closet, the builder should also consider using hinges on the closet plywood doors so that they swing outward. Holes cut in the plywood doors at each meter will allow meter reading without having to open the doors (Figure 3.22). Furthermore, this will also provide coverage and protection for the switch gear. If someone needs to work on the meters, the closet doors are simply swung open.

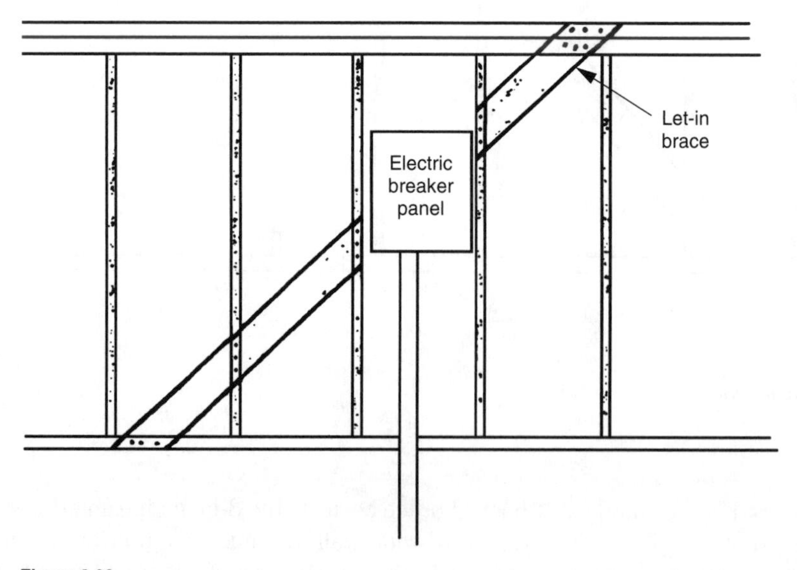

Figure 3.20

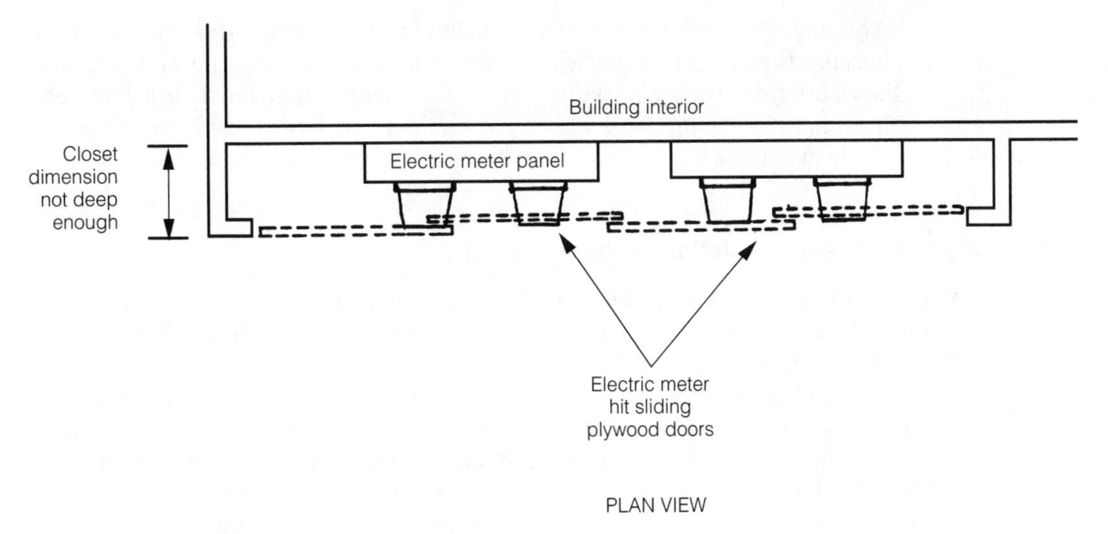

Figure 3.21

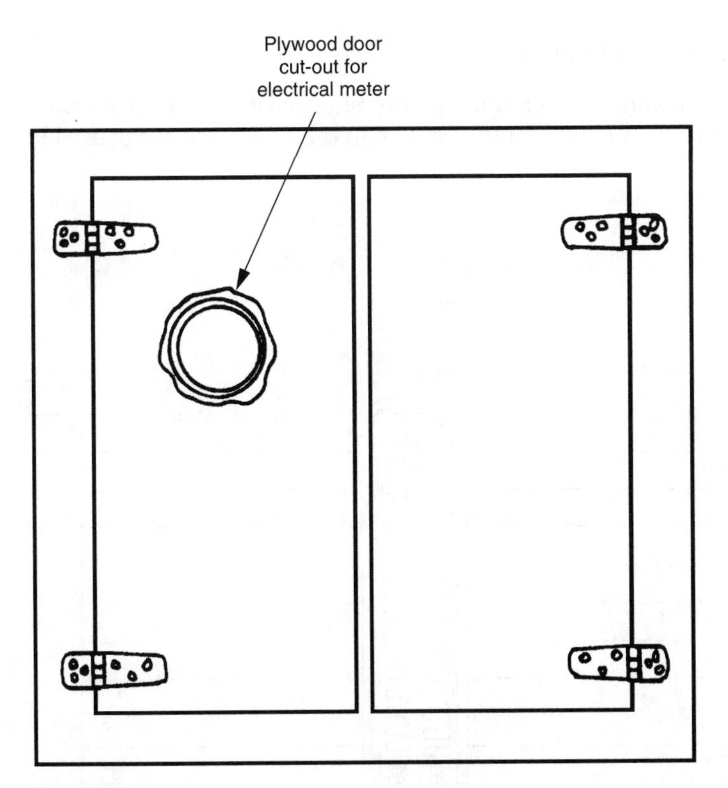

Figure 3.22

The situation to avoid is not to mix a shallow meter closet with sliding plywood doors because the meters sticking out through the holes cut in the plywood will prevent the doors from sliding back and forth. The meter closet must either be deep enough so that the sliding doors miss the meters or the doors must be on hinges so that they swing outward.

Preliminary Drywall and Insulation during Framing

When a minor amount of preliminary drywall or insulation is needed during rough framing, such as at a partywall with shear panel plywood and stairs on both sides of the wall, the builder should include this work in the framing contract. This allows the framer to insulate or drywall as the framing requires without having to wait for another subcontractor to show up for 15 minutes or a half an hour of work.

In the stair example shown in Figure 3.23, the shear panel must be installed and nailed off before the stairs can be built. Because the partywall in this case has both, shear panel and stairs on each side and must be insulated, it is easier for the framer to stuff the insulation into the wall prior to closing up the second side with shear panel plywood. The stair builder and the shear panel installers are not held up then waiting for the insulating contractor.

Grid-Lines on the Structural Plans

Illustrated in Figure 3.24 is a simplified version of a grid-line system used for a set of structural plans for a house. It helps both in the reading of the plans and in the vertical layout of the structural members to have the structural floor plans divided into

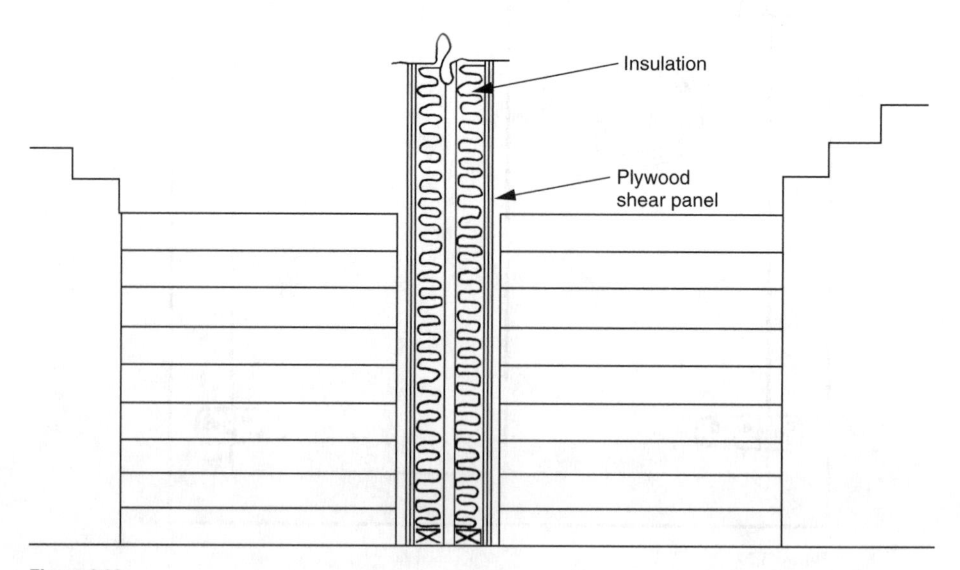

Figure 3.23

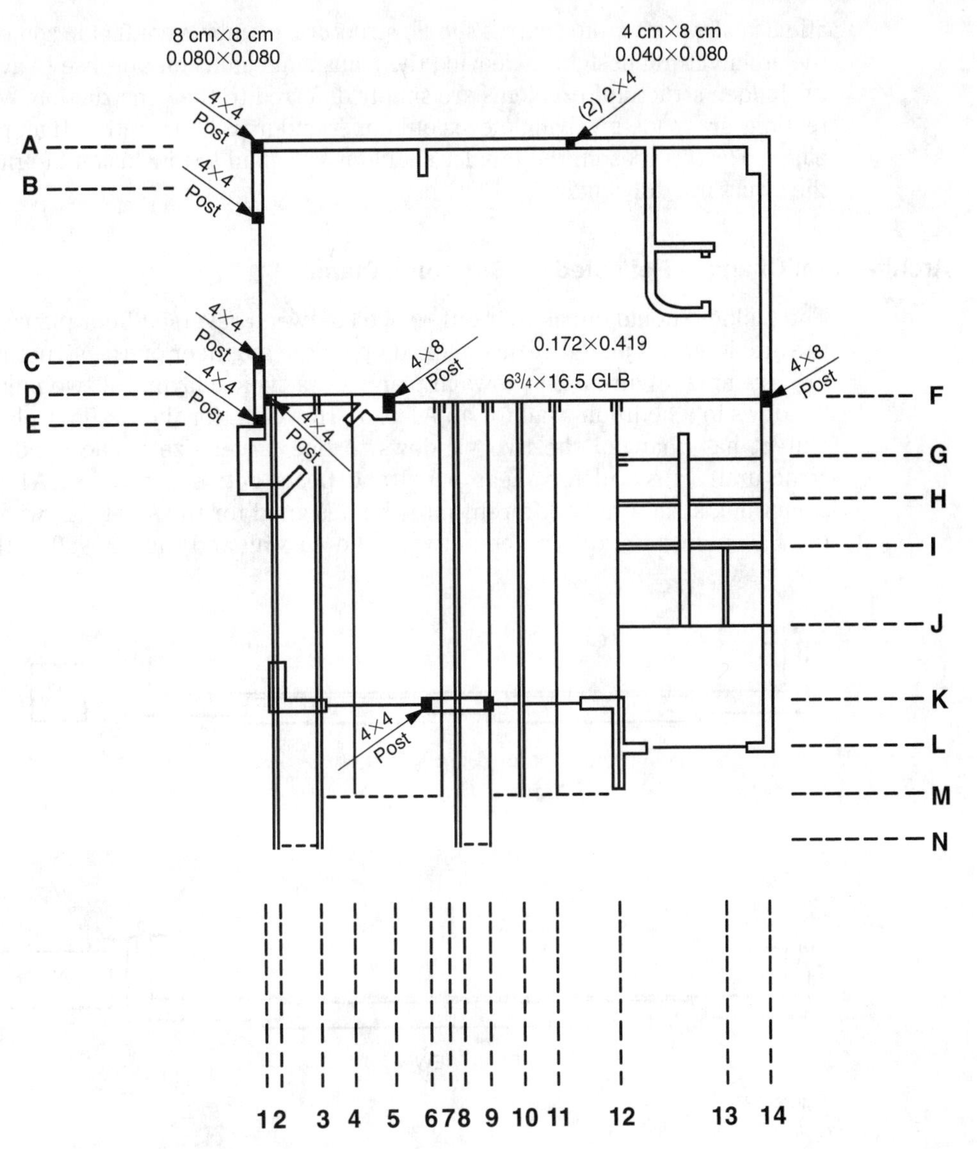

Figure 3.24

grids. Labeling the horizontal and vertical grids with letters and numbers helps in tracing a structural item such as a post and post anchor from the first floor, through the second floor, and up to the roof. Conversely, by finding the grid letter and number for a structural member in the roof, the structural support can be traced down through the building with greater clarity and ease.

Some structural engineers do not like to use grid-lines. They feel using grid-lines places the responsibility for getting everything to line-up upon themselves, thus creating liability for framing extras should structural members conflict in some way with the architectural design. By leaving the plans ambiguous on purpose to avoid liability, hidden structural problems are simply deferred to the construction, where corrections are made involving pieces of lumber and man-hours rather than pencil and paper. For this reason, the builder should insist upon the inclusion of grid-lines on the structural drawings.

Architectural Changes Reflected on Structural Plans

The builder should check that differences between reversed floor plans shown on the architectural plans are also picked up by the engineer on the structural plans. For example, Figure 3.25 shows in plan view a typical layout of two medium-size windows in a bedroom wall for an A1 plan. However, for the A2 floor plan, the architect has changed the two windows to one larger-size single window. If the structural plans call for shear panel and posts between the two A1 windows, something structurally different must be designed for the single A2 window. Figure 3.26 shows the difference between the A1 wall and the A2 wall in the elevation view.

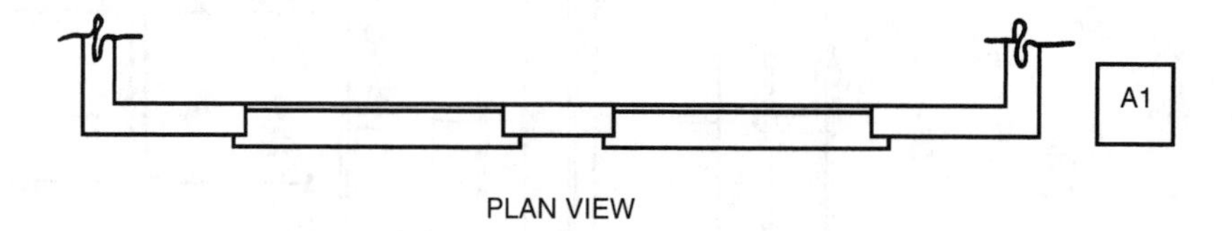

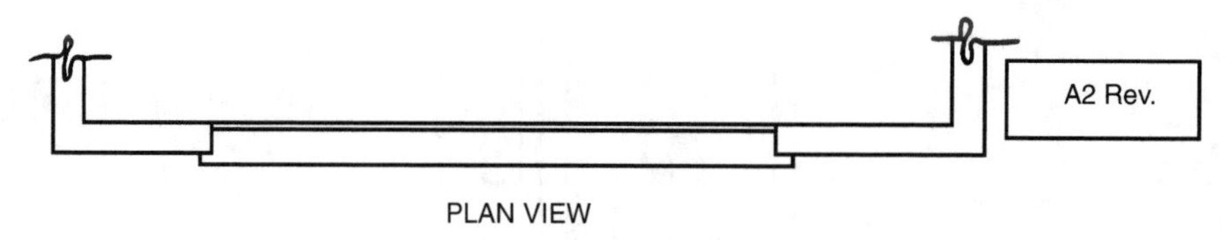

Figure 3.25

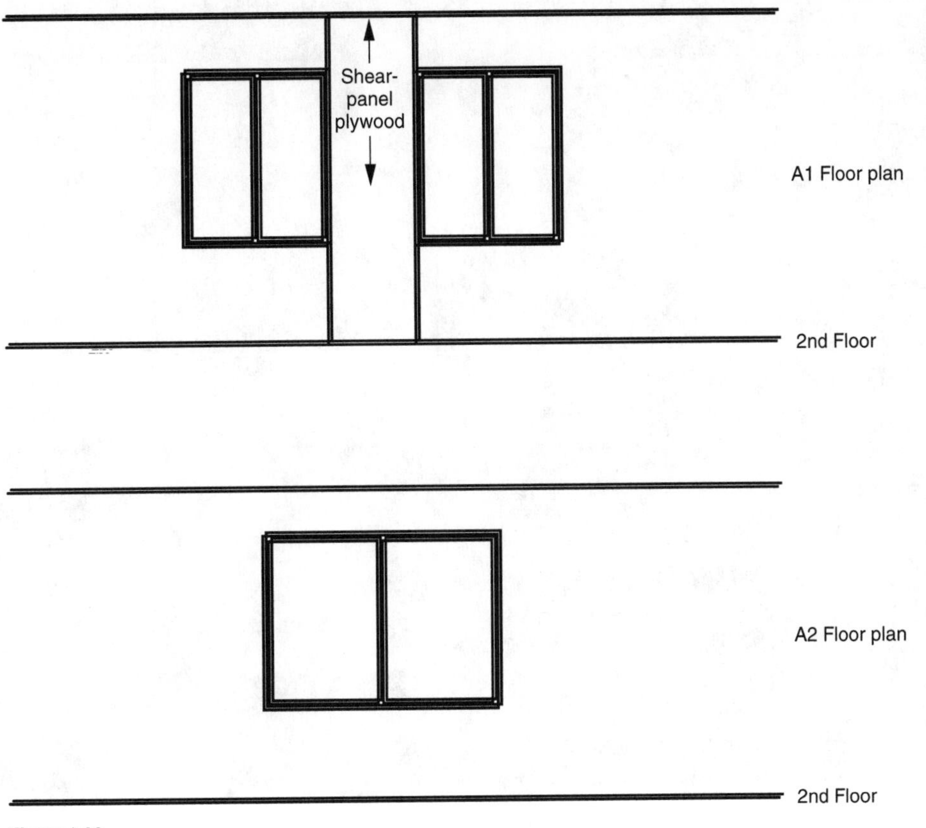
Shear-
panel
plywood
A1 Floor plan
2nd Floor
A2 Floor plan
2nd Floor

Figure 3.26

Chapter

4

Framing Floors and Ceilings

Joist Layout and Blocking

It is important that floor and ceiling joists be framed on their correct spacing layout. Joists that are off layout make the installation of plywood sub-flooring and ceiling drywall more difficult and time-consuming. Joists usually become off layout when cross-sectional joists, which are solid or mid-span blocked with precut blocking, bow. A series of bowed joists, adding an ⅛ to ¼ of an inch at each joist, can quickly throw the layout off (Figure 4.1).

All runs of floor or ceiling joists should be checked every 8 feet to ensure a true layout.

Joist Blocks

Illustrated in Figure 4.2 is the end view of a first floor wall below second floor joists. In this example, solid blocking is nailed between the floor joists and on top of the wall framing. If the joist blocks are cut out of a joist that is slightly deeper in dimension than the other floor joists, then the blocks will stick up above the joists because of the wall framing below. Even a slight projection above the other floor joists can result in a hump in the floor.

Layout Joists and Beams for Flush-Lights

The plan view in Figure 4.3 shows a powder room containing a pedestal sink and a toilet. A flush-light is centered directly above the pedestal sink. In this example, the center of the pedestal sink is placed 16 inches from the side wall. Because the flush-light is centered over the sink, its center is also 16 inches from the side wall.

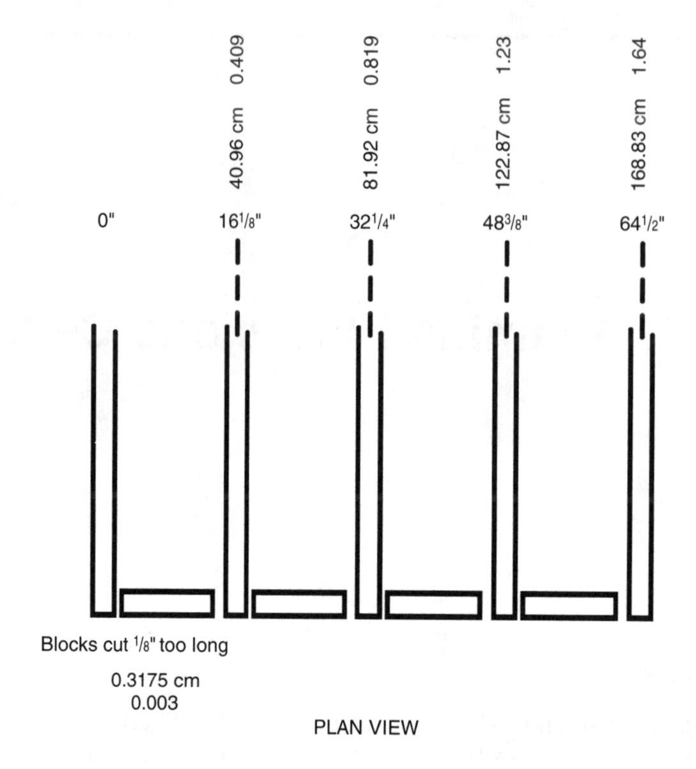

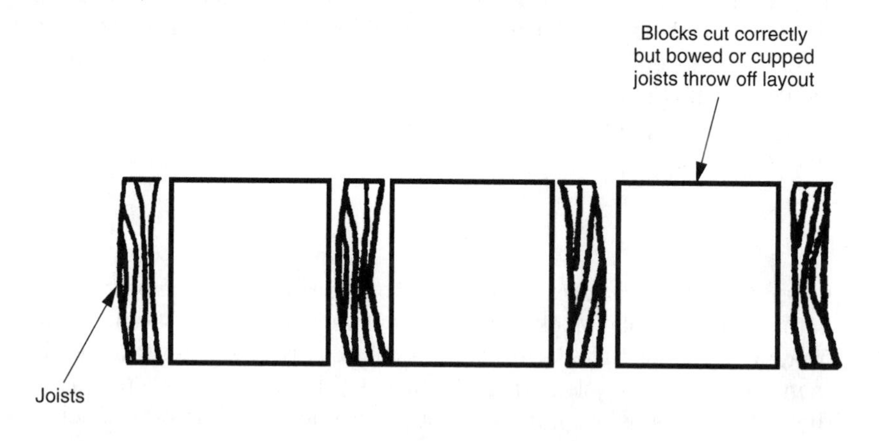

Figure 4.1

The problem to avoid is laying out and framing the ceiling joists on a normal 16 inches, on-center from the side wall without knowledge of or consideration for the flush light that must be installed later. With this joist layout, the first joist will land exactly at the center of where the flush-light must go. The joist layout in this case should start with a clear joist bay centered with the flush-light, then begin its 16-inch, on-center layout in both directions from there. This saves coming back later, during the rough electrical wiring, and having to cut-in a head-out for the flush-light when it is discovered that a joist is directly in the way.

Ceiling Joist Overlap

Ceiling joists are required to overlap a certain amount over bearing walls. This overlap should not exceed the required length. If joist ends extend a few feet past the

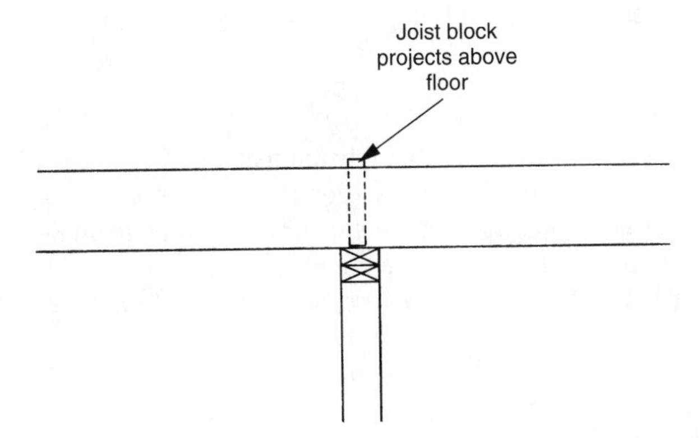

Figure 4.2

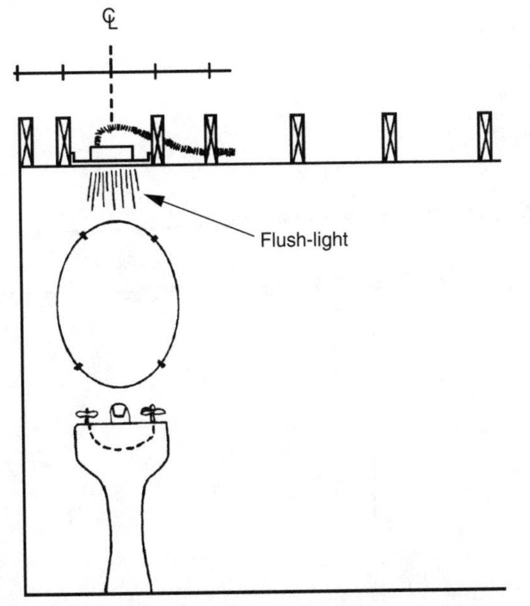

Figure 4.3

bearing wall, movement at the joist end may cause drywall cracks. A slight deflection in the ceiling joists results in a teeter-totter effect, raising the joist ends enough to cause nail pops or cracks as shown in Figure 4.4.

Glulams Next to Joists

Floor joist systems combining glulam beams with truss floor joists should not have a cambered glulam beam parallel to a floor joist. Figure 4.5 illustrates a cambered glulam beam placed in the center of a living room floor. A standard camber glulam has about ¼-inch to ⅜-inch upward crown in 20 feet of length. The adjacent parallel truss floor joists are manufactured straight without camber. The crowned member placed in the middle of flat joists will cause a hump in the floor (Figure 4.6).

Deflection Problems

The steel I-beam or a structural wood glulam beam featured in Figure 4.7 is placed in the center of a room, connected to cantilevered floor joists that support a second floor roof above. In this example, the steel I-beam or the wood glulam beam must be designed to be straight after all the other members are connected and the loads applied. Any downward deflection due to overspan or upward camber not equally counterbalanced by the loads will transfer from the center of the room through to the cantilevered ends of the joists, up to the roof line.

If the steel I-beam deflects downward, for example, the floor joists connected to it will also go downward at the center of the room, causing a teeter-totter upward effect at the opposite cantilevered ends of the joists. This then creates an upward

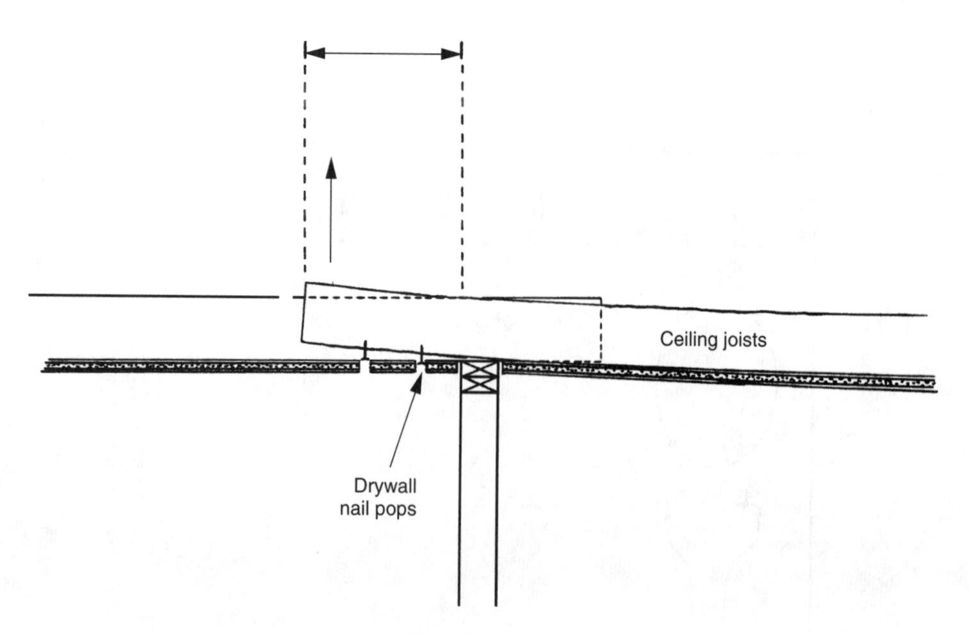

Figure 4.4

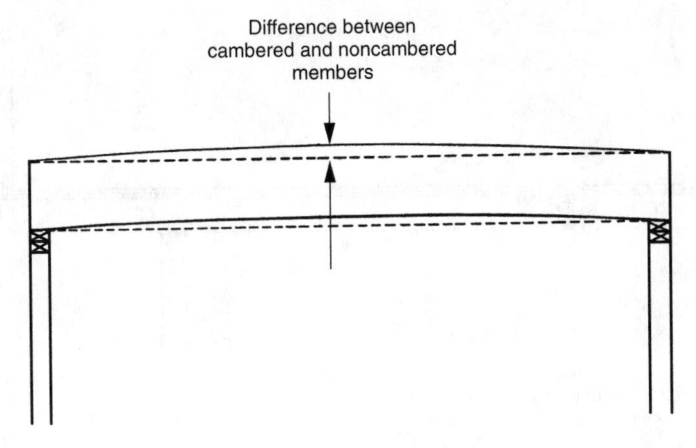

Figure 4.5

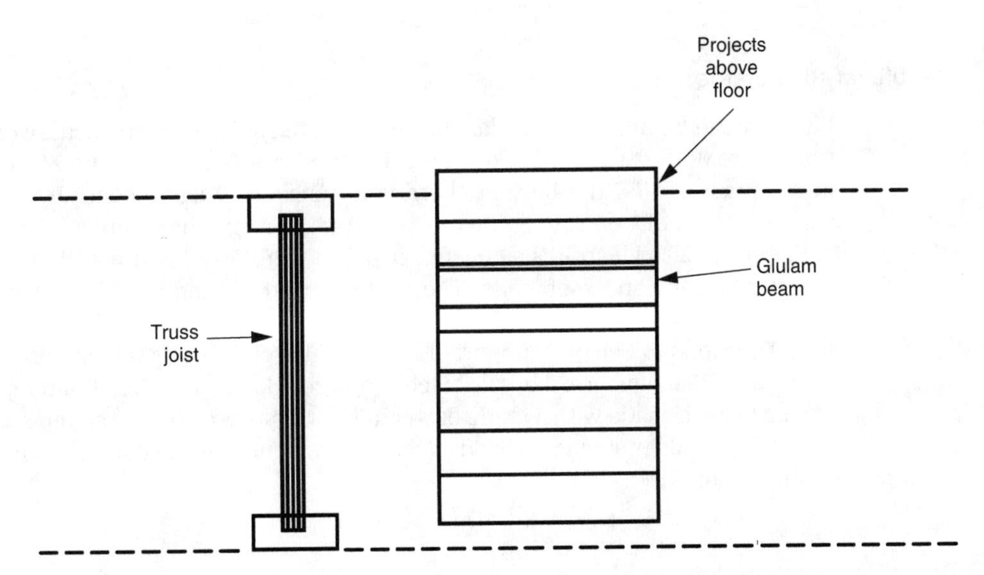

Figure 4.6

crown in the exterior floor line and transfers through the wall framing to the double top plates supporting the rafter tails and facia board. Depending on the amount of deflection and the length of joist cantilever, this can result in an upward or downward variation at the roof line for each room that has a deflecting I-beam or glulam beam.

This situation can create havoc for the framing carpenter who is trying to achieve a straight rafter tail and facia board roof line. If one second floor bedroom has a floor with an I-beam that deflects downward, and adjacent bedrooms have cambered glulams that do not deflect, as actually occurred in one tract of houses, the double plate line at the roof can end up uneven.

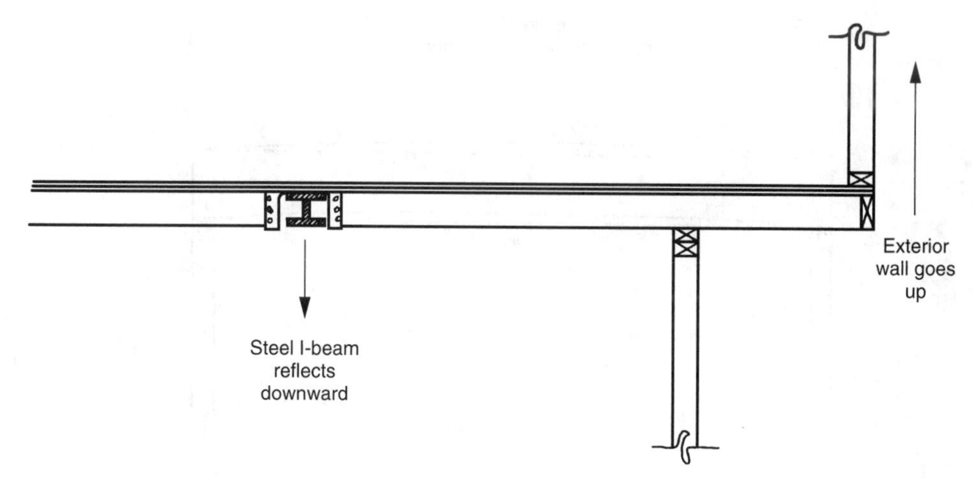

Figure 4.7

Cantilevered Glulams

Illustrated in Figure 4.8 is a glulam beam supported at both ends. Because of the natural desire of the beam to sag downward in the center, the top chord of the beam is in compression and the bottom chord is in tension. Glulam beams are made with varying grades of wood along the top or bottom to resist these forces.

Figure 4.8 also illustrates a glulam beam that cantilevers past a wall to support an exterior balcony and roof loads. The center portion of the beam still wants to sag downward but the tension side changes from bottom to top as the beam changes from floor joist to balcony support. The curvature changes as the end of the glulam beam is pushed downward by the balcony and roof loads. For this situation, a special glulam beam is made with a grade of wood that resists tension on the top and bottom chord. The builder should check during procurement that the correct type of glulam beams are ordered.

Don't Use Large Beams in Floors

During the plan review, preconstruction phase, or even prior to the start of the design work, if the opportunity arises, the builder should check that large beams, such as 10×12s or 12×12s, are not used in the floor joist system.

Large cross-sectional lumber has a tendency to twist as it dries out. When a 10×12 or 12×12 beam is used in a floor system, it only takes a slight twist in the beam to rock the 9½-inch or 11½-inch flat top surface of the beam out of level, forcing one of its edges up anywhere from ⅛ to ½ of an inch. This results in a hump in the floor. When a smaller size beam twists, such as a 4×12 or 6×12, the flat surface on top of the beam is only 3½ or 5½ inches wide, so the amount one edge of the beam can lift is less.

Because none of the surrounding members, such as plywood subflooring, blocking, or rim joists, are strong enough to resist the twisting force of a large beam, it will twist. The best defense is to take these large dimensional pieces of lumber out of the structural floor design. The builder should ask the architect or structural engineer to substitute a glulam or microlam beam in place of the solid, monolithic wood beam. These structural members are designed to equal the strength of solid wood beams but without the unpredictable twisting and bowing that is common to solid wood beams (Figure 4.9).

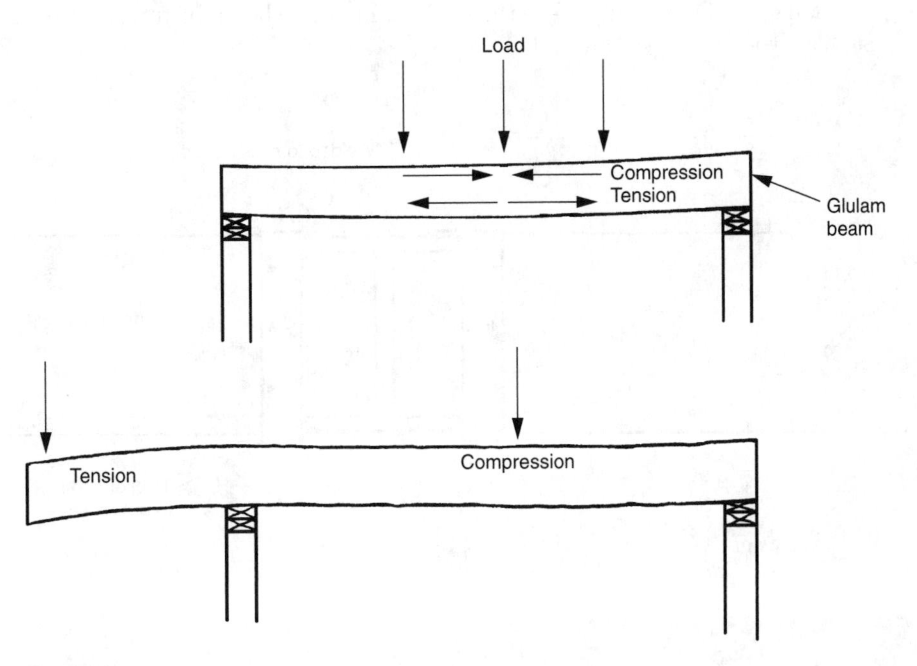

Figure 4.8

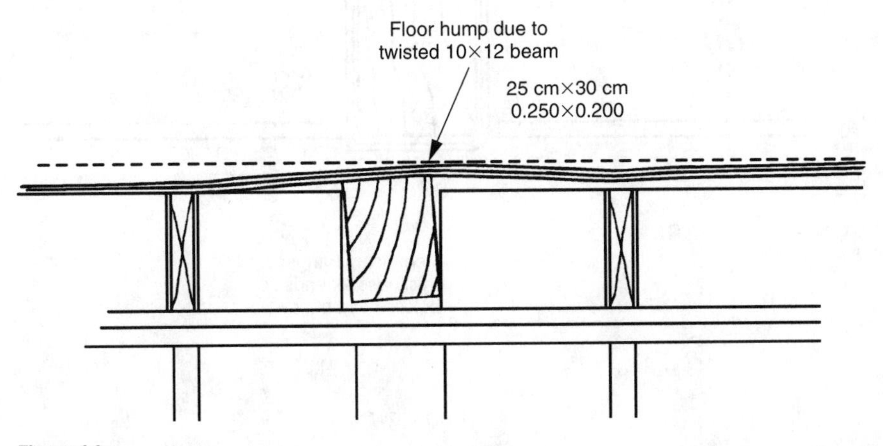

Figure 4.9

Thick Joist Hangers

When steel I-beams are used with a wood joist floor system for a second floor, the metal joist hangers are often a thicker gauge material because they have to be welded to the I-beam (Figure 4.10). Even when the bottom of the hangers are a manufactured brand with form-fitting tight corners, the thickness of the metal will still cause a downward bulge in the underside of the first floor ceiling. To eliminate this bulge, a ceiling having these thicker joist hangers should be furred with a 1×4 or similar wood strips (Figure 4.11).

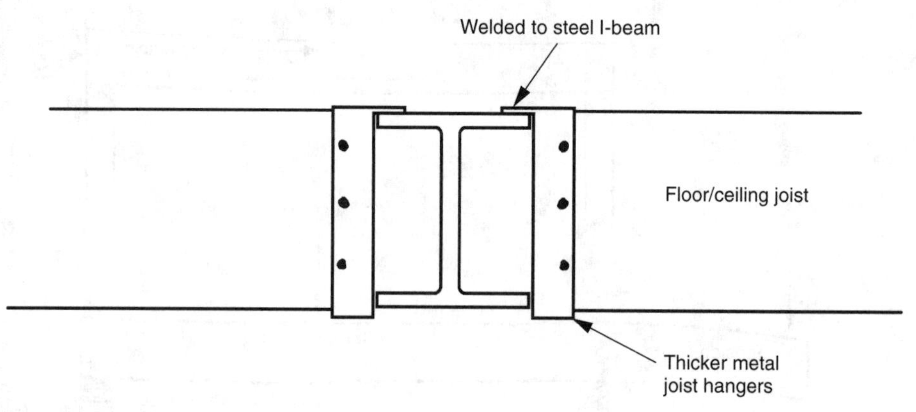

Figure 4.10

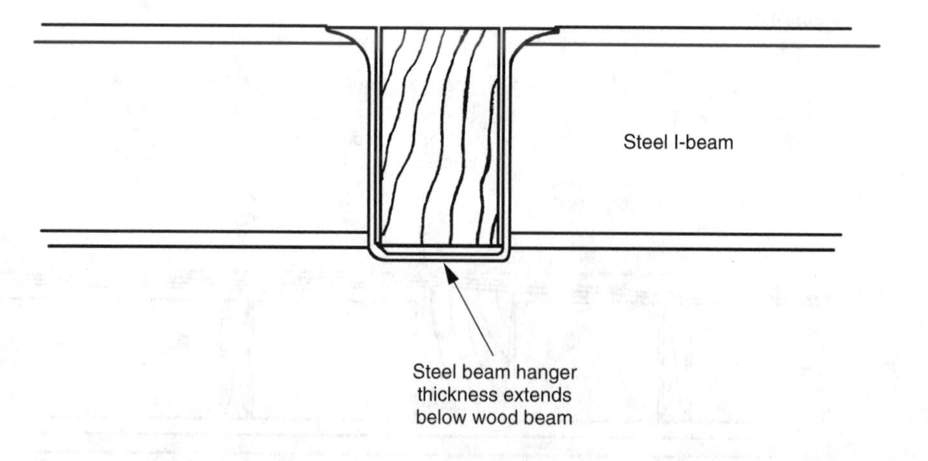

Figure 4.11

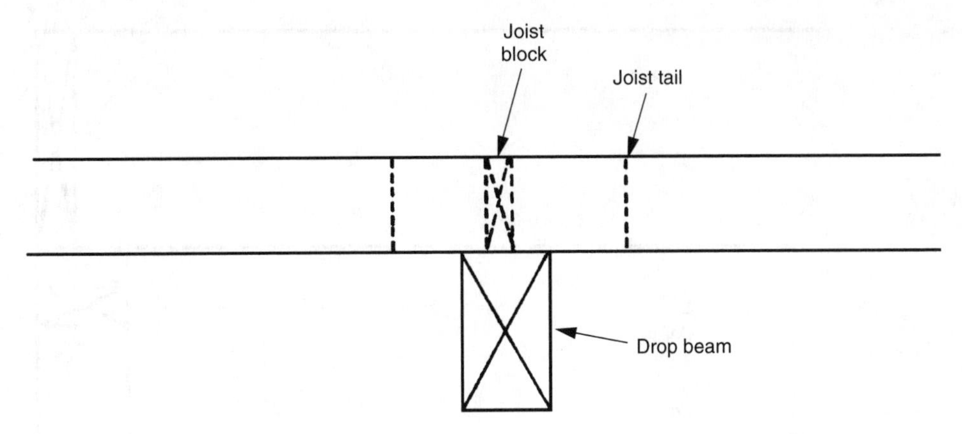

Figure 4.12

The builder should anticipate this extra furring in the framing contract. If not anticipated, the builder will get stuck having to pay for the furring of a number of ceilings at a more expensive time-and-material rate rather than being included more cheaply in a lump-sum bid.

Drop Beams vs. Flush Beams

A structural wood beam found in a garage is pictured in Figure 4.12. Because it is below the floor joists, it is called a drop beam. Drop beams can save money during framing because the joists and beams above rest on top of the drop beam, thus eliminating joist and beam metal hangers. On a large project, this can be a substantial cost savings. Before a drop beam can be used instead of a flush beam, it first must be determined that there is enough headroom clearance between the finish floor elevation of the garage slab, and the bottom of the proposed drop beam.

Microlams as Garage Headers

Illustrated in Figure 4.13 is an end view of a microlam beam used for a garage door header. Because microlam beam construction is similar to plywood, the edges of microlam beams reveal exposed wood veneer edges and glue joints. In this instance, if the garage door rough header height is 7 foot, 1 inch above the floor, then the microlam beam should be framed ¾ or 1½ inches higher. This allows the microlam to be furred out on the bottom edge with 1- or 2-inch finish trim material, thereby hiding the rough looking veneer edge (Figure 4.14).

Bolt Microlams

When two or more microlam beams are combined into one structural beam, the engineer usually asks that they be bolted together, top and bottom, at some staggered

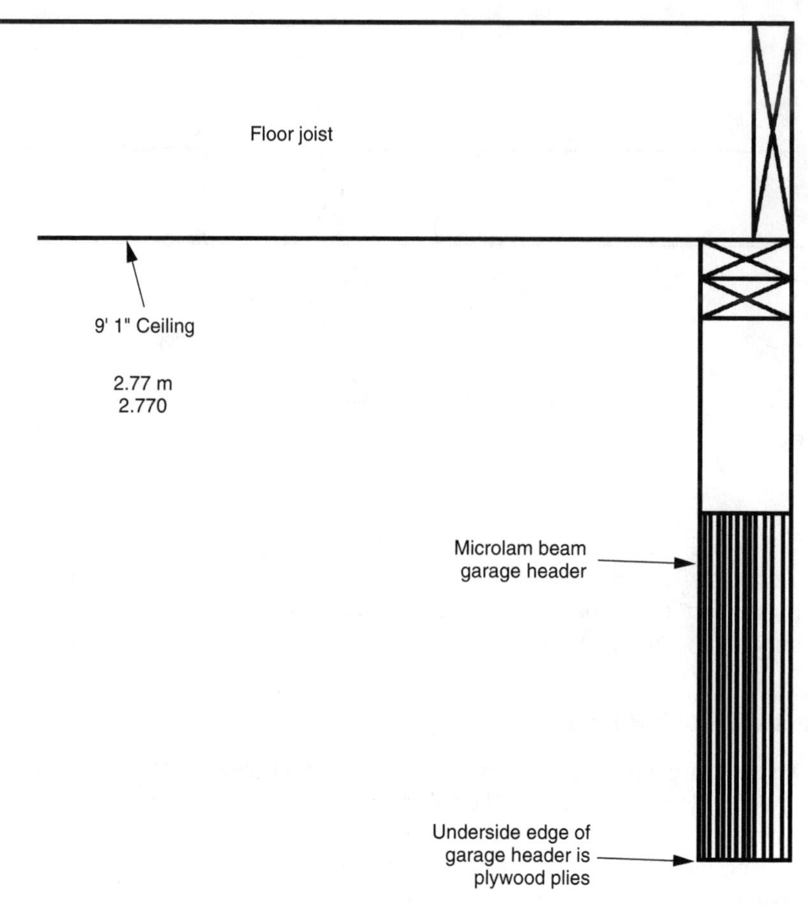

Figure 4.13

increment such as 18 or 24 inches on-center. Figure 4.15 shows one such microlam combination between two floor joists. Due to the lack of working space around the microlams once they are in place, they should be bolted together on the ground and before installation.

Fur Bolted Microlams

Figure 4.16 shows the side view of a three-piece microlam beam used in a garage. If the three beams are required to be bolted together so that they become one structural member, each side must be furred out to cover the bolt heads and nuts.

If the underside of the ceiling and microlam are drywalled, such as in a garage with living space above, this furring should be anticipated from the plans and spelled out in the contract. This furring can then become part of the lump-sum framing bid and not an expensive time and material extra.

I-Beam Flange Connections

Occasionally a structural detail will call for an end connection between a steel I-beam and a wood beam by means of a metal flange with bolt holes. Figure 4.17 shows this connection from the side and from the top. In this example, the saw kerf opening in the end of the wood beam to admit the metal flange must fit snugly around the metal flange. If the cut is carelessly made too wide, one side of the beam may break as the flange bolt nuts are tightened.

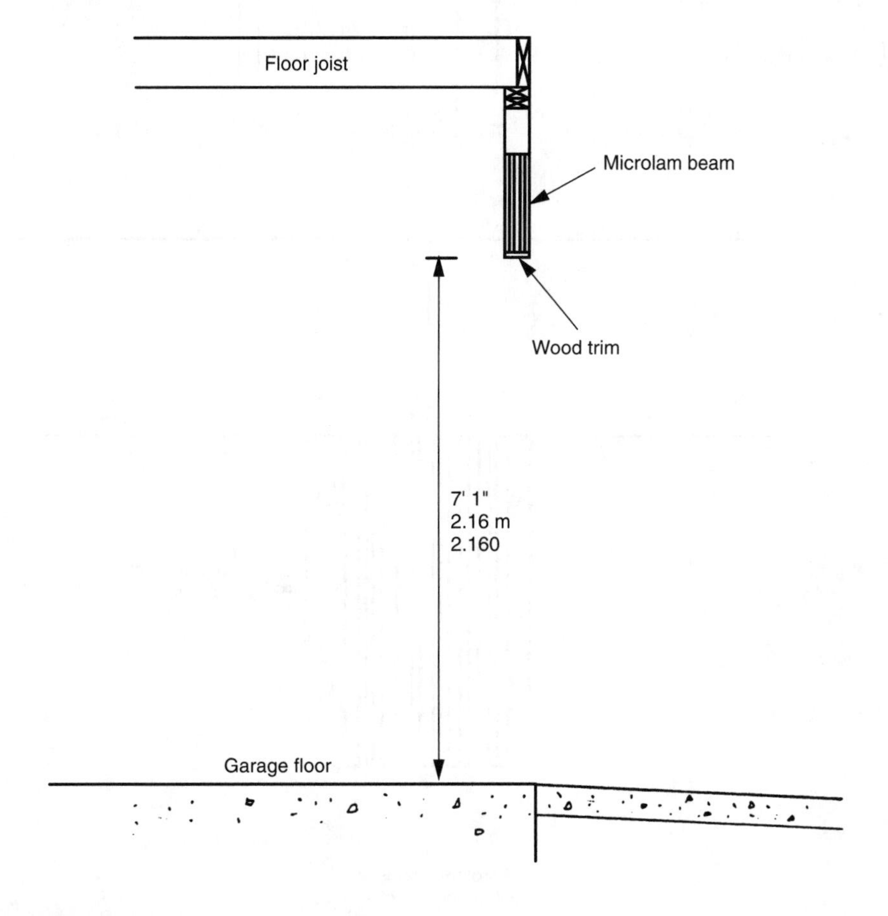

Figure 4.14

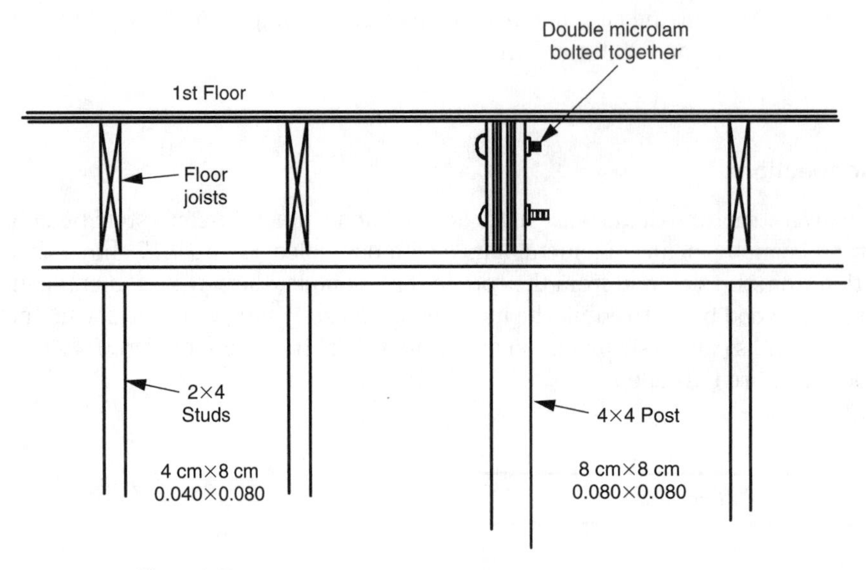

Figure 4.15

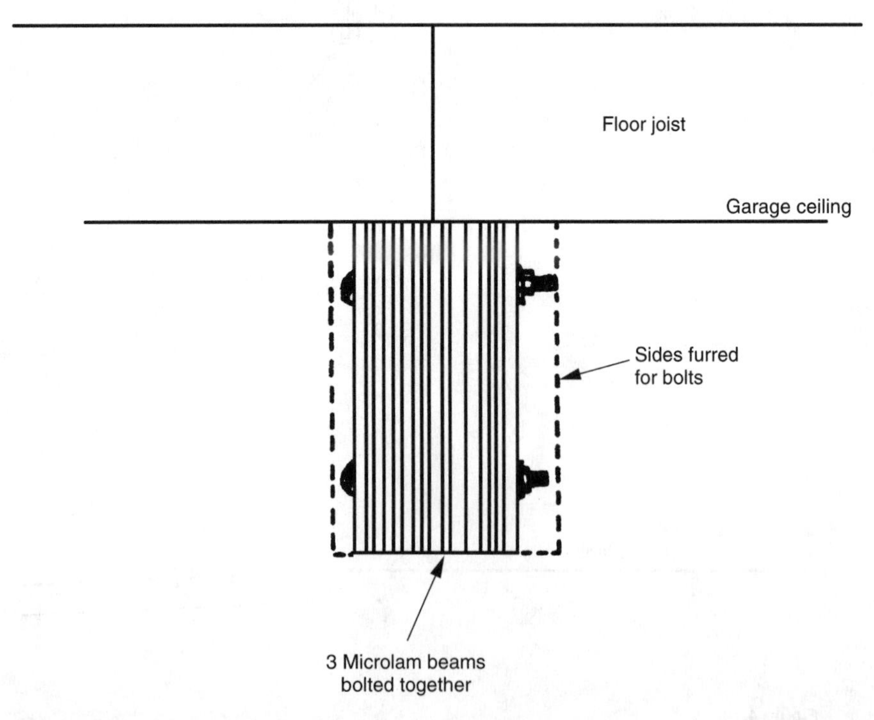

Figure 4.16

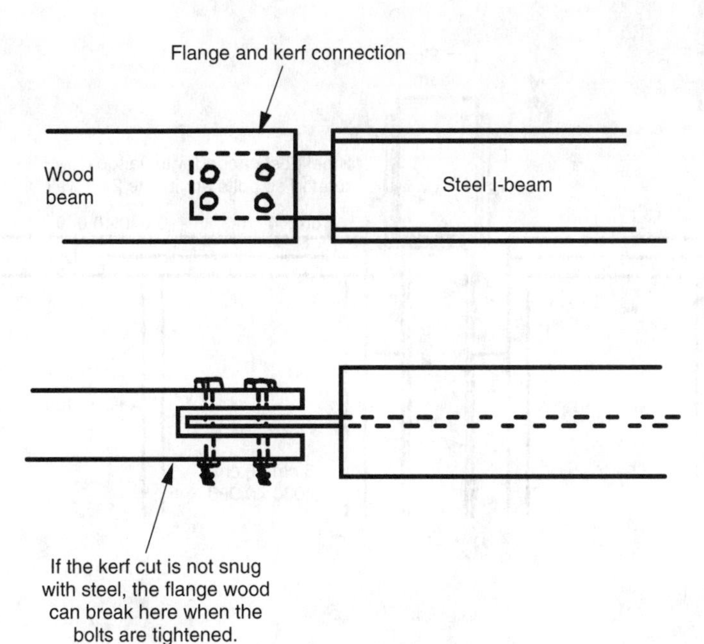

Figure 4.17

Bolts Inside Walls

When special post-to-beam steel connections are designed by the structural engineer, the builder should check that the direction of the saddle flanges place the bolts within the wall framing.

Figure 4.18 illustrates a beam saddle supported by a 4×4 post. The bolts in this example should be parallel with the wall so the bolt heads and threaded ends are within the wall framing, not projecting out into the room or hallway.

This problem becomes more difficult when 4×6 or 4×8 posts are needed. The temptation for the engineer is to orient the saddle to the narrowest dimension of the post, thinking he or she is doing the carpenter having to drill the bolt holes a favor. However, this leaves the builder with the problem of having to hide the bolt ends by using either furring or two layers of drywall, which creates additional interior finish problems such as having to order different size door jambs.

Balcony Deck Slope

Figure 4.19 shows the side view of an exterior balcony deck floor joist that has its top cut to provide slope for water runoff. Because floor joists can vary in depth from 11¹⁄₁₆ to 11½ inches, if a template is used to mark the slope on each joist before cutting, the template should be used from the bottom up rather than from the top down. By

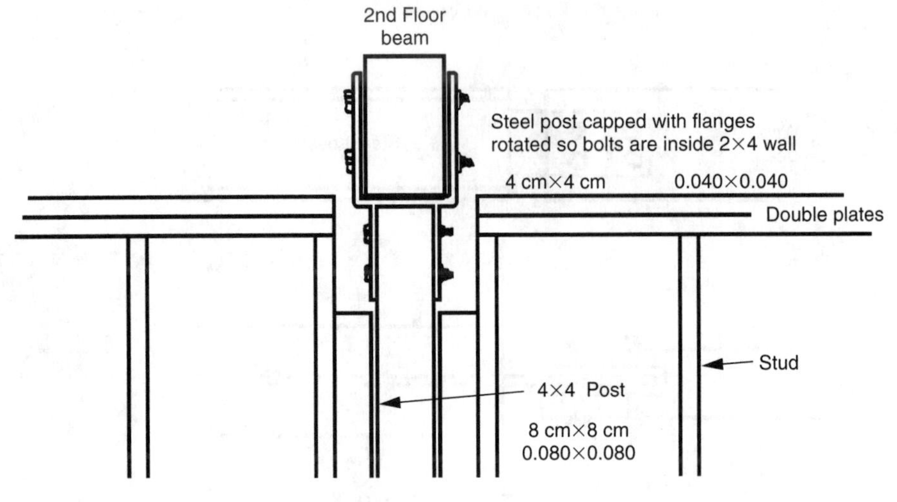

Figure 4.18

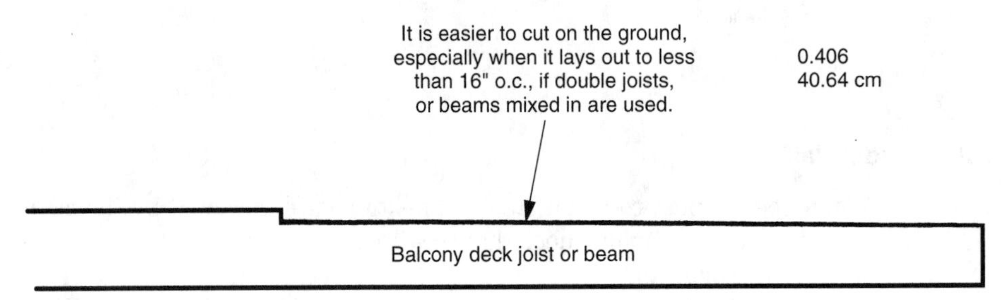

Figure 4.19

marking each joist from the bottom up, the tapered depth of each joist remains uniform while the throwaway top cutoff pieces vary with the different depths of each joist (Figure 4.20). If marked from the top down, the throwaway cutoff pieces will all be equal but each joist will vary. Joists with varying depths results in balcony deck surfaces that have humps and valleys.

Truss Joist Layout for Steel

Truss joists are floor joists fabricated from plywood that are similar in shape to steel I-beams. A plywood center web is glued to plywood top and bottom flanges or chords. They are becoming more popular as compared to standard lumber floor joists because of their relative straightness and ability to prevent floor squeaks.

When the structural plans say that the layout and spacing of truss joists will be determined by the truss joist manufacturer, but approved by the structural engineer, the builder should commence this process before the fabrication of any structural steel I-beams. The steel fabricator welds joist and beam hangers to the steel I-beams used in the floor system (Figure 4.21). For the steel fabricator to have the current information, the truss joist layout must be designed and approved before shop fabrication of the steel members can begin.

The problem to avoid is the time crunch between getting a truss joist layout from the manufacturer, getting the structural engineer to review and approve the design, and then getting the steel fabricated in time for construction. If the manufacturer's layout design is used by the steel fabricator before approval from the structural engineer, any changes to the floor system required by the engineer can result in expensive field welding of additional hangers and buckets at the job site.

Figure 4.22 shows an example of a typical truss joist layout design supplied by a manufacturer. The question can be asked, why doesn't the structural engineer provide the floor system design to start with, thus saving review and approval time? The answer is that truss joists are manufactured by several companies and some structural engineers will not specify something that can vary from manufacturer to manufacturer. By requiring the manufacturer to provide the design and engineering calculations, the engineer merely reviews and approves the design, while the liability for the product stays with the manufacturer.

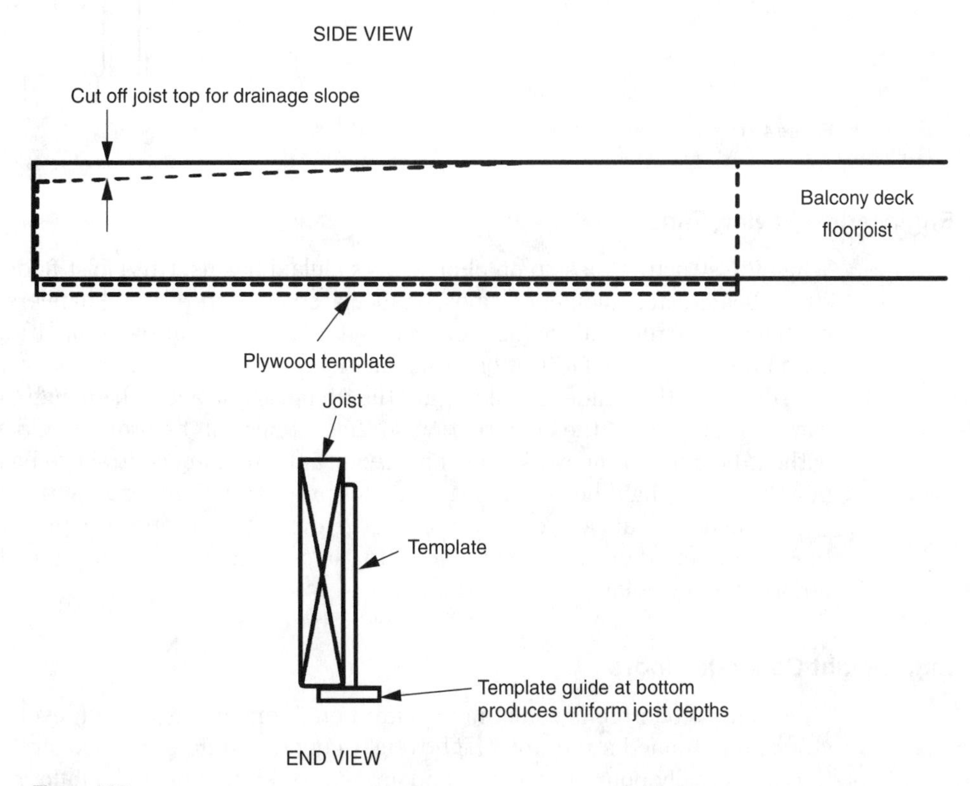

Figure 4.20

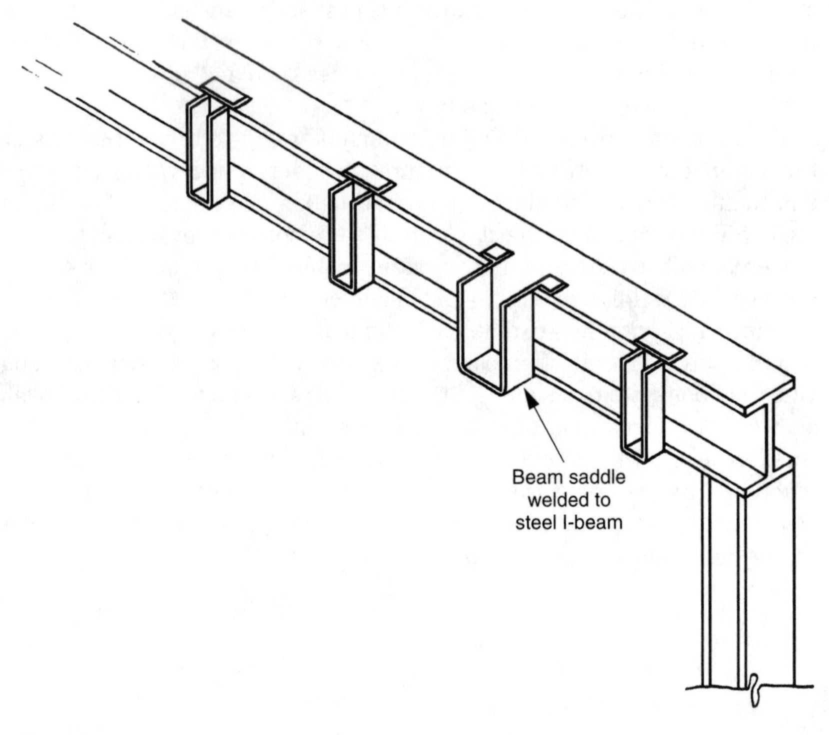

Figure 4.21

Engineering Review Time

When the structural design of a house uses glulam beams, truss joist floor systems, steel I-beams, steel moment frames, or special steel wood-to-wood hangers and connections, the structural engineer or architect must often approve shop drawings and layout designs before fabrication or installation.

In this case, the builder should award the framing contract with enough time in advance of the start of the construction so these items can be ordered and approved without holding up the work. This may require the framing contract to be set a few months earlier than the normal lead time for projects without engineering review.

The builder should also closely monitor the submittal and approval process so that the review period does not carry over into the ordering lead time for the above materials, thereby holding up construction.

Lightweight Concrete Floors

Lightweight concrete is sometimes poured on floors above ground level for multistory wood framed structures. Lightweight concrete differs from concrete slabs in that it is usually poured on top of and around wood framing and subflooring that is already in place. This creates a subtle problem within interior doorways that does not exist for slabs on grade.

Walls are normally laid out and framed with continuous bottom plates. The portions of bottom plate remaining within the doorways are cut out after the walls are stood up, during the door plumb-up phase (Figure 4.23).

Wall framing on floors with lightweight concrete have two bottom plates instead of one. The lowest of the two plates is buried in the concrete after it is poured (1½ inches thick), while the upper plate remains above the concrete and acts as a normal bottom plate for nailing drywall and baseboard.

If both bottom plates are left within the doorways at the time the lightweight concrete is poured, the concrete finisher may trowel off the concrete either too high or too low in relation to the lowermost plate. After the uppermost plate is cut out of the doorway, a high or low area is left in the floor between the doorway.

To avoid this problem, the uppermost of the two bottom plates should be cut out before pouring the lightweight concrete. The remaining plate can then be used as a screed to finish off the concrete even and flush within the doorways.

On one particular condominium project, this problem was not caught until some of the units were carpeted. The homebuyers complained of stepping on sunken areas in the floor when walking through the doorways. The carpet had to be cut out between the doorways, and these low spots filled-in with pour-stone cement. The unsold, uncarpeted units also had to have their doorways similarly repaired.

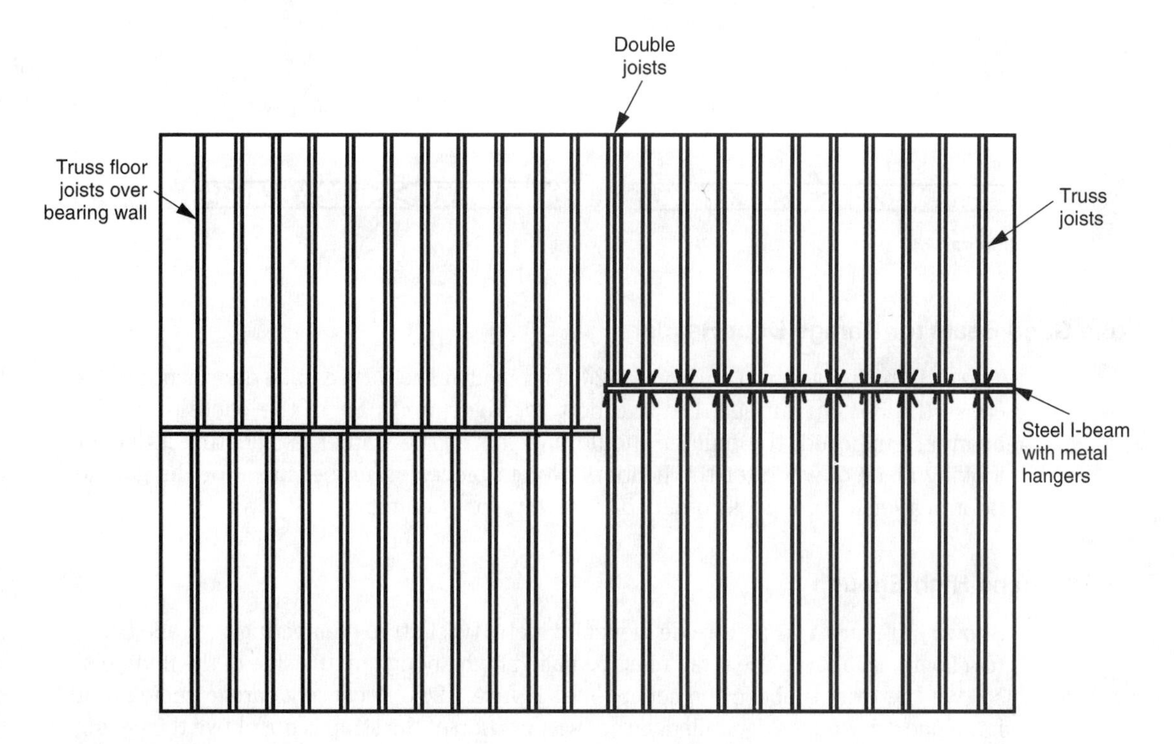

Figure 4.22

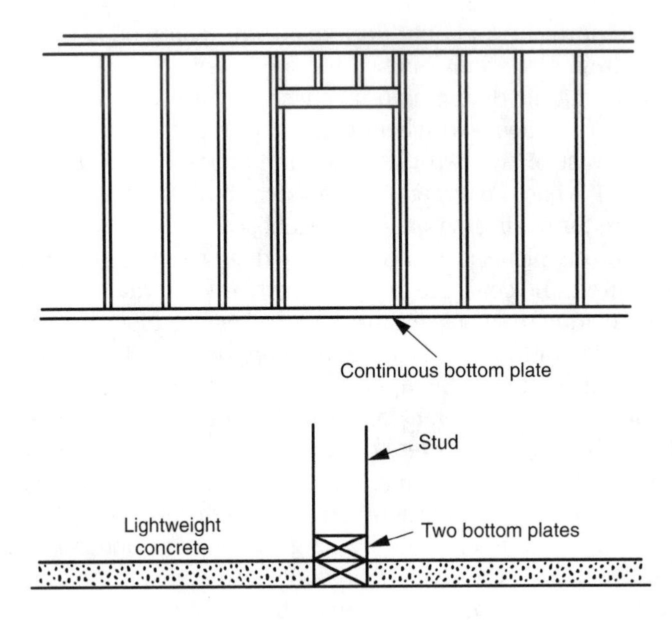

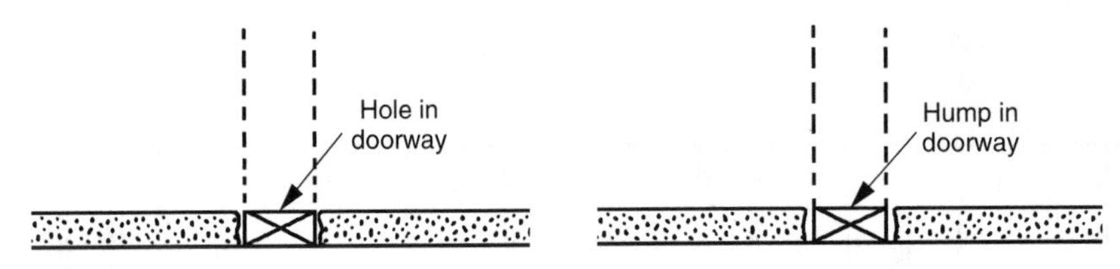

Figure 4.23

Use Good Beam for Garage Door Header

The 4×12 beam pictured in Figure 4.24 is typically used for a garage door header. Because the bottom surface, the two bottom edges, and part of the backside of the beam are exposed, the builder should check that the framer is selecting a decent looking piece of wood for the headers. Rough pieces of lumber used for the garage door header will still look rough after painting or staining.

Twist Straps High Enough

Shown in Figure 4.25 is the side view of a metal twist strap connecting a garage beam to a floor joist above. The strap must be nailed high enough up the side of the floor joist so that the twist is above the ceiling line (Figure 4.26). Because a garage ceiling that has a floor above gets drywalled, if the twist portion of the strap is not above the ceiling line, than that part of the strap will be exposed after the ceiling and beam are finished.

Select Cabinets before Drop Ceilings

The dimensions of kitchen cabinets must be known before drop ceilings can be framed so that the drop ceilings can be deep enough to accommodate the upper cabinets and the varying types and thicknesses of cabinet scribe molding (Figure 4.27). It is important to address this issue before it is brought up by the framing carpenter who is building the kitchen drop ceilings. Because the amount of drop ceiling soffit reveal gives a definite architectural feel to a kitchen, this is something the builder wants to look at prior to when the question is raised out in the field.

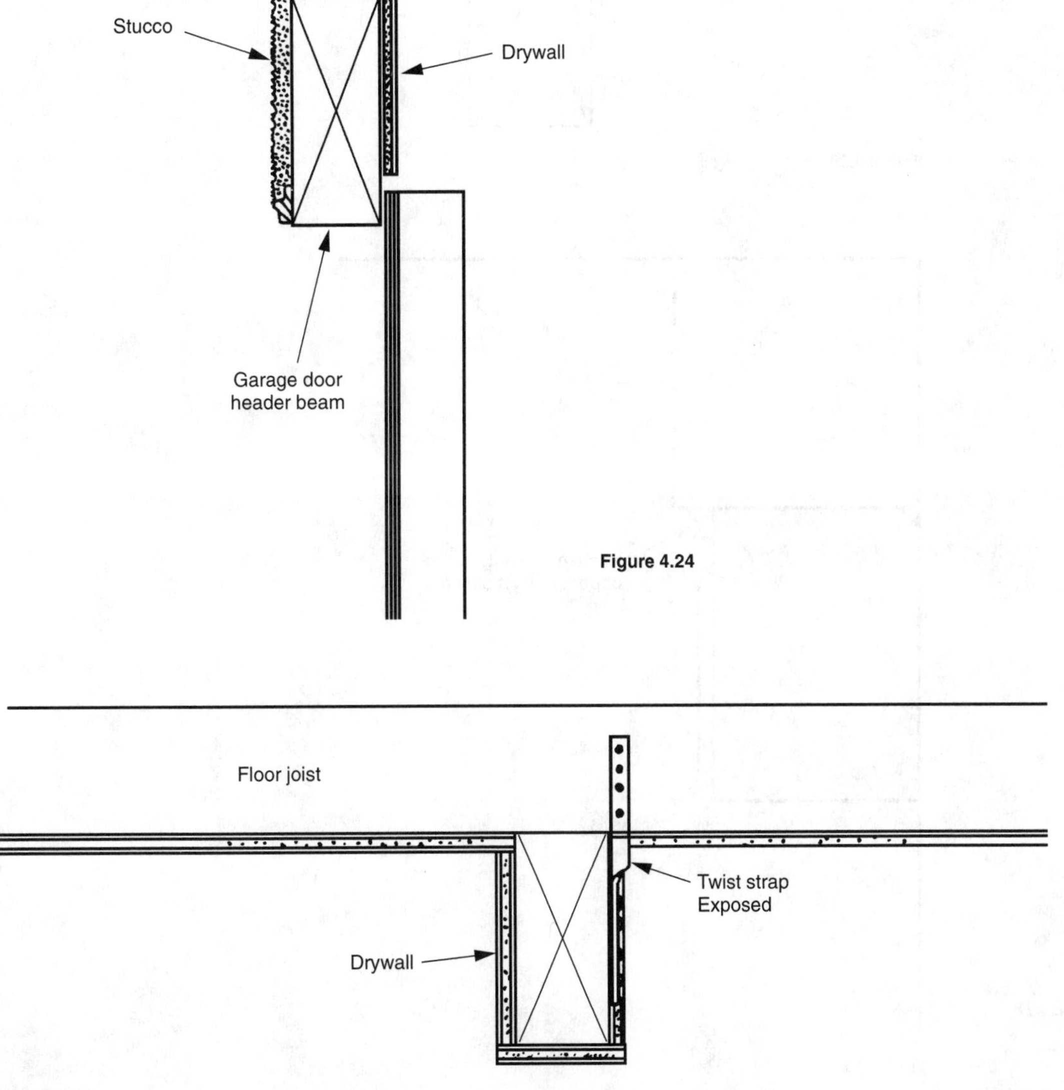

Figure 4.24

Figure 4.25

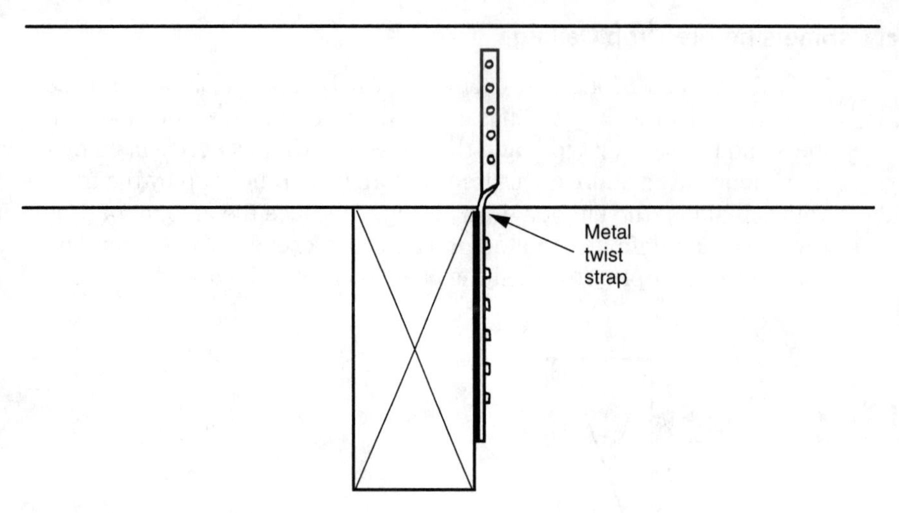

Figure 4.26

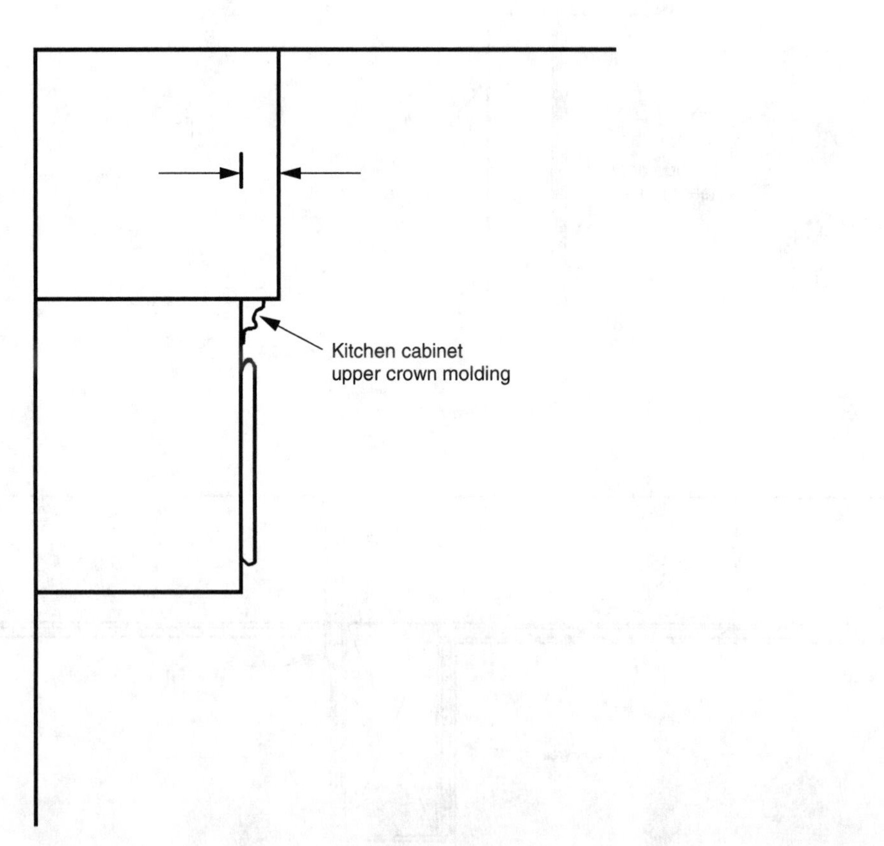

Figure 4.27

Split-Level Floors

A detached house with a conventional wood joist floor has a crawl-space beneath the floor that is vented to the outside. Air circulates under the floor but does not enter the interior of the house above.

Illustrated in Figure 4.28 is an unusual problem that occurred when an on-grade concrete slab was mixed with a wood joist floor to create a raised split-level floor. In this example, air drafts from the underfloor crawl-space flowed around and through the gaps between the electrical cover plate, the outlet, and the outlet box, creating quite a strong draft of air.

With this type of floor design, the crawl-space side of the wall separating the two floor levels should be closed-off with plywood or drywall to block-off any air flow from the crawl-space into the house.

Slope Gas Meter Ceilings

Gas meters for apartments and condominiums, which are placed within an open and vented exterior room, should have headerless door and window openings. The ceiling should also be sloped upward toward the exterior (Figure 4.29). This design helps prevent the possibility of gas rising from the meters, pooling within a ceiling space, and creating an unsafe condition.

Nail Floor Sheathing Quickly

When floor sheeting plywood or composite sheets are glued to the floor joists, they should be nailed soon afterwards. Depending on how long it takes for the glue to dry, the best bond between the sheathing and the joists occurs when the entire floor is nailed off while the glue is wet.

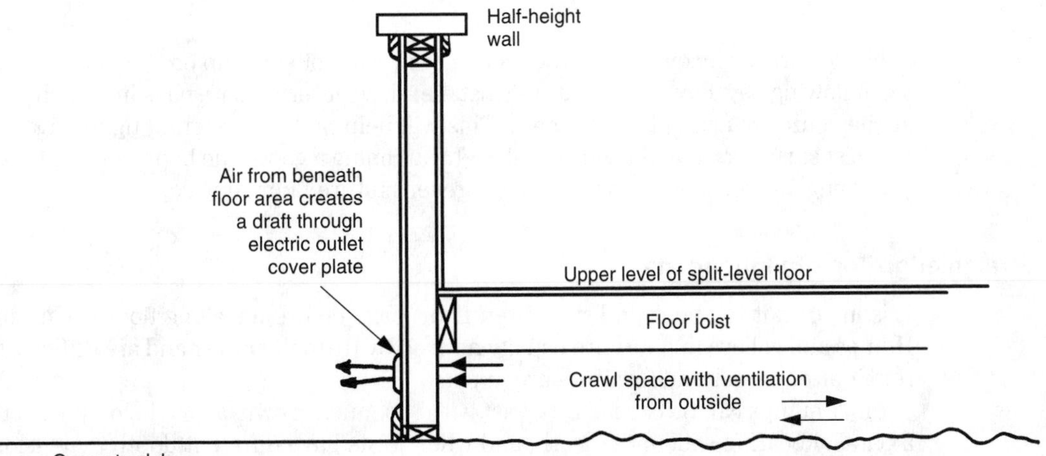

Figure 4.28

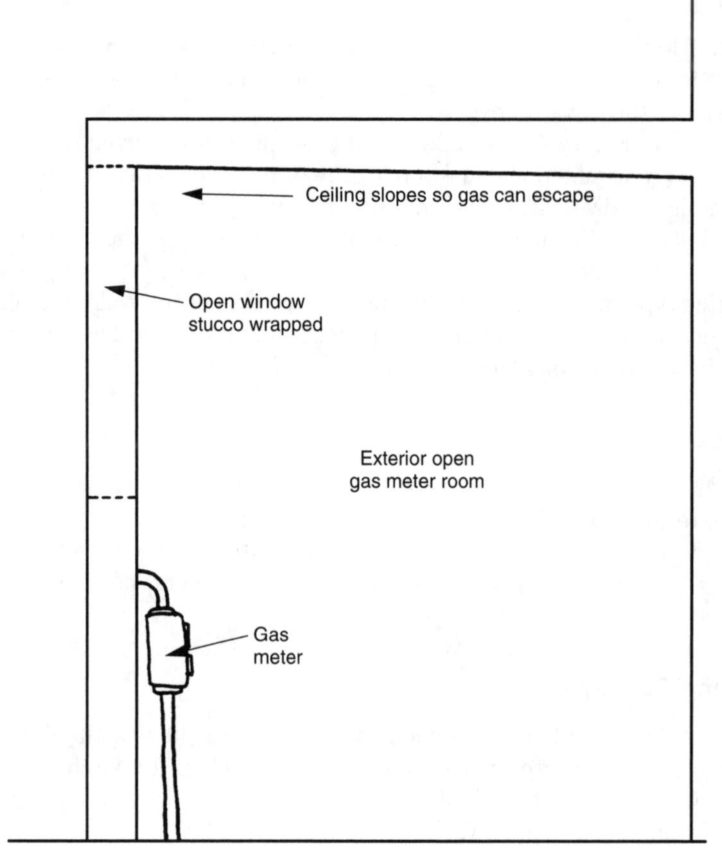

Figure 4.29

If the work is piecework and the nailing crew will not show up on the job site until the following day, have the sheathing installer drive intermediate nails into each joist at the center of each plywood sheet. This will help pull the sheeting tightly against the joist surfaces while the glue is still wet. Obtaining a good glue bond between floor sheathing and joists is one way to help prevent future floor squeaks.

Straightedge Top of Floor Joists

It is important to check and straighten floor joists before installing floor sheathing. Humps and valleys in floors are usually noticed by the homebuyer and are difficult to repair after carpet and furniture are in place.

Floor humps can be caused by a joist with too much crown, a joist held up rigid by a parallel bearing wall underneath, and when joists have different depth dimension. To avoid floor humps, the builder should check the floor joists with a long straightedge just prior to floor sheathing and have the framer correct any problems then.

Drip Kerfs on Exterior Beams

Figure 4.30 shows the side view of an exposed exterior wood beam that has had a kerf notch cut on the underside of the beam. This kerf notch prevents water on the bottom of the beam from reaching the adjacent wall surface. When water hits the interrupting notch on this beam, it drips downward off the beam before reaching the wall. This prevents water that has interacted with stain or paint on the beam from dribbling down the wall and staining the wall surface. This also reduces the chances of water leaks at the beam penetration through the wall.

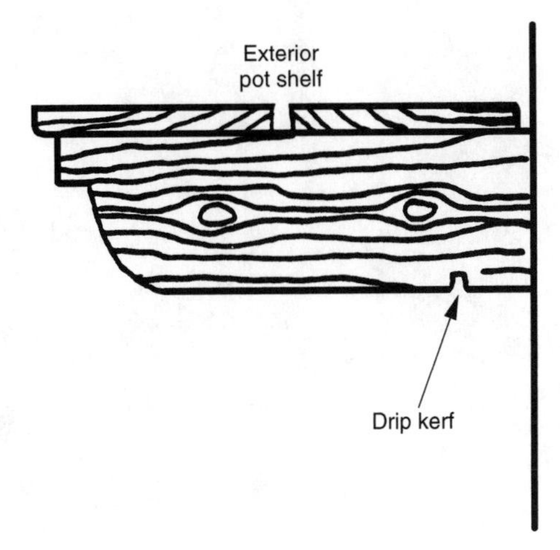

Figure 4.30

Chapter 5

Layout and Framing for Doors

Keep Door Sizes Uniform

Confusion can be kept to a minimum on large housing projects when interior door sizes are kept to a few standard widths. This permits easy exchange of doors to complete each unit in sequence, should a particular shipment of doors to the job site be short one size, have the wrong swing direction because of a reverse floor plan, or have one or two damaged doors.

For instance, suppose one particular size and swing direction of a prefit door and jamb is missing in one of the deliveries to the job site. If the door widths and swings vary in each floor plan, the next similar door size and swing may not occur until some future shipment of materials. In this example, because a suitable size cannot be "robbed" from houses up the line, the house is otherwise completed except for the area around the missing door. This creates future door hanging, baseboard, and painting pickup work once the door is obtained.

For another example, suppose floor plan A has a 2-0 left-hand no-bore closet door, which is not found in any of the other floor plans. A delivery of finish carpentry materials for lot numbers 21 through 30 contain one A plan and one A reverse plan. If the delivery mistakenly delivers two 2-0 left-hand no-bore doors, then the closet door for the A reverse plan cannot be installed because it has its hinges on the wrong side. Because no other floor plans have this door size and type, a door cannot be robbed, for example, from lot number 30 to complete lot number 23. Lot number 23 remains uncompleted around the closet opening until the next delivery of finish carpentry materials for lots 31 to 40. At that time a door can be robbed from lot 37 to complete lot 23.

If there are only two, or at the most three, interior door widths used, then a door can be borrowed from the end of the shipment to complete the unit with the missing or damaged door. Other phases of the construction will not catch up to this last or end unit until the next shipment of materials, so the **robbed** door can be installed. This helps to place the pickup work at the leading edge of the construction rather than buried at different places back throughout the project.

Floor-to-Ceiling Doors

Interior doors and bypass wardrobe doors that are floor-to-ceiling in height can cause problems during the finish stages of construction. Because of their size and weight, they are difficult to maneuver up stairways and elevators without damaging the surrounding walls. Once inside the unit, they are difficult to stand upright and move around without scratching or gouging the ceilings. Floor-to-ceiling entry doors in condominiums cause a problem because there is typically very little space between the top of the door and the ceiling for door closers. Furthermore, the door closer arms often drag across the ceilings. From a construction standpoint, therefore, standard-height doors are preferred to floor-to-ceiling height doors.

Door Clearance

Doorways sandwiched between two walls should show a minimum of 6 inches added dimension to the net door opening between the two walls for all the door parts to fit. For example, if the plans call for a 2-foot, 6-inch wide bedroom door, the hallway leading to the bedroom should be at least 3 feet wide for the door, door jamb, and casing to fit without having to rip the casing down to a narrower width. Hallway coat closets are often given less inside dimension than is needed for full pieces of casing on each side (Figure 5.1).

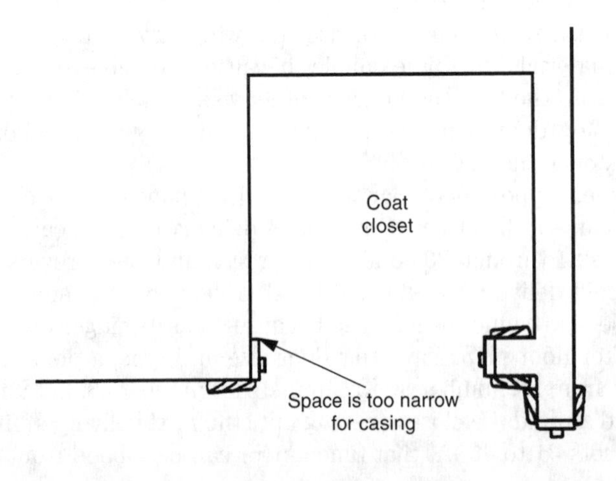

Figure 5.1

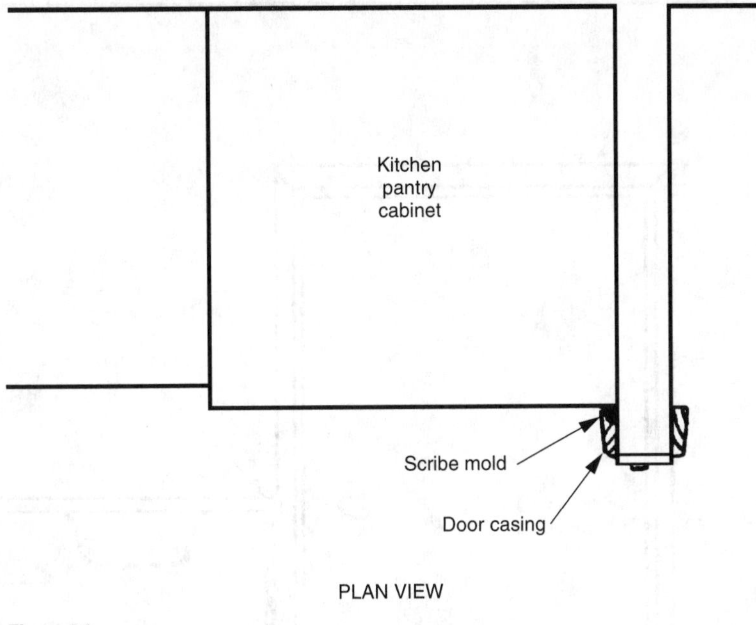

Figure 5.2

Clearance for Casing

Figure 5.2 illustrates in plan view a kitchen door next to a pantry cabinet. In this case, the pantry cabinet is slightly deeper than the other cabinets and the wall return dimension to the door opening must be enough to accommodate the width of the door casing, the cabinet scribe mold, and some in-between empty wall space. The builder must check during the framing phase that the dimension from the wall corner to the kitchen door is wide enough.

Clearance for Cabinets and Doors

During the rough framing, the builder should check that bathroom and kitchen side walls are deep enough for cabinet tile, marble lavatory tops, and door casing. Figure 5.3 illustrates a door opening that is too close to the bath cabinet. The door casing must be cut around the imitation marble lavatory top because the side wall did not have enough dimension for the door opening, casing, and cabinet lavatory top.

Bath Door and Toilet

Illustrated in Figure 5.4 is the plan view of a small bathroom in which the door opens across the front corner of the toilet. In this particular case, the dimensions of this bathroom were such that the bathroom door barely cleared the toilet, and if the closet ring of the toilet was not exactly 15 inches from the back wall, the door would hit the toilet seat.

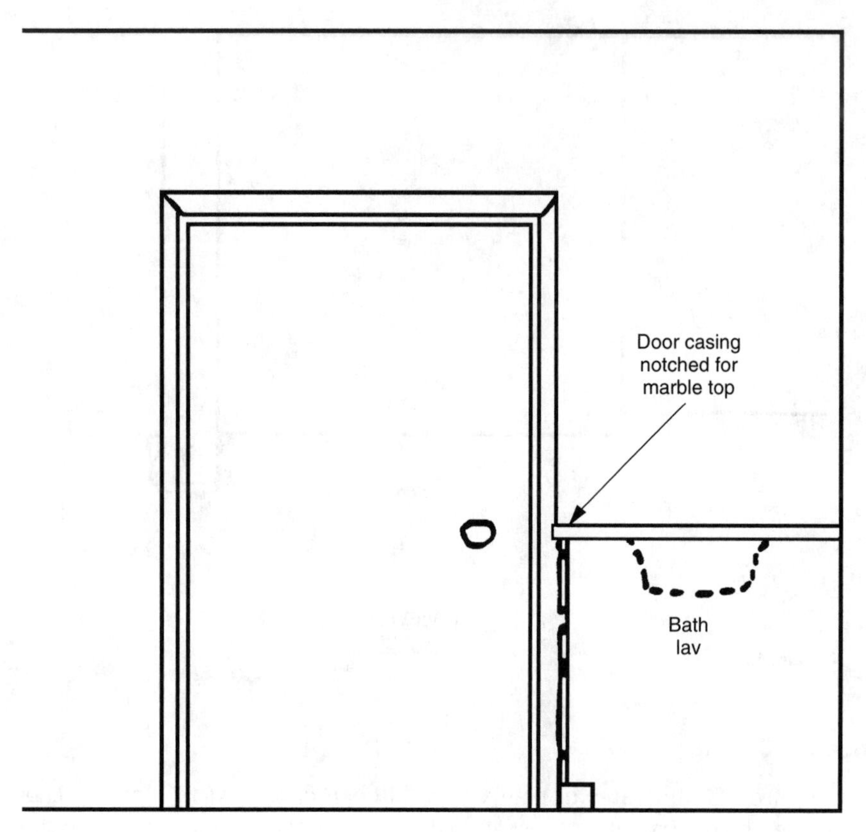

Figure 5.3

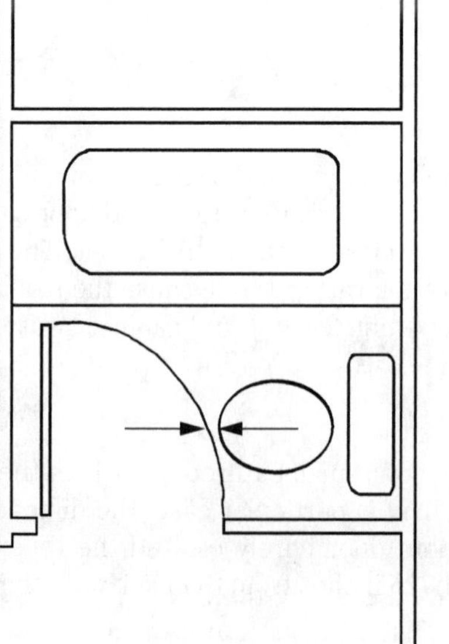

Figure 5.4

During the preconstruction plans review period, the builder should check whether there will be a possible conflict between the door and the toilet. This can be done by using the dimensions given on the specification sheet of the toilet, checking the distance the toilet closet ring is set from the back wall, and looking at the layout of the door opening. If it appears that there may be a conflict between the bath door and the toilet, the builder should consider reducing the size of the door. A 2-foot-6-inch door can be reduced to a 2-foot-4-inch door without any major difference in appearance or function, providing an additional 2 inches of clearance.

If the design results in a situation where the door barely clears the toilet and the design cannot be changed, then the builder must check all of the factors involved during the construction, such as the door opening framing, the layout of the closet ring, and so forth.

Door Openings

Both vertical sides of rough-framed door openings must be plumb. With conventional wood jambs, each jamb unit is installed plumb within the door opening, independent of the plumbness of the surrounding rough framing. When the drywall and wood casing trim are installed later, if the wood jamb and drywall surfaces are not flush, the casing will be tilted. This slows down the finish carpenter who is installing the casing trim because he or she has to flatten or mash-in the projecting drywall with a hammer to make the drywall flush with the jamb so the casing will sit flat against the wall.

With prefit or prehung interior doors, an additional problem presents itself when rough-framed openings are not plumb. The door stop trim and casing for prefit doors come prestapled to the jambs, so there is a limit to how far one side jamb piece can be pushed or bent over to match the opposite side jamb so that the door closes evenly against the jamb stop (Figure 5.5). Beyond a certain point, the staples connecting the casing to the edge of the jamb will split the casing, not to mention the fact that the awkwardly tilted casing looks bad.

When the door does not close evenly against the door stop, the predrilled hole and mortise for the door striker plate often will not engage the door catch, so the striker plate must be relocated. In some cases, the door stop must be removed from the prefit jamb and renailed in place to match the door, which cannot be adjusted over anymore by tilting the casing trim (Figure 5.6).

Ripping Casing

The clearance between the edge of a door and the corner of a wall must be wide enough to allow for finish door casing. This requires at minimum a king stud and trimmer around rough-framed openings (Figure 5.7).

If the door header is framed directly into the adjacent wall, with only the trimmer separating the edge of the rough opening from the corner, the finish door casing must be ripped to a narrower width in order to fit. This quick-fix will not satisfy the demands of custom housing if the homebuyer chooses a wide, detailed wood casing and trim. Few things look more awkward than a detailed, 3-inch or wider casing above a doorway, with the vertical pieces having to be ripped to a narrower width.

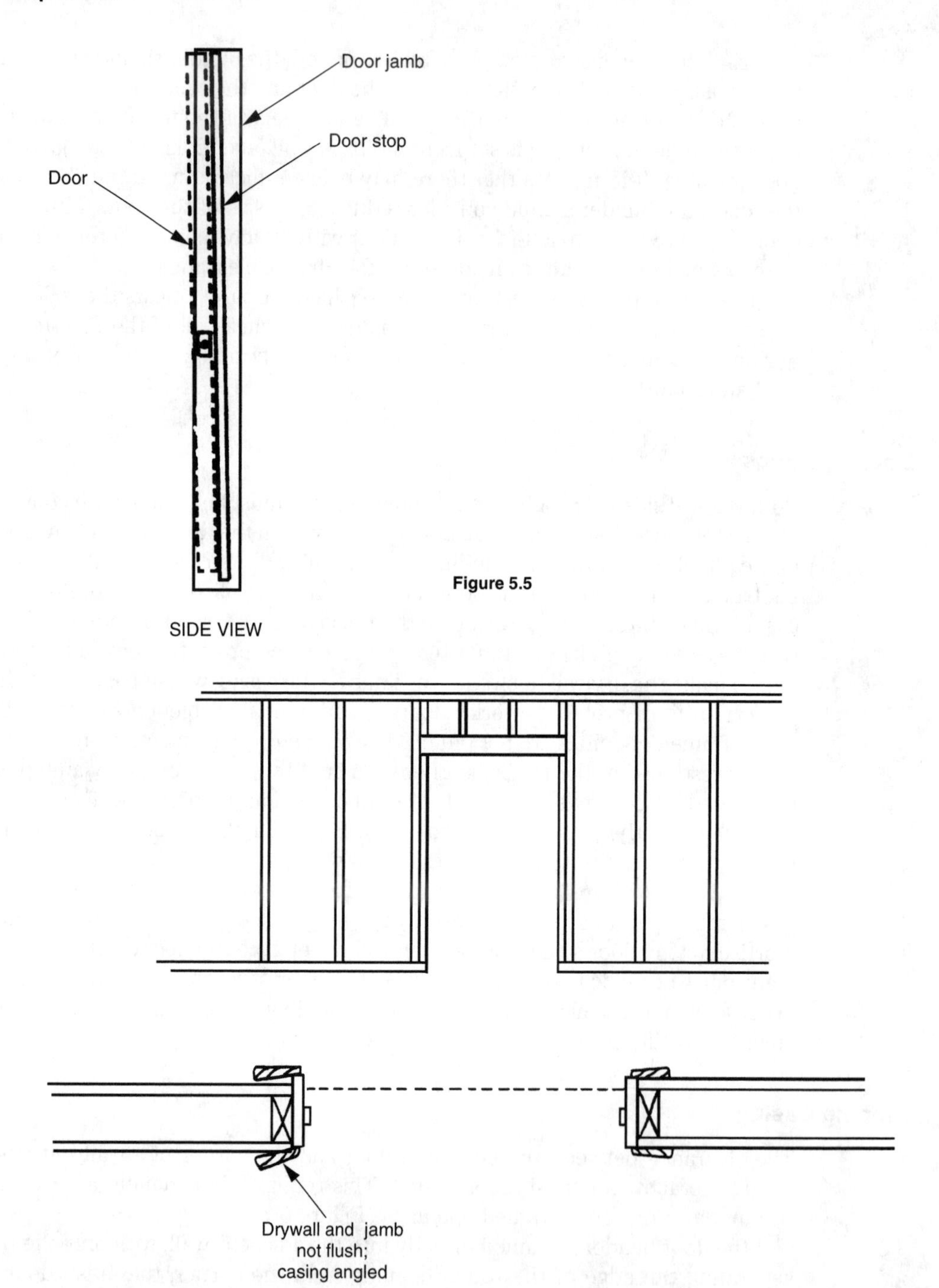

Figure 5.5

Figure 5.6

Double Doors

The builder should check the two vertical sides of rough-framed double-hung door openings. If one or both sides are out of plumb or not vertically aligned, the finish carpenter can correct this during the interior door installation. This correction, however, is limited to the amount the bottom of each door jamb can be bent sideways in the opening. The unwanted result is door casing trim that is angled awkwardly inward or outward rather than flat against the wall surface (Figure 5.8).

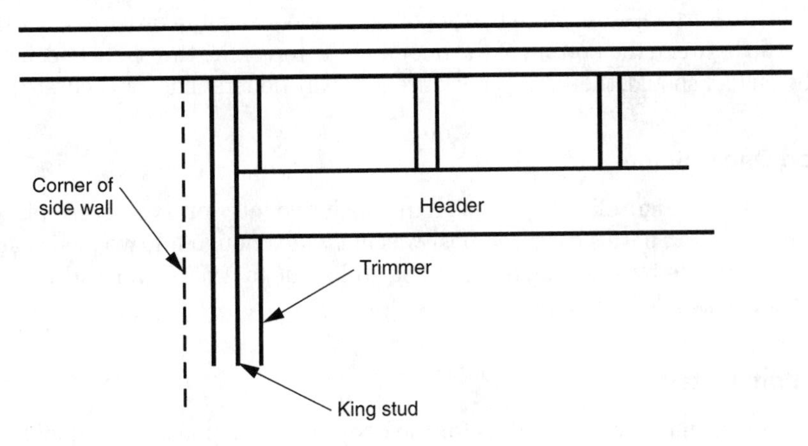

Figure 5.7

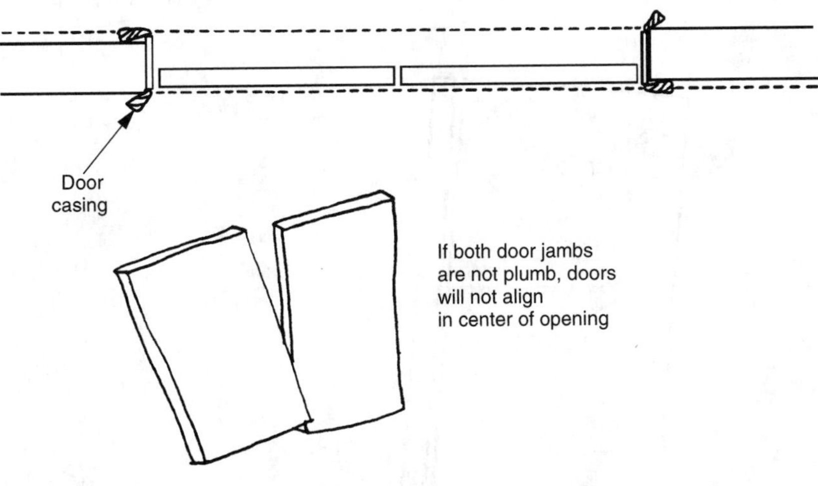

Figure 5.8

Corridor Firedoors

Multistory condominium and apartment buildings have firedoor partitions within hallways and corridors to prevent the spread of fire and smoke during a fire. In many cases, a low metal threshold is installed within the doorway, which is even with or barely above the height of the corridor carpet. The bottom of the firedoor is cut to match the threshold. This results in a small amount of clearance between the door and the carpet.

If the partition wall containing the firedoor is out-of-plumb and leaning toward the direction the door opens, the leading bottom corner of the door will get lower and lower as the door opens (Figure 5.9). If the door rubs on the carpet, it will not pass fire inspection. Furthermore, if the bottom of the firedoor is cut in order to clear the carpet, the gap between the bottom of the door and the top of the threshold become too large. The builder should, therefore, check firedoor partition framing for accurate plumbness.

Two Doors on One Column

Illustrated in Figure 5.10 is a design in which two interior doors were placed on the same 4×4 post. In this case, the post was not wide enough for two pieces of casing to be placed side-by-side. Each piece had to be cut to a narrower width. This design should be avoided when possible.

Doorway Bottom Plates

In checking interior door widths for the door plumb-up phase, the builder should not only tape off the width at the top and middle of the door opening, but also at the bottom plates. Illustrated in Figure 5.11 is a door opening in which the two ends of the bottom plates are not cut perfectly flush with the 2×4 trimmers. Even though the

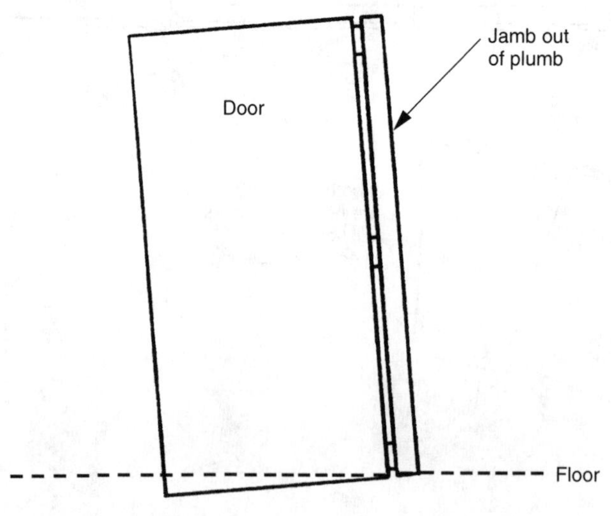

Figure 5.9

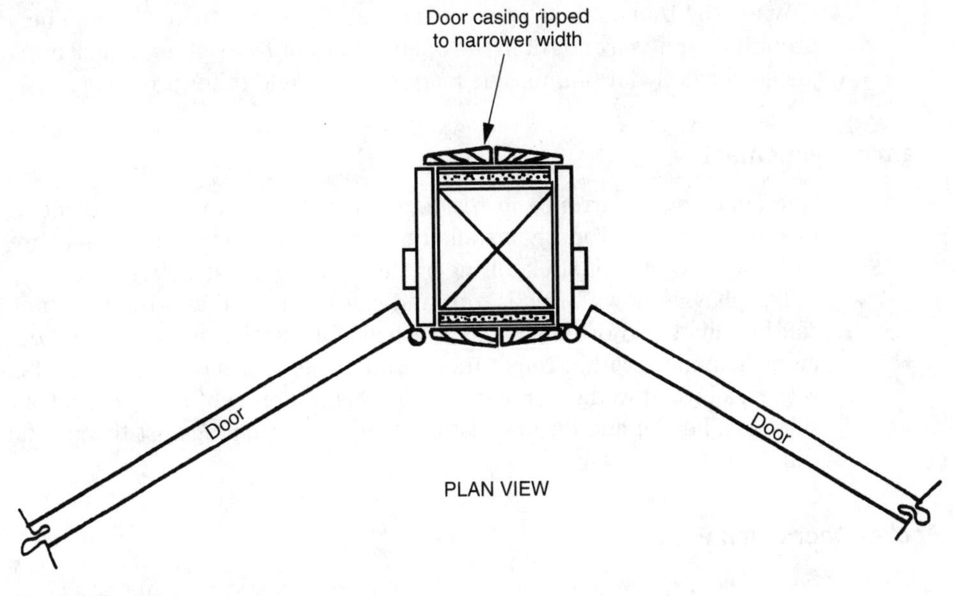

Figure 5.10

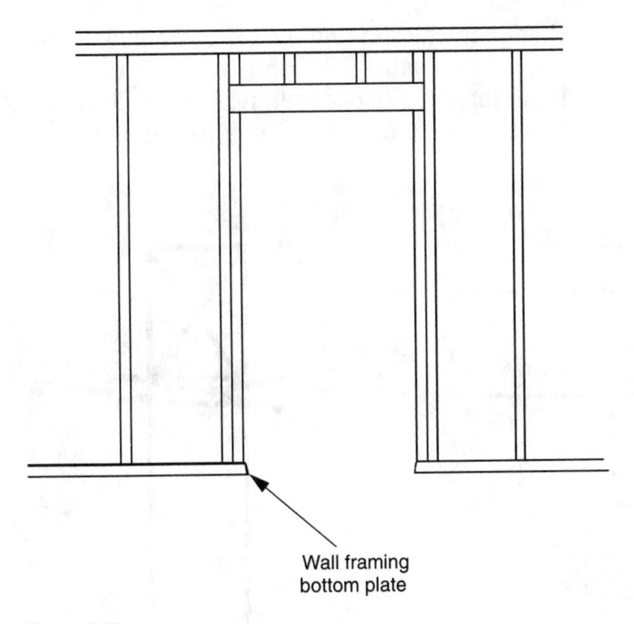

Figure 5.11

added framing allowance to the net opening is 2 inches in the middle of the opening, this necessary allowance is reduced to 1¾ or 1⅝ of an inch between the bottom plates. A 2 foot, 6 inch bedroom door that needs a 2 foot, 8 inch framed opening for installing the door would now be undersized.

When the framed door opening is smaller in width than it should be, the interior prefit door unit may not fit in the opening, or not have enough slack or play to be adjusted correctly for plumb, side margins, gaps, and other problems.

Framing Settlement

One annoying occurrence in residential construction is the settlement of framing members around properly installed doors and windows. The results are doors and windows that bind or stick during or shortly after construction.

The shaving down of door tops and sides that bind and the resultant repainting can be minimized or eliminated by checking the tightness of rough opening trimmers. Trimmers within rough-framed openings that have gaps should be shimmed with small wood wedges (Figure 5.12). This will prevent the lowering of one or both sides of a header and ensure that the trimmer is tight against the ground or sill and will not settle.

Pocket Door Jamb Plumb

Pocket door jambs and frames are installed in the same manner as conventionally hung door jambs. The vertical striker side and head jamb must be plumbed, leveled, and straightened using wood shims. Pocket door jambs should be checked for plumbness along with other door jambs during the rough-framing phase.

In Figure 5.13, the pocket door is adjusted (using the header track hardware) to an out-of-plumb jamb. This causes the door to tilt, making the leading bottom edge of the door high, and the back of the door low. The problem is that the door may

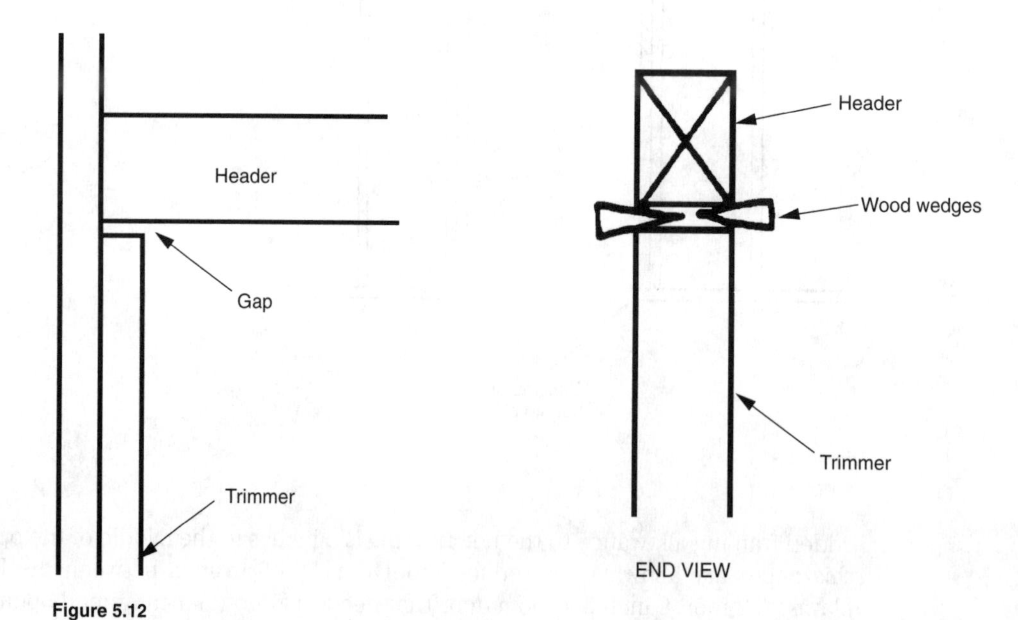

Figure 5.12

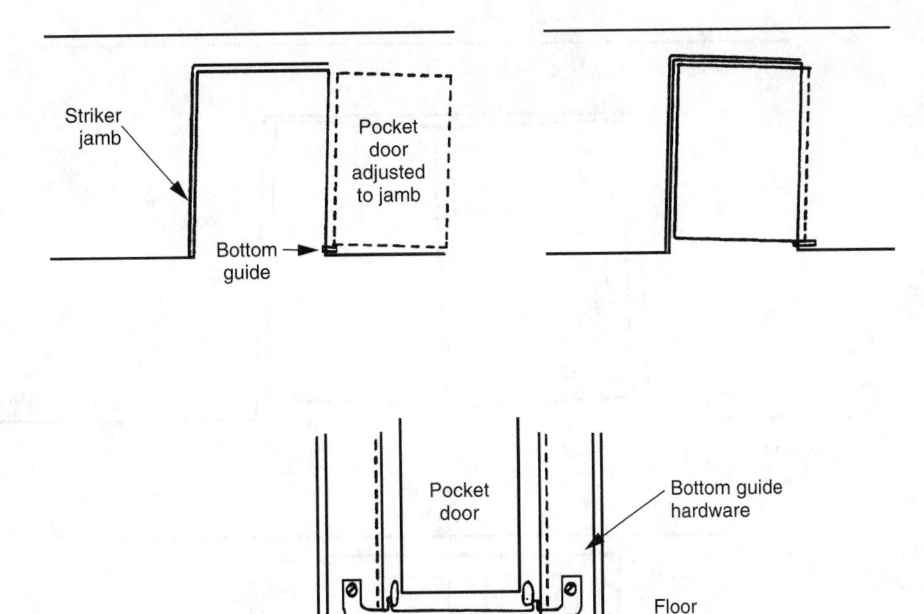

Figure 5.13

barely rest within the bottom guide hardware when open, then touch, drag, or stop on the guide before being fully closed. If the jamb and pocket door are both plumb, the bottom of the door will glide at the same level within the guide.

Wardrobe Closet Sides Plumb

The degree that interior walls, windows, and doors are noticeably out-of-plumb often depends on how close they are to other vertical lines of comparison. Wardrobe closet bypass doors (Figure 5.14) have the unique and singular quality of being able to slide from one side of the closet opening to the other, thereby offering an immediate comparison to the relative plumbness of both vertical sides.

By adjusting each wardrobe door to match its side wall when in the closed position, some of the out-of-plumbness can be hidden in the center where the two doors overlap. If one of the doors is transferred to the opposite side or the order of the doors is mistakenly reversed by the homebuyer, the difference in plumbness becomes immediately apparent. Wardrobe side walls should be among wall framing openings checked for plumbness.

Washer Closet Sides Plumb

In condominiums, sometimes clothes washers and dryers are placed in each individual unit in a hallway, kitchen, or bathroom closet as shown in Figure 5.15.

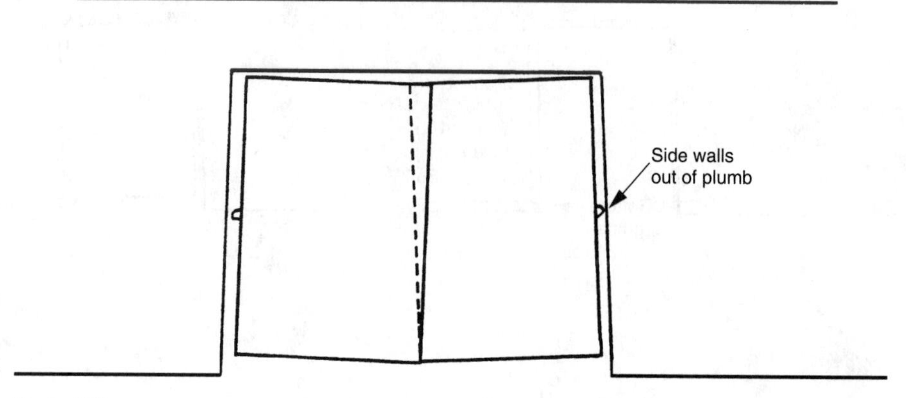

Figure 5.14

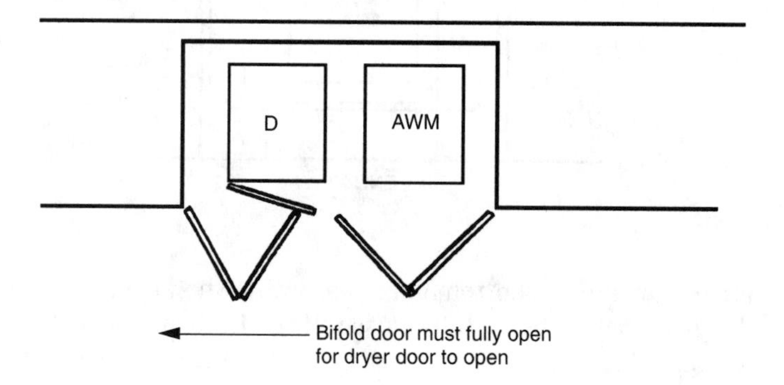

Figure 5.15

When bi-fold doors are used to enclose these closets, both vertical sides of the opening must be plumb. The reason is that the top active or sliding corner of a bi-fold door will bind before being fully opened, if the doorway opening is not plumb, level, and square. A bi-fold door that does not fully collapse toward the side wall, in this instance, restricts the width of the opening of the laundry closet, reducing accessibility to the washer or dryer.

Spreading Door Jambs

Exterior door jambs are often delivered to the job site and stored in an empty garage of an unoccupied house in a previous phase or in a storage bin.

These jambs should be spread from their storage areas to the houses for installation only as they are needed each day. If too many or all of the jambs are spread to the houses, the ones that are not immediately installed, but left to lean up against walls inside garages or the houses, will be damaged or broken by careless tradesmen or by the delivery of other materials. Wind blowing through the houses will knock over the door jambs onto the floors where they can be stepped on and broken.

Install Jambs Late

Conventional wood jambs for exterior door openings should be installed as late as possible during the rough framing. This helps prevent the jambs from being dinged up and dented by the variety of tradesmen that carry materials through them during rough construction.

Shim Entry Door Jamb Gaps

Illustrated in Figure 5.16 is an entry door that has been shimmed over by the door hanger on the striker (latch) side of the opening. This is done to obtain the correct margin or space between the door and the door jamb during the door hanging operation. Small wood angled shims are used after the door is hanging on its hinges. This is the easiest and best way to adjust the margin around the door prior to the installation of door casing.

Avoid shimming the door jamb away from the 2×4 trimmer without placing shims in the gap where the deadbolt striker plate will be installed later. Any gap or space between the jamb and the trimmer, which has not been filled in solidly with a shim, can be sucked up together again by the extra long screws (2½ to 3 inches) that normally come with deadbolt hardware (Figure 5.17). These long screws are designed to reach all the way over to the trimmer and engage the structural part of the door opening. If there is a gap between the jamb and the trimmer, the jamb is pulled over toward the trimmer by these long screws. The door hanging margin may be widened at this point, spoiling the door margin reveal and causing the door knob latch to barely catch the door knob striker plate.

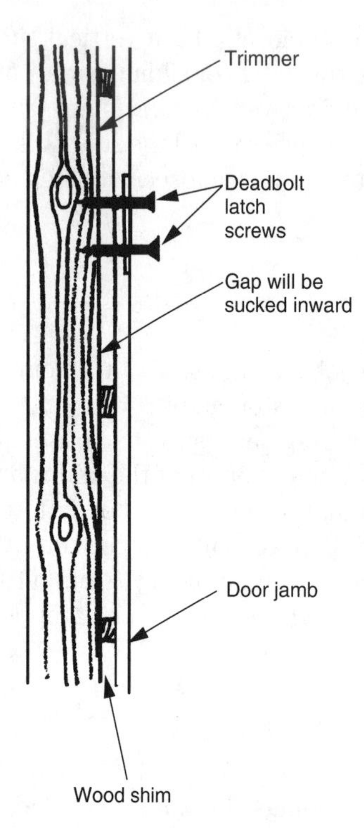

Figure 5.16

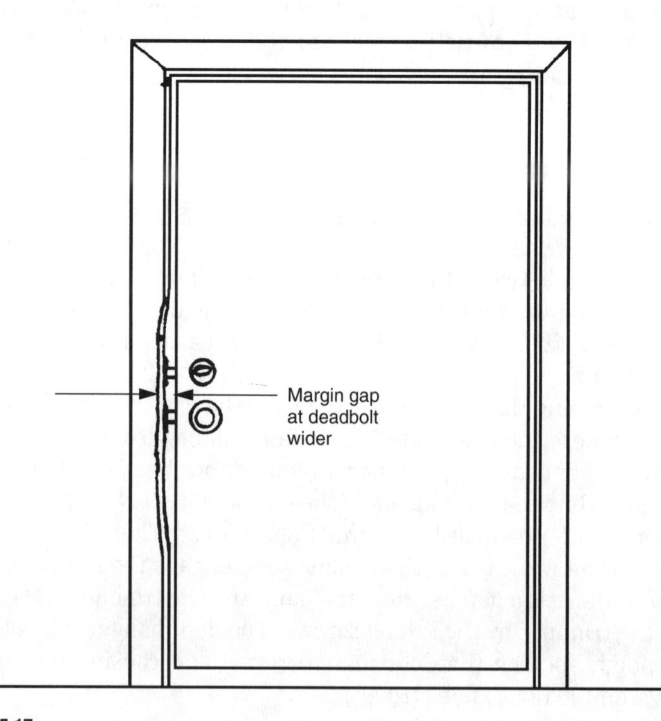

Figure 5.17

Garage Man-Door

When a garage slab has a raised footing curb around its perimeter, an opening must be left in this footing for the garage man-door. This allows the man-door opening to have the same floor elevation as the garage slab.

Because this opening is formed a few inches wider than the rough door opening width to allow some extra clearance for the door installation, a gap results between the footing concrete and the door jamb, as shown in Figure 5.18. This gap should be filled-in by the concrete contractor as part of the chipping and patching pickup stage prior to the exterior lathing or wood siding installation. If this gap is not filled-in, the backside of the lath paper is exposed and visible, which can look unsightly when viewed from the interior of the garage.

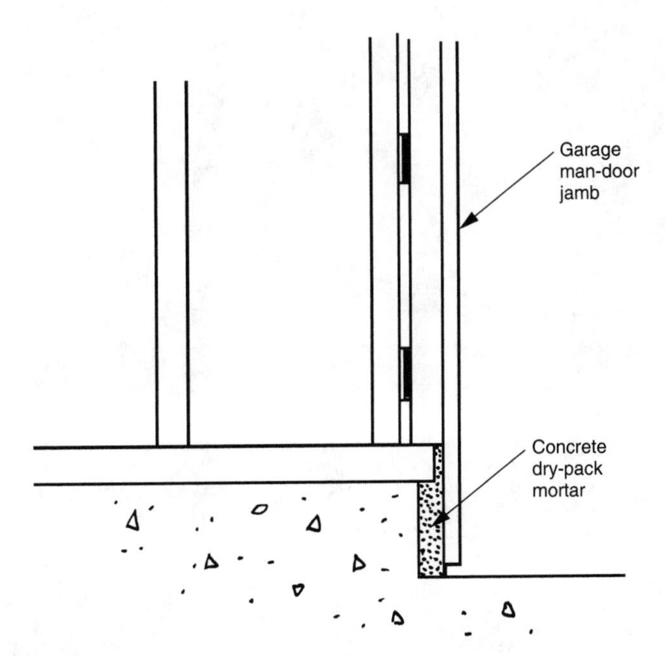

Figure 5.18

Chapter

6

Layout and Framing for Windows

Keep Window Sizes Similar

There are many factors considered in determining the size of windows for each room in the house include building code requirements, heat transfer, insulation requirements, and design aesthetics.

One factor that should also be considered is simplicity of construction. Window sizes that seemingly vary for no apparent reason add confusion to the job site. Metal window frames may sometimes vary between 3-4, 4-3, 3-5, 5-3, 4-5, 5-4, 4-6, 6-4, and so forth, for essentially the same square footage of window area required. This subtle variation in window sizes often merely complicates the task of getting the correct size window frames into the right houses. For example, when a delivery is short one or two windows, it becomes more difficult to borrow the missing sizes from houses further up the line in order to complete each house in the correct sequential order.

Check Windows as They Are Off-Loaded

It is a good idea to check the condition of windows as they are off-loaded from the delivery truck, especially when the window frames come with glass already in place. This helps identify scratched metal window frames, bent window track flanges, scratched glass, and cracked glass at the time of delivery. The builder does not then get mistakenly charged for repairs or glass hack-outs in windows that were defective upon delivery.

Store Windows in Container Bin

One method of protecting window frames, sliding glass-door frames, and glass sash is to require the window subcontractor to maintain a storage bin on the site. When windows and sliders are delivered to the job site, they can be delivered in large enough quantities to satisfy the window company, but spread to only the number of houses that can be quickly set by the framing contractor. This eliminates window frames from sitting around inside the garages of houses for days waiting to be installed by the framer, during which time some are stolen, disappear, or are knocked over and stepped on by tradesmen.

Because the window company installs the sliding glass door frames and glass panels, this also provides them with an on-site storage bin for materials.

No Backorders

It is important that window frame suppliers send out complete units of window frames with each delivery, without shortages or back orders. Wood siding, shear panel, and lath installers need house exteriors with all of the window and sliding glass door frames installed. A window supplier that repeatedly sends out units of frames with one or two frames missing and backordered per house can quickly disorganize production. Workers jump from complete house to complete house, skipping incomplete houses or leaving portions undone, until the backordered frames finally arrive and are installed.

The builder should not tolerate the added confusion and supervision problems that are created by incomplete deliveries of materials. This should not be considered a normal occurrence or to be expected. The window contract should spell out that all window frame shipments are to be complete with no backorders and that trucks containing backorders will be sent back at no charge to the builder, with a complete shipment brought out that same day.

Eight-Foot Sliders Too Large

For installation and customer service reasons, 8 foot wide sliding glass door openings should be avoided. Half the width of an 8-foot wide opening leaves a 4-foot wide by either 6-feet, 8-inch or 8-feet, 0-inch high piece of glass, the weight of which makes it difficult for even two or three people to unload off a truck and lift and carry up stairways. The heavy weight of 4-foot wide doors also makes it more difficult for the homebuyer to open and close these doors, even when the bottom track guide is clean and straight, and the bottom wheels are lubricated. If the bottom track has the slightest kinks in the side flanges or the track or wheels are not clean, the 6 or 8 pounds of pressure that should be the maximum effort required to push the doors becomes double or triple that amount.

Six-Foot Sliding Windows Too Large

Shown in Figure 6.1 is a 6-foot by 6-foot sliding window. At 3.5 to 4.0 pounds per square foot for dual-glaze windows, the weight of this size window is too heavy for some people to open easily. A 6-foot by 6-foot window is approaching the square foot size and weight of a sliding glass door, but has wheels that are not as hefty as those used underneath sliding glass doors.

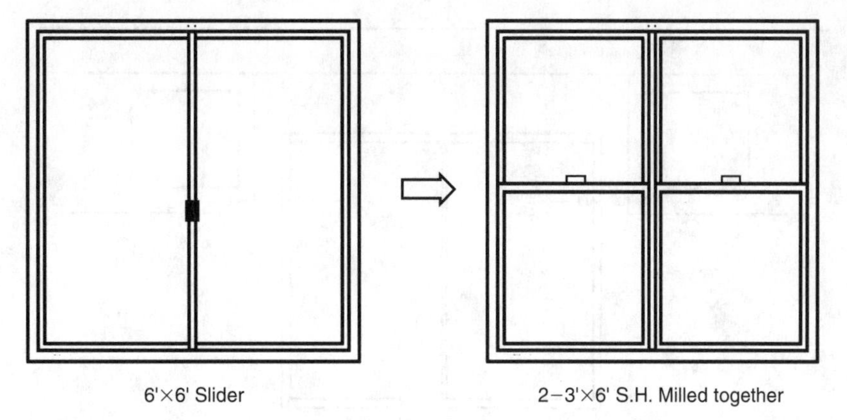

Figure 6.1

As a customer service and quality consideration, the builder should examine the possibility of breaking up a 6-foot by 6-foot sliding window into two 3-foot by 6-foot single hung windows, joined together in the center by a vertical mullion strip. Although this may cost 10 to 15 percent more than a solid 6-foot by 6-foot sliding window, the look and feel of the pair of smaller size windows joined together may be a better choice.

Simplify Tempered Glass Window Installation

If, when building a house, tempered glass is called for in only one window of a group of similar-size windows, the builder should make all the similar-size windows tempered glass so as to not confuse the framer who is installing the window. For example, suppose there is a 4-0 × 5-0 fixed glass window at a stair landing, which is required by code to be tempered, and there are two other fixed windows elsewhere in the house that are also 4-0 × 5-0. All three fixed windows should be tempered glass. The slight increase in cost eliminates the potential of the framer mixing up the tempered and nontempered windows, and mistakenly installing a nontempered window where a tempered window should go.

Window Reveals around Upper Cabinets

Shown in Figure 6.2 is the window above the kitchen sink that has upper cabinets to either side of it. The upper cabinets should be designed and installed so that the wall space between each cabinet and the window is equal for better appearance.

Center Windows below Gable Roofs

A window below the peak of a gable roof should always be centered from the exterior view (Figure 6.3). An occasional mistake on the design plans will show the window centered within the interior of the room or kitchen, but not centered on the exterior wall and therefore off-center with the roof.

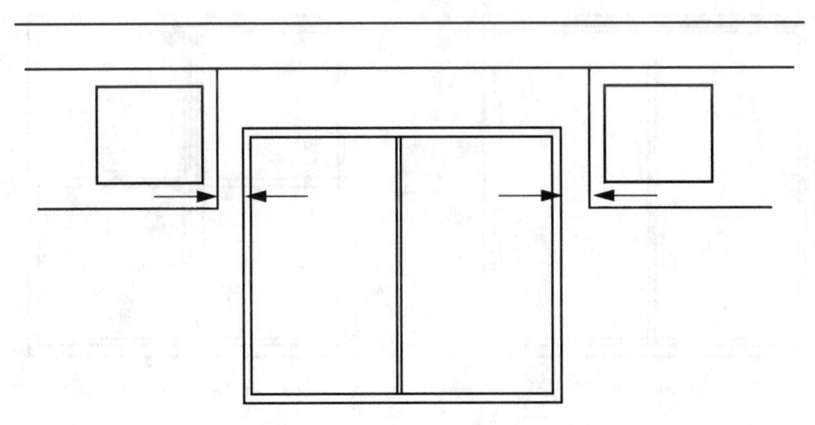

Figure 6.2

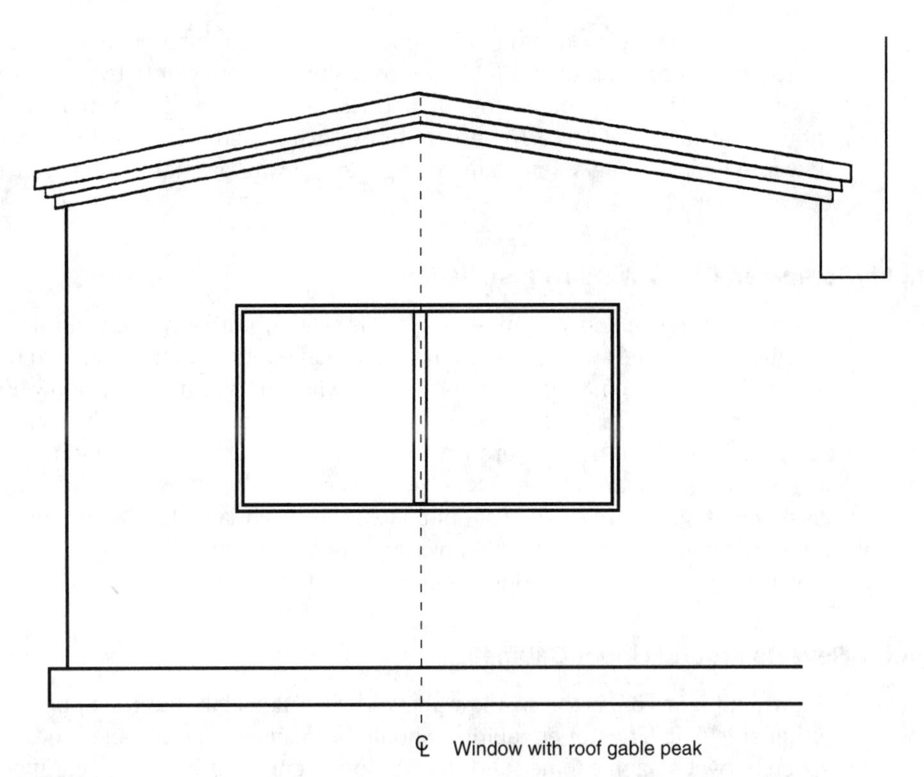

Figure 6.3

Sometimes the floor plan will also mistakenly be dimensioned from the outside of the exterior wall to the inside face of the interior wall, and then give a centerline dimension for the window (Figure 6.4). An outside to inside dimension, however, will not yield the center, but be 3½ inches off-center. The dimension should be from outside to outside, or, from inside to inside. The plans should be checked and a decision made as to which is more important—interior or exterior centering of windows, before surrounding structural posts, shear panel, plumbing pipes, and the window frames are installed.

Corner Windows

Although commonly done in tract housing, it is difficult from the standpoint of drywall cornerbead alignment and taping to use a single 4×4 post as the trimmers for two corner windows. The reason why joining two corner windows at a 4×4 post is difficult is that 6 wall surface planes with 3 lines of sight must all come together at both the top and bottom window sills (Figure 6.5). Two horizontal cornerbeads at 90-degree angles at the window sills must come to a point joining the vertical post cornerbead and the taped inside corner of the two walls. A cross in 6 planes must be aligned at a single point to be aesthetically correct.

By cutting the window headers a little longer and framing trimmers 3, 4, or 5 inches away from the 4×4 post on each side, the intersecting planes are separated, leaving one floor-to-ceiling inside taped corner (Figure 6.6).

Roofs and Windows

Illustrated in Figure 6.7 is a roof over a garage that comes up to join with a second floor wall. In this example, because the roof pitch is 3½ feet in 12 and the run is 8 feet, there

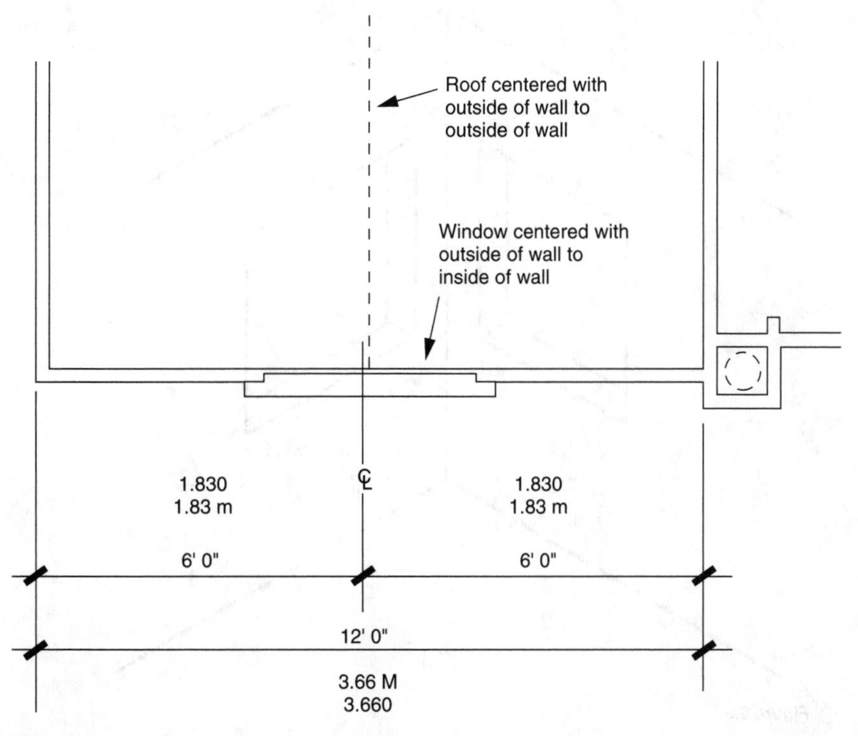

Figure 6.4

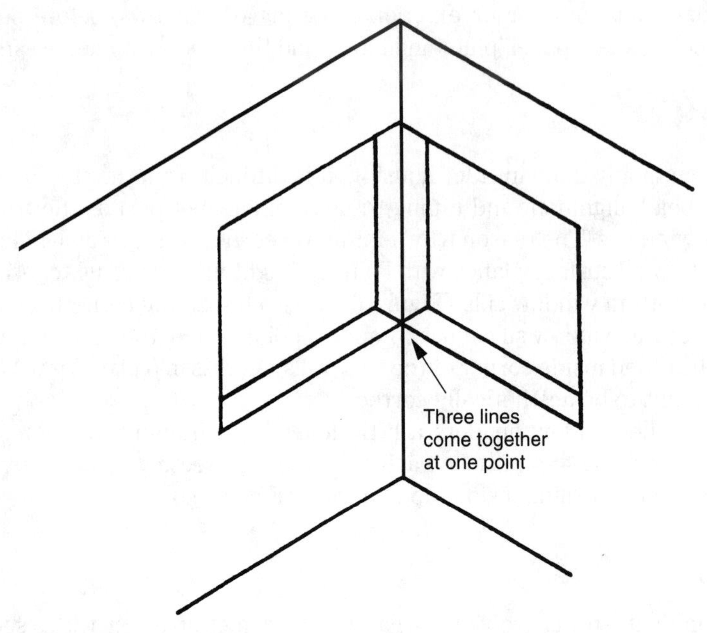

Figure 6.5

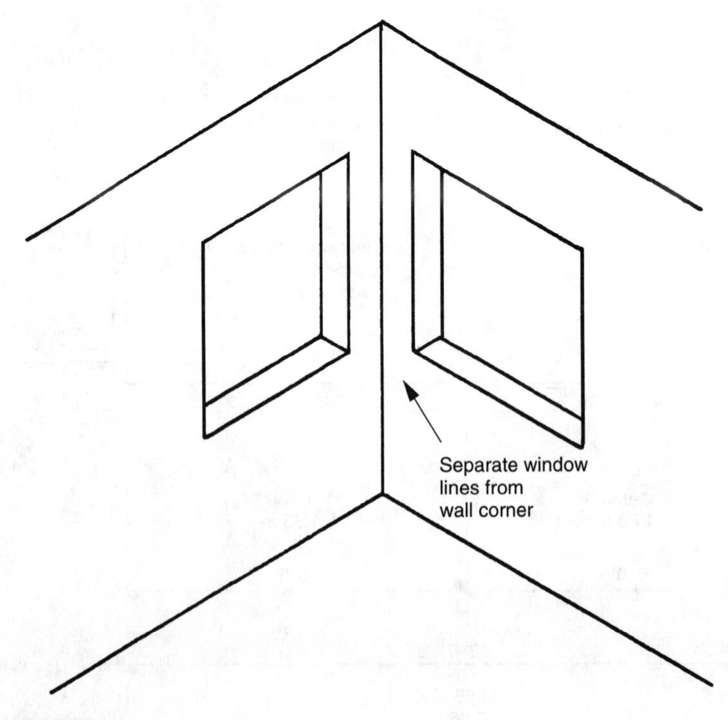

Figure 6.6

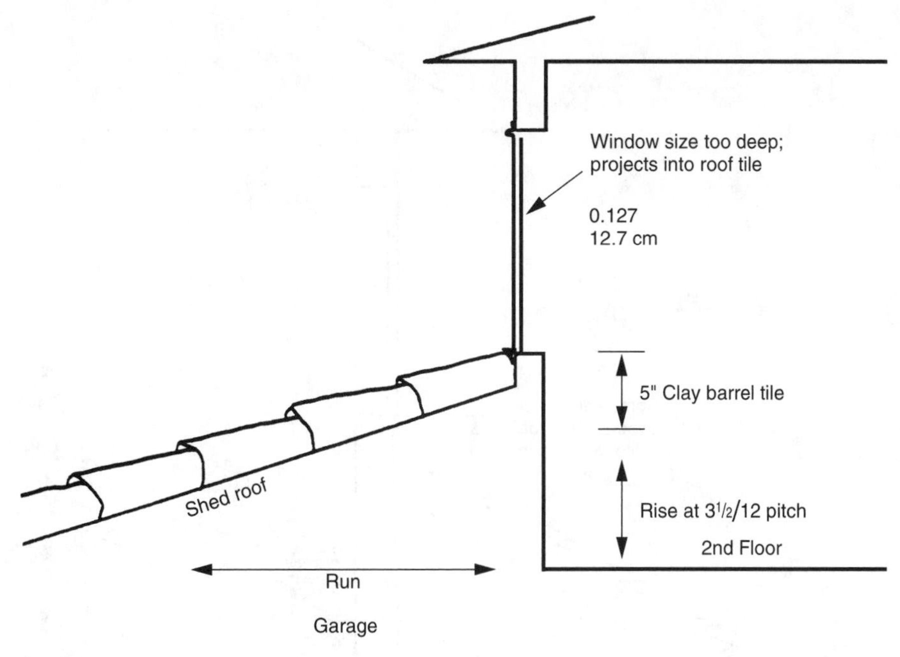

Figure 6.7

is not enough room for the 5-inch, S-shaped clay roof tiles and the 6-0 × 4-0 window. The builder should check the plans for similar situations prior to the start of the construction and the ordering of the windows. Figure 6.8 shows this garage roof designed low enough below the second floor to provide adequate clearance for the roof tile.

Cabinet Countertop and Windows

The kitchen cabinet countertop overhang, pictured in Figure 6.9, projects into an adjacent window. In this example, the location of the window was not coordinated on the plans with the cabinet pass-through countertop, which had an overhang of 14 inches. This is another item the builder should check during the plans review activity.

Check Windows for Squareness

The builder should check the window frames for squareness soon after they are installed by the carpenters. This can be done using a framing square or a smaller metal square. The builder does not want to wait until after the window frames are fixed in place by exterior stucco or wood siding, to be told by the glass sash installer that some windows and sliding glass door openings are out of square. The time to find out that fixed glass panels won't fit, sliding windows bind, and sliding glass door center latches will not catch because the openings are out of square, is not when the repairs involve tearing out stucco, wood siding, or masonry brick veneer.

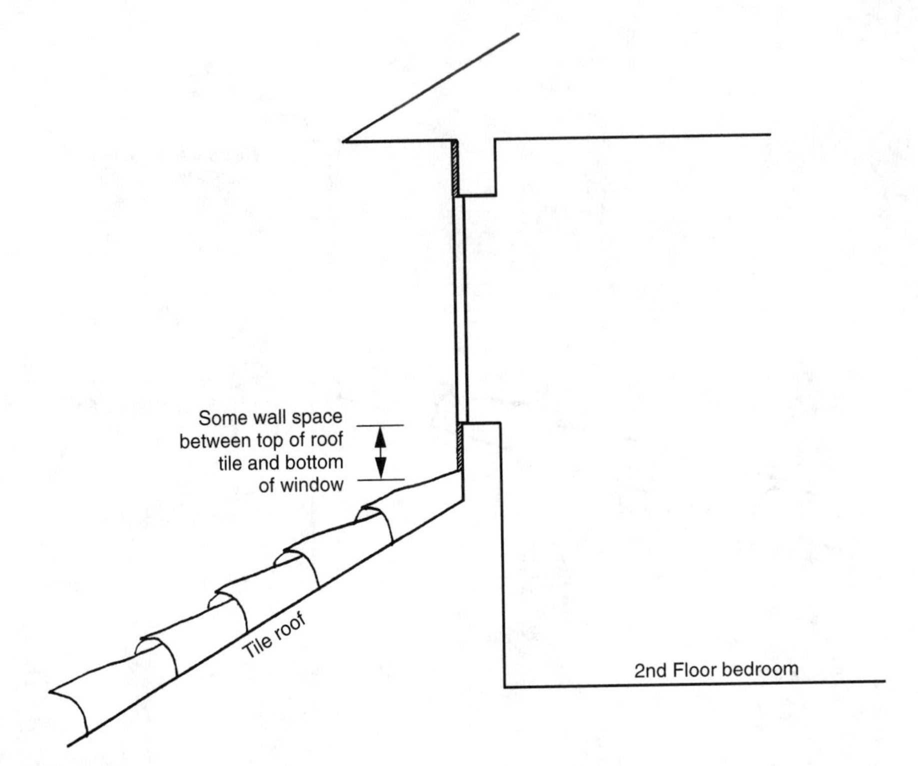

Figure 6.8

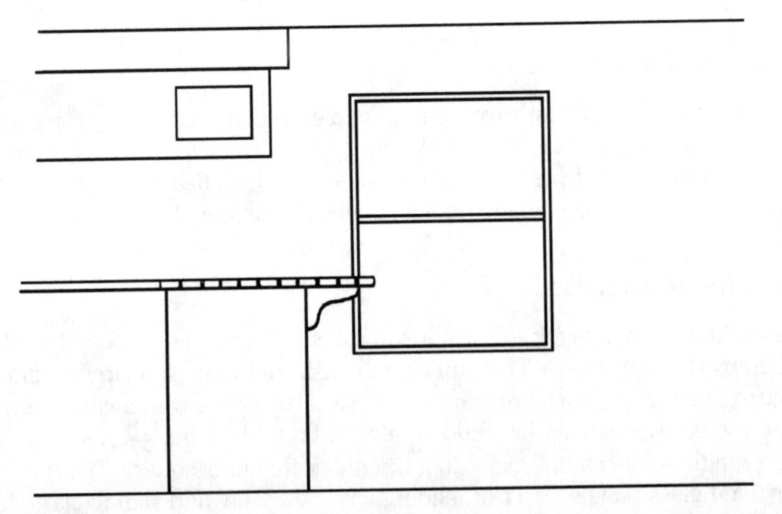

Figure 6.9

The problem of out-of-square windows becomes more crucial with buildings over one-story high. Second and third floor windows surrounded by stucco plastering require scaffolding just to get up to the windows to make the repairs. If an out-of-square window is not discovered until after the homebuyer moves in and complains about it on a customer service letter, then additional problems, such as landscaping and access to garages blocked off by scaffolding, make a difficult repair even more difficult.

Check Corian on Window Sills

Figure 6.10 shows the side view of a kitchen countertop and window sill, both covered with Corian. Although the Corian countertop worked out well for the homebuyers, the Corian window sill at the kitchen sink created framing problems for the builder.

On this particular project, the window openings were framed with a vertical dimension that produced a window reveal at the window sill. The window sill was not deep enough to allow for stool and apron to be added later. If stool and apron was added, or in this case Corian, the thickness of the stool would extend up and over the lip of the window frame. This would make it impossible to remove the window because the window sill became higher than the window frame itself. In this instance, the Corian was about ½ inch above the window frame, preventing the window from being lifted up and over the window sill to be removed.

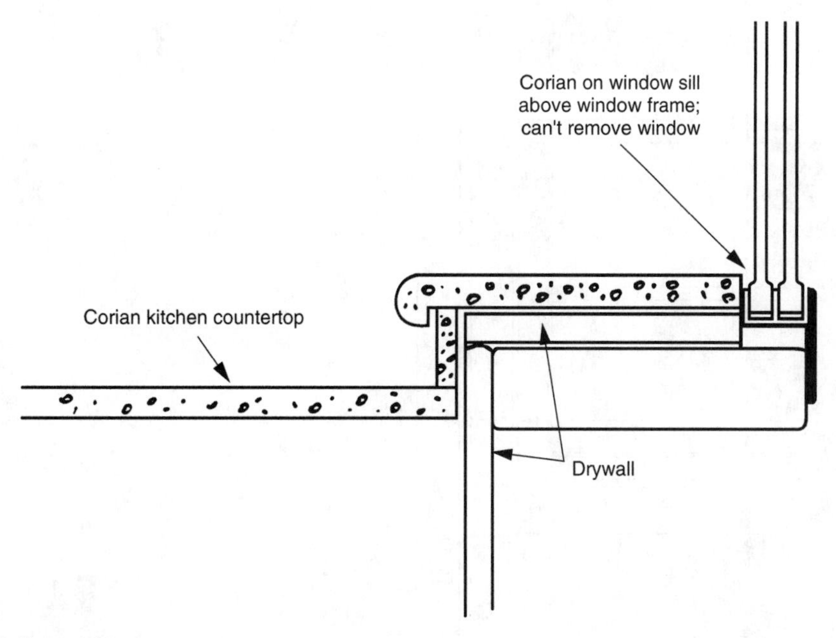

Figure 6.10

During construction, the Corian installer did not recognize that by adding a Corian stool to the window sill, the window was now irremovably stuck in place. The builder also did not realize that a Corian window stool came with the Corian kitchen countertop option. Because the sale of this particular house with the accompanying homebuyer's selection of Corian countertops did not occur early enough in the construction, the builder did not know to lower the framing of the window sill or leave the drywall off the window sill to accommodate the later addition of the Corian stool.

In this case the Corian stool had to be removed, the drywall torn off the window sill, then the Corian stool reinstalled. This lowered the stool to the point where the window could then be removed.

Window Reveals

The metal window frame pictured in Figure 6.11 has drywalled sills on all four sides. In order for the leftover, exposed portions (reveals) of the window frame to be uniform and even, the framing around the window frame must be straight, square, and have the correct rough opening dimensions. Because this is difficult to do with wood, the framer can tighten up the uniformity during the rough framing by attaching strips of building paper, roofing paper, thin wood strips, or thin tapered wood shingles, around the sides of the window sills that need correction. This is something the builder should insist upon and check.

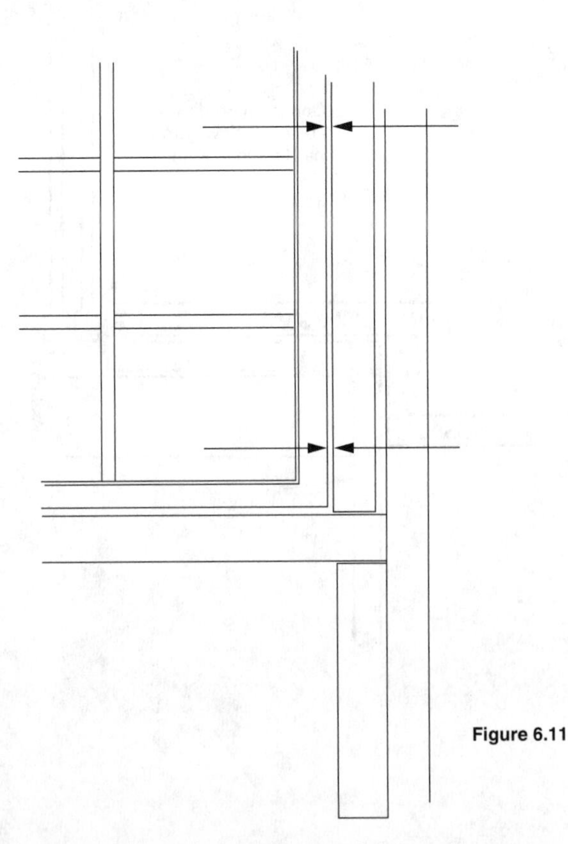

Figure 6.11

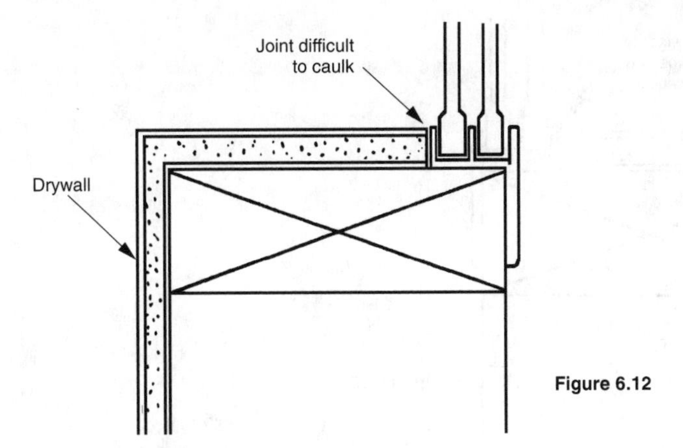

Figure 6.12

Get Framing Correct for Window Reveals

Illustrated in Figure 6.12 is a situation that occurred in which the openings for the window frames were correct in terms of width, but were the wrong height. The drywall at the top and bottom of the window sill met the window frame at the same elevation, resulting in a ugly crack at this joint on 90 percent on the windows. Because these cracks were on a flat joint between the drywall and the window sill, they were difficult to caulk and have them look right.

Figure 6.13 shows drywall at the window sills that butts into the window frame. In the example, this is what actually occurred at the sides of the window frames, which had the correct width framing dimension. When a crack forms at this type of joint, caulking can be applied to the 90-degree inside corner that now exists between the drywall and the window frame. The bead of caulking can be thicker at the actual crack because it is round on the front side, but square at the joint. The caulking can also adhere to both the drywall surface and the window frame, which cannot occur at a joint on a flat, 180-degree surface, without looking bad.

Windows Aligned for Stool and Apron

The two windows shown in Figure 6.14 are designed to be close together at a corner and have window sill stool and apron trim. The two windows in this example should be installed to be individually square, level, and plumb, as well as be positioned at exactly the same elevation. The wood stool and apron trim at the two window sills must align equally at the wall corner where they come together in order to look right. The windows themselves must therefore be at the same elevation, or the window frame reveals at the stool and apron will either be unequal or slightly out of level in order for the corners to match. For the window reveals to be equal, the stool and apron to be installed level, and the corners of the two stools to match, the window frames must be installed level and at the same elevation.

Align Windows for Plantons

The row of windows shown in Figure 6.15 should not only be installed so that they are individually square and plumb, but they should also align as a group if the win-

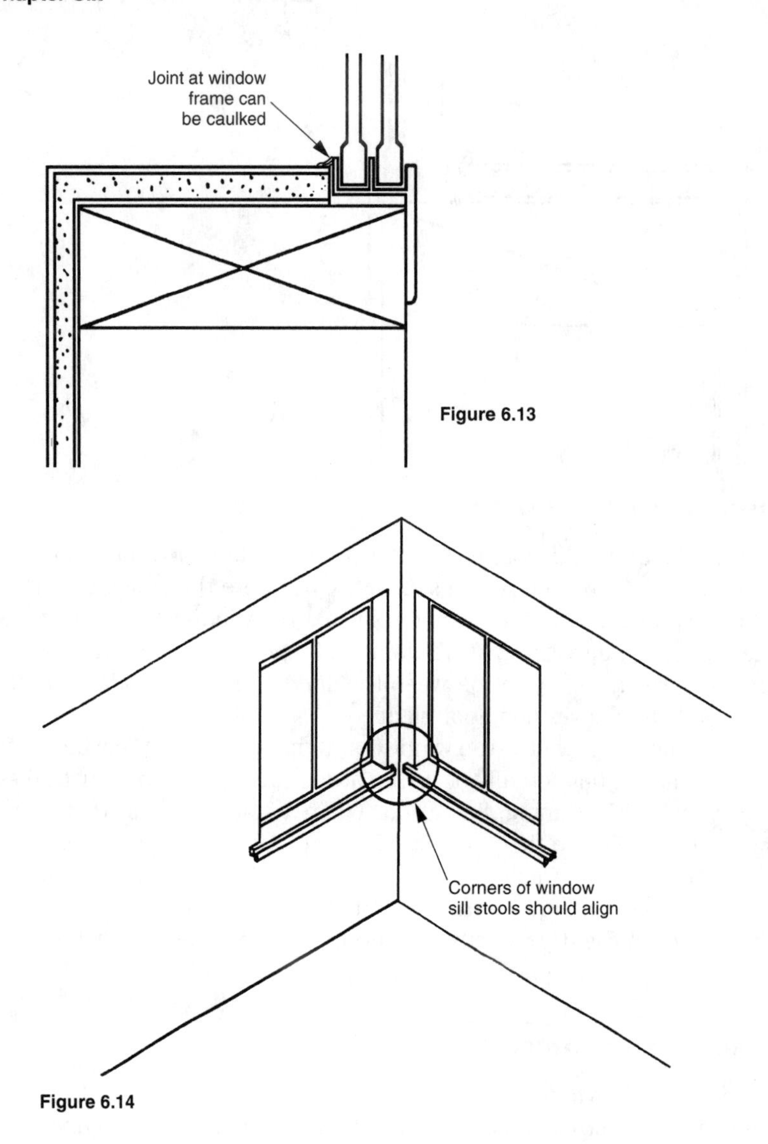

Figure 6.13

Figure 6.14

dows are to be trimmed out with wood plantons. By aligning all the windows, uniform reveals or gaps are formed between each window frame and planton.

Tile Over Window Sills

Illustrated in Figure 6.16 is the end view of a kitchen cabinet and window sill at the kitchen sink. When kitchen countertop tile continues over the window sill and butts into the window frame, the window sill should be framed the same height above the floor as the cabinet rough top. This enables the tile setter to apply a uniform thickness of cement on the cabinet roughtop and the window sill in preparation for the tile.

The cabinet roughtop height dimension (usually 35–36 inches) must be obtained from the cabinet manufacturer before framing layout in order to plan the correct kitchen window sill height.

Window Arches

The framing for an arched window is pictured in Figure 6.17. The header is framed into the wall 2 inches higher than normal so that the plywood arch can be one continuous piece, in this example 2 inches wide at its narrowest point. This makes ob-

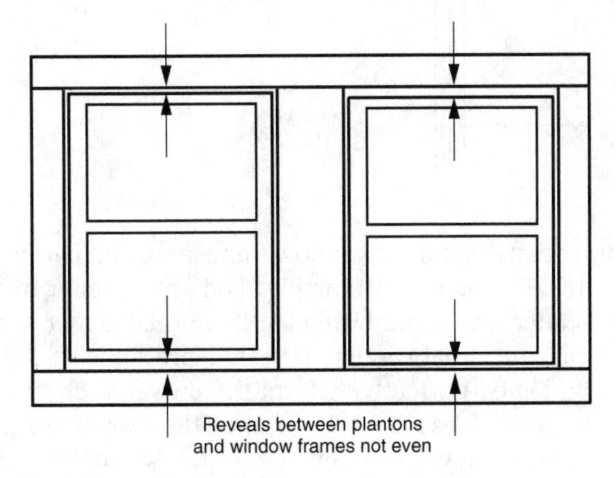

Figure 6.15

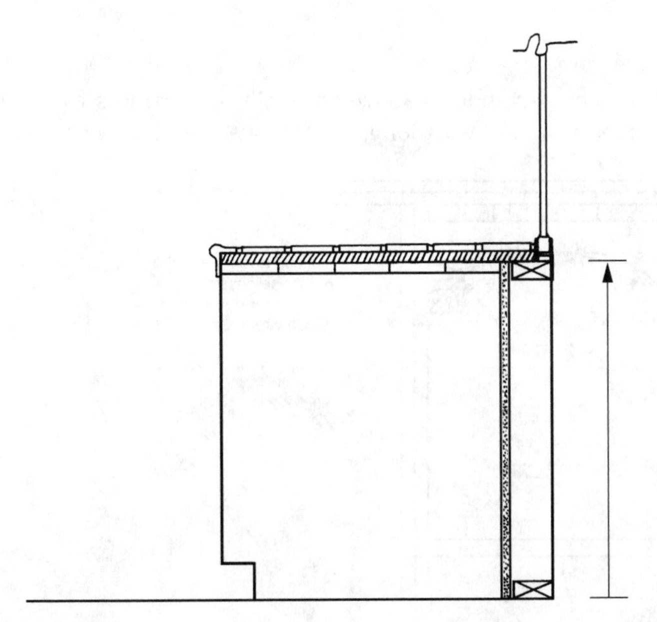

Figure 6.16

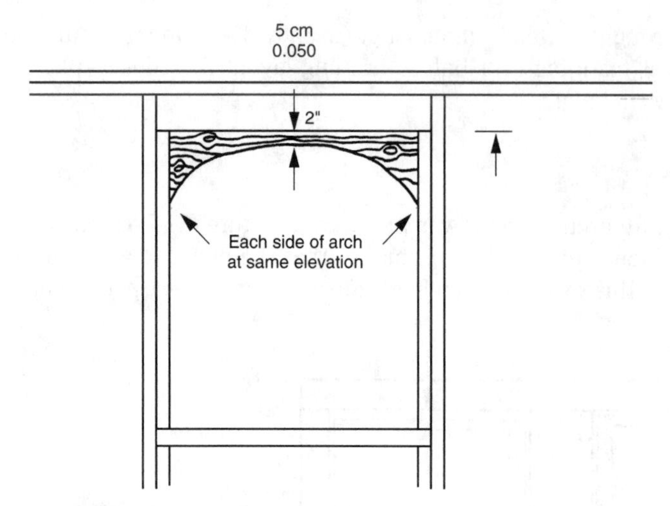

Figure 6.17

taining an even and equal reveal around the window frame easier and more accurate.

Figure 6.18 shows an arched window with the plywood arch framing in two separate pieces because the header was either framed at its normal height or could not be raised. This makes achieving equal window reveals more difficult, because the two half pieces must to be joined perfectly to form the curved arch to match the window frame. If one of the two halves is slightly off, then the reveals around the top arched portion of the window frame will be uneven. A solid plywood arch eliminates this problem, but the header must be framed higher in anticipation of the one-piece plywood arch.

Align Jambs for Casing

Entry doors with transom windows above and hallway corridors with rows of windows are sometimes trimmed with long, continuous pieces of wood casing rather

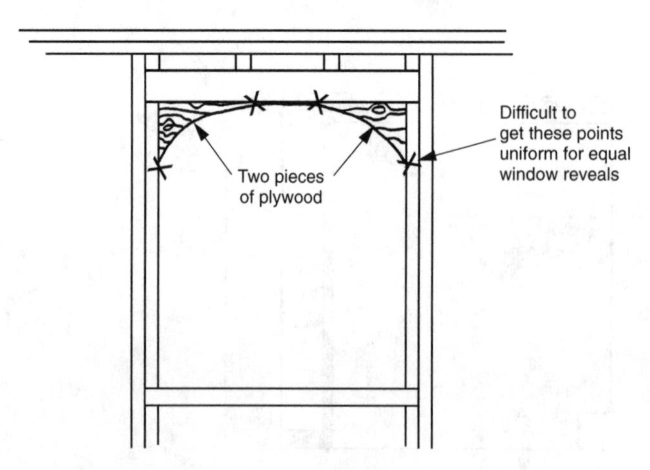

Figure 6.18

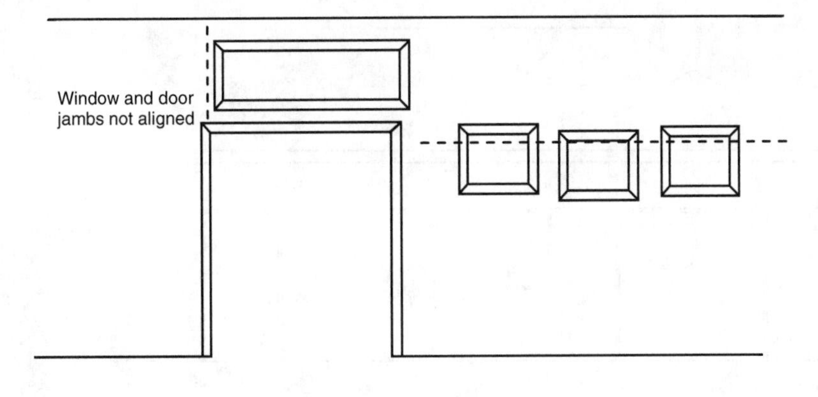

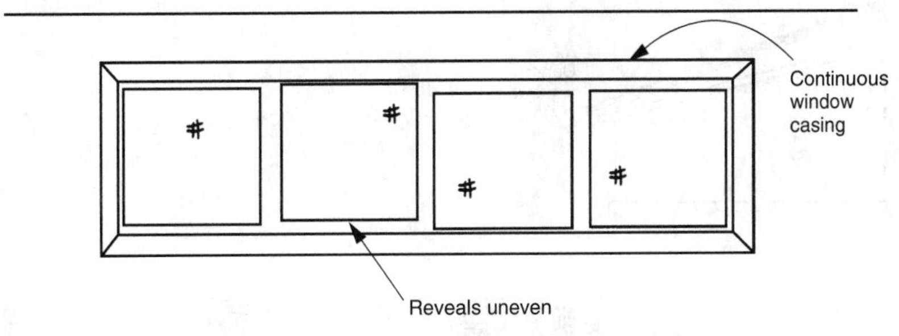

Figure 6.19

than individual pieces (Figure 6.19). In this example, the wood jambs in each opening should be aligned if the jambs are to correspond correctly with the casing trim. Jambs that are not aligned in a straight line will result in different reveals when the casing is installed.

Sliding Glass Door Header Heights

For apartment and condominium projects, the builder should check the architectural floor plans versus the exterior elevation plans when both 6-feet-8-inch and 8-feet sliding glass doors are called out. The floor plan may have a 6-feet-8-inch slider for the dining room balcony and an 8-feet slider at the living room balcony on the same floor level. The exterior elevation drawing, however, may have both sliders and any other windows drawn with all the header heights equal and on a straight line (Figure 6.20).

The problem is that different draftspersons may be working on different pages of the plans. The architectural intent may be to have either all the window and slider headers align on each floor level or have the headers at different heights. The person

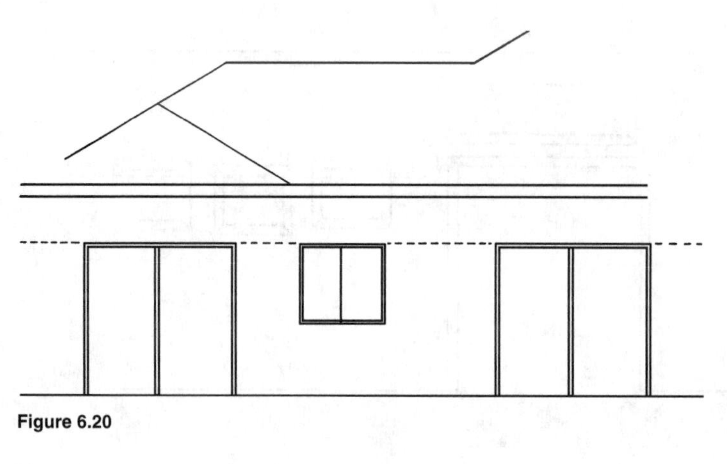

Figure 6.20

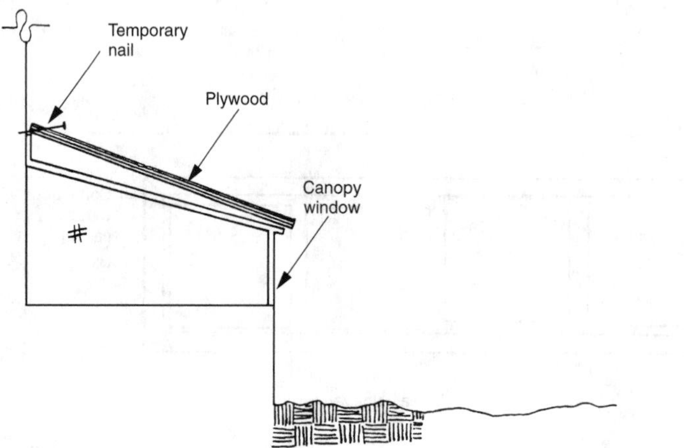

Figure 6.21

drawing the floor plans may not communicate with the person drawing the exterior elevation plans on this one point. Another scenario is that the building exterior elevations may be an earlier generation of plans that were not revised to reflect floor plan changes. The framer may cut cripples and frame the walls using the floor plans without looking at the building exterior elevation plans. In any case, this should be on the builders checklist of things to look at on the plans (Figure 6.20).

Protect Canopy Windows

Canopy windows, like the one shown in Figure 6.21, are usually installed as complete units, with glass panels in place. Because they are vulnerable to damage from things dropped from above during the course of the construction, the builder should include plywood panels, cut to fit over the top of the canopy windows, in the framer's contract. The plywood panels can be held in place by nails driven into the top edge of the plywood and hooked over the top flange of the canopy window (depending upon the window design). The plywood panels can then be easily moved from house to house or building to building as the construction progresses. The cost of these plywood panels will be much less than the cost of glass hack-outs for canopy windows with broken glass.

Chapter

7

Framing Stairs

Stair Skirtboard

Figure 7.1 illustrates the typical framing of a stairway without stair skirtboard that can be found in housing construction. To build stairs without the stair skirtboard, fur out the stair stringer or jack ¾ inch from the wall framing and hold back the stair treads and risers an equal distance. Then slip the drywall between the stairs and the wall framing. By doing the job in this fashion, the drywall does not have to be cut saw-tooth to match the stair steps (Figure 7.2).

If the builder decides to include a stair skirtboard, the decision must be made prior to the framing of the stairs. By introducing the thickness of an additional 1×12 skirtboard between the stairs and the wall framing, the stair stringer, treads, and risers must be furred out 1½ inches. This allows for both the drywall and the skirtboard to slip into place without having to spend time cutting the skirtboard to match the stair steps. Figure 7.3 shows the end view and side view of this condition.

Plywood Over Stairwells

The exposed edges of plywood subflooring on open second floor hallways, on stair landings that are open on one or both sides, and for open lofts, should cantilever past the rough framing an amount equal to the thickness of the interior drywall (Figure 7.4).

During the finish carpentry phase of the construction, materials such as doors, shelving, and bundles of baseboard are leaned against and lifted up and over these open floor areas as they are spread throughout the house interior. Having the plywood project over the drywall protects the edge of the drywall from being mashed or broken.

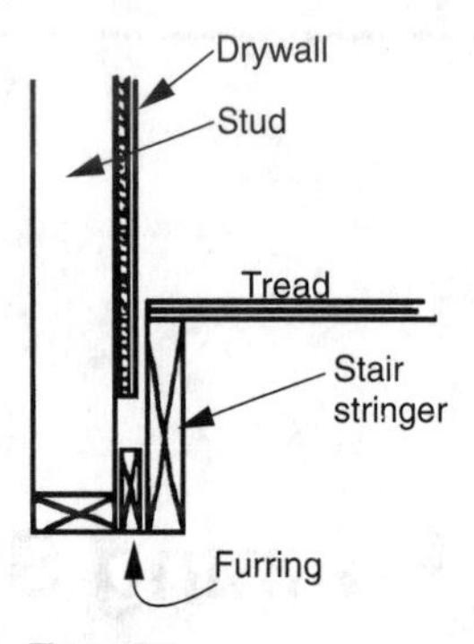

Figure 7.1

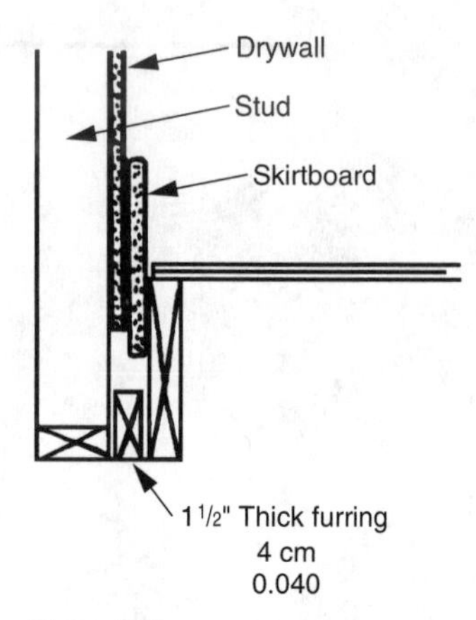

Figure 7.3A

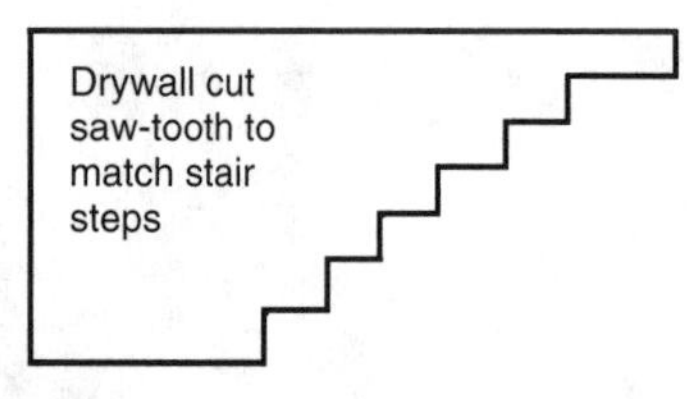

Figure 7.2

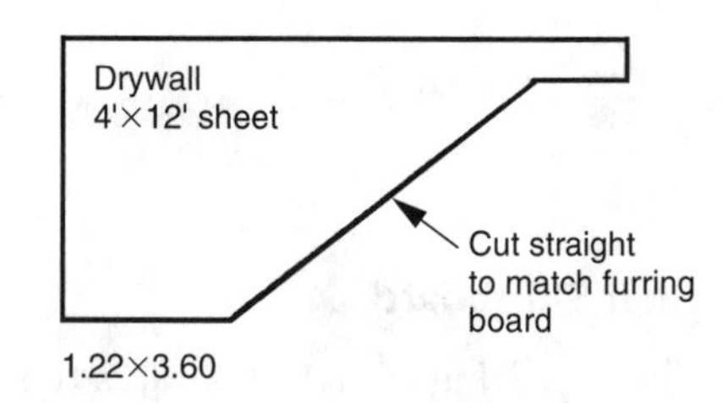

Figure 7.3B

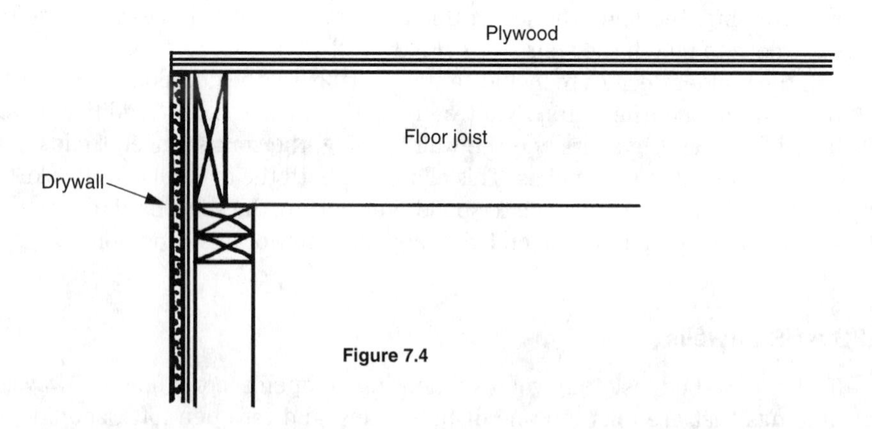

Figure 7.4

Thicker Stair Tread Plywood

Figure 7.5 shows a curved stairway in plan view. The span required of the plywood for the treads at the outside of the stair curve, where the treads flare out, is sometimes greater than what is allowed for the thickness of the tread material used for straight, noncurved steps. Depending upon the width of the staircase, ¾-inch, 1-inch, or even 1¼-inch plywood should be used at these wider curved treads in order to pass inspection and to prevent spongy stair treads. The same result can also be accomplished by solid blocking underneath each tread to reduce the plywood span, thus allowing the use of the original thinner plywood.

No Curved inside Corners

In the curved stairway pictured in Figure 7.6, the outside curve of the stairs should be straightened out to become a flat wall. This eliminates the awkwardly tight inside corner that is difficult to frame, drywall, tape, texture, baseboard, paint, and carpet. For production tract housing, the architect should throw away this curved stair wall detail.

Veneer at Curved Stairs

The plan view (Figure 7.7) and front view (Figure 7.8) of a curved stair step shows how the curved portion of the stair tread is supported by short length cripple studs that are spaced several inches apart. The problem with this design is that the spacing of the cripple studs at the curved portions of the steps leaves empty areas between the studs. The carpeting going up the stair risers at these areas has nothing solid for backing other than the cripple studs. The carpet then becomes wavy and indented at the voids between the cripple studs. The solution is to have the stair builder nail a door skin or Masonite veneer over the cripple studs at the curved portions of the stairs. A thin wood veneer will bend to match the contour of the curved stair step, yet still be stiff enough to provide a solid backing for the carpeting.

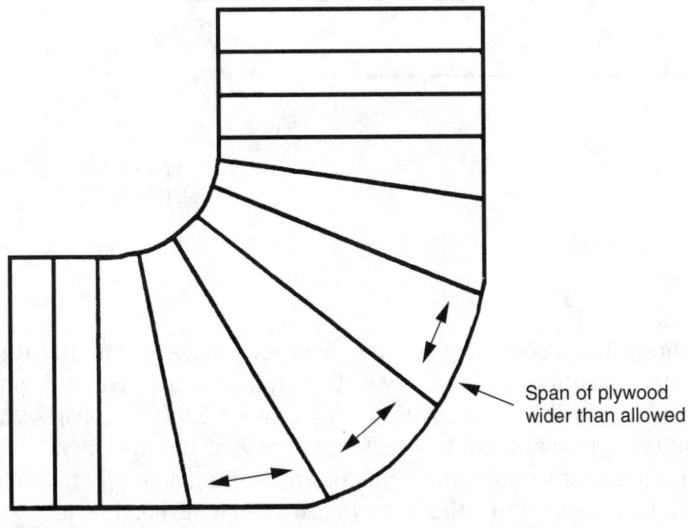

Figure 7.5

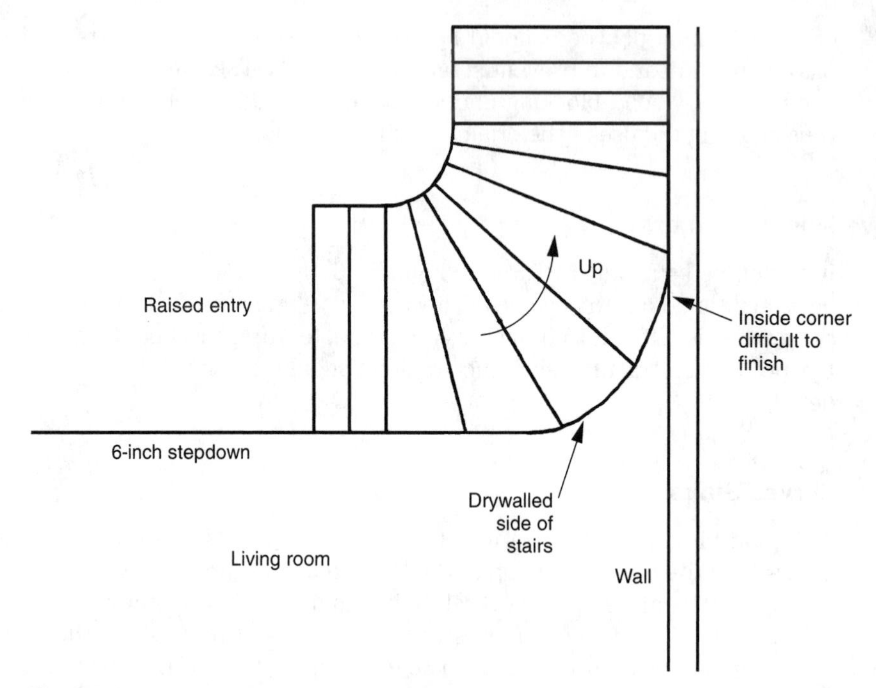

Figure 7.6

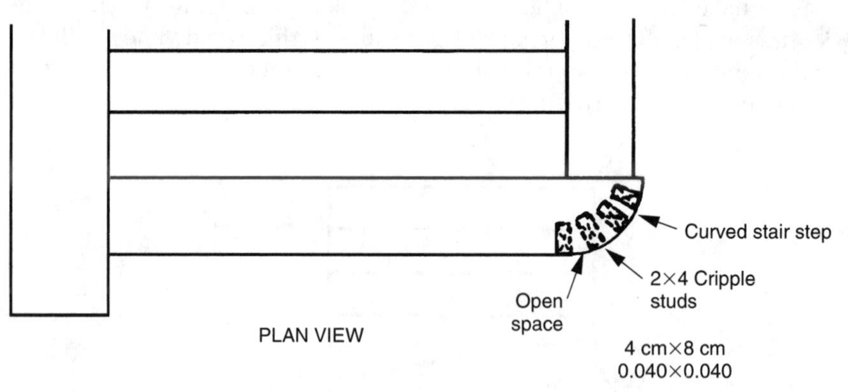

Figure 7.7

Stair Handrail Post

Figure 7.9 shows a 4×4 post anchored to a floor joist in a stair landing at the end of a half-height stair partition wall or pony wall. In this example, the stair partition wall will be framed, drywalled, and covered with a wood cap. A similar situation could also occur at the termination of this half-height wall at the first floor.

By running the post through the subfloor and bolting it solidly to one of the floor joists, the independent and otherwise unattached half-height stair wall is made structurally more solid and stable.

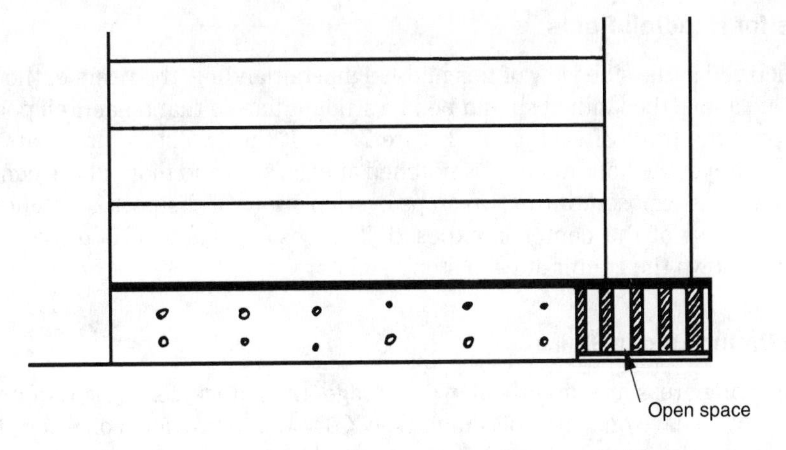

Figure 7.8

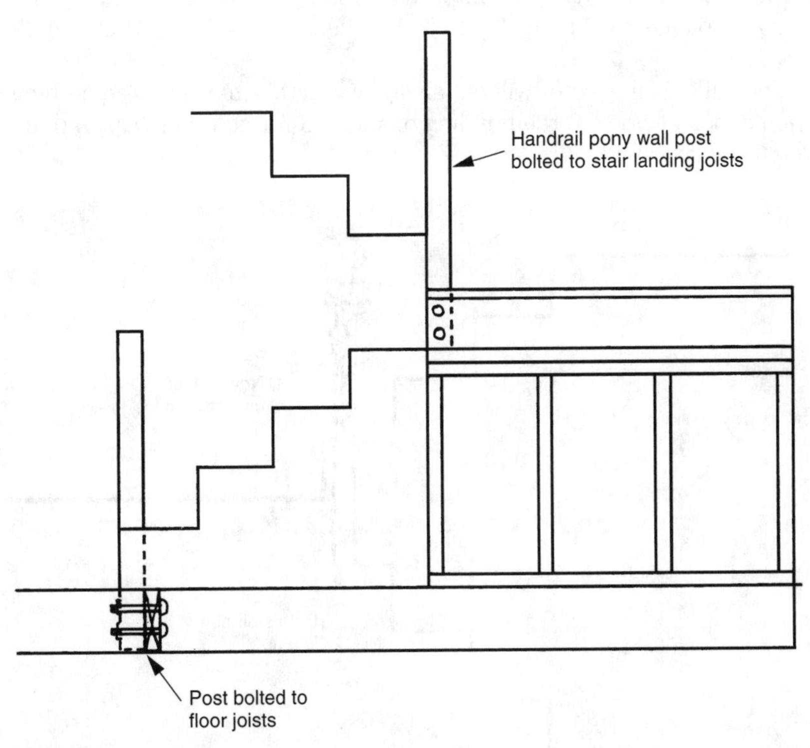

Figure 7.9

Line Up Risers for Handrail Posts

As pictured in the side view of this mid-height stairlanding, the risers of the steps below and above the landing should be in a straight line so that a handrail post will fall half over the front of each step (Figure 7.10). If the two steps are apart by a few inches, less of the handrail post is notched at each step and more of the central meat of the post is exposed. If the two steps overlap, more of the post is notched at each step and less of the center is exposed. The best appearance is to have the post notched down the center at each step height.

One-Hour Fire Rating Under Stairs

When living areas are directly above a garage, the garage ceiling is required by the UBC code to have one layer of ⅝-inch type X drywall to provide a one-hour fire rated separation. The garage ceiling joists for a one-hour fire rating cannot be spaced farther apart than 16 inches, on-center.

Figure 7.11 shows the underside framing of a flight of stairs from the garage level to the second floor, which is part of the garage ceiling. The staircase is a minimum 36 inches wide, with three stair stringers supporting the treads and risers. The spacing between the stringers is wider than the one-hour fire rating allows for framing on the garage ceiling. If the staircase were 42 inches wide, the spacing of 3 stringers would be 21 inches on-center—well over the 16-inch spacing allowance.

The building inspector will not always catch this, but this is something the builder should look at prior to the building of stairs, if for no other reason than to avoid future liability should a fire occur.

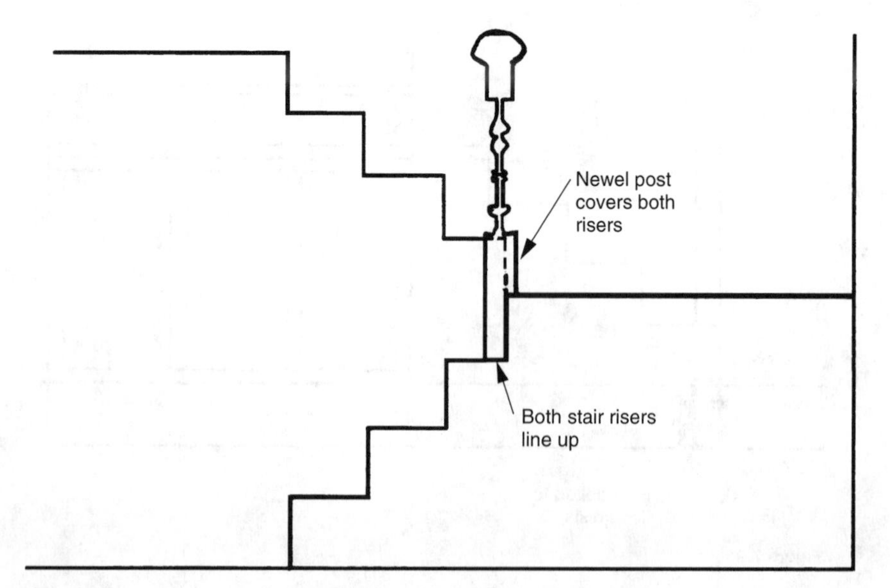

Figure 7.10

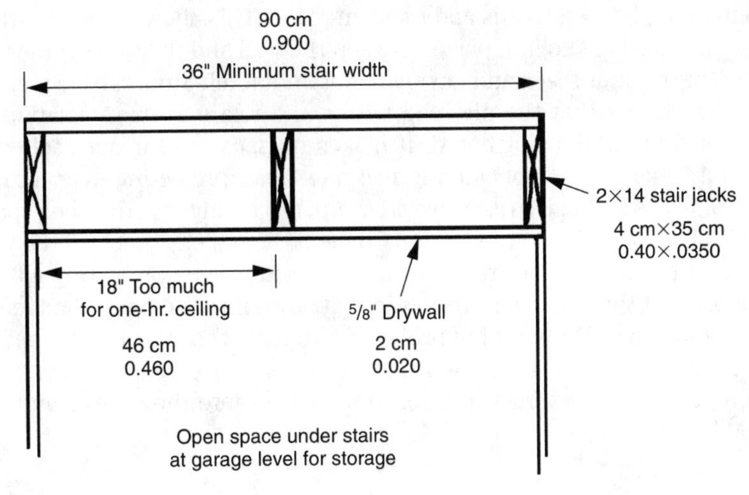

Figure 7.11

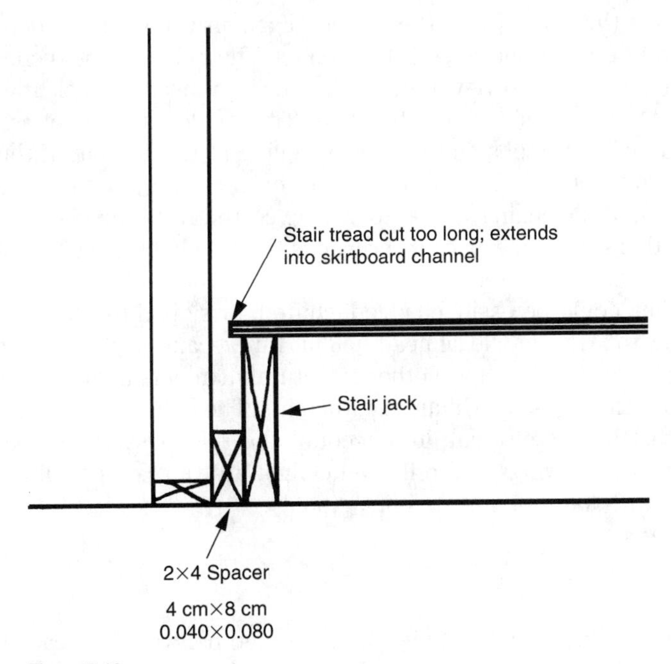

Figure 7.12

Check Stair Tread and Riser Clearance

Shown in Figure 7.12 is the front view of a flight of stairs that are framed to receive both drywall and a wood stair skirtboard between the wall framing and the stair jack. Check to make sure the stair jacks are framed at least 1½ inches away from the wall framing and the stair builder remains consistent with this clearance when cutting and nailing the plywood treads and risers in place. This allows space for the drywall and the skirtboard to slip into place between the wall and the stairs rather than resting on top of the stair steps and having to cut and notch around each step.

Be sure to check that the plywood treads and risers are not violating the open channel for drywall and skirtboard. It makes no sense to fur out each side of the stairs the 1½ inches, then accidently nail one tread projecting over into the open space on one side and nail a riser, two steps up, projecting into the open space on the other side. The drywall and the skirtboard need the full clearance all the way up the stairs, and it takes only one tread piece or one riser piece projecting into the open space to prevent the skirtboard from slipping down into the open channel.

One way to solve this potential problem is to have the stair builder cut the tread and riser pieces a ¼-inch short, then split the difference on each side of the stairs so that each piece is ⅛ of an inch short of running over into the drywall and skirtboard channel.

Stair Head Height Clearances and Beams

Figure 7.13 shows the side view of the bottom of a stairwell with a structural beam placed at the front edge of the sloped stair ceiling. The builder should check during the preconstruction design review period whether legal headroom clearance exists in stairways as they are designed on the plans. The builder should not assume that the rise and run of stair steps and a hallway ceiling at the bottom of the flight of stairs, coordinate to provide a legal head-room height simply because the bottom step and the edge of the ceiling appear to line-up correctly on the plans.

What makes this situation especially crucial is that the hallway ceiling at the bottom of the stairs was not merely ceiling joists, but a structural beam. If it was only ceiling joists, they could be easily changed, shifted, or sloped to accommodate the stair steps below to arrive at a legal head-height. But because a structural beam in a ceiling or floor has a relationship to other structural members above, it cannot simply be shifted over for the sake of stair head-heights. The builder cannot also assume that the architect accurately took into account the thickness of this beam when drawing the section drawings and reflected ceiling drawings. A pencil line labeled 4×12, 6×12, or 8×12 does not always fit in actuality.

Stairway Arches

Illustrated in Figure 7.14 is a front view of an arched hallway that leads into a living room at the top of a flight of stairs. In this particular case, the arch was not centered because the width of the hallway did not work out with the ceiling height of the stairs and the arch could not begin on one side until the stair ceiling died, about 18 inches past the start of the hallway. The three fixed points—the height of the stair ceiling, the width of the hallway, and the high point of the arch—could not be changed in order to center the arch in the hallway in this particular case. The off-centerness of the arch was accentuated when a wall sconce light fixture was centered on the stair wall in the hallway at the top of the stairs. The flat plane

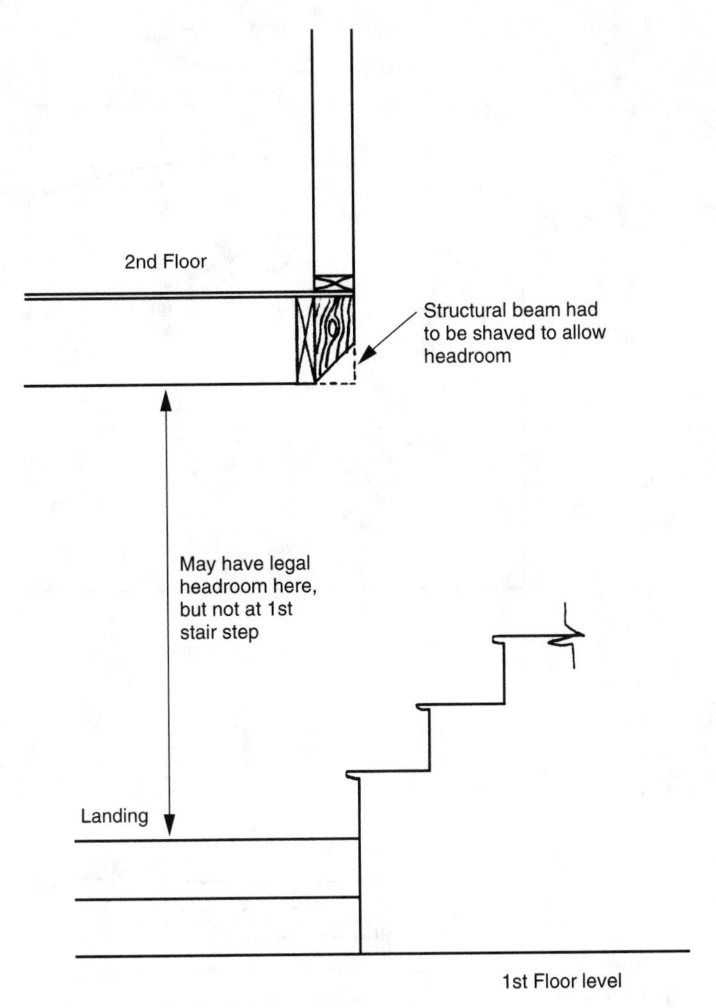

Figure 7.13

of the stair ceiling, the point at which the arch began, and the curvature of the arch itself, was all now noticeably off-center with the addition of the sconce light.

Stair Handrail Wall High Enough

The central sloping pony wall that separates two flights of stairs with a wrapping around handrail is depicted in Figures 7.15 and 7.16. In this example, the builder should check that the wall is accurately framed according to the plans and high enough in dimension to allow for a handrail at the legal height above the stair steps. For example, if the pony wall is given a height dimension on the plans somewhere between 30 and 36 inches, which is also roughly the interval at which a handrail is legal,

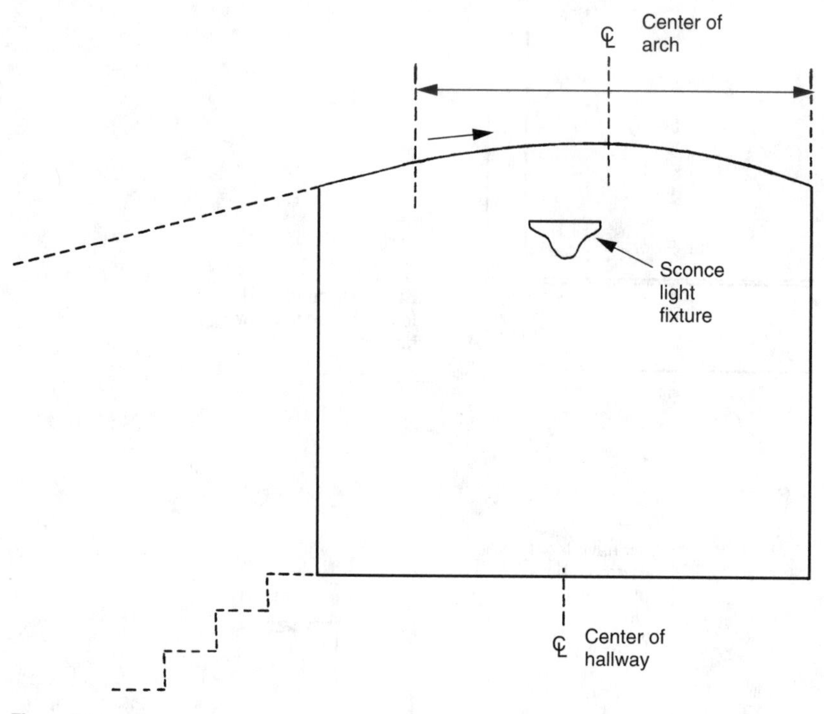

Figure 7.14

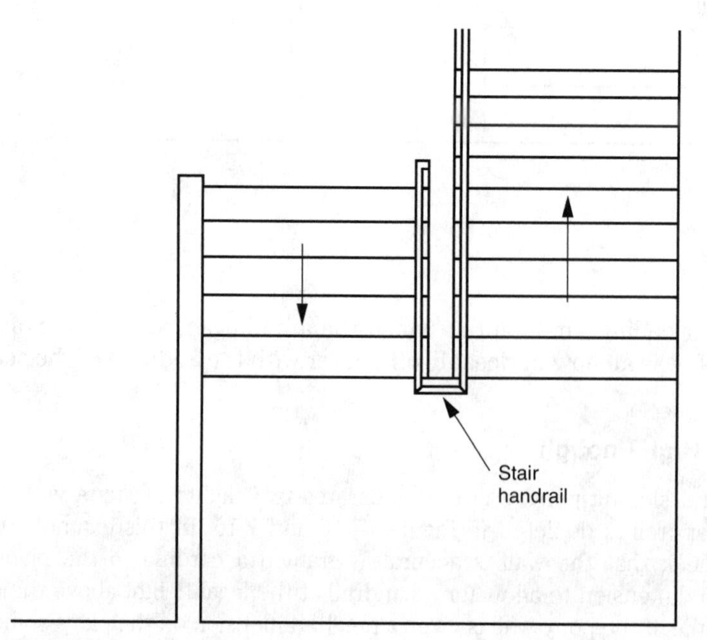

Figure 7.15

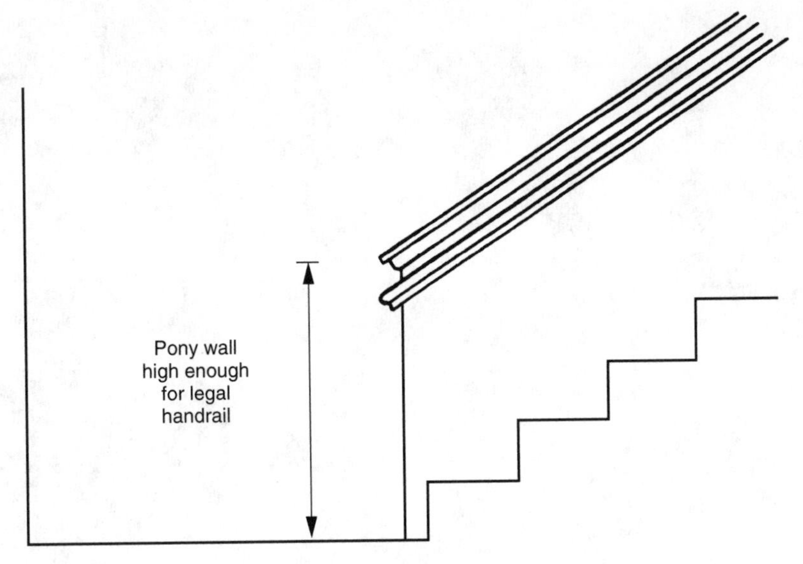

Figure 7.16

then the builder should draw to scale all of the pony wall parts to see at what elevation the handrail ends up. If the top of the pony wall is supposed to be at 36 inches above the floor, by the time you work downward from there with a ¾-inch thick wood cap, a 1½-inch or wider wood apron, and metal handrail support brackets, the handrail itself may be too low.

The time to discover that the handrail is not at the legal height, due to a cap, apron, handrail bracket, the handrail itself, and a wall height that wasn't correct to begin with, is not during the final building inspection. It is much easier and cheaper to analyze the whole thing on paper prior to the start of the construction.

Chapter

8

Shear Panel

Layout for Shear Panel

Plywood panels are often installed on interior and exterior wall sections to provide shear strength to the building. Wall framing should be laid out whenever possible so that the first sheet of plywood can be nailed up without having to be cut. Figure 8.1 illustrates a long main wall with one side running without interruption and the other side partitioned with a closet that is to be shear-paneled. If the wall framing layout starts at one end and continues the entire length of the wall, there is little chance that the location of the closet partition walls will accommodate the plywood installation. The wall framing layout should be stopped at the closet, then a new 16-inch, on-center layout started within the closet. This will enable the first 48-inch wide sheet of plywood to be nailed up without having to be cut to a narrower width (all vertical edges of shear-panel plywood must be on a wall framing stud).

By changing the framing layout each time to accommodate shear-panel plywood, only the last sheet in a run of shear-panel needs be cut. Not having to measure and cut the first sheet can save a lot of time on a large housing tract that has a lot of shear panel.

Shear Panel in Garage

Sometimes the front corners of garages, on each side of the garage door, are structurally designed to be shear-paneled with plywood on both sides of the wall (Figure 8.2). Be sure to inspect hold-downs and anchor bolts inside the wall before both sides of the garage walls are covered with the shear-panel.

There are two ways of ensuring that the hold-downs and anchors are inspected prior to the wall being covered. The first way is to cut and nail on the shear-panel to

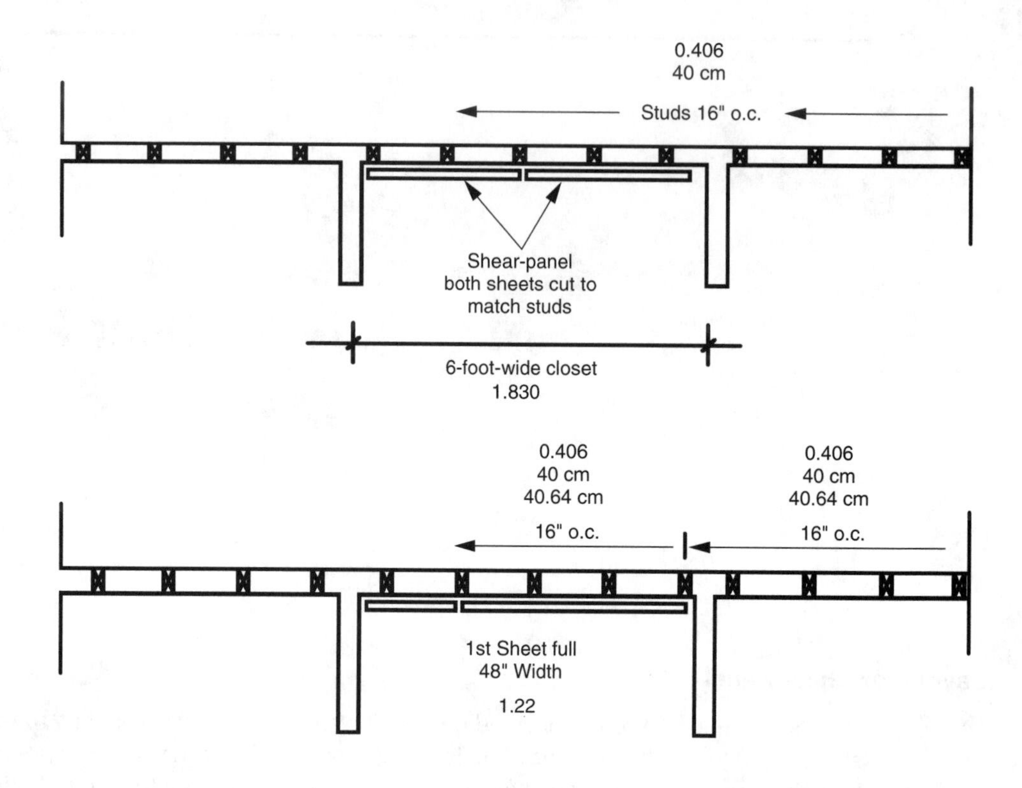

Figure 8.1

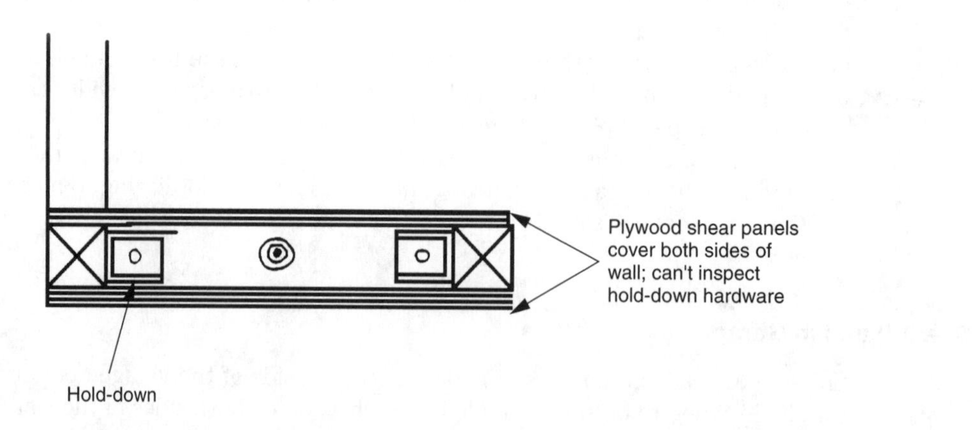

Figure 8.2

one side of the wall, then cut and tack the second piece of plywood to an adjacent wall. During the shear-panel inspection, the one installed side is inspected along with the open and exposed hold-downs and anchor bolts. The second piece of plywood is then nailed on the second side of the garage wall, and inspected later during the combination framing inspection.

The second solution is to install the hold-downs and anchor bolt nuts early, talk the inspector into checking these items separately from the shear-panel inspection, get them signed off on the inspection card, and then shear panel both sides of the garage wall.

It is easy for the shear-panel plywood installers, in production housing, to cover both sides of the garage wall without thinking about the required inspections. One side of the shear panel must then be torn off so that the inspector can see the hold-downs and anchor bolts.

Cantilever Second Floor Walls for Shear Panel

Illustrated in Figure 8.3 is a second floor wall that has been laid out and framed past the first floor in anticipation of the installation of the first floor shear-panel plywood. By using a cantilever effect or offsetting the wall an amount equal to the thickness of the shear-panel plywood, the second floor wall ends up flush with the plywood. This eliminates having to add furring strips to the second floor wall.

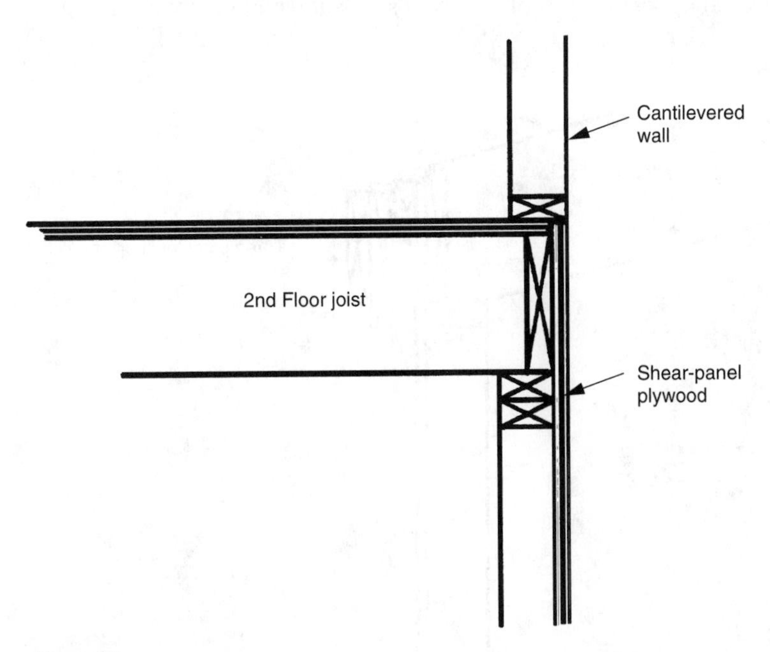

Figure 8.3

Divide Up Shear-Panel Inspections

When shear-panel plywood must be inspected at different stages during the construction, but there is only one line on the inspection card on which to sign off, it is a good practice to list the various inspections on a separate piece of paper or the back of the inspection card. The items can then be signed off as they are inspected at each stage. For example, the top of shear panel at the roof line must be inspected before the nailing is covered up with zero overhang facia board, yet the installation of the facia board should not have to or often cannot wait for all of the other shear-panel plywood to be completed and inspected (Figure 8.4). Shear connections and nailing from the attic walls through the roof may want to be inspected before the remainder of the shear panel is completed, yet before the roof is covered with felt paper. Attic shear panel may be completed and inspected before preliminary attic drywall can be installed, yet ahead of the entire building shear inspection.

By breaking up the inspections to reflect the reality of the construction more closely, the progress of the shear-panel inspections can be noted and initialed, and therefore recorded, should inspectors be changed or memories lapse. When the shear panel is inspected and approved for the entire building, then the single line on the inspection card can be signed off.

Upgrade Shear Panel One Side

In two- and three-story condominiums and apartments, shear-panel plywood is often designed on opposite sides of partywalls, as shown in plan view in Figure 8.5. When

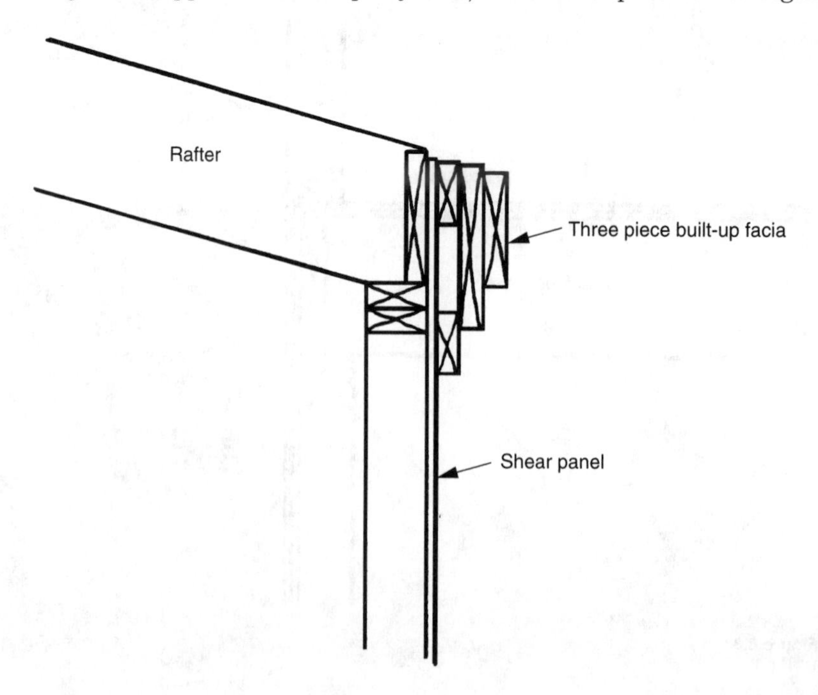

Figure 8.4

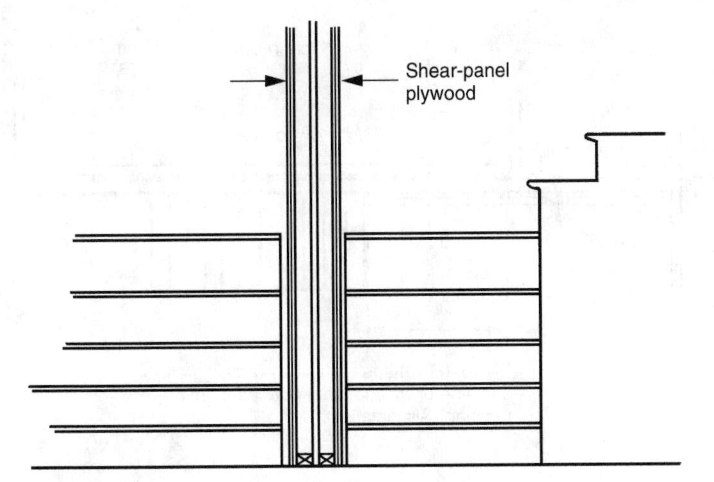

Figure 8.5

this is the case, the builder should try to get shear-panel sections of minimum design, such as, for example, ⅜-inch plywood with 6-inch, on-center perimeter nailing, upgraded to ½-inch plywood with 4-inch or 3-inch, on-center perimeter nailing. This can shorten the length of shear-panel sections on each side of the partywall, reducing or eliminating overlapping shear panel. This enables all of the shear panel to be installed at one time, without worrying about enclosing both sides of a partywall section, closing off access to the mechanical trades such as plumbing and electrical.

When plywood shear-panel sections do overlap on two sides of a partywall, one side of the wall cannot have shear panel installed until all of the mechanical work is in place and inspected. This creates two shear-panel installations, one for each side of the wall, which complicates the construction scheduling.

In one particular case, a section of shear panel was upgraded on a wall inside a powder bath, which then eliminated two sheets of shear panel extending out into the stairway outside the powder bath. This not only eliminated the overlap with shear panel on the opposite side of the partywall, but also saved us the expense of having to fur out the remainder of what would have been the nonshear paneled areas of the stair wall (Figure 8.6). These are issues about which the structural engineer and the architect do not always think.

Attic Shear Panel

Some structural designs for houses call for shear-panel plywood to extend from the floor line of a wall all the way up to the underside of the roof sheeting. These shear sections are designed to resist earthquake forces. The architect and engineer should also consider the code requirements for access to attic spaces when enclosing attic areas with shear panel sections.

There are only so many desirable locations within a house where an attic access in the ceiling can be placed. Walk-in closets, large coat closets, and hallways are the best locations for ceiling attic accesses because of their less than attractive appear-

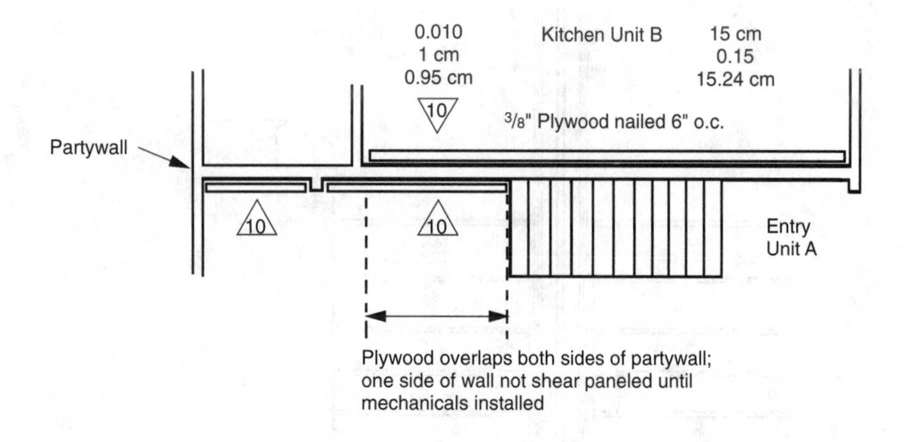

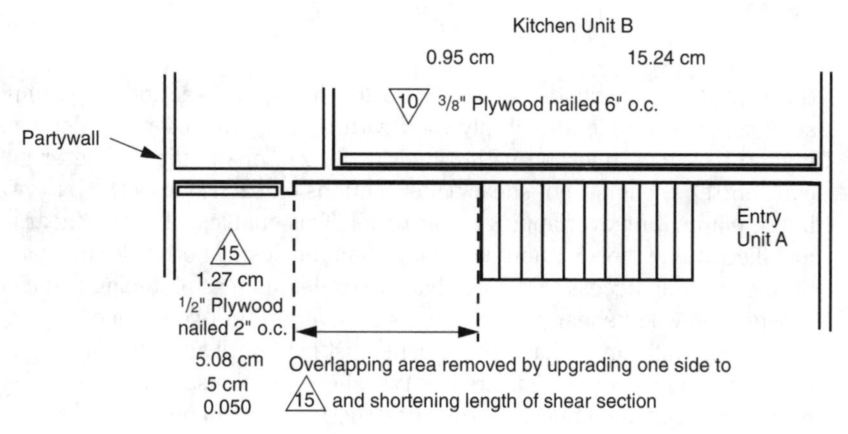

Figure 8.6

ance. If attic space above living rooms, dining rooms, entry hallways, and kitchens are blocked off from the normal closet or hallway attic access due to shear-panel plywood in the attic, then additional accesses may have to be placed in these ceilings.

Crawl space holes are allowed through shear-panel walls in attics to provide access to other areas within the attic, but these openings must be anticipated in the shear-panel design and shown on the structural plans.

The time to consider and solve problems with shear-panel design is during the design phase of the project, not during the framing inspection or, worse yet, the final inspection.

Check Plans for Two-Inch Shear Nailing

Prior to the start of the construction, the builder should check the plans for shear-panel plywood nailing that is called out to be 2 inches, on-center (Figure 8.7). This close of nail spacing in plywood requires that the studs and plates around the perimeter of each plywood sheet be thicker than the 1½ inches of a standard 2×4. This means that the bottom plate mudsill must be a 3×4 or 3×6, and that the studs in this section also be 3×4 or 3×6. This also requires that the anchor bolts for the 3×4 mudsill project up above the concrete an inch higher than for a 2×4 mudsill.

This is a requirement that should be colored or highlighted on the builder's set of plans and on any concrete and framing layout sketch sheets so that it is not missed during concreting and framing. This is especially easy to do when concrete anchor bolts are wet set during the pouring of the concrete, as the person setting the anchor bolts may not know that some areas have to have the anchor bolts placed higher for thicker mudsill.

Shear Panel behind Fireplaces

Illustrated in Figure 8.8 is a situation in which the architect or structural engineer has placed shear-panel plywood on a wall behind a fireplace. This particular section of shear panel should be installed, nailed-off, and inspected prior to the installation of the fireplace, ahead of the rest of the shear-panel plywood installation. If not anticipated and completed before the fireplace is installed, the framing carpenters are then stuck with having to fit and nail-off sheets of plywood with little or no working space around the fireplace unit. Rows of nailing may even be missed because there is not enough room to swing a hammer or to use a nail gun.

The builder should look for this special situation as a check-off list item during the plans review period. If this situation does occur, the builder can either get the architect to revise the structural design so that the shear-panel section stops before entering the fireplace area or the builder can break up the shear-panel installation into enough phases on the construction schedule to handle this particular design.

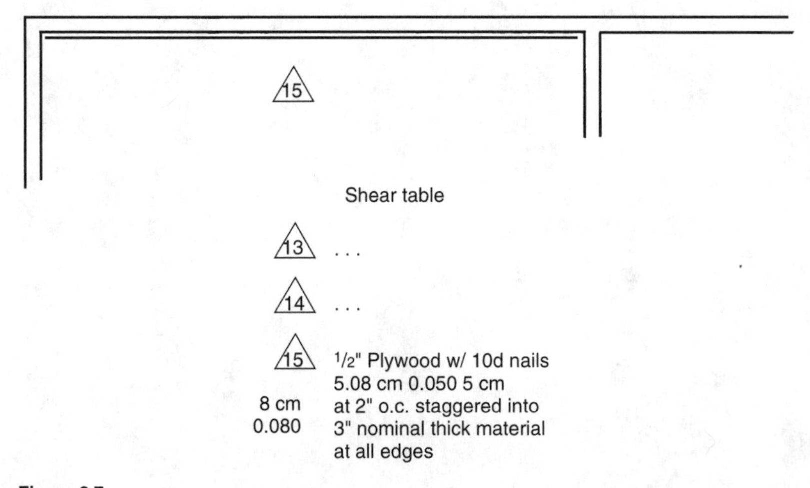

Figure 8.7

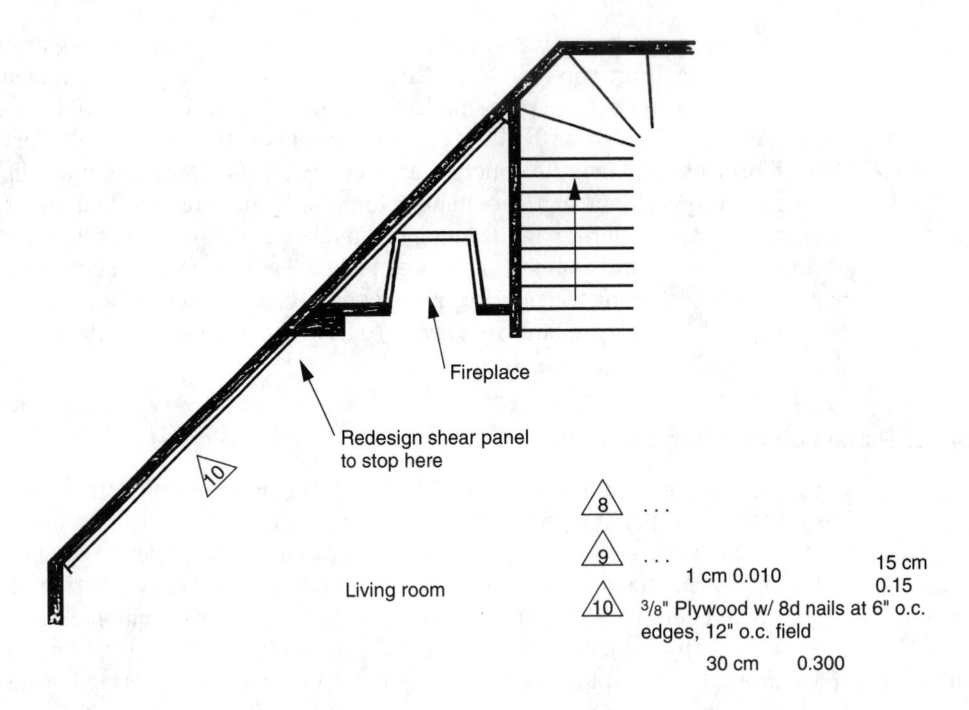
Fireplace
Redesign shear panel to stop here
10
8 . . .
9 . . .
1 cm 0.010
15 cm
0.15
10 3/8" Plywood w/ 8d nails at 6" o.c. edges, 12" o.c. field
30 cm 0.300
Living room

Figure 8.8

Chapter

9

Fireplace Framing

Fireplace Fronts

Decorative fireplace fronts with wood plantons and plywood siding, as shown in Figure 9.1, require specially placed backing at the edges where the plantons and plywood meet. If the planton material is 2×12, the joint between the planton and the plywood will be several inches beyond the normal stud framing for a corner (Figure 9.2). To provide solid backing for nailing and to prevent these edge joints from coming loose and curling outward later, backing must also be placed at the planton and plywood vertical and horizontal joints.

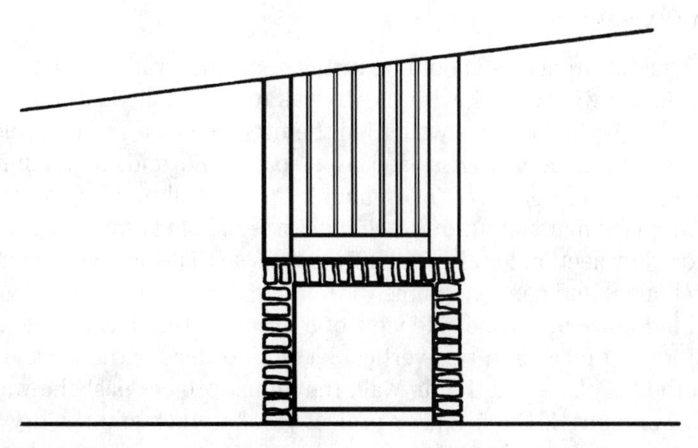

Figure 9.1

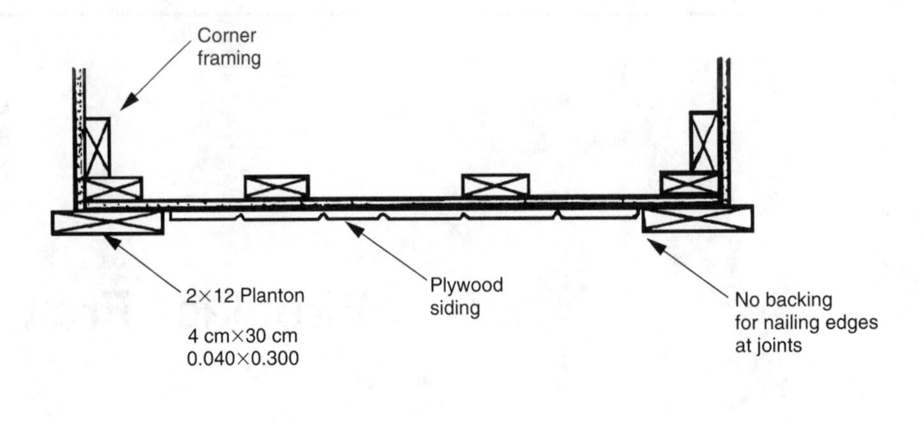

Figure 9.2

Design Fireplaces around Tile Size

When designing the framing dimensions for drywalled mantles and side pop-outs around fireplaces, the size of the decorative tile or marble around the fireplace should be considered. Illustrated in Figure 9.3 is a fireplace with a framed and drywalled mantle and side pop-outs, with 12-inch × 12-inch ceramic tile squares. The placement and dimensions of the firebox opening, the width of the black fireplace metal to remain showing, the tile size, the thickness of the grout joint, and the thickness of the drywall, must all be calculated into the design of the fireplace in order to arrive at the correct dimensions for the framing of the mantle and side pop-outs. Otherwise, the tile setter must cut tiles.

Fireplaces Flush with Drywall

Prefabricated metal fireplaces should be installed against wall framing so the front face of the fireplace will be flush with the drywall to be installed later. If ½-inch drywall is installed over the fireplace wall, then the fireplace face should project ½ inch past the framing. If ⅝-inch drywall is used, as in apartments and condominiums, then the fireplace face should project ⅝ of an inch from the wall framing. A word of caution—some fireplace manufacturers place their side nailing flanges to provide a ⅝ of an inch standard projection for all of their fireplaces so that they will be equally suitable for both houses and condominiums.

Illustrated in Figure 9.4 is the side view of a fireplace that has marble surrounds consisting of a head piece and two vertical legs. In order for the marble to lie flat against both the fireplace and the drywall, they must project past the wall framing studs an equal amount. If the fireplace projects ⅝ of an inch and the drywall thickness is only ½ inch, the marble surround will lie flat against the fireplace, but not touch the drywall. The gap at the drywall must then be caulked. If ceramic tile, clay tile, or marble squares were used instead of the marble surrounds in our example, the grout joint between the tile and the drywall would simply be wider.

To minimize the thickness of this caulking or grout joint, the fireplace manufacturer's specifications should first be checked. If necessary, the thickness of the drywall on the fireplace wall can simply be changed from ½ inch to ⅝ of an inch.

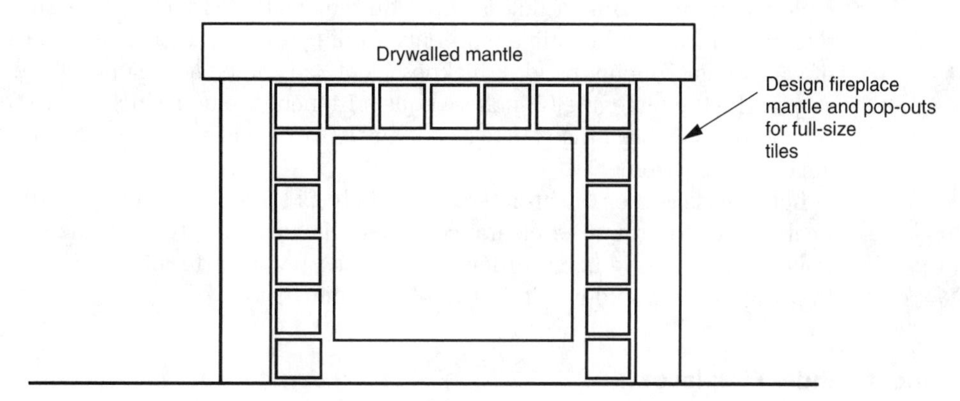

Figure 9.3

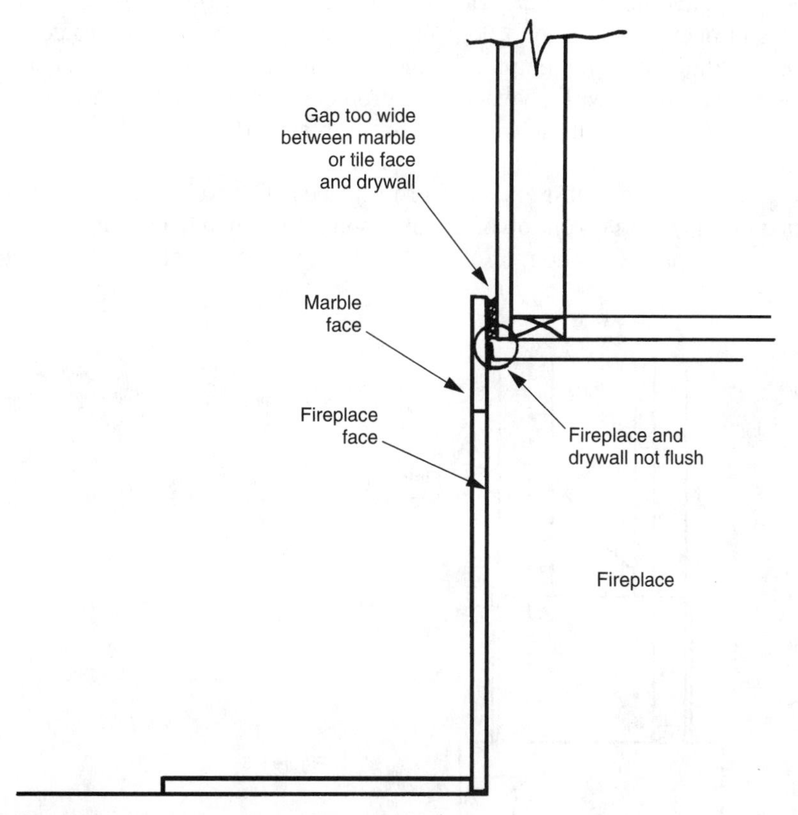

Figure 9.4

Fireplace Flue Clearance

Figure 9.5 illustrates an occasional conflict between design and construction. Shown is a double-sided fireplace, open to both the living room and a family room. The fireplace space separating the living room from the family room is dimensioned on the plans to be 24 inches deep. The metal fireplace flue is 14 inches in diameter.

Stamped on the metal flue is the requirement that 2 inches clearance must be maintained from combustible materials. Adding up the dimensions of 2 thicknesses of wall studs (7 inches) and 2 thicknesses of 2×4s on edge to support the flue draft stop at the top plate line (3 inches) equals 14 inches, plus 10 inches for the flue and the entire 24 inches is used up without any leftover space for the 2-inch combustible clearance requirement.

In this case, either the architect did not check the fireplace manufacturer's specifications for the required clearances when dimensioning the fireplace walls or the builder's purchasing agent ordered the wrong brand of fireplace without checking first whether it would fit within the 24-inch dimension.

Support under Fireplaces

A prefabricated metal fireplace located on the second floor should be supported by floor joists and not just a 2×4 rim with plywood filling the opening. Figure 9.6 illustrates in plan view a second floor fireplace and chimney stack projecting from the main building wall. The temptation for the framer is to rim this open space with 2×4s, then fill it in solidly with plywood. The problem with this method is that the fireplace box will not extend all the way back to the rear of the framing because of required combustible clearances or because it simply doesn't fit the architectural allowance space given for the different sizes of fireplaces. The weight of the rear of the fireplace will then end up on top of the plywood rather than being supported directly by the 2×4 rim pieces. If the rear of the fireplace causes the plywood to deflect slightly

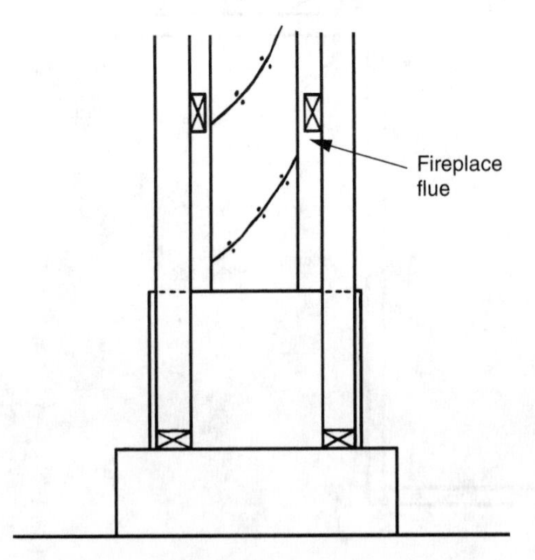

Figure 9.5

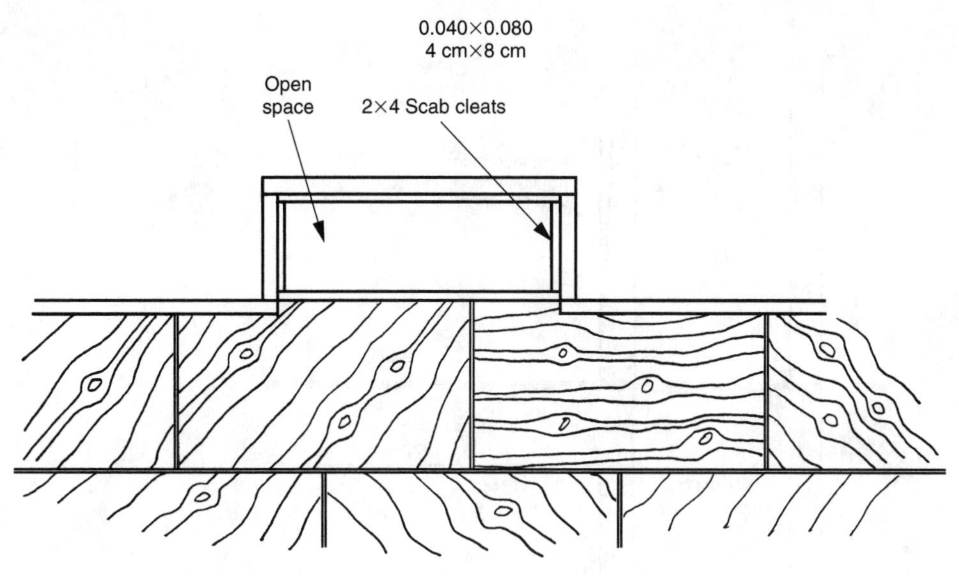

Figure 9.6

downward, the top front of the fireplace will move toward the building exterior and the face of the fireplace will no longer be plumb (Figure 9.7). When drywall and decorative tile or marble are installed later, this out of plumb fireplace front will result in an uneven gap between these finish materials and the fireplace.

To avoid these problems, small floor joists should be framed across the fireplace floor to provide solid structural support underneath the fireplace.

Plywood over Fireplace Chimney Rakes

Illustrated in Figure 9.8 is a fireplace chimney that reduces in size at a raked angle at mid-height. If the chimney and the surrounding walls are to be plastered with exterior stucco, the raked angle surface should be covered with plywood or scrap boards before being lathed. This will prevent the plasterer's foot from going through the lathing later when this area is used as a step to reach higher wall areas during the scratch, brown, or even the final color coat plastering phases.

Fireplace Chimneys

Shown in Figure 9.9 is the framing for a fireplace chimney. A 1×6-inch let-in brace is used for keeping the chimney square and straight until exterior plastering.

A better method is to use plywood instead of the 1×6-inch brace. Scrap pieces of plywood can be nailed to the inside or outside faces of at least two sides of the chimney. This will provide a stronger brace than the 1×6s, enabling the chimney to remain straight and true until the plastering stucco holds the fireplace chimney structure rigidly in place.

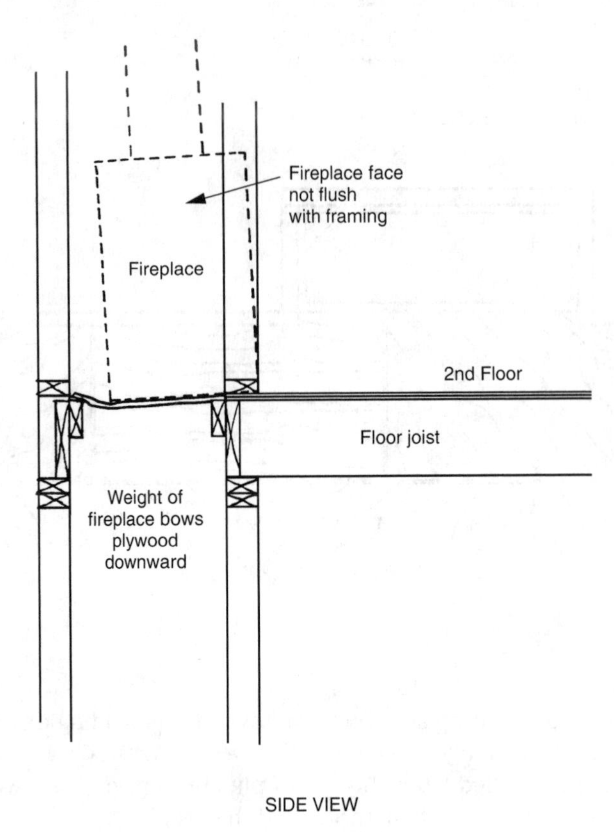

Figure 9.7

Fireplace Flues and Framing

The fireplace flue, shown in Figure 9.10, snakes its way up through a second floor chase to the roof. In this particular case, the fireplace manufacturer and brand was not selected by the builder at the time the plans were drawn, so the architect did not have the benefit of the manufacturer's spec sheet to design the location and size of the chase. It was discovered during the construction of the models that because of several large beams in the floor above and a bathroom wall that placed the chase a few feet offset from the fireplace, the flue could not bend and twist at the angle required to reach the chase. This proves the point that there is no substitute for drawing all of the conflicting members to scale in three-dimension prior to construction. By drawing the width and depth of the 8×12 floor beam to scale at its exact location and adding the 15-inch wide vent pipe that can only tilt at a 37-degree angle and twist off-center only so many inches in so many feet of rise, the builder could have quickly discovered whether or not the fireplace flue vent would fit as designed on the plans.

In this example, the fireplace flue could not get to the mechanical chase that was framed above on the second floor. The beam design in the second floor above the fireplace had to be changed and the fireplace moved over several inches for the chosen type and brand of fireplace to fit. Several thousand dollars worth of framing changes could have been avoided simply by having someone examine this situation in more detail prior to the start of the construction, while the whole thing was still on paper.

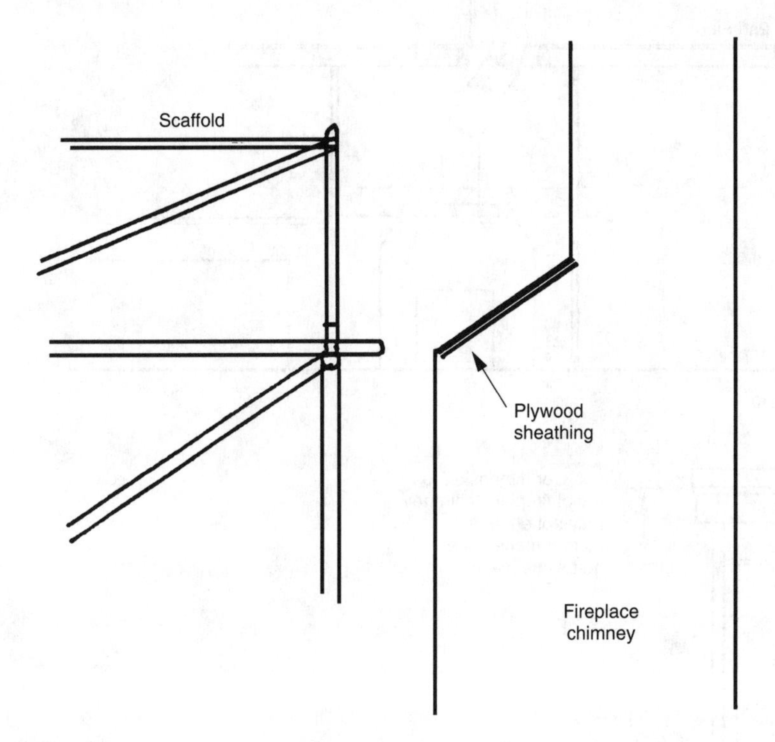

Figure 9.8

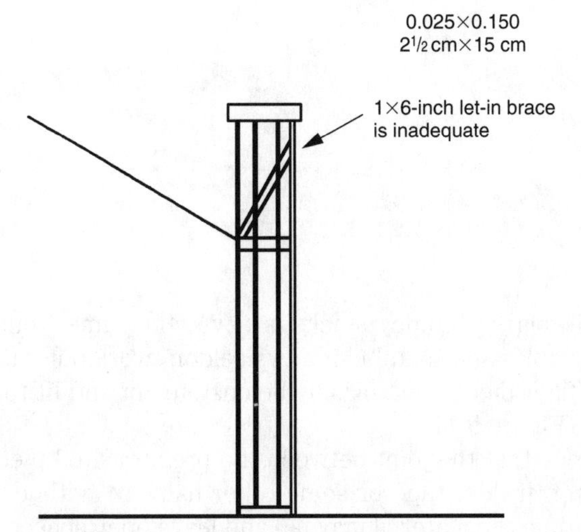

Figure 9.9

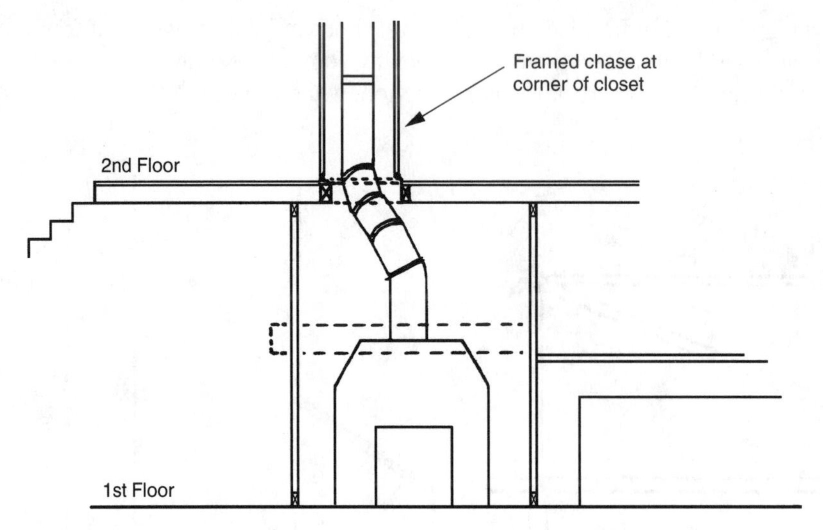

Figure 9.10

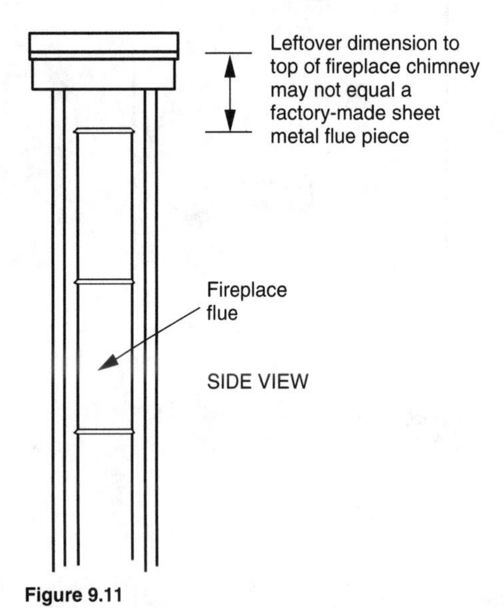

Figure 9.11

Prefab Fireplace Chimneys

Because the height of fireplace chimney stacks is never the same from house to house, the lengths of prefabricated metal chimneys seldom work out with an even number of pieces. A top filler piece must usually be custom cut and fit to reach the top opening of the stack (Figure 9.11).

The builder should check that the joint between the prefabricated piece and the top filler piece is fireproof. If duct tape or some other halfway method is used to cover over joint holes, in time the material may fail and leave open spaces for sparks to get through and cause a fire.

Chapter

10

Roof Framing

Ridge Height Restrictions

For view lot houses, a city or county may establish maximum roof height elevations to protect the views of the surrounding existing houses. Some houses may have a tolerance of plus or minus a few feet. Other houses may only have a tolerance as little as one or two tenths of a foot.

When ridge heights are crucial, the builder should have the framer slightly lower the roof pitches. A 3½ by 12 pitch roof can be cut and stacked at 3¼ by 12 without any noticeable difference, yet gain several inches of peace of mind. Bird mouth seat cuts in rafter tails can also be slightly deepened in an effort to lower the entire roof.

These small measures can help offset ridge heights established unrealistically by people unfamiliar with building construction, and save the builder from having to argue with the city building department or planning commission over a few houses that are 5 or 6 inches too high.

Layout Rafters for Flush-Lights

Shown in Figure 10.1 is the side view of a rafter that has been mistakenly framed at the location of a ceiling flush-light, which extends up into the roof attic about 6 to 8 inches. The layout of flush-lights that are centered above a bathroom sink should be considered prior to the rough framing stage. Roof rafters can then be positioned to miss flush- lights (Figures 10.2 and 10.3).

Rafter Tails at Fireplace Chimney

Roof rafters are not laid-out to coincide with fireplace chimneys. For this reason, a small gap of 6 inches or less sometimes results between a rafter tail and the chimney

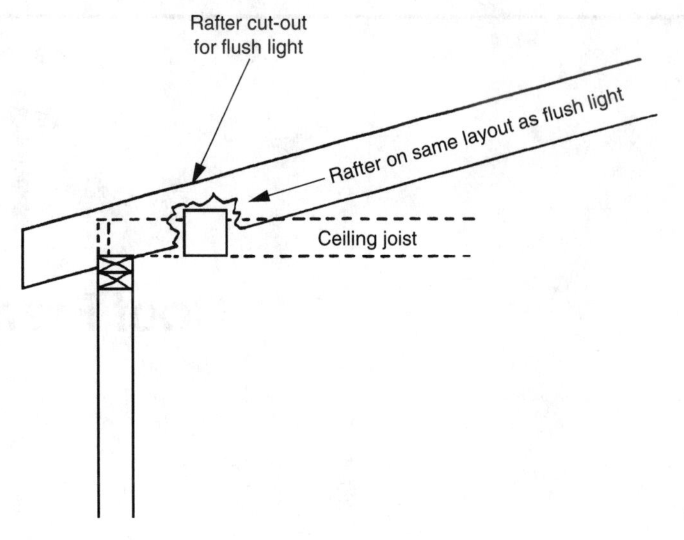

Figure 10.1

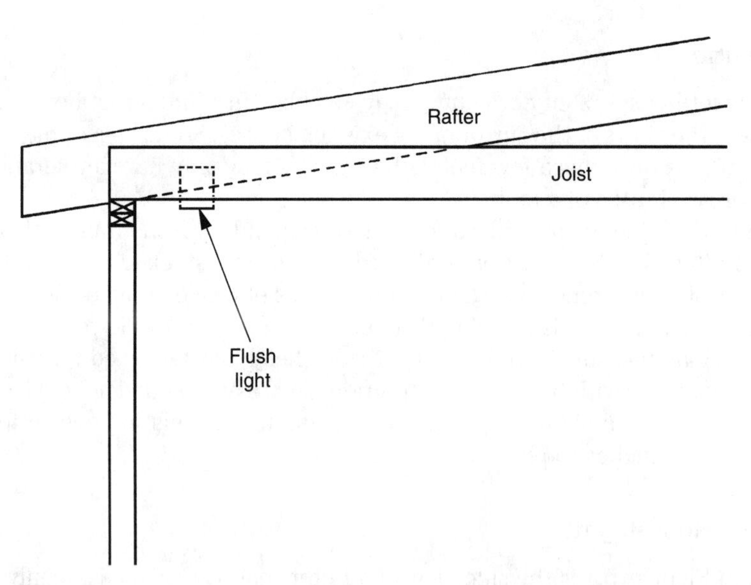

Figure 10.2

framing (Figure 10.4). This small gap between a rafter tail and the fireplace chimney presents a problem for the lather and the stucco plasterers. For example, lath cannot be nailed to the chimney framing higher than the rafter tail because there is not enough room for swinging a hammer or using a nail gun. Similarly, the stucco plasterers also have trouble spraying scratch and brown coats and hand troweling the color coat into this narrow space.

To solve these problems, false rafter tails can be nailed to each side of the chimney framing, with the lathing and plaster stopping at the bottom edge of the false rafter tails (Figure 10.5). If the space between the rafter and the chimney is less than 3½ inches, a 2×4 block can be nailed in flat between the bottom of the rafter and the chimney, with the lath and plaster stopping at the block (Figure 10.6).

Steel Connections on Wood Members

Illustrated in Figure 10.7 is a ceiling beam, post, and roof hip with a steel post cap and a special steel hanger connection. In this particular example, the structural engineer went to all the trouble of designing large roof members supported by posts

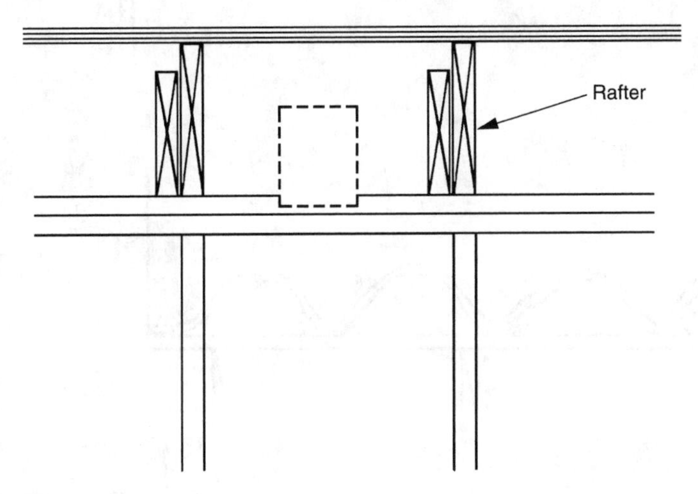

Figure 10.3

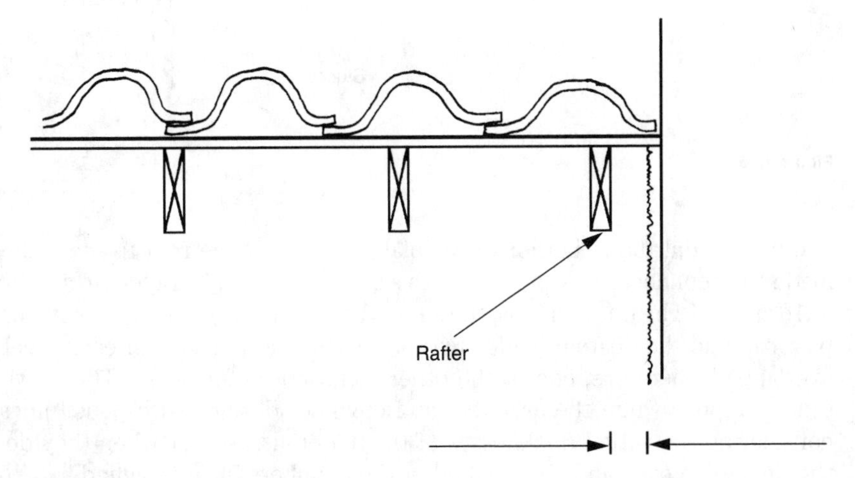

Figure 10.4

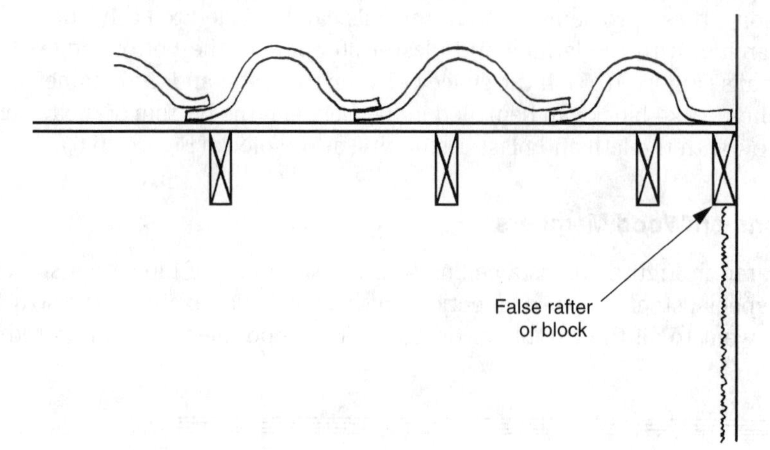

Figure 10.5

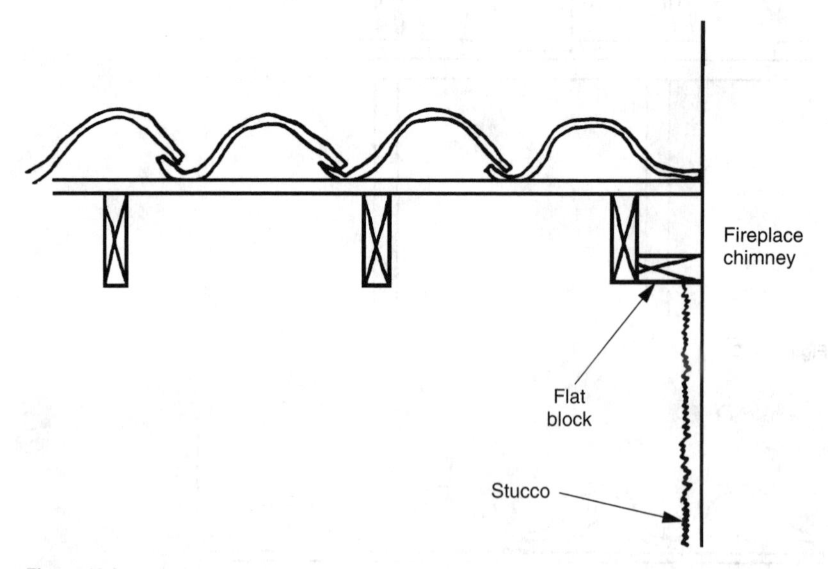

Figure 10.6

and beams, but didn't bother to calculate the rise of the roof to see if the custom-made steel connections would fit. With a span of only 10 to 14 feet in some cases and 4×10 and 4×12 beams and hips, the supporting post was not long enough for both the post cap and the custom-made stirrup. The tongue portions of each steel member overlapped; therefore, one or the other could not be installed. This is what sometimes happens when the architect and engineer designs with pencil lines without considering the actual thicknesses of the materials used or when they do not take the time to go back and check whether the members fit as designed.

Ridge-to-Hip Steel Hangers

Figure 10.8 shows a wood-to-wood steel connection between the end of a roof ridge board and two incoming hip boards. The builder should check the steel fabrication shop drawings to see if the steel connection designed by the structural engineer provides for a method of attaching king rafters and jack rafters to the ridge at this location. If not, the king and jack rafters will butt into this steel hanger without any

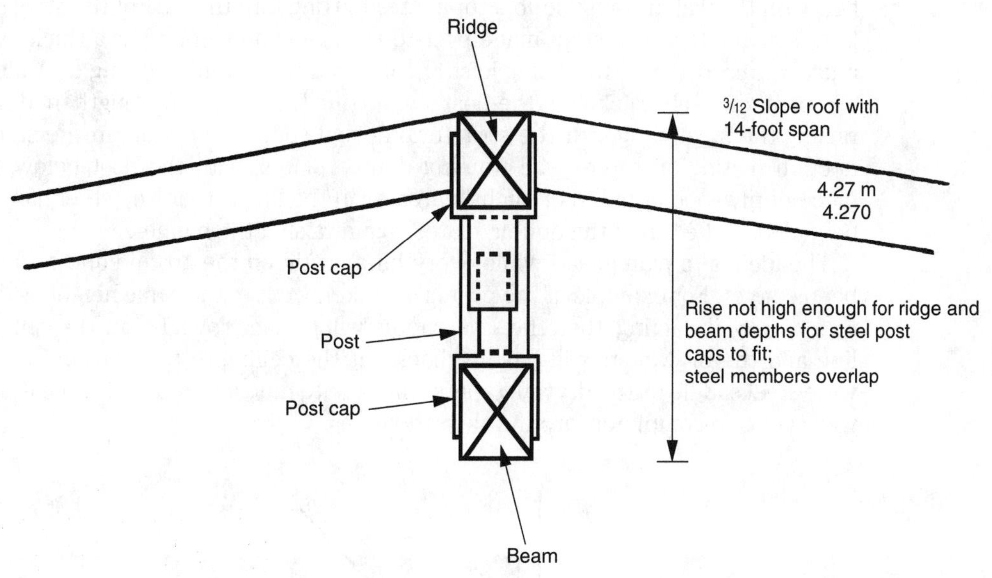

Figure 10.7

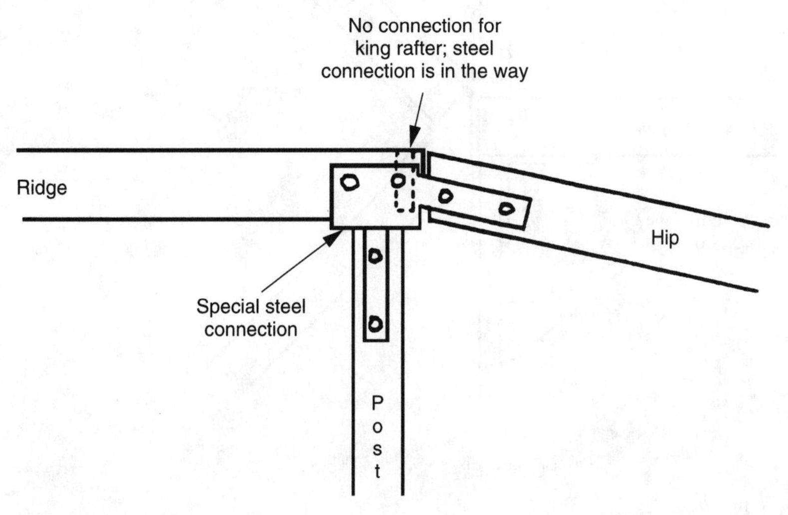

Figure 10.8

means of nailing or bolting them in place. Because it takes time to fabricate special custom steel connections, the builder should examine and resolve potential problems before the start of the rough framing. Figure 10.9 shows the added flanges required for attaching the king rafters and hip boards.

Raise Steel Hangers at Ceiling

The post in Figure 10.10 runs full height to the bottom of the roof framing hip member, with the wall framing double top plates butting into the side of the steel saddle. In this example, the custom-made post-to-hip steel connection has a thickness that must be raised above the ceiling joist line in order to prevent a ceiling drywall bulge. To correct for this thickness, the post supporting the hip can be lengthened slightly, raising the steel hanger. If the structural design somehow results in the steel connection resting on top of wall framing double plates, with the post below, then a piece of plywood equal to or slightly greater in thickness than the steel hanger can be added to the top of the double plates, again raising the hanger.

The idea is to plan ahead while everything is still on the ground and nothing has been cut yet. For example, if the post is mistakenly cut to the same height as the surrounding wall framing, then the steel saddle will project down below the ceiling joist line and the drywallers will have to back-cut the ceiling drywall around a 3½-inch wide steel saddle or the drywall tapers will have to mud and tape this hole if the drywall is cut up completely around the saddle.

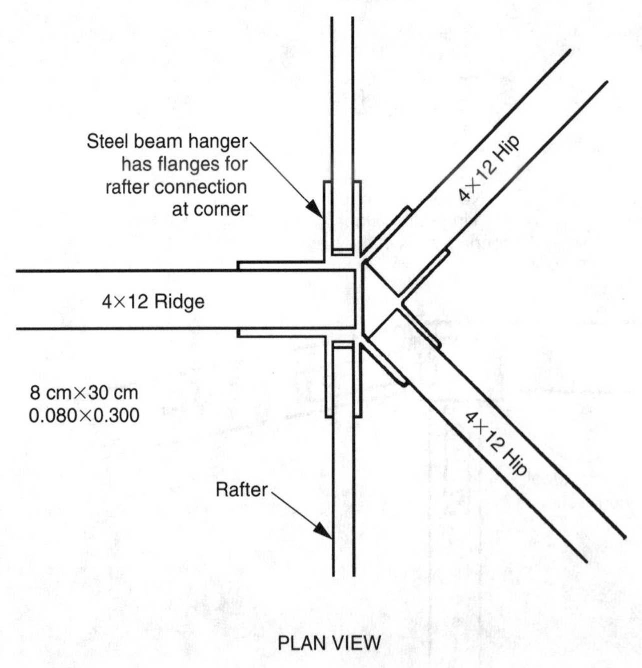

Figure 10.9

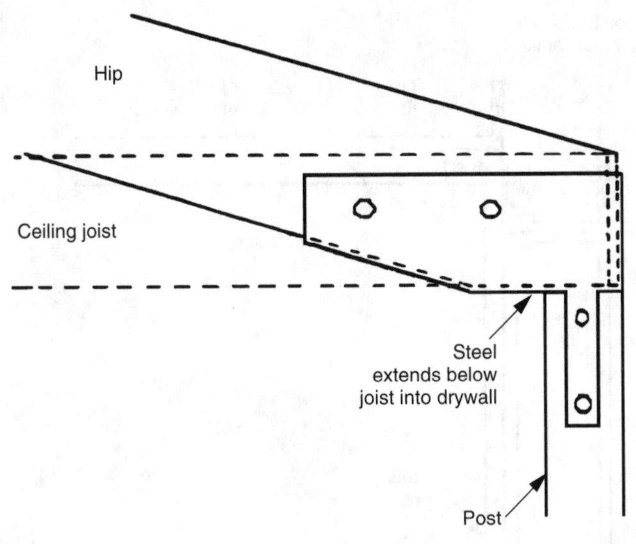

Figure 10.10

Steel Post Next to Joist and Rafter Tails

Shown in Figure 10.11 is a steel post used to support a ceiling beam on one side, with a metal flange for connection to the double top plates of the wall on the other side of the post. In this example, because the metal flange was mistakenly placed at the top of the wall (at the 8-foot-1-inch height), both the ceiling joist and the roof rafter at this point had to be cut and notched out for the thickness of the metal flange and bolt heads. This was necessary so that the roof line would remain straight at this point, instead of rising upward due to the metal flange.

A better way to design and build this connection is to place the metal flange at the underside of the double top plates so that they are completely out of the way of the ceiling joists and rafters. Because the flange supports and connects the top of the wall at this point, the wall stud that is bolted to the steel post acts merely as backing for drywall nailing. It can thus be easily cut slightly shorter in length to fit underneath the metal flange. This is a lot less work than cutting and notching the bottoms of joist and rafter tails (Figure 10.12).

Roof Eave Nail Shiners

Roof sheathing nail shiners are nails that have missed the rafters and are therefore exposed and visible from underneath the roof sheathing (Figure 10.13). At roof overhangs, nails that have missed the rafter tails are openly visible (Figure 10.14). Few things look worse at the exterior entry than to look up and see dozens of green, vinyl coated nails sticking through the tongue-and-groove siding starter board at the roof eave overhang.

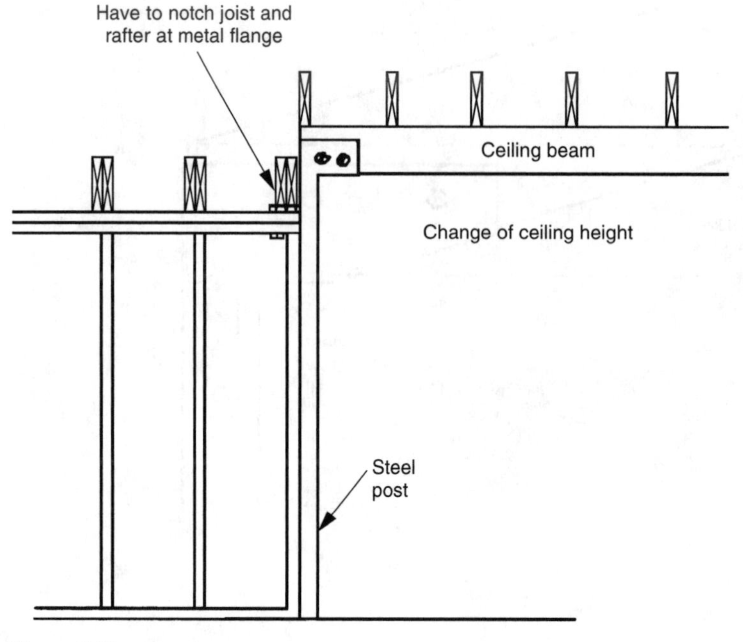

Figure 10.11

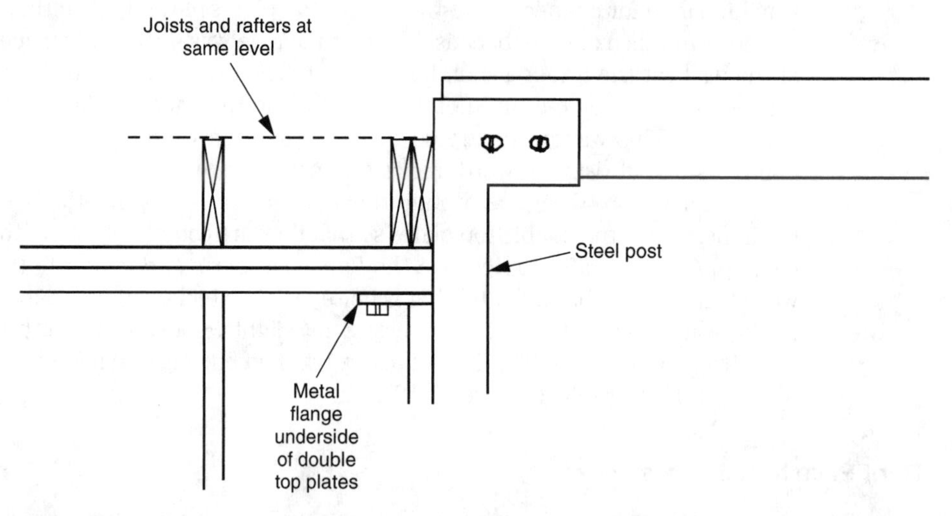

Figure 10.12

The time to discover and correct roof sheathing nail shiners is during the roof sheathing phase, while the sheathing crew is still on the job site and the roofs have not been papered yet. The builder should walk each house with the framing foreman, and with the help of a carpenter on the roof, push up or pull out nail shiners as they are spotted.

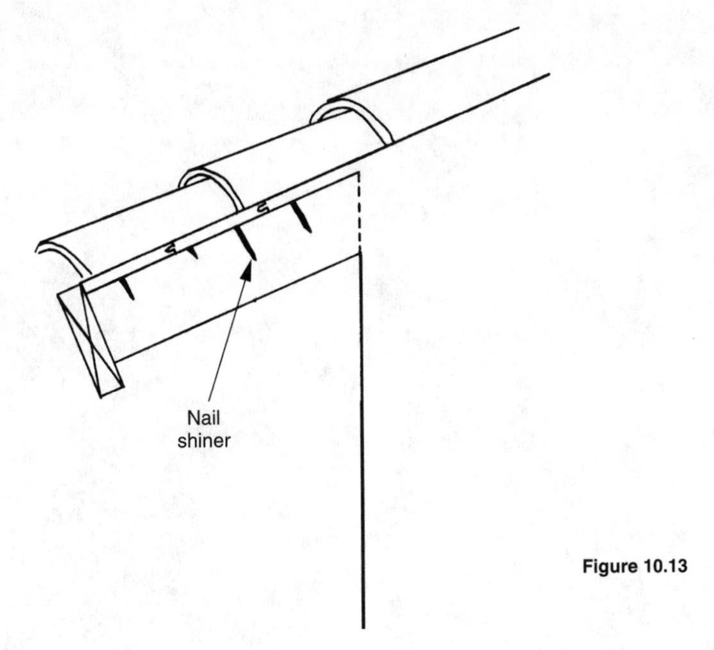

Figure 10.13

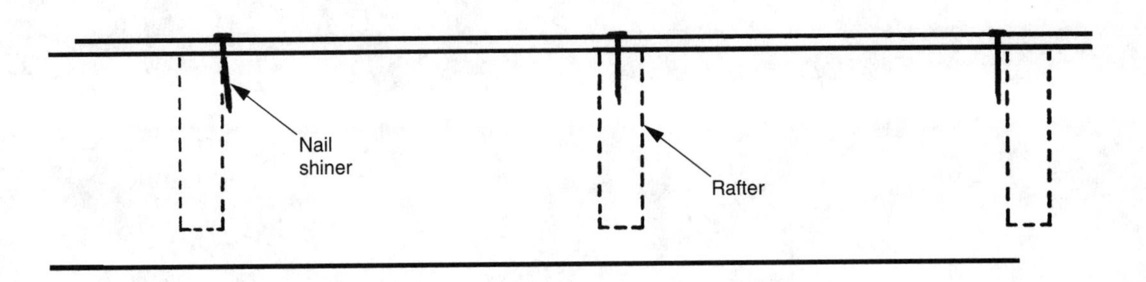

Figure 10.14

Chapter

11

Moment Frames and Steel Posts

Tap Ends of Threaded Bolts

Illustrated in Figure 11.1 is a shop fabricated steel moment frame. The moment frame is either made out of hollow, rectangular tube-shaped steel or an I-beam. It is bolted to the concrete foundation and usually has threaded bolts on the inside and outside edges for attaching wood ledgers. A moment frame is used where the standard wood post, beam, and shear-panel assembly is not structurally adequate for the design. It can also be used where there is an opening in the wall that is too large to provide enough leftover wall space for shear panel. A garage door opening is a typical example of this.

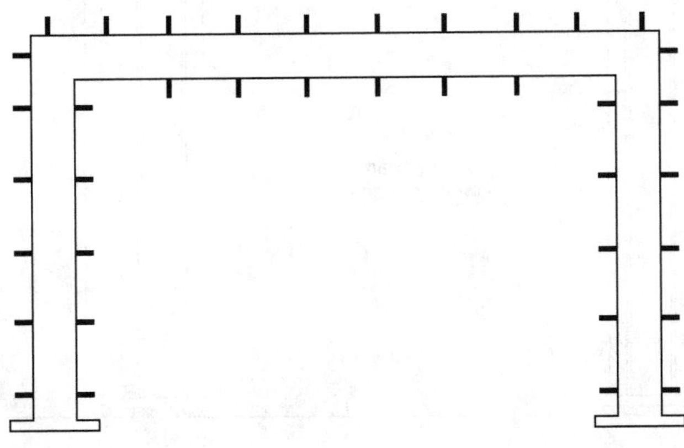

Figure 11.1

The threaded bolts on the inside and outside edges of the moment frame are sometimes cut out of continuous threaded rod and then welded to the moment frame. This method results in threaded ends that are difficult to put nuts on. The builder should therefore specify in the steel fabrication contract that all bolt ends be tapped or that nuts be already started and installed. This will save the framer a lot of time and frustration when installing the ledger boards.

Set Moment Frames and Nailers before Framing

Moment frames are steel arches placed within houses to add shear strength to resist earthquake forces. Illustrated in Figure 11.2 is a typical square-shaped moment frame with threaded bolts attached for wood backing nailers. Moment frames and other structural steel posts should be installed right after the concrete forms are stripped, prior to the framing and standing up of any surrounding wall framing. This allows the wood nailers to be drilled and bolted to the moment frames while there is working space around the moment frames.

The scheduling, fabrication, delivery, and installation of the steel moment frames and posts must be coordinated closely with the concrete and framing schedule. If the moment frames do not come to the job site on time, the framer is forced to leave out part of the framing around the moment frames so that they can be inserted once they arrive. This can delay floor joists from being put in above a garage, hold up beams and joists from being positioned at a large sliding glass door opening, or halt the framing altogether. The movement of a truck-mounted crane around the building pads and the placement and installation of the moment frames themselves is also much easier and faster for the steel fabricator when done right after the concrete cleanup, prior to the start of framing.

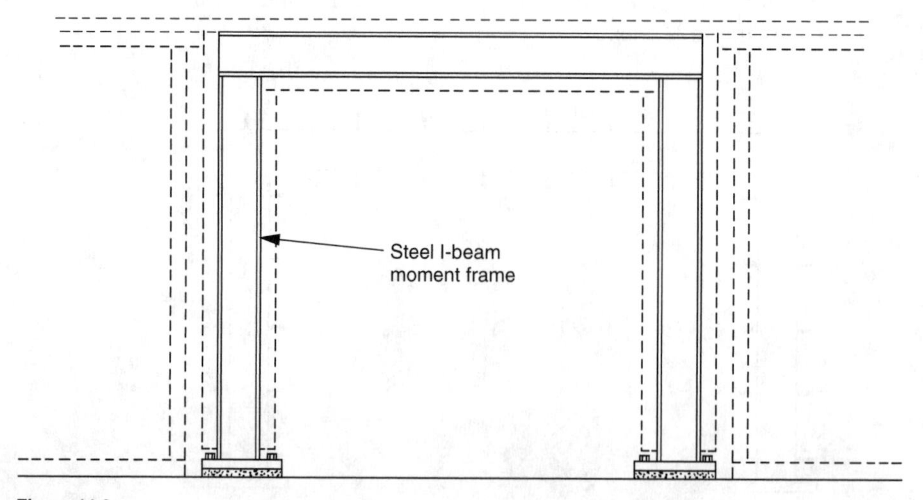

Figure 11.2

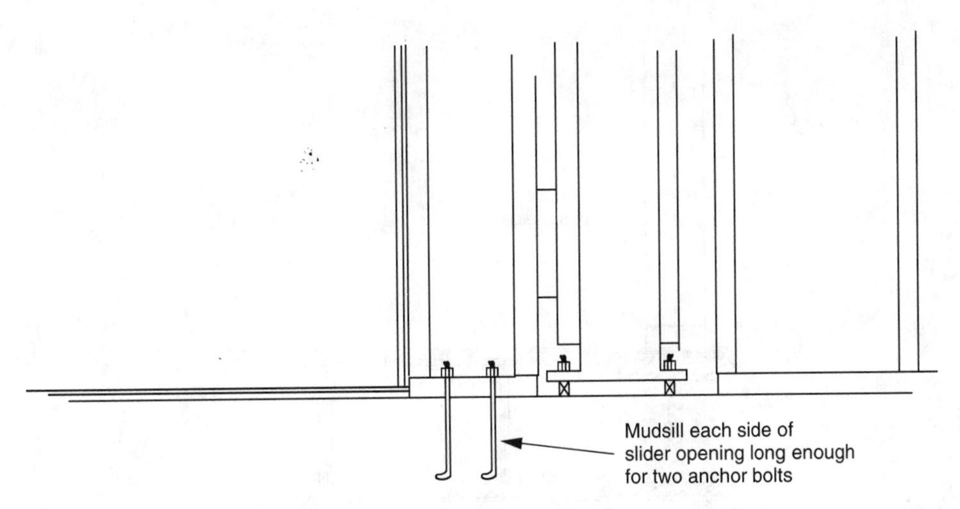

Figure 11.3

Moment Frame Width

When possible, the moment frame should be designed wider than the wall opening it is trying to structurally reinforce. For example, a moment frame around a sliding glass door opening, when designed a few feet wider than the opening, allows a mudsill bottom plate to be anchored to the concrete floor on each interior side of the opening (Figure 11.3). This allows for easier and stronger wall framing around the sliding glass door because the mudsill is firmly secure. If the moment frame is designed only slightly wider than the opening to allow for wood nailers and some amount of framing margin, then the mudsill will be short in length and too small to be attached firmly to the concrete.

Steel I-Beam and Framing

Illustrated in Figure 11.4 is the end view of a steel I-beam that is supporting a second floor wall in an open stairway. The ceiling below the second floor is open to the stairway, which is to the left of the I-beam and the wall in the figure. This is a typical example of what can happen when steel members are mixed with wood in the structural design. If the beam in this case was a solid wood glulam or microlam beam, it could easily be moved over to line up with the wall above. The steel I-beam, however, is supported on each end by steel posts bolted to the concrete and cannot be moved over. The structural engineer and the architect were thinking in terms of the width of a pencil line, not taking into account the actual widths of the different members. The fix in this case was to fur out the second floor wall above the I-beam with a second framed wall that would be flush with the beam. The original wall could not be moved because it was a weight-bearing wall. The new wall resulted in a 4-inch bump between the second floor hallway handrail and the bedroom door opening, which was not intended on the architectural details.

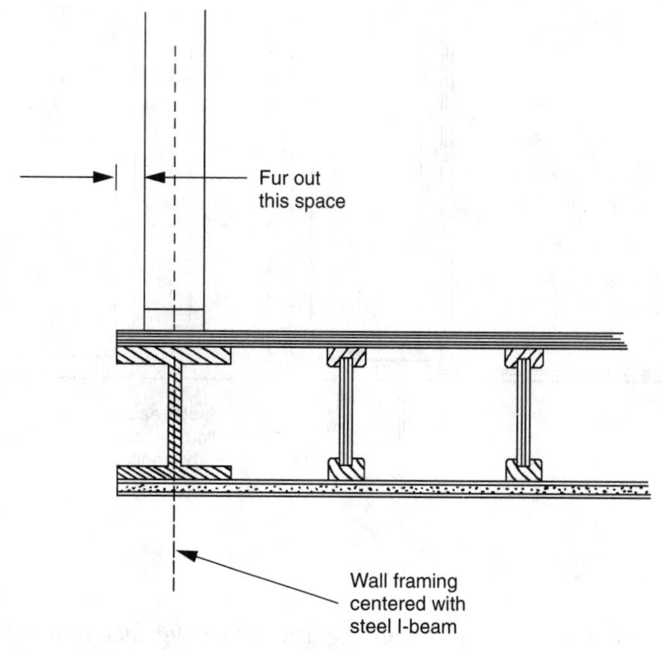

Figure 11.4

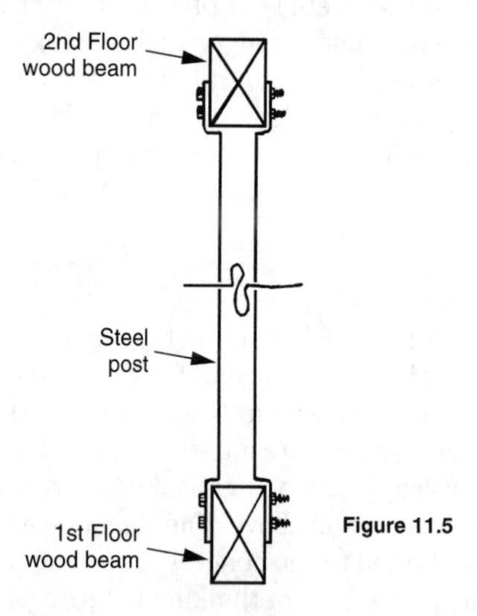

Figure 11.5

Steel Posts, Saddles, and Beams

Shown in Figure 11.5 is the side view of a steel post that is sandwiched between the first and second floor beams. In this particular case, the steel post provides mid-span support to the second floor beam, which runs the full-length of the room, from wall to wall.

Do not assume that if the steel fabrication runs behind schedule and the post is not delivered and installed during the second floor framing, that the post can be installed later as a one-piece unit. The problem is that a full-length post with an attached bucket and saddle cannot be installed between two beams that are already in place because the two opposite corners of the upper and lower buckets will bind and dig into the beams before the post can be stood upright. The diagonal dimension from the opposite corners of the upper and lower buckets is greater than the dimension through the centerline of the post; therefore, the post will not fit (Figure 11.6).

When the steel fabrication runs behind schedule and the wood framing cannot continue without structural problems, as in this example, the builder needs to rethink the steel installation. The steel post and buckets may have to come out to the job site in pieces, to be assembled and welded into place. It does no good to have a one-piece, full-length unit, that was fabricated in the shop according to the structural details, show up two weeks late on the job site, only to discover that it does not fit.

Moment Frame Nelson Studs or Holes

Figure 11.7 illustrates a steel moment frame that is located at the garage door opening. It has threaded Nelson studs welded to the inside and outside surfaces of the two vertical sides, but holes drilled out through the I-beam flanges along the top horizontal member. This moment frame design allows the framer to choose predrilled holes spaced along the steel I-beam that miss the layout of large-size wood beams in the garage ceiling. This eliminates having to cut and notch the ends of wood beams around threaded steel bolts that are sticking up

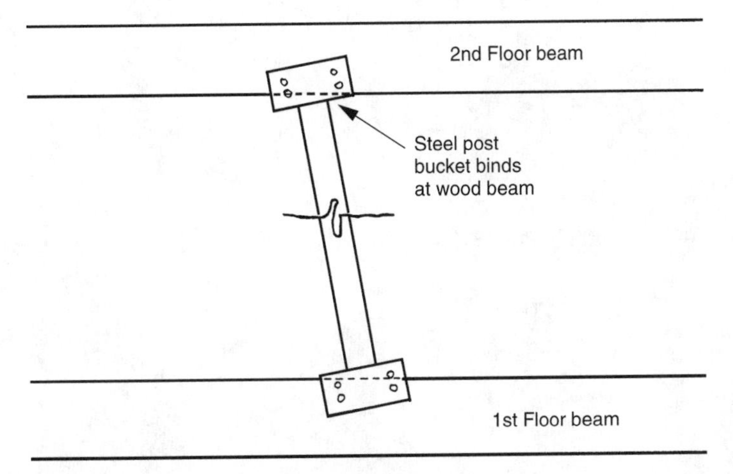

Figure 11.6

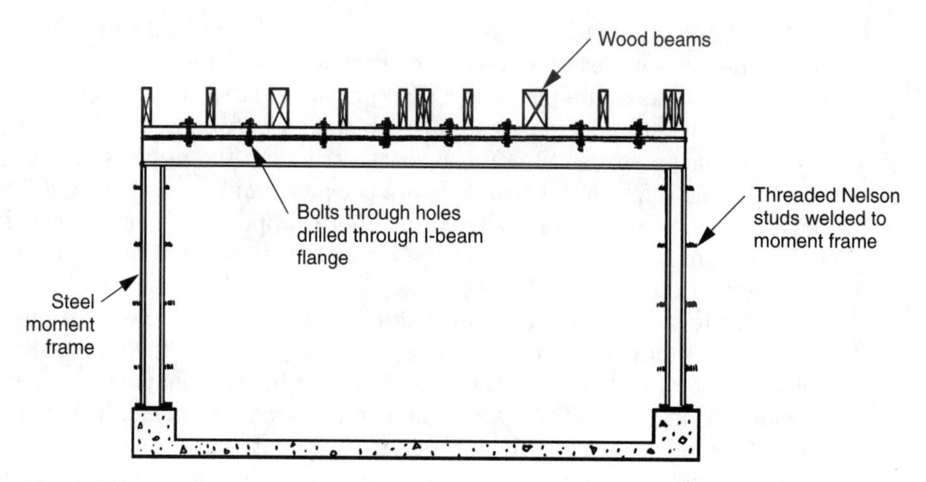

Figure 11.7

from the moment frame. The framer can lay out the location of the wood beams, double joists, and floor joists, then attach the wood nailer board to the top of the moment frame using holes that do not interfere with these floor members.

Because there is no conflict with beams or joists at the sides of the moment frame, the Nelson studs that are welded to the vertical sides of the moment frame provide a quick and easy means for attaching wood nailers.

Chapter

12

Exterior Framing

Banding and Weep Screed

Illustrated in Figure 12.1 is the framing for a lower banding pop-out on an exterior wall. The banding is framed with 2×4s that are nailed flat to the exterior wall at the bottom plate line and window sill line. The center portion is filled-in with 2×4s that run horizontally, or with short vertical cripples nailed to each stud. The exterior wall is then lathed and plastered. The banding pop-out provides an architectural break up of the wall surface.

The bottom 2×4 of this banding should be kept above the wall framing mudsill at least 4 to 6 inches. This provides clearance for installing the stucco weep screed.

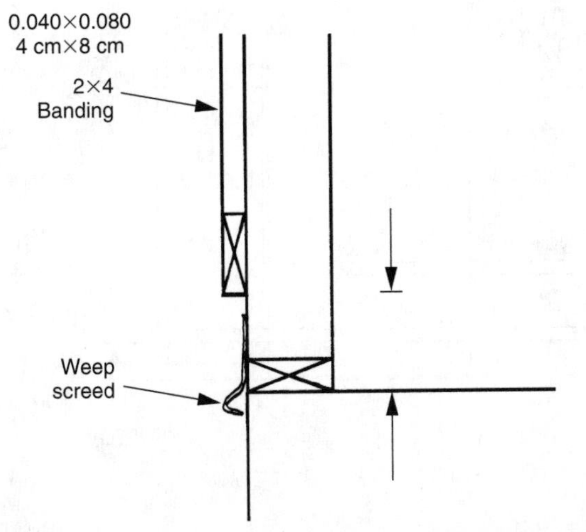

Figure 12.1

Fur behind Banding

Figures 12.2 and 12.3 illustrate the front and top views of architectural wall banding. This banding is usually a 1½-inch or 3½-inch pop-out from the wall framing, which breaks up large plastered wall areas for better appearance.

When exterior walls at banded areas get shear-panel plywood, the spaces between the banding and the wall framing without shear panel should be furred out to the same thickness as the plywood. This prevents the banding from bowing inward when nailed to wall areas without shear panel.

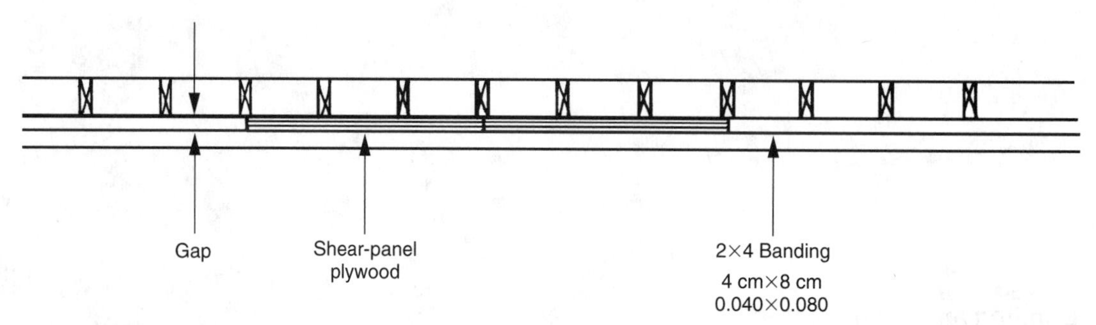

Figure 12.2

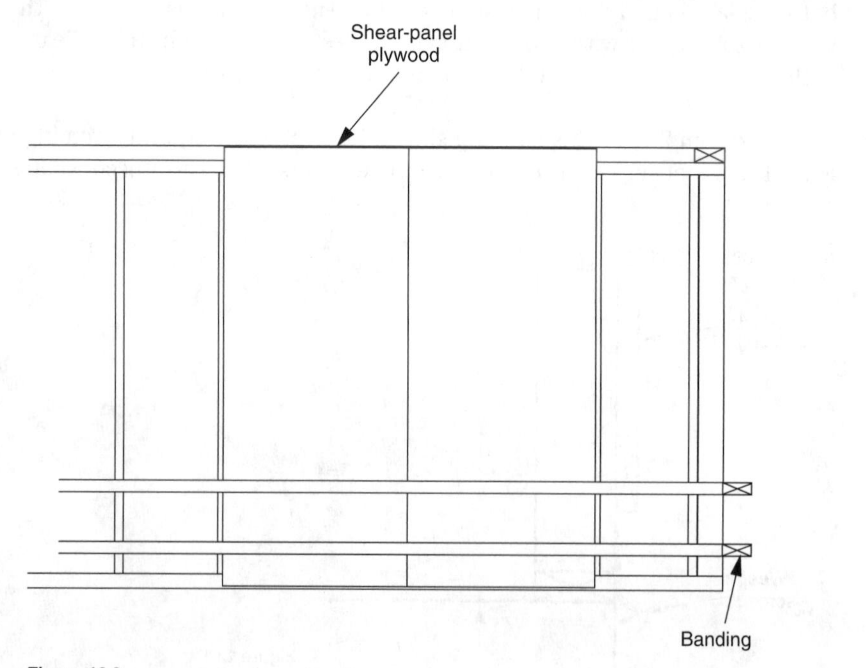

Figure 12.3

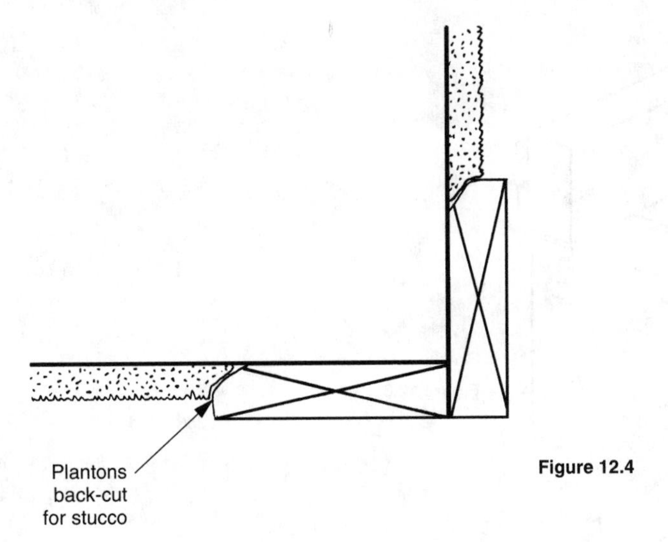

Figure 12.4

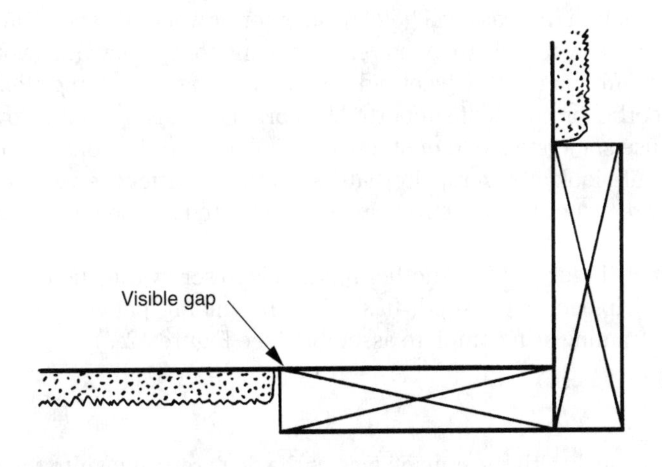

Figure 12.5

Back-Cut Plantons

The inside edges of vertical wood plantons should be back-cut with a bevel as shown in Figure 12.4. This allows the stucco to fill in behind the wood planton, preventing future shrinkage cracks from showing at this joint (Figure 12.5).

Column Tops

Things are often visible from a second floor window that cannot be seen from ground level. Figure 12.6 illustrates the top of a column that is wider than the beam it supports. When visible from a second floor window, the exposed rough framing at the column top should be trimmed with finish wood material.

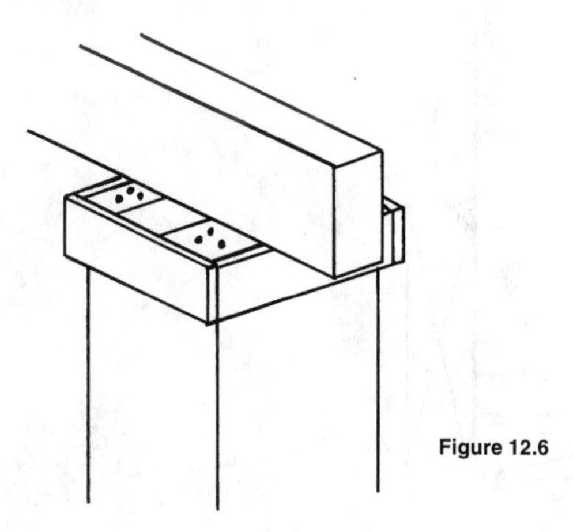

Figure 12.6

Secure Corbels

Illustrated in Figure 12.7 is the side view of a wood corbel projecting from an exterior wall. Two wood corbels in this case will hold up an exterior wood pot shelf. Oftentimes tradesmen will use these corbels to stand on or rest one foot upon while working on exterior lathing, scaffolding, lath corner aid, or wood siding. Few things look worse than an exterior corbel or pot shelf supported by corbels that is drooping downward, out of level, because someone stood or stepped on it earlier in the construction.

The builder should look at several alternate ways to construct the wood corbel so that it ends up more structurally secure. One method is to nail the end of the corbel inside the wall framing to a solid block, with possibly an additional ledger board underneath the corbel (Figure 12.8). Another method is to screw the end of the corbel securely to a small piece of plywood, flashed with building paper, which is then nailed to the wall framing as a complete assembly (see Figure 12.7).

Recess Deck into Wall

An exterior balcony deck with the outside face of the deck returning into the building corner is pictured in plan view in Figure 12.9. If the building has a belly band at the same elevation as the deck, the transition between the deck and the belly band can be eliminated by building the deck recessed in from the building corner (Figure 12.10).

Whether the deck has an exposed wood rim joist or a built-up wood facia, as was the case in this instance, by designing and building the deck recessed in from the building corner, the deck will return into a flat surface and not have to join with a belly band of a different width or thickness.

Knots in Wood Decks

On one project, a particular developer decided to use a marginal grade of lumber for the exterior decks. This lumber contained a lot of knots. One homeowner tripped over a knot that had worked itself free and was projecting above the deck surface. The homeowner consequently broke her ankle and sued the developer. Because of the resulting problems, the developer's customer service crew went back through the entire project and sanded down every knot so that they were flush with the deck surfaces.

This was an incredible hassle in that the decks could only be reached by going through the condominiums, which were all occupied. Deck furniture had to be moved

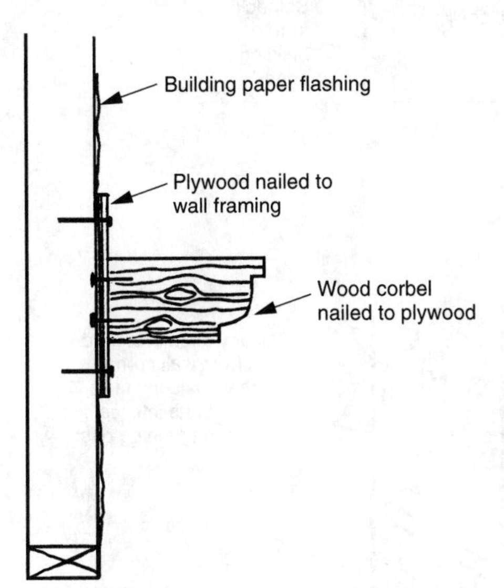

Figure 12.7

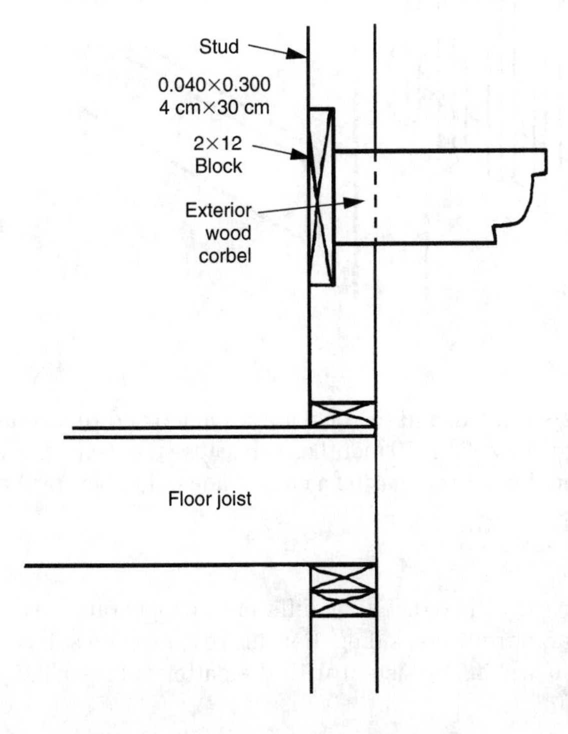

Figure 12.8

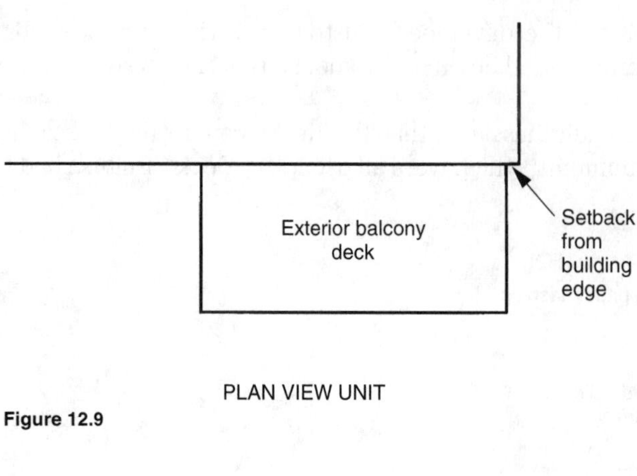

Figure 12.9

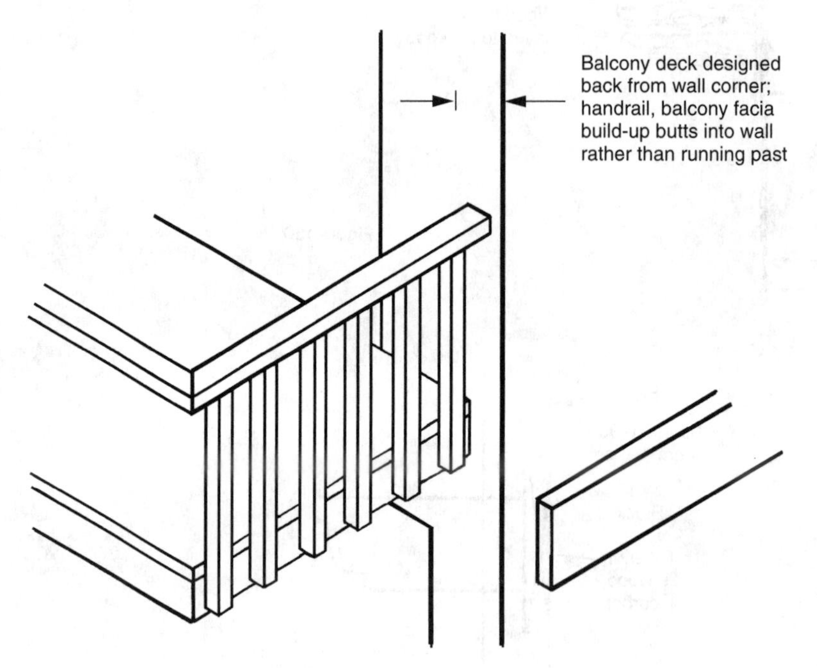

Figure 12.10

somewhere else, the knots sanded, and the decking restained and re-lacquered. This took the customer service crew about 10 months, all because the developer wanted to save a little money upfront by not purchasing a clear grade of lumber for the decks.

Exterior Siding Boards

Exterior wood siding boards can be distinctly different in appearance from piece to piece, such as pine, cedar, or redwood siding. For this reason, such siding should be purchased in lengths that will be consistent with the pattern of installation. Figure

12.11 shows an exterior wall containing windows on which the vertical siding boards on both sides of the windows are continuous pieces. Each successive board may be entirely different from the adjacent piece—different in shade of color, grain pattern, or number of knots, but a definite pattern of continuity is established by having each board run from top to bottom. When coming to the windows, however, there is the choice of purchasing shorter lengths of unrelated pieces to mean that random and unrelated pieces would be above and below one another, contrasting with the pattern of vertical continuity. Siding boards for this section of the wall (between the windows) should be purchased in lengths which are both shorter than the full length pieces, yet long enough when cut to fill in the three spaces above, between, and below the windows.

Galvanized Nails and Rust Stains

For nailing exterior wood siding boards, the correct type of galvanized nails must be used. The smooth surfaced, electro-galvanized nail will rust when exposed directly to moisture, causing a brown rust streak running downward below each nail on the wood siding. The rough surfaced, hot-dip coated galvanized nail will not rust; therefore, it should be used for wood siding where the nailing is exposed.

If the blighted, bombed-out look is the one the architect wants for the exterior wood siding—fine. It is a shame, however, to see expensive redwood or cedar siding ruined simply because someone used the wrong type of galvanized nails.

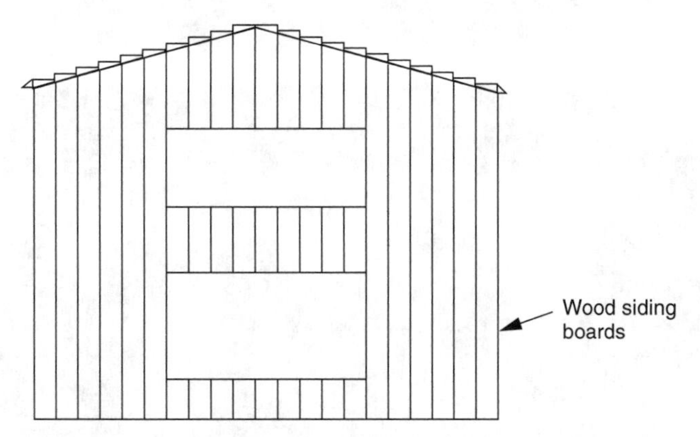

Figure 12.11

Chapter

13

Miscellaneous Framing

Plywood on Plant Shelves

Shown in Figure 13.1 is the end view of an interior plant shelf, placed on the top portion of an 8-foot high soffit in a room with a higher ceiling. The framer should install plywood on top of such shelves to provide support should heavy potted plants or other items be placed on them. Drywall over finger joists spaced 16 inches on-center is not strong enough should the pots be placed on the drywall alone, between the joists.

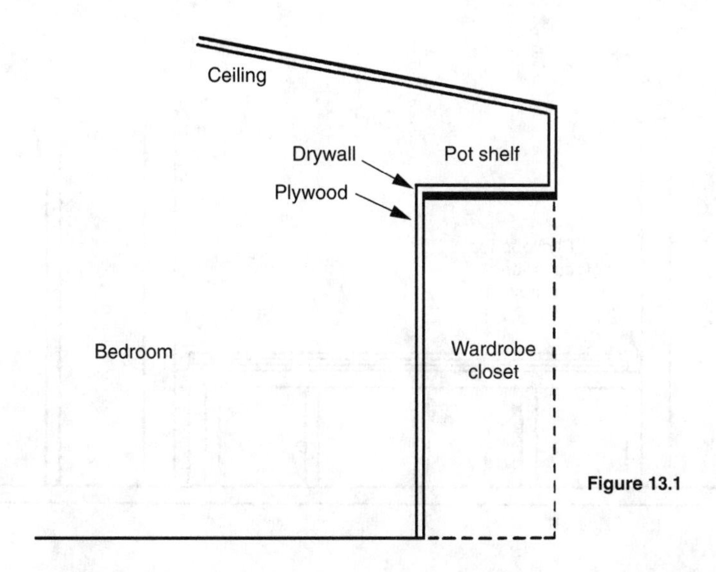

Figure 13.1

Water Heater Platform Edges

Figure 13.2 shows the corner of a water heater or HVAC platform typically found in many single-family homes. The plywood top projects ⅝ of an inch past the front cripple studs so that drywall can be installed later with no need for a protective metal corner bed—the plywood protects the drywall edge.

Water Heater Platform Plywood and Drywall

Illustrated in Figure 13.3 is a water heater and HVAC furnace platform in the garage of a house. The platform area is recessed in the garage wall, and therefore has two side walls and a rear wall to surround the water heater and furnace. The plywood top has been cut and tacked into place, to be removed later for the ⅝ inch drywall, and then reinstalled over the drywall. The drywall gives the platform a one-hour fire rating.

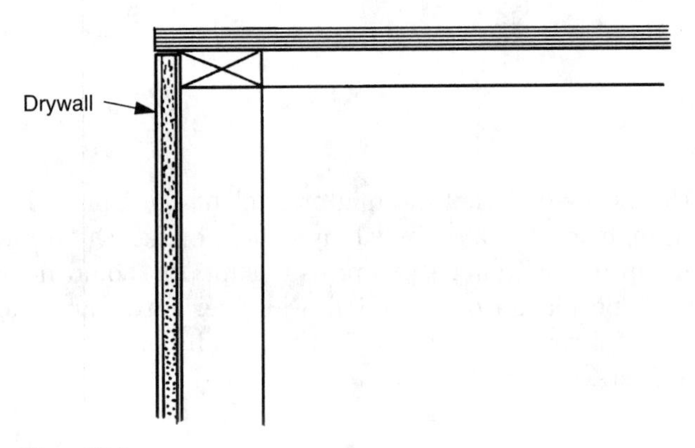

Figure 13.2

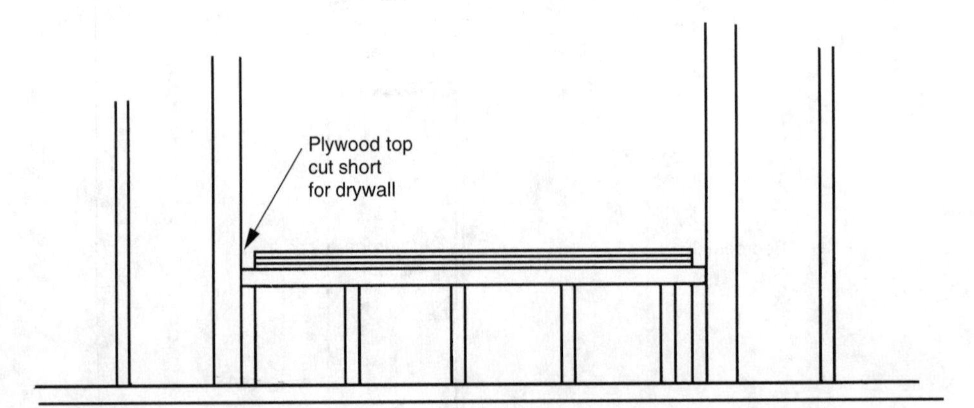

Figure 13.3

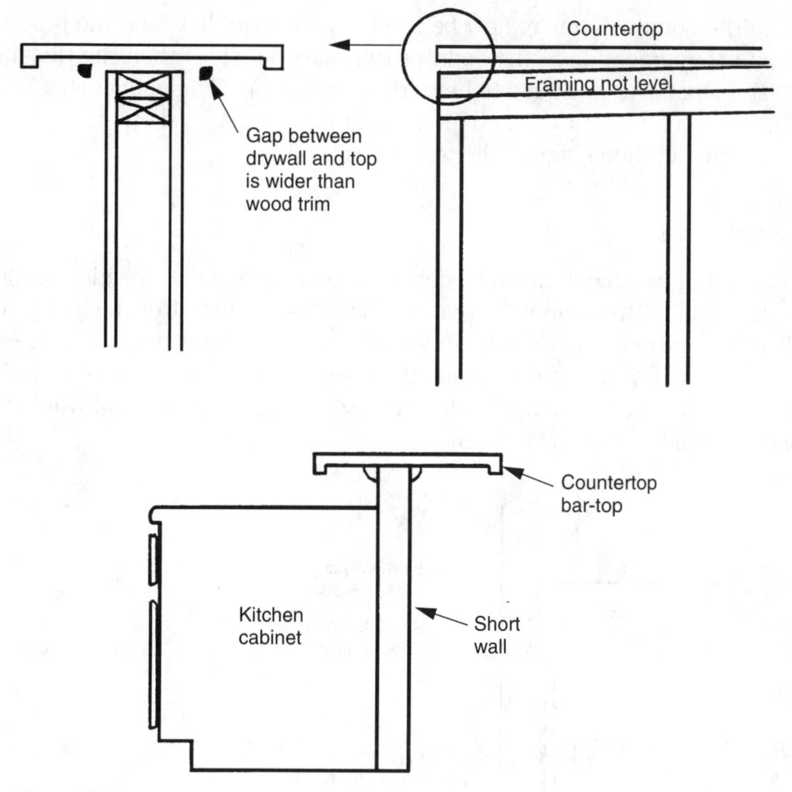

Figure 13.4

The problem to avoid is if the framer cuts the plywood from side wall to side wall when nailing the drywall to the wall framing around the water heater platform, this area is reduced by the thickness of the drywall and the plywood will not fit. The drywall crew will invariably pull off the tacked-in-place piece of plywood, install the drywall on the platform and the surrounding walls, discover that the plywood top does not fit, and then leave it on the garage floor. The cleanup crew mistakes the plywood top for trash and throws it away.

In this example, the plywood top to the water heater platform should be cut 1½ inch less than the dimension between the two side walls. This will allow for the thickness of the layers of drywall on each side wall; therefore, the plywood top will fit after the drywall is installed.

Framing under Bar-Top Counter

The formica laminate countertop pictured in Figure 13.4 is supported by a mid-height pony wall. A problem arises when the pony wall is not level, either because it is not framed accurately or the floor is not perfectly flat and level. Because the countertop must be installed level, one of the ends of the counter is shimmed-up. Although the countertop may now be level, a gap exists between the top of the pony wall and the underside of the counter that may be wider than the thickness of the base-shoe that is to be added to help trim out this gap. The builder should check this during the rough framing punch-list phase by placing a level on top of the pony wall. The framer can then shim up the top plates level.

Garage Door Jamb Size

Shown in Figure 13.5 is the side view of a 4×12 garage door header supported by a 2×6 door jamb. The jamb is 2×6 rather than 2×4 so that after the garage door is installed, the extra 2 inches of jamb width will prevent a gap at the side edge of the door. Because the top of the garage door rests against the door header, if the same width of material as the door header was used as a jamb, the result would be an open gap at both sides of the garage door.

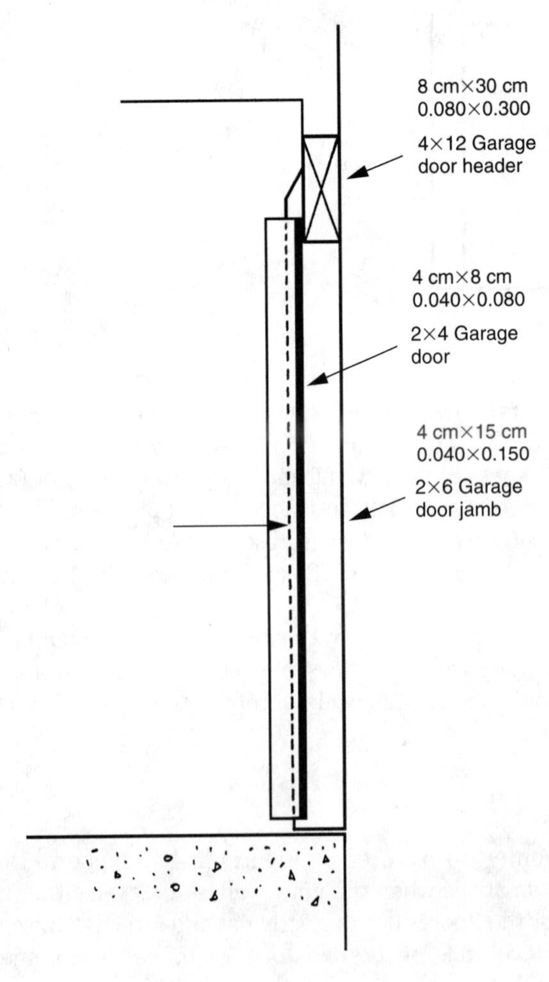

Figure 13.5

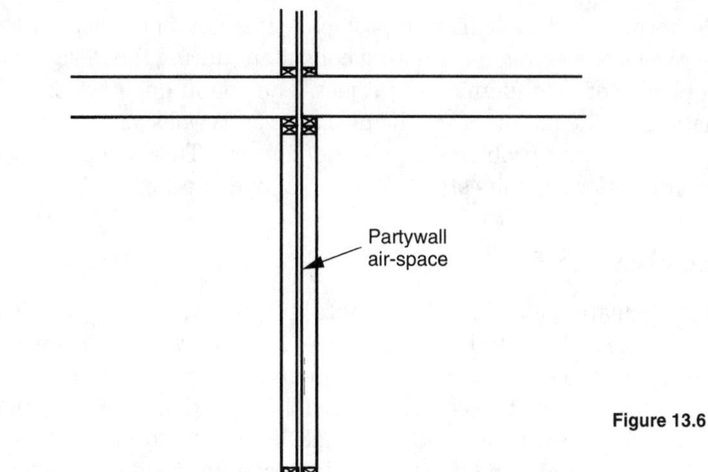

Figure 13.6

Sound Control

There are several construction quality-control measures that can be taken to help prevent the transmission of sound through apartment and condominium partywalls. Using Figure 13.6 as an illustrative example, these measures include the following:

1. Keep the integrity of the air-space between the partywalls free of anything touching from one partywall to the other—pipes, structural members, plumbing pipe straps, wall framing braces, loose debris, and so forth.
2. Keep the plumbing for one apartment in its own framing wall—do not cross over from one partywall into another.
3. Direction changes in copper pipe water lines should be 45 degrees rather than 90 degrees in partywalls to eliminate the cavitation noise from water making a sharp 90-degree turn.
4. Holes drilled through plates, blocks, and corners for plumbing pipes should be slightly larger than the diameters of the pipes, then cushioned with acoustical rubber foam padding or filled with acoustical caulking. This eliminates the acoustical bond created when plumbing pipes fit too snugly within wall framing, yet at the same time prevents pipes from rattling around and creating noise when they are free to move within holes that are drilled too large.
5. Consider increasing a 1-inch wide partywall air-space to 1½ inches. This eliminates the possibility that bowed framing members will touch and gives the construction an additional margin of error. No one will miss a reduction of ¼ inch on either side of the partywall, as long as the construction layouts are changed accordingly.

Because the transmission of sound through partywalls can be a source for lawsuits and unhappy homebuyers and because most construction people are not aware of the subtleties of sound transmission problems, it is not a bad idea for the builder to hire the services of a sound engineering consultant during the design and early construction phases of a condominium project. The sound engineer should make recommendations to the architect for the design of partywalls, and possibly inspect the construction for sound problems on a periodic basis. This is especially true of condominium projects with units stacked on top of one another.

Polyseal and Window Reveals

Polyseal, an insulating foam, is injected into the gaps around the window frame and the wall framing as illustrated in Figure 13.7. The builder should check the installation of the polyseal to insure that it is not applied excessively around window frames and window sill corners. If too much polyseal is applied, it can ooze out and expand in such a way as to affect the window reveals between window frames and drywall. A thick bead of polyseal, placed in the inside corner between the window sill framing and the window frame can throw off the uniform window reveals by not allowing the drywall to lay flat against the window sill framing. The result is that windows framed with perfect reveals can end up, after the drywall is installed, with reveals that are uneven and unequal.

No Hot-Mop Tar on Subfloor

Illustrated in Figure 13.8 is the side view of a bathroom shower dam with hot-mop tar applied to its inside, top, and front sides, all the way down to the plywood subfloor. The builder should avoid getting hot-mop tar on the plywood subfloor. This is a prime example of sloppy workmanship. The hot-mop should end at the corner between the shower dam and the floor. This will then be covered after the shower dam is cement floated and the tile installed.

Hot-mop tar on the subfloor creates the same problems for the vinyl flooring installer as leftover paint overspray does for the painter—except the coat of hot-mop tar is thicker than paint. If the hot-mop tar spreads past its intended stopping point, it must be cut-out and scraped off the plywood subfloor. This task is not only difficult to do, but results in tar on the fingers and tools of the person removing it.

Chipping Concrete

Figure 13.9 illustrates foundation concrete that has projected slightly past the exterior wall framing. This problem happens when concrete forming is slightly off in dimensions or square. It only becomes apparent after the walls are laid out and framed.

The projecting concrete must be chipped or bush-hammered flush with the wall framing before the installation of exterior wood siding or lath. Wood siding boards usually extend below the mudsill plate line; therefore, they require a flat transition from the framing to the concrete for the metal weep screed that overlaps both the mudsill and the top of the foundation. Concrete chipping in this case is done by the concrete contractor.

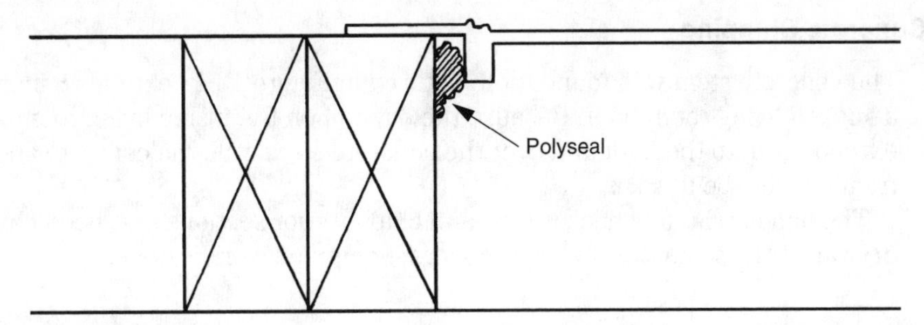

Figure 13.7

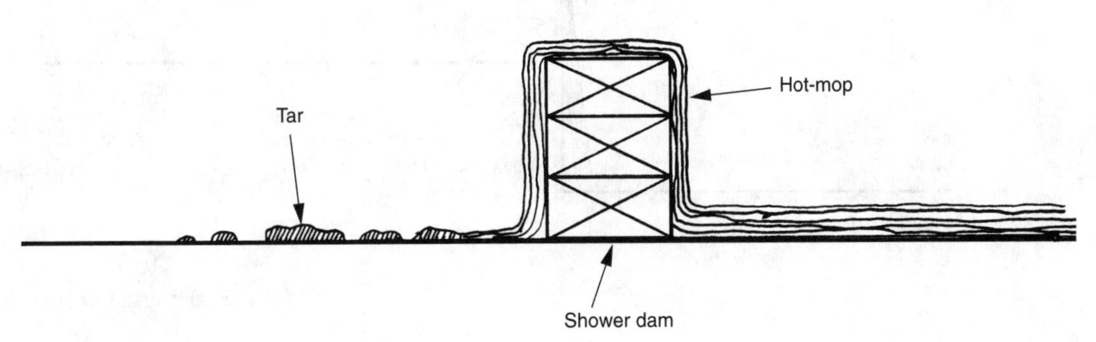

Figure 13.8

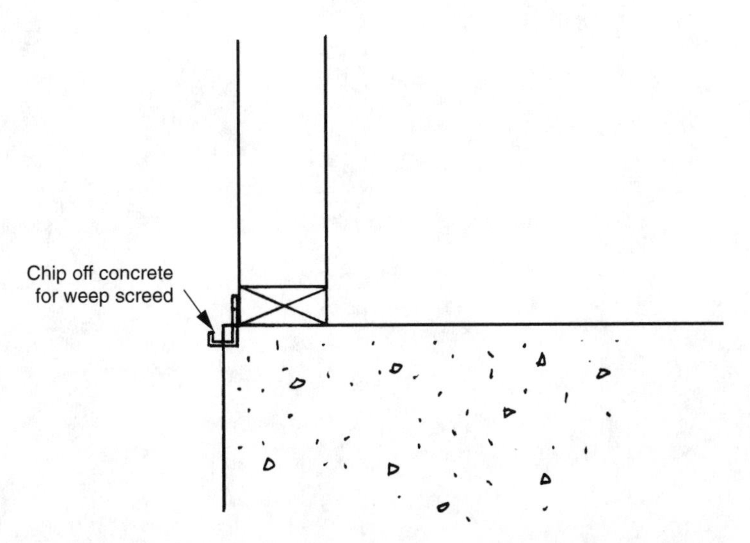

Figure 13.9

Interior Concrete Chipping

The concrete stem wall foundation pictured in Figure 13.10 extends 8 inches above a sunken living room floor. Because the drywall on the inside living room wall must extend down to the floor and over the concrete stem wall, the concrete and the wall framing must be flush.

The builder should check, mark, and chip interior sections that need chipping for drywall at the same time exterior concrete chipping work is done.

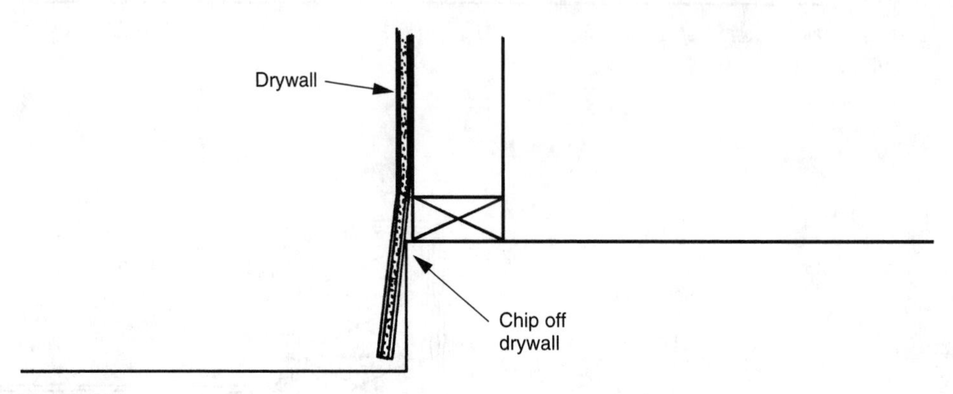

Figure 13.10

Chapter

14

Framing for Bathtubs

Framed Shelves around Bathtubs

A wood-framed, ceramic-tiled shelf that sometimes surrounds the perimeter of a bathtub is pictured in Figure 14.1. Whenever there is some choice in the dimensioning of these type of shelves around bathtubs and showers, the framing should be designed to eliminate rows of cut tiles. Shelves that can be built in widths of one or two rows of full-size pieces of tile, simplify and speed up the installation of ceramic tile later in the construction.

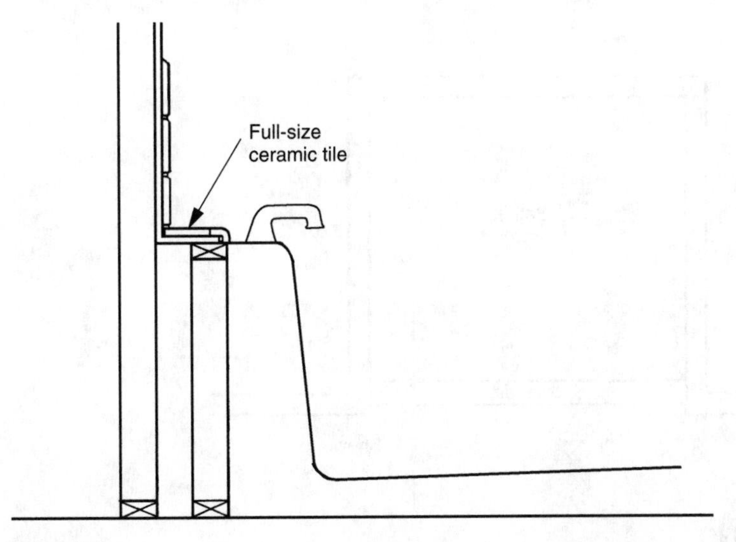

Figure 14.1

Hold Back Shower Dam

The bathroom shower stall, pictured in plan view in Figure 14.2, can either be built flush with the two adjacent side walls or be recessed an inch to allow for ceramic tile. For example, if the shower dam is framed flush and in line with the wall framing, after the shower pan hot-mop and ceramic tile mortar floating is added, the tile will also end up flush with or extend beyond the ⅝-inch thickness of the drywall. If the tile is flush with the drywall, then the baseboard must be awkwardly back-cut at the edges of the tiled shower pan. If the tile extends past the drywall, then tile cornerbead and small cut pieces of tile must be used to return the shower pan back to the drywall.

If the dimensions of the shower stall permit, a better method is to frame the shower dam recessed in from the two side walls at least 1 inch. This places all of the finish surfaces of the shower tile inside the three walls of the shower stall, without projecting past the two side walls. The floor baseboard can then make a 90-degree turn at the shower stall corners, and die neatly into the ceramic tiled shower dam.

Length of Bathtub Spout

Bathtubs are sometimes designed by the architect with a small shelf at the faucet end of the tub (Figure 14.3). The builder should check that this shelf is not deeper than the bathtub spout before framing around the tub. Sometimes this shelf is designed and built to accommodate two full widths of ceramic tile for better appearance and tile-setting ease. While this may look good, often the specified standard tub spout may not extend past this shelf into the tub.

Fur behind Tubs and Showers

Fiberglass shower and tub enclosures have contours that result in gaps between the wall framing studs and the shower sides (Figure 14.4). In order to provide a

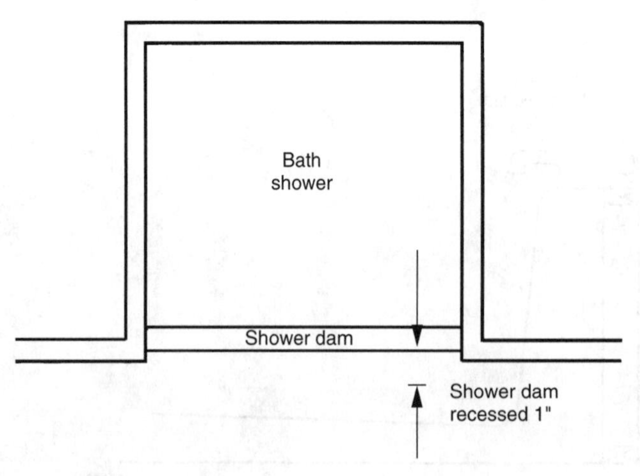

Figure 14.2

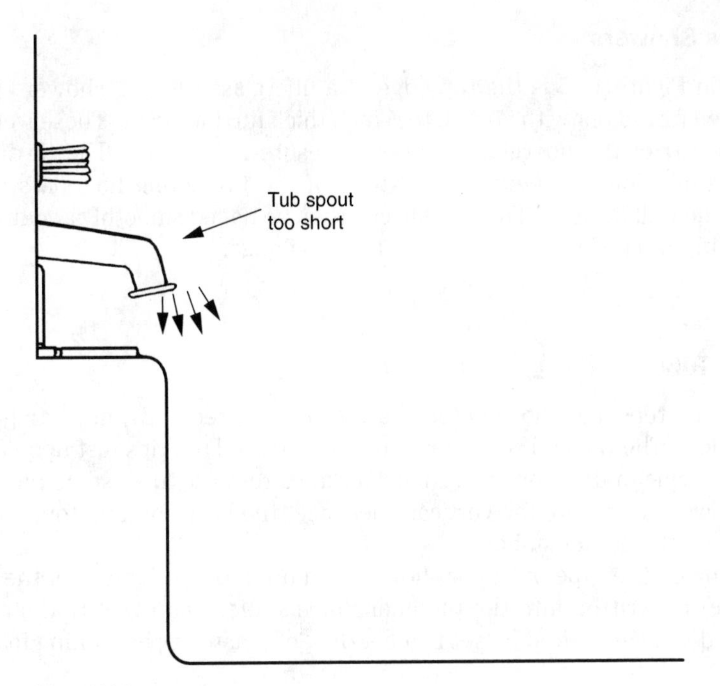

Figure 14.3

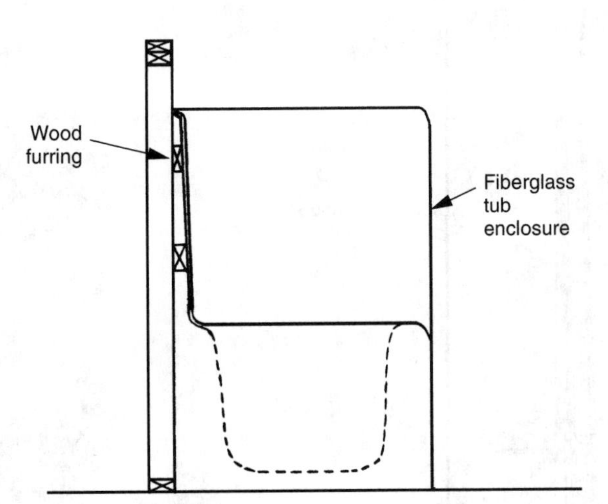

Figure 14.4

solid feeling to the sides of the shower unit, the bathtub installers should be required to fill in these gaps at one or two mid-heights. This can be done by gluing 2×4 blocks, shim shingles, or furring strips within these spaces using panel or subfloor adhesive. This should be spelled out and required in the fiberglass tub contract.

Fur above Fiberglass Showers

Illustrated in Figure 14.5 is the top edge of a fiberglass tub and shower enclosure that has been furred out with $\frac{1}{16}$-inch to $\frac{1}{8}$-inch thick furring strips. These were nailed in place as part of the fiberglass shower enclosure installation. The furring strips soften the transition between the thickness of the projecting fiberglass tub edge flange and the wall framing. The drywall can then be nailed smoothly and firmly over this joint without kicking out abruptly at the tub flange.

Support underneath Tubs

Most fiberglass tubs and showers today are manufactured with supporting ribs underneath them. The tub and shower rest on the ribs, and the ribs rest upon the floor. After a short time in use, however, small hairline stress fractures sometimes appear along the lower rim, where the vertical sides meet the horizontal bottom, and in the middle, where the bather stands.

Depending on the type of tub or shower, and upon the judgment of the builder, a clause can be written into the plumber's or installer's contract that states that the tubs and showers are to be set in a bedding of plaster or hardening foam (Figure 14.6).

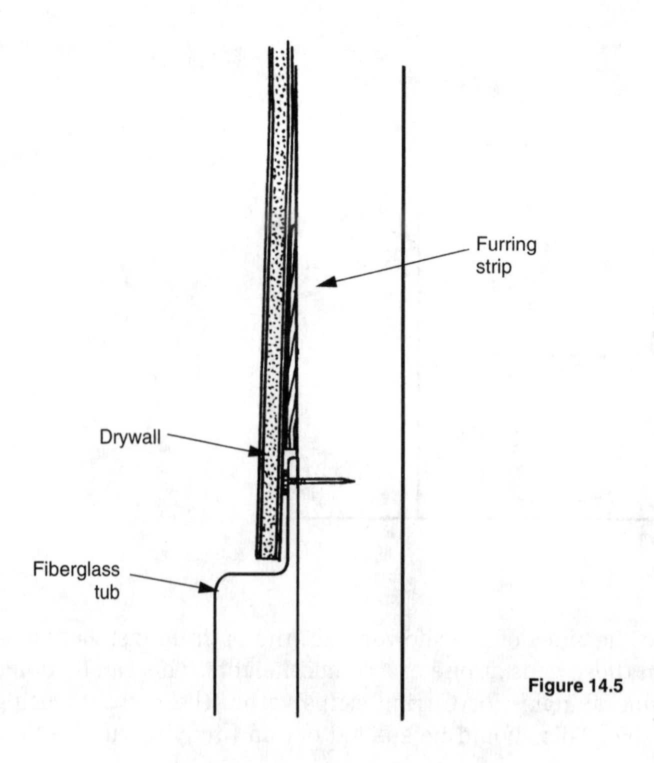

Figure 14.5

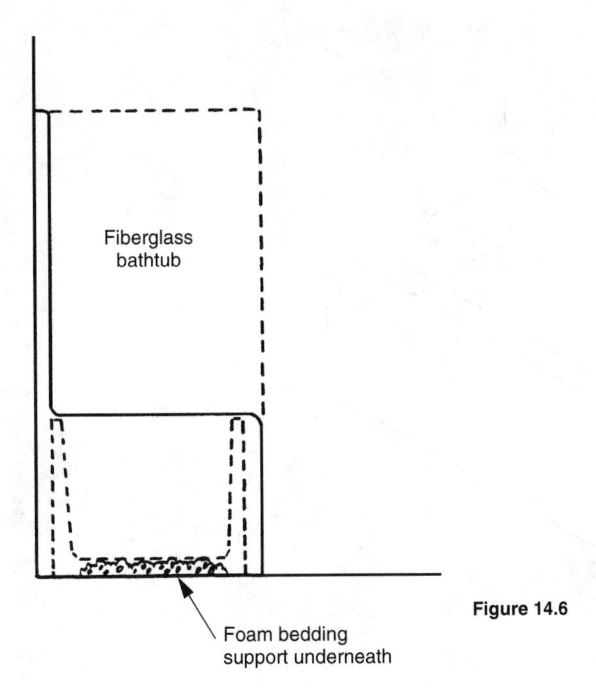

Figure 14.6

Fiberglass Tubs without Overflows

Sometimes the builder wants to use bathtubs and combination tub/shower units that do not have overflows, in order to avoid having to place an unsightly access panel to the overflow on a bedroom side of the bathtub wall. When the builder specifies in the fiberglass tub/shower contract that units are to be without backflows, the builder should not accept units that are delivered to the job site with holes cut out for overflows (Figure 14.7). Sometimes the fiberglass tub manufacturer, by mistake, will send out tubs that have overflow holes, and say that these holes will be patched in the field, after the tubs are installed. The problem is that these patches seldom match the exact color and sheen of the rest of the tub surfaces. The builder is then stuck with having to fast-talk his way through homebuyers who question the dull area in the fiberglass tub where an overflow would normally go.

When tub/shower units with overflows are delivered to the job site that are supposed to be without overflows, the builder should simply tell the truck driver to turn around and take the tubs back and return with the correct ones.

Don't Install Motors with Jacuzzi Tubs

Jacuzzi tubs are installed early in the rough framing, at the same time as any other type of bathtub. Because of this, they often remain open and uncovered until the bathtub is enclosed by drywall, cabinets, ceramic tile, or a bathtub skirt, which is

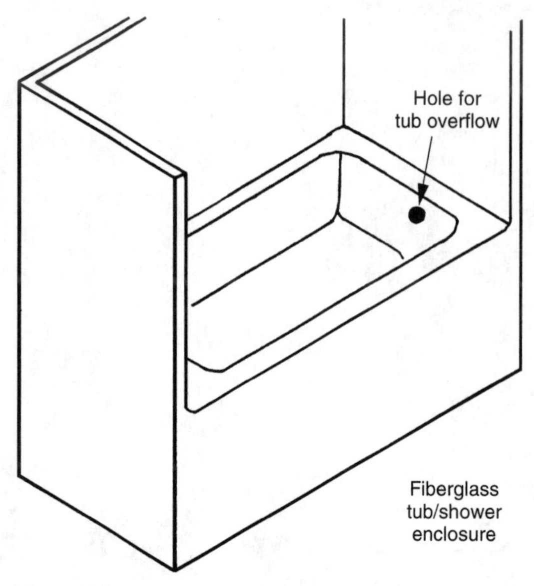

Figure 14.7

usually not for several weeks or even a few months later. For this reason, the builder should not have the Jacuzzi motors installed along with the tubs because this leaves the motors vulnerable to theft during the remainder of the construction or until the tubs are covered up.

The motors should either be delivered separately by the Jacuzzi tub manufacturer or, if that is not possible, the motors should be removed from each tub, and stored in a safe place on the job site until they can be installed later. The Jacuzzi motors should be installed just prior to the time that the bathtubs are enclosed by surrounding drywall, cabinets, tile, or a tub skirt.

Chapter

15

Backing

Entry Wall Ceiling Backing

Figure 15.1 illustrates an entryway into a house or apartment with a short partition wall to the right of the front door. If the entry ceiling is drywalled and the ceiling joists run parallel with the partition wall, the entry wall is not long enough to warrant placing two full-length ceiling joists alongside the wall for drywall nailing (Figure 15.2). Instead, a single 2×8 or two 2×4s are nailed flat to the top of the wall framing

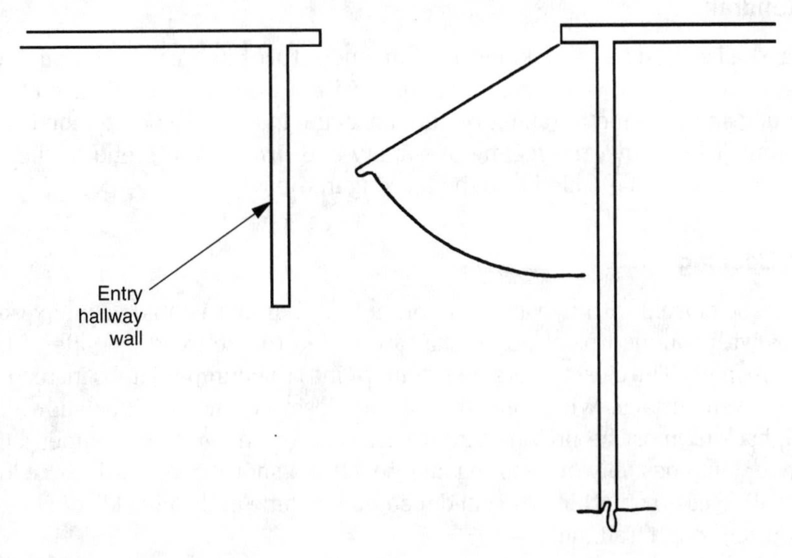

Figure 15.1

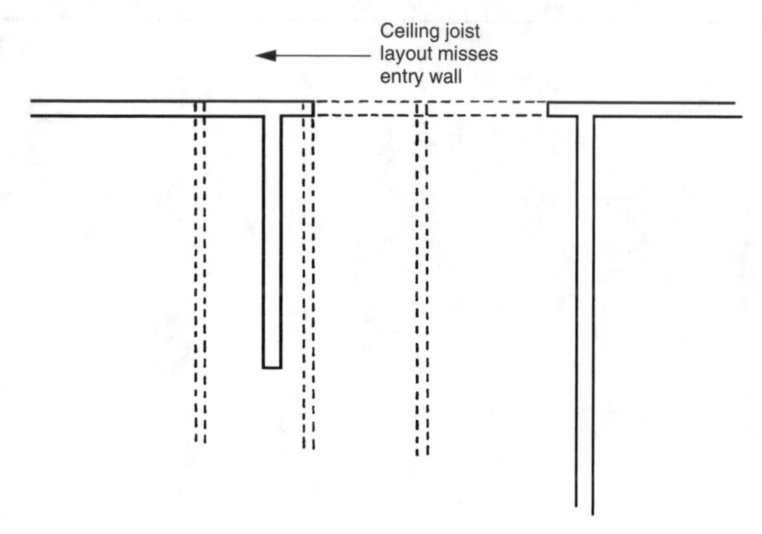

Figure 15.2

double plates, with a portion cantilevered out over each side of the wall to act as backing for drywall nailing.

The problem to avoid is casually nailing these short pieces of backing in place with only a few nails instead of securing them properly. The thing to remember is that the drywall is going to be nailed to these short pieces of backing on the ceiling, and if one end or a portion of the backing should come loose with the upward force of a hammer blow, the entry ceiling will bow as a result (Figure 15.3).

Backing for Handrail

A handrail should have backing where an end returns or dies into a wall. Figure 15.4 illustrates a typical wrought-iron handrail with a wood cap that is used for production housing. The metal railing requires backing in order to be attached securely to the wall. To improve the feeling of solidity and firmness, the end of the wood cap piece should also be nailed into the backing in the wall.

Bumper Jamb Backing

Wardrobe closets with bypass doors often have bumper jambs—1-inch×4-inch plain jambs with rounded front edges that are nailed to each vertical side of the closet door opening. The closet doors then "bump" into the bumper jambs instead of hitting the drywall surface. When one side of the closet is a flat wall, as shown in Figure 15.5, backing must be provided in order to nail on the vertical bumper jamb. If the bumper jamb has only drywall to nail into, after a short time, it will come loose from the wall. This is something the builder should include in the checklist of items to look for during rough framing.

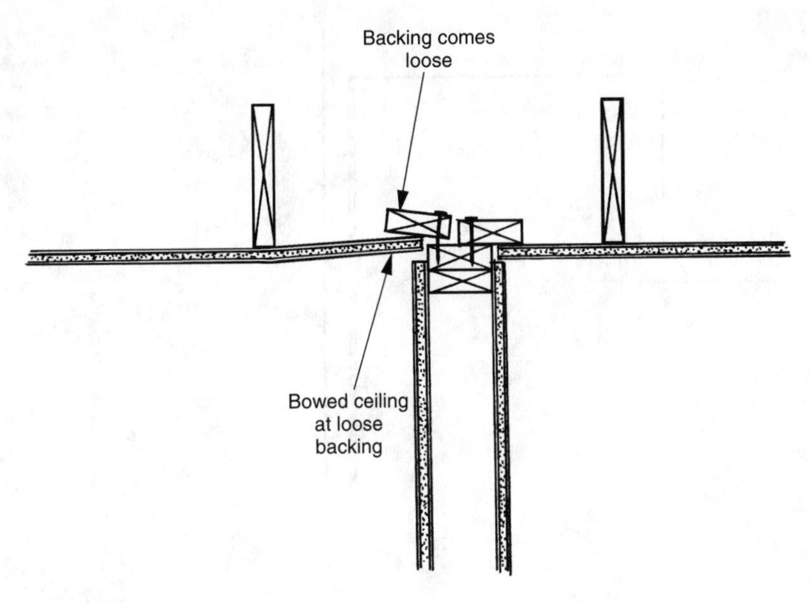

Figure 15.3

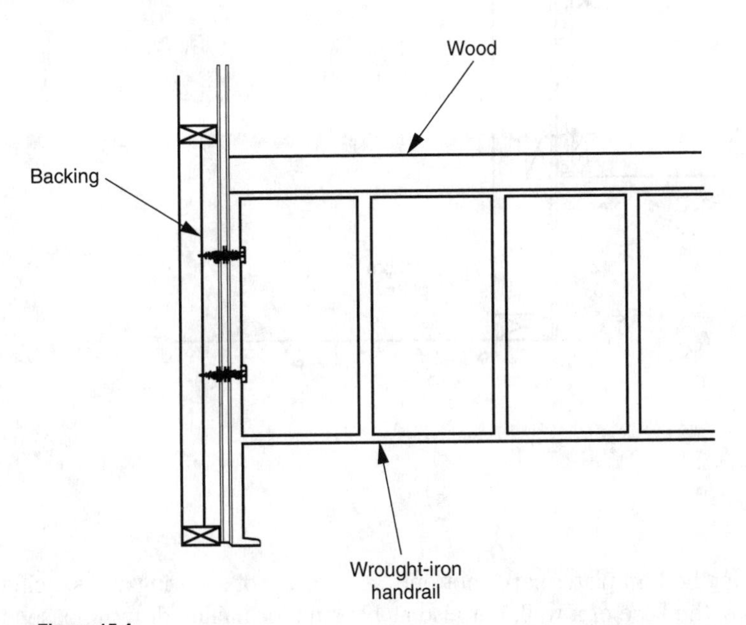

Figure 15.4

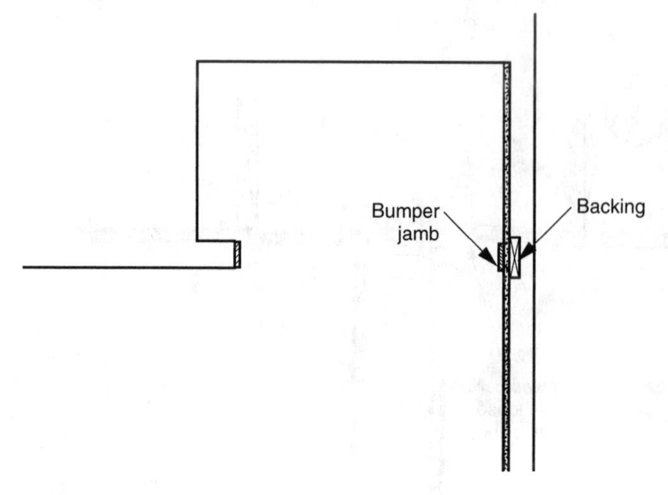

Figure 15.5

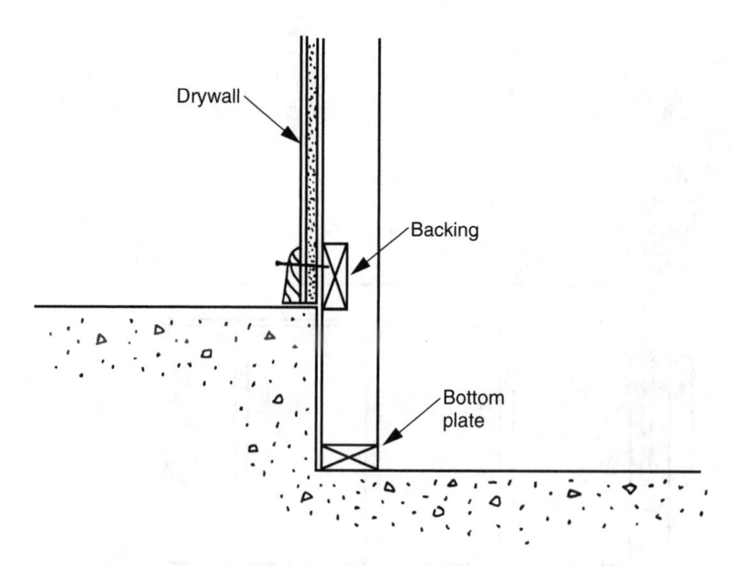

Figure 15.6

Split-Level Floors

Wall framing bottom plates serve not only as a means of securing and spacing framing studs along the base of a wall, but also as backing for nailing drywall, baseboard, and baseshoe trim. When a single wall serves both sides of a split-level floor, as shown in Figure 15.6, backing should be added at the higher floor level to serve as the bottom plate.

Never assume that wall framing studs spaced 16 inches apart are adequate nailing on the upper floor level for drywall, baseboard, and baseshoe. Baseboard and baseshoe often require nailing that is closer together than 16 inches in order to eliminate gaps at inwardly bowed wall areas. Nailing into drywall alone does not have the pulling power than nailing solidly into wood. Also, the upper floor wall surface may be painted enamel (e.g., as in a bathroom or kitchen) before the installation of baseboard trim, thus hiding the location of the wall framing studs.

Backing for Wardrobe Track

Illustrated in Figure 15.7 is a wardrobe closet with floor-to-ceiling doors. Because there is no solid header spanning the opening, solid blocking must be added between the ceiling joists in order to be able to attach the wardrobe door metal track. The ceiling joists alone cannot be considered adequate backing because the metal track comes with uniformly spaced predrilled holes that will not correspond to ceiling joist layout. Without solid blocking, each ceiling joist must be located and new holes drilled in the metal track (Figure 15.8).

The builder should especially check that solid backing is provided at each end of the wardrobe track. Here the ceiling occasionally bows or curves slightly upward, and a gap will result between the track and the ceiling if the track cannot be pulled up tightly with a screw.

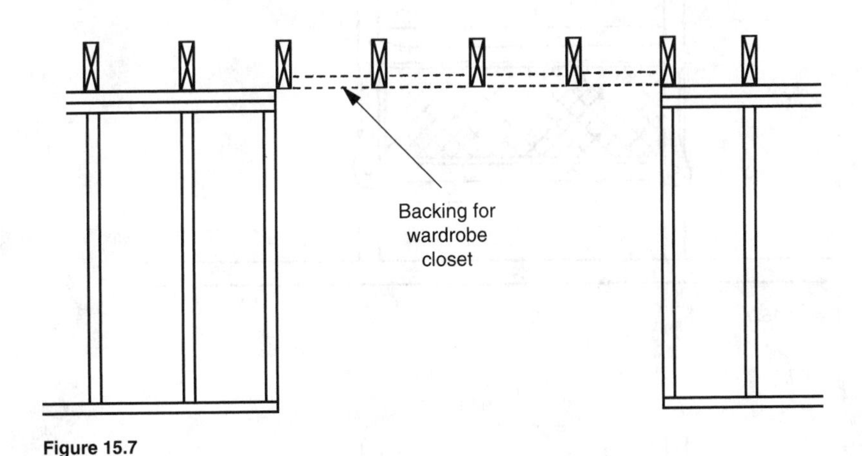

Figure 15.7

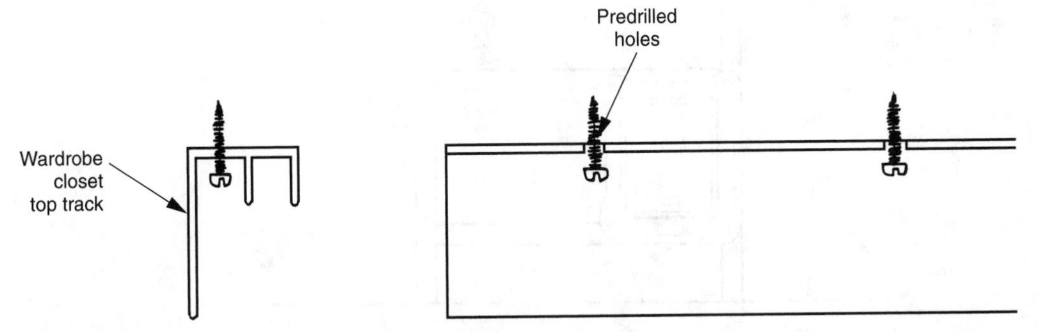

Figure 15.8

Garage Vent Backing

It is good practice to install blocks above and below a garage vent (Figure 15.9). The blocks provide a solid backing so that stucco lath can be nailed around all four sides of the garage vent.

Utility Panel Backing

Telephone and television utility panel boxes that are located on exterior walls should have wood blocking around all four sides of the panel. Wood blocking at the top and bottom provide solid nailing for exterior stucco lath (Figure 15.10).

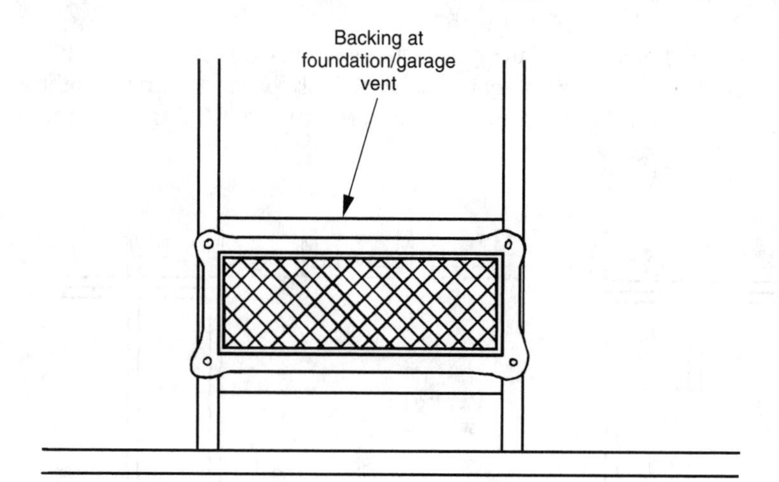

Figure 15.9

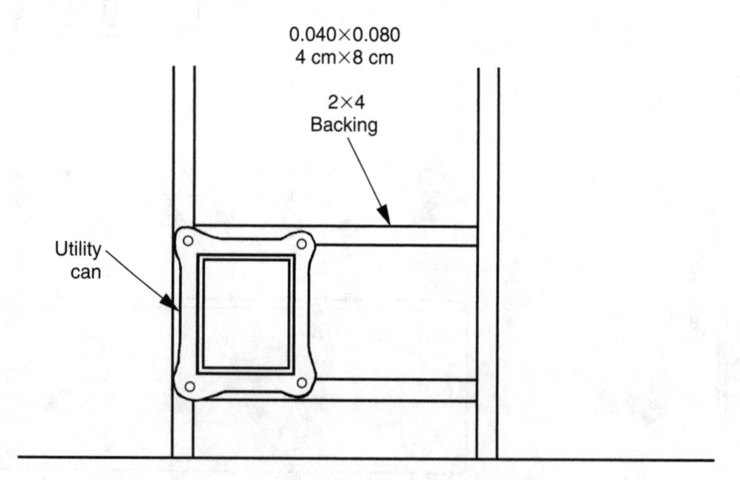

Figure 15.10

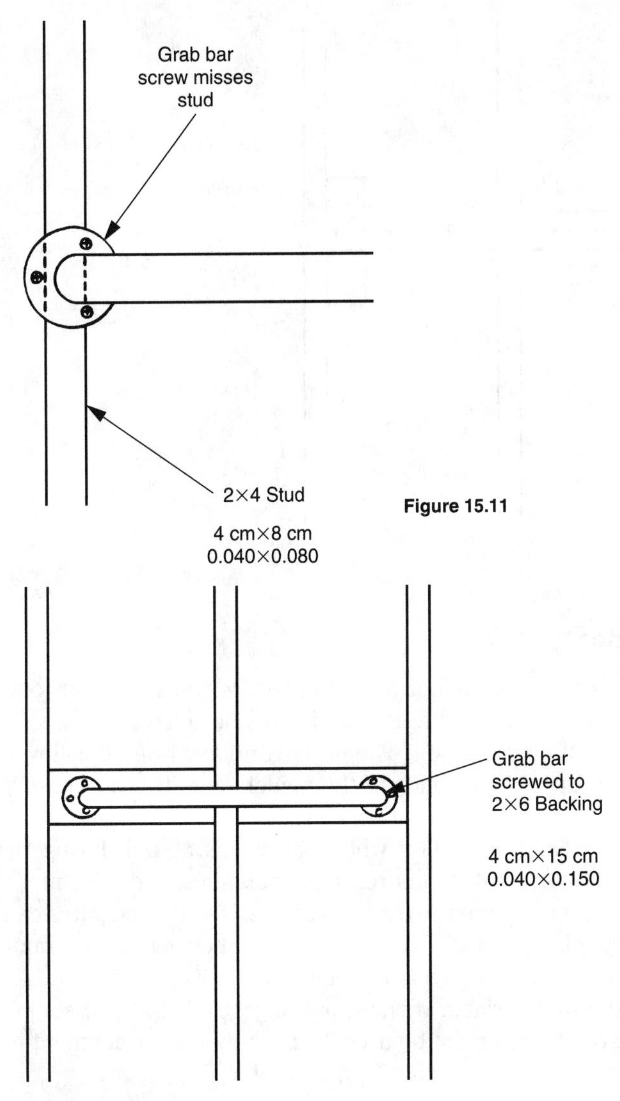

Figure 15.11

Figure 15.12

Backing for Grab Bars

Metal grab bars are sometimes installed in bathrooms around toilets, bathtubs, and shower stalls. If wall framing is to be used as backing for attaching the grab bars, solid blocking must be added at each grab bar end. As illustrated in Figure 15.11, grab bar ends have three or four holes for screwing the bar to the wall. The 1½-inch wide edge of a 2×4 stud is not wide enough to pick up all three screws, even when the grab bar end is centered over the stud. In order to provide adequate support for the grab bar, either a 2×6 stud extending along the back of the grab bar or pieces of blocking at the end of the grab bar ends should be added for solid nailing (Figures 15.12 and 15.13).

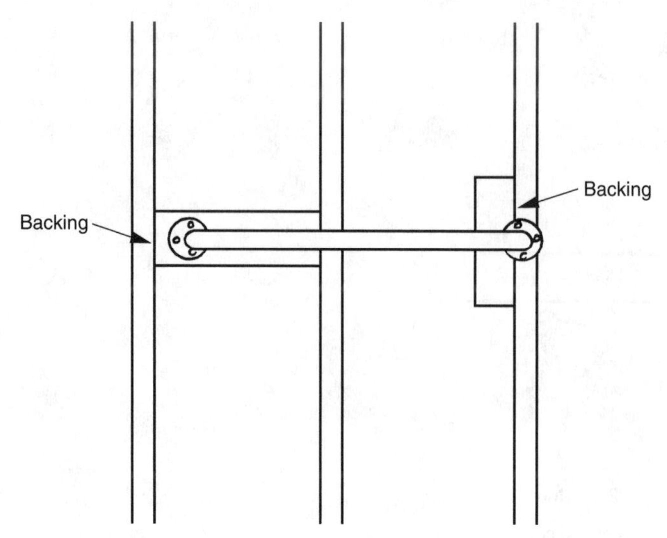

Figure 15.13

Backing for Fences and Gates

In tract housing, wood backing is not placed in garage walls or other exterior walls for the future attachment of wood fences and gates because their exact location is unknown at the time of construction. The design and location of walkways, fences, and gates are usually the responsibility of the homebuyer after purchase.

For other types of housing construction, the attachment of fences and gate posts to exterior walls is greatly simplified when backing is added during the framing stage. Condominium/apartment structures, custom homes, and sales models and remodels are all examples of where the exact location of fences and gates can either be determined from the plans or decided upon by the owner before the framing stage.

Fence or gate posts that have wood backing within the walls can be lag-screw bolted directly to the wall surface. Without backing, toggle bolts or some other hollow-wall means of attachment must be used. This method of attachment is generally more complicated and time-consuming (Figure 15.14).

Firedoor Magnetic Catches

Exit corridor firedoors must close automatically when the fire alarm system is activated to prevent the spread of smoke within exit corridors. These doors are held open by a magnetic catch, half of which is on the door and the other half on the wall. Upon activation of the fire alarm, the electric circuit to the magnetic catch is opened, the magnet loses its force, the two catches disengage, and the door closer closes the firedoor.

These magnetic catches are installed late in the construction, after the walls are covered up with drywall. The half of the catch mounted to the wall must be attached to wood backing for adequate strength. If backing is not placed in the wall near the outside edge of the door, the catch must be mounted to the next closest stud, somewhere within the middle span of the door. Because of the angle the firedoor makes with the wall, the doorknob may hit the wall before the two catches can engage. In one case, the doorknob was mashed into the drywall so that the two catches could touch (Figure 15.15).

This problem can be prevented in one of two ways. Backing can be placed in the wall during the rough framing period to correspond with the top corner of the firedoor, allowing the magnetic catch to be placed on the edge of the door and the doorknob (Figure 15.16). The other solution is to frame the hinge side of the doorway partition only 4 inches in projection away from the main wall. This allows the firedoor to open parallel to the main wall and enables the magnetic catch to be placed anywhere within the span of the door without causing the aforementioned problem (Figure 15.17).

Corridor Handrail

Corridor handrail is usually supported by metal brackets that are spaced 4 to 5 feet apart and screwed into wood backing inside the wall (Figure 15.18). Because handrail is sometimes installed after the corridor walls are painted, the location of wall framing studs and backing may be hidden by the paint. For this reason, studs and backing locations must be marked in some way prior to painting.

One common method of installing the handrail in the proper place is to mark the location of studs and backing on the floor with spray paint before the walls are drywalled. A dot is used to indicate a stud and a line or dash is used for blocking. When the time comes for handrail installation, the carpenter reads the markings on the floor and attaches the brackets accordingly.

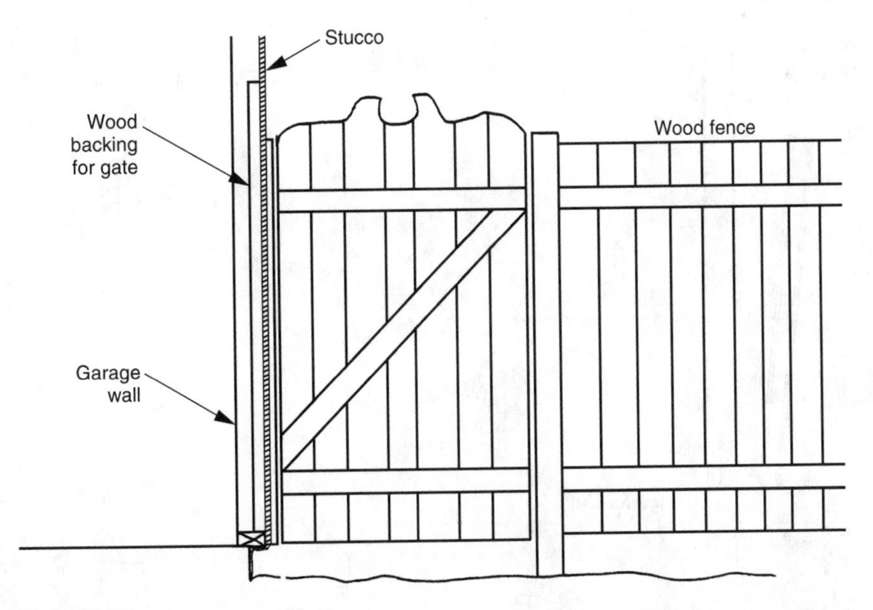

Figure 15.14

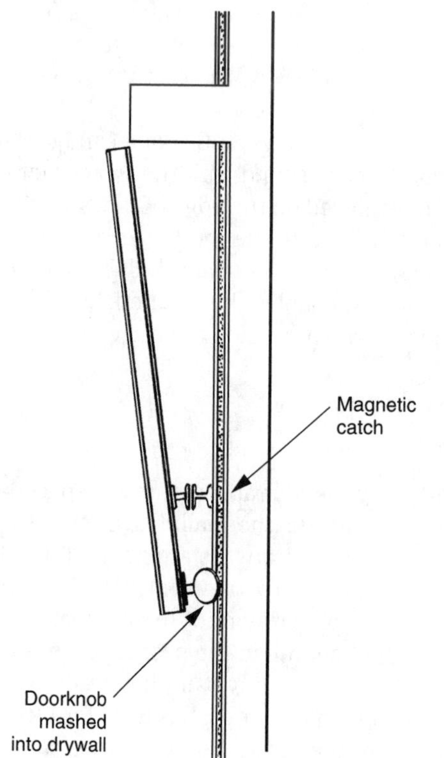

Figure 15.15

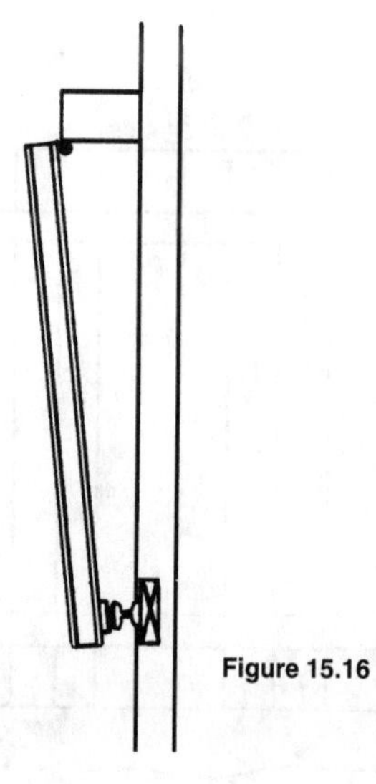

Figure 15.16

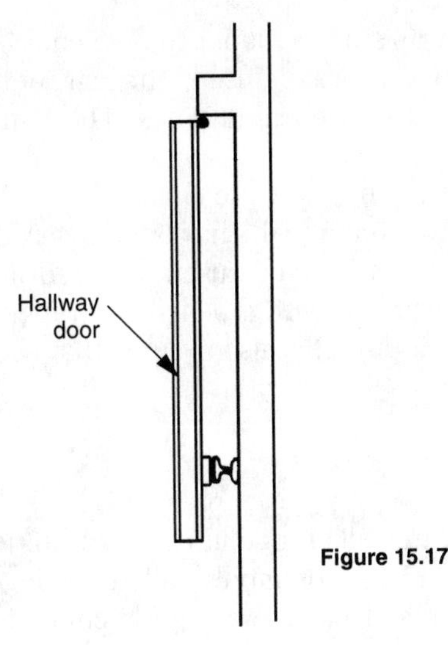

Figure 15.17

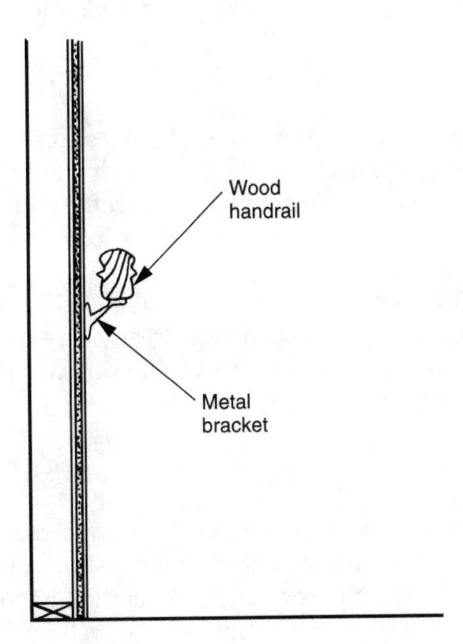

Figure 15.18

A better method is to have the carpenter chalk a line with white chalk at bracket height on the corridor wall after they have been drywalled, taped, and textured. Because the walls have not yet been painted, the drywall nailing pattern is clearly visible and each stud is easy to find. Using one bracket as a template guide, the carpenter goes through all the corridors and predrills the holes where the brackets are to be installed later. By drilling through the drywall and into the studs or backing, a hole is left in the drywall surface paper that will not be covered up or masked by painting. After the walls are painted, the carpenter simply screws the brackets into the walls using the predrilled holes. The handrails can then be attached to the brackets.

This second method has two advantages. First, a white chalk line can be used to establish a straight line for the bracket attachment, which is later covered by painting. Second, the exact location of backing is determined during the predrilling phase, enabling holes that missed backing to be circled with a pencil and repaired during drywall pickup and before the walls are painted. This also ensures that each bracket will indeed be attached to backing.

Herring-Bone Siding

Wood siding boards should have backing in walls for solid nailing at all joints. The pattern of siding boards shown in Figure 15.19, commonly called herring-bone, requires backing at the vertical mitered joints. This allows slightly cupped or bowed boards to be straightened to match the corresponding mitered edge of the opposite board. It also helps keep the ends from pulling away from the wall in the future.

Light Soffit Frames

A dropped light soffit, such as the one illustrated in Figure 15.20, is sometimes constructed out of rough-sawed or resawed 2-inch thick lumber with all of the pieces cut, assembled, and installed on the job site. This type of light soffit requires additional backing in the walls and ceiling in order for the nailing to adequately draw the wood members tightly and permanently up against the ceiling. If the light soffits were prefabricated out of thinner and lighter materials at a cabinet shop, the extra backing would not be needed. With the prefabricated and already assembled units, the light soffits can usually be attached anywhere to the ceiling with a combination of nailing into ceiling joists and the use of toggle-bolt fasteners.

Don't Use Plywood for Hot-Mop Backing

Illustrated in Figure 15.21 is one method sometimes used to provide solid backing for the hot-mopping of shower pans. It involves nailing 12-inch-wide, $\frac{3}{8}$-inch-thick pieces of plywood around the base of the shower pan. This method is not recommended because the plywood adds $\frac{3}{8}$ of an inch to the framed wall surface. This takes $\frac{3}{8}$ out of the $\frac{3}{4}$ of an inch that the wall should be mud floated for tile.

The better method is nail 2×12-inch solid blocking between the studs at the elevation just above the shower pan. This gives the tile setter a flat surface with which to start and allows three plys of hot-mop to be applied tightly to the wall framing of the shower stall (Figure 15.22).

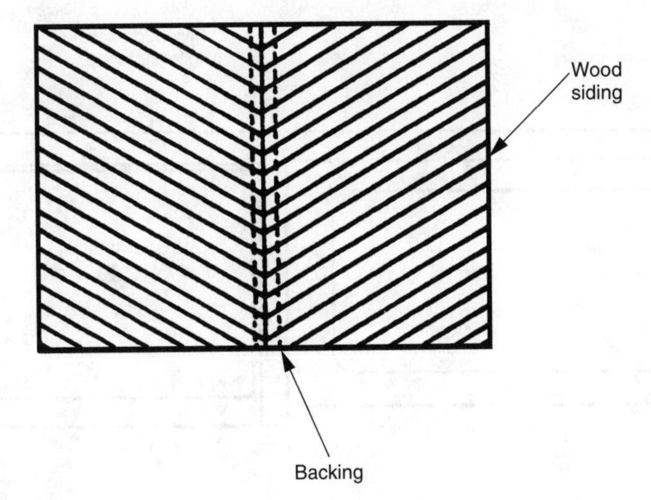

Figure 15.19

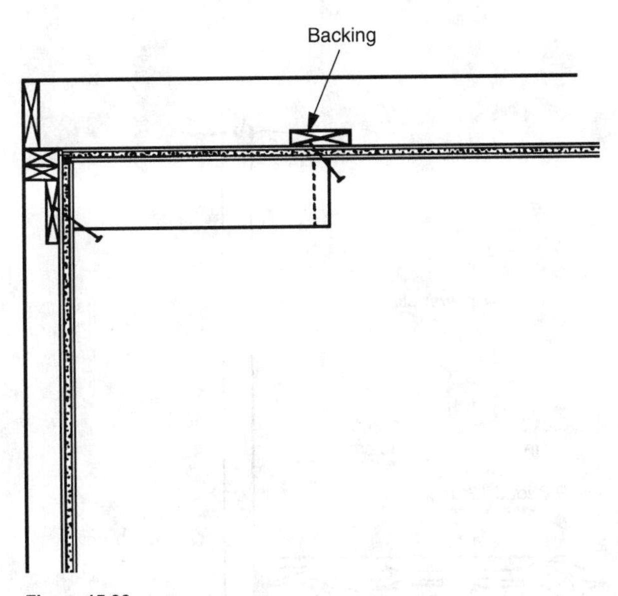

Figure 15.20

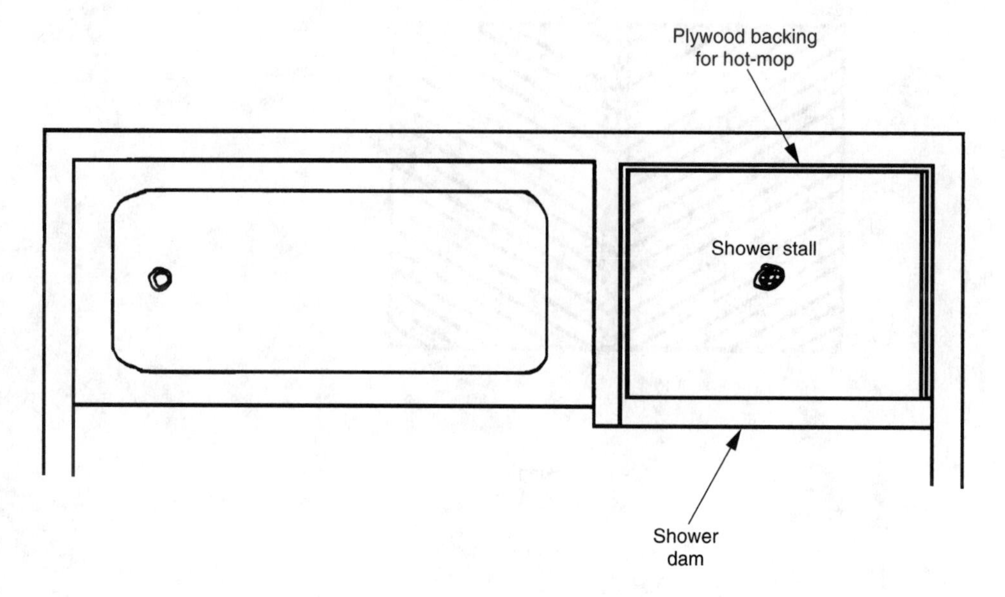

Figure 15.21

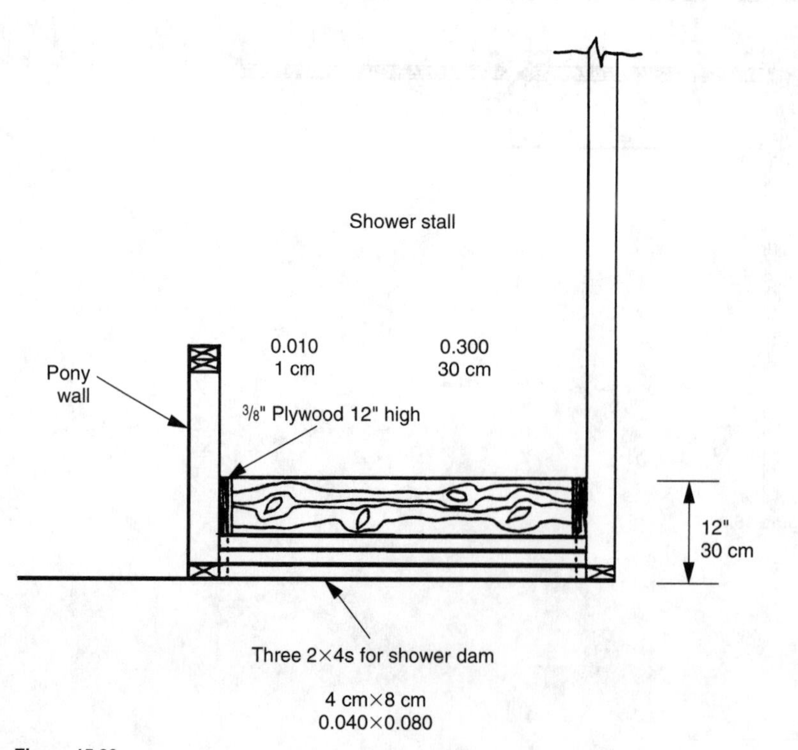

Figure 15.22

Chapter

16

Straightedge

Mirror Walls Straight

The builder should check during the framing straightedge phase that bath walls on which mirrors are to be hung are straight. This is especially important at the top of the mirror where a bow in the wall can result in a gap between the wall and the mirror either at the corners or in the center (Figure 16.1).

If a large bow exists on a wall receiving a mirror, the mirror installer sometimes has to pull or suck in a corner for the mirror clip screws to reach and engage a stud. Bending a mirror causes distortion in the mirrored image. Furthermore, too much tension placed on a mirror may cause a mirror to break. If mirror clips are not used, the gap between the mirror and wall caused by a bowed section in the wall may still be unsightly.

Tile Walls Straight

The builder should check that walls receiving tile are straight during the framing straightedge phase. This is especially important on long runs of splash tile in kitchens that are in a direct line of sight (Figure 16.2). Equally important are inside corners of bathtubs and showers where square tiles and grout joints readily reveal vertical corners that are out of plumb and horizontal corners that are out of square. This problem becomes more crucial when tile is glued directly to drywall greenboard rather than floated with a base of mortar.

Bar Light Walls Straight

The bar light is a currently popular light fixture used in bathrooms. They are placed above the bathroom mirror and can range from as short as 2 feet to as long as 6 or 8

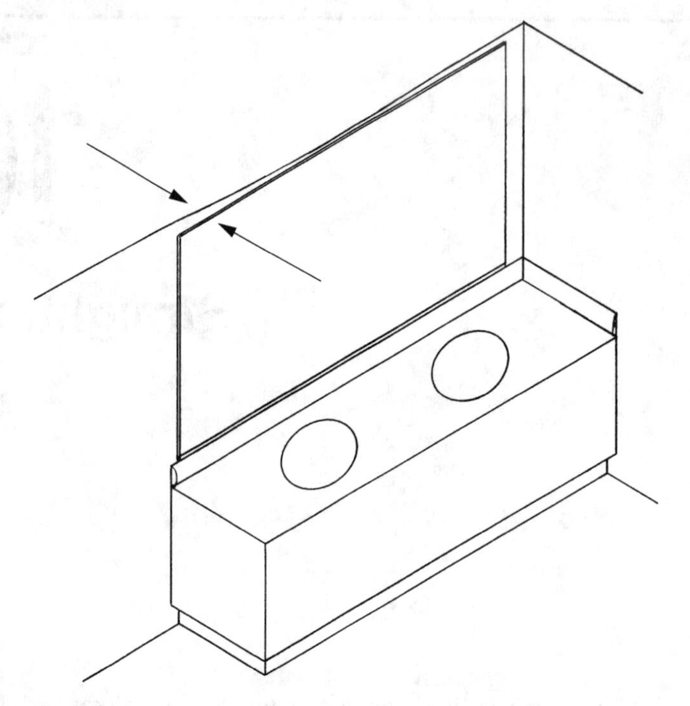

Figure 16.1

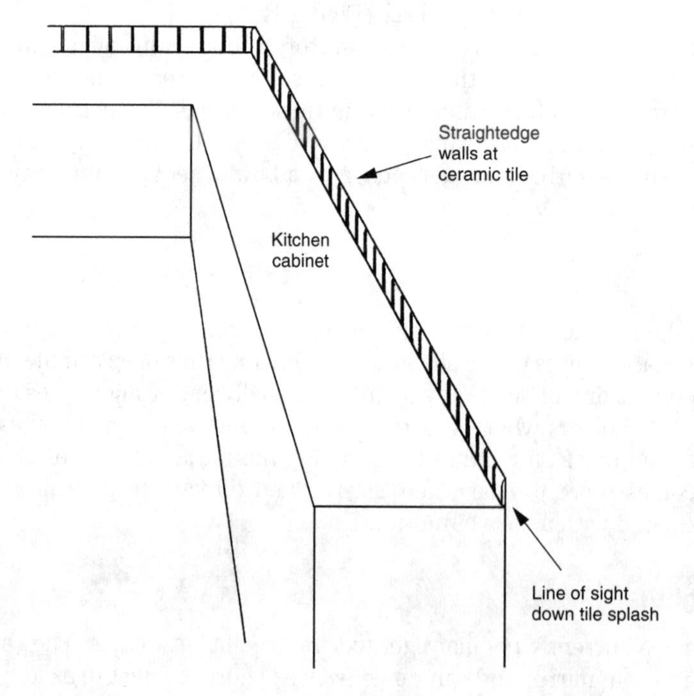

Figure 16.2

feet. Because the bar light fixture is rigid, the wall above the bath mirror must be very straight. The builder must pay special attention to getting these walls straightedged and true (Figure 16.3).

A bar light fixture on a bowed wall will not fit evenly to the wall. If the builder tries to remedy the situation by pulling the end or middle of the fixture in with a screw, nail, or toggle bolt, the side of the fixture may buckle or crease.

Lav-Top Walls Straight

One-piece imitation marble bath lavatory tops are rigid and inflexible. The builder, during the framing straightedge phase, should check that the walls at the lavatory tops are straight in order to get a tight fit between the lavatory top and the wall. When a lavatory top is placed against a crooked wall, the gap between the top and the wall is uneven and can become large. Although the bath mirror placed on top of the lavatory top splash covers some or all of this gap, the ends of the lavatory top splash are exposed. On a bowed wall having a 5-, 6-, or 8-foot-long lavatory top, an exposed gap between the end of the lavatory splash and the wall of an ½ inch or more is difficult to caulk neatly and may be unsightly.

Wall Corners Plumb for Mirrors

Illustrated in Figure 16.4 is the typical result when bathroom walls with 45-degree angle corners are not properly straightedged for mirrors. In this example, the wall corner where two mirrors come together is out of plumb because one of the 45-degree angle walls was not straight. Because the two mirrors are cut straight and must

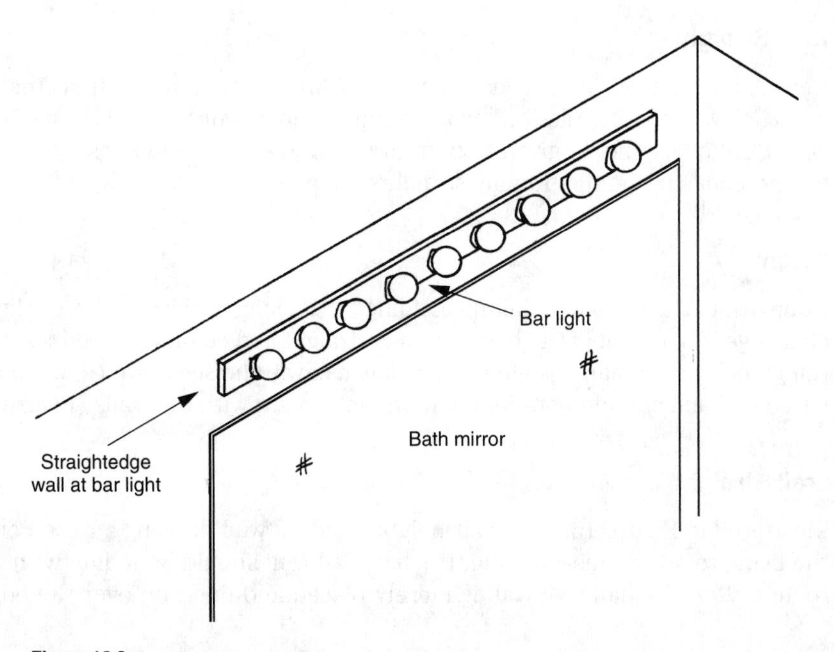

Figure 16.3

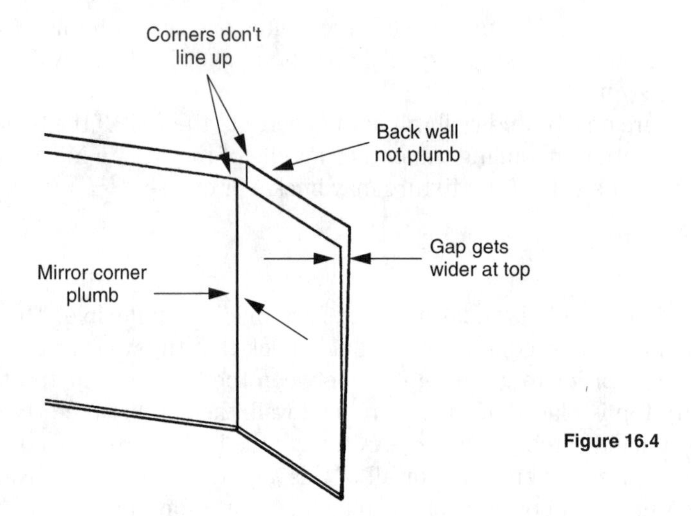

Figure 16.4

join together in a straight, vertically plumb inside corner, a gap is created behind one of the mirrors that gets larger at the top because, in this example, the one wall is falling away as it goes up. The top corners of the two mirrors also do not line up with the wall corner, which is also falling away backwards as the corner goes up.

In this particular instance, the 45-degree angle wall was about ½-inch out of plumb in 4 feet, resulting in a ½-inch gap at the top of the smaller mirror between the mirror and the wall. This is something you do not want to discover late in construction when the mirrors are to be installed.

Flooring Walls Straight

Bowed, crooked, or out-of-square walls show up more readily in areas that have ceramic, clay, or marble tiles and wood parquet squares or planks (Figure 16.5). Special attention should be given to bottom plates in entries, kitchens, bathrooms, and family rooms during the framing straightedge phase.

Hallway Straight

In apartment and condominium buildings with long hallways, the quality-control challenge is to get all of the hallway walls straight and column pop-outs square. Few things look worse than to sight down a long hallway and see bowed and crooked floor baseboard and column pop-outs that are not square with the walls (Figure 16.6).

Entry Handrail Straight

Illustrated in Figure 16.7 is a half-height handrail wall dividing a raised entry from the living room. In this example, the handrail wall should be in line with the living room wall. If the handrail wall is merely placed and lined up over the edge of the

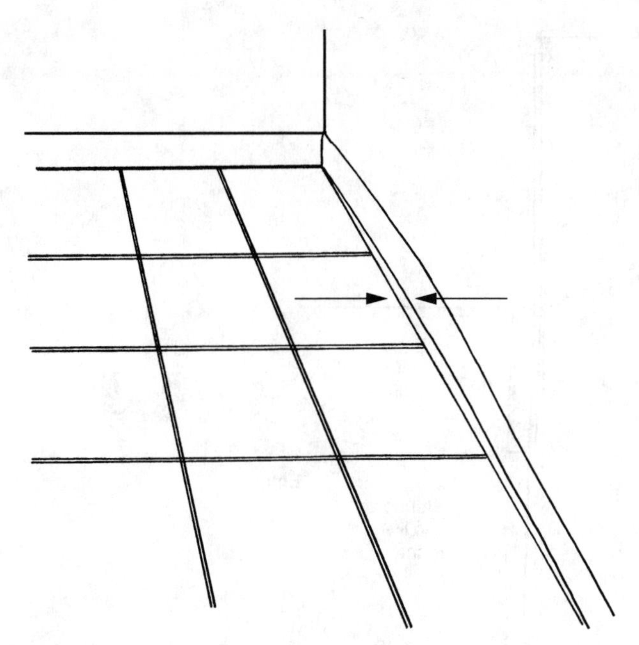

Figure 16.5

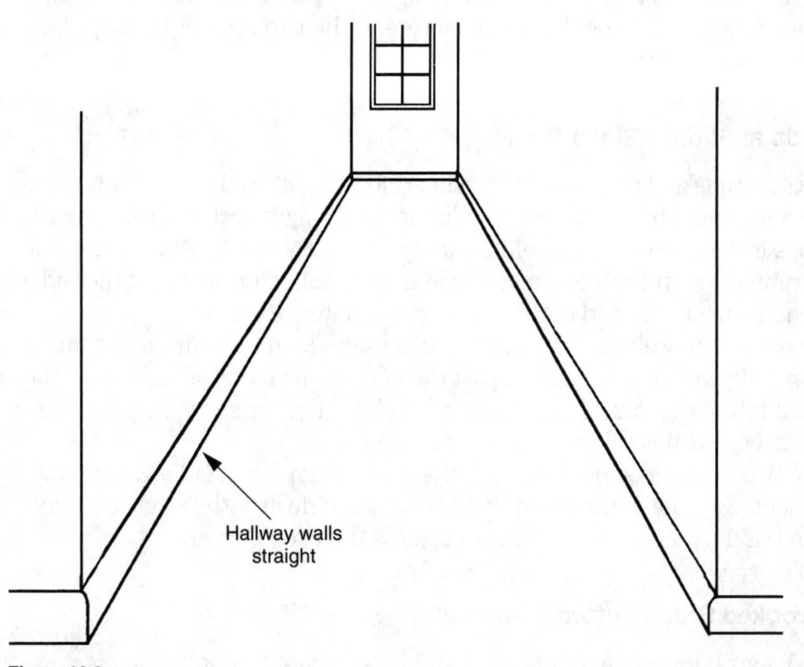

Figure 16.6

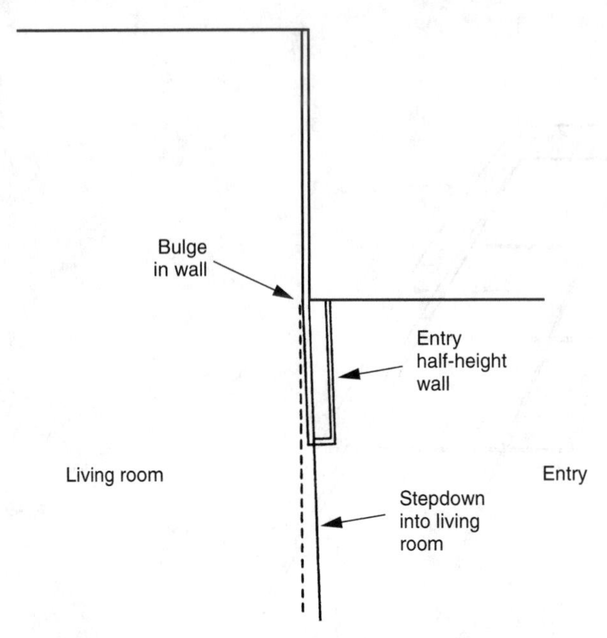

Figure 16.7

entry raised slab, the straightness of the handrail wall and the living room will depend on the alignment of the concrete. If the two are not in line, a bow or bulge will show where they join.

Bowed Studs at Sliding Glass Doors

King studs and trimmers at the sides of sliding glass doors must be straight and true. They should be checked in production housing before the sliding glass door frames are installed or at least before the walls are drywalled. Bowed or crooked studs and trimmers next to sliders may cause the drywall to project past the slider jamb, which can be unsightly and difficult to repair (Figure 16.8).

Once the walls are drywalled, finish taped, and textured, not much can be done with drywall that bulges out past the edge of the metal slider frame. Mashing or beating-in the drywall, then adding drywall taping "mud" will sometimes correct small bulges, but this seldom works with larger ones.

Wall framing should also be checked where the header, king stud, and trimmer meet. Any one of the three that is not nailed flush with the others may cause a bulge in the drywall at one of the sliding glass door top corners.

Remove Crooked Studs before Electrical

The straightedging of crooked and bowed wall framing studs is done just prior to the installation of drywall. However, there is a preliminary phase in the rough framing when the removal of badly bowed and crooked material is of benefit to the straightedging later. This phase is just prior to the start of the rough electrical work. Once the electrician drills holes and pulls wiring through wall framing studs and posts, the framer is pretty much stuck with this material. By going through and replacing badly twisted or bowed 4×4 posts, along with any crooked studs, prior to the start of the electrical work, the straightedging work is made easier for the framing carpenters later.

Straightedge Stair Skirtboard Walls

Another area within the house that should be checked for straightness is the stair walls that receive skirtboard. The top edges of skirtboard provide a line of sight down a stair wall. Any waviness in the wall framing transfers directly to the skirtboard, which is then easily seen by anyone looking down the stairway (Figure 16.9).

Ceiling Springlines

Illustrated in Figure 16.10 is the joint between a flat section of a bedroom ceiling, and a raked or sloped section of the ceiling, called a springline. The rafters that form the sloped half of the ceiling must be designed for zero deflection, as any downward bow in the center of the rafter spans may transfer to the springline. This becomes crucial when the roof is loaded with heavy clay or concrete roof tiles, as the center of the room ceiling will tend to deflect, as a whole, more than the edges of the room. The rafters may bring the entire springline downward, showing a noticeable downward bow in the flat portion of the springline.

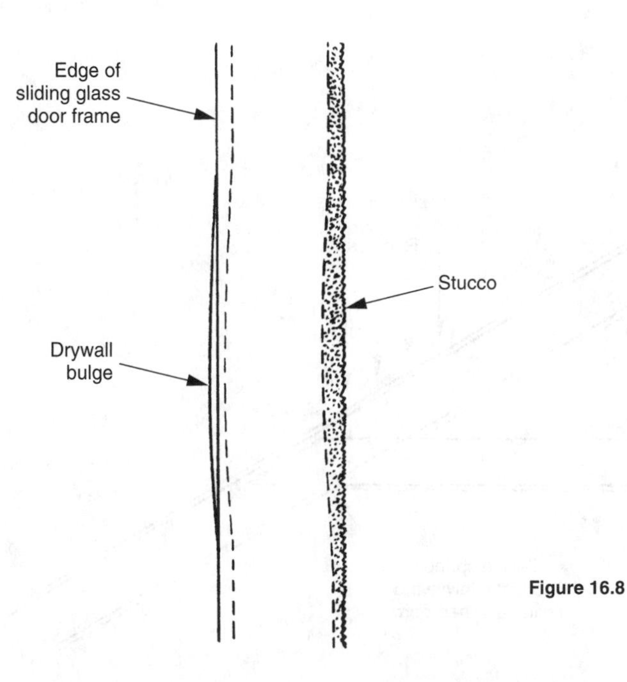

Figure 16.8

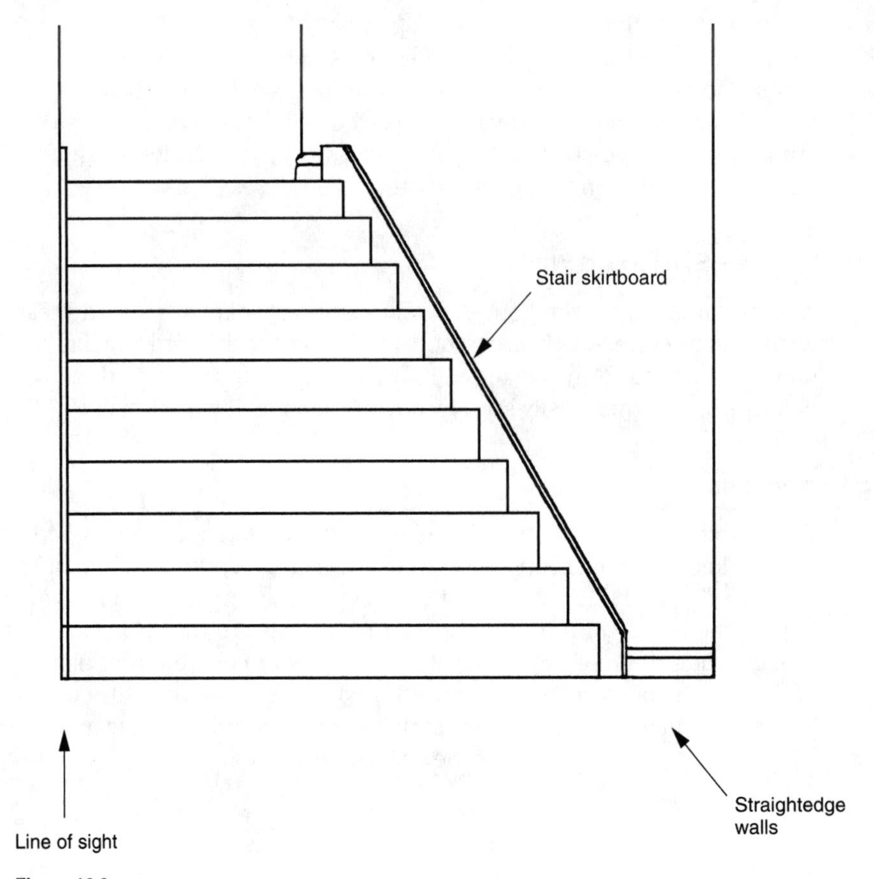

Figure 16.9

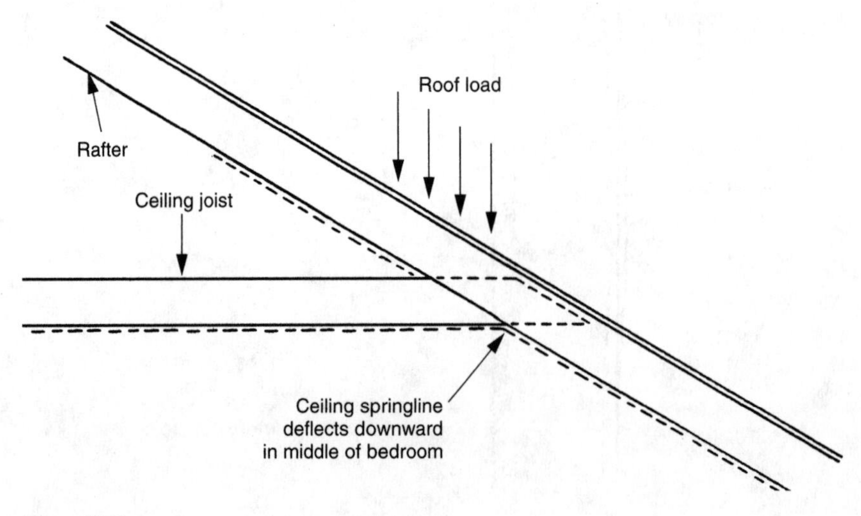

Figure 16.10

Chapter

17

Plumbing

Space for Toilets

Illustrated in plan view in Figure 17.1 is a recessed space in a bathroom for a toilet. The builder should check that at least 30 inches of clear space remains after any shear panel and drywall is installed on the two side walls. A toilet by code must have 15 inches clearance on each side from its centerline. The architect sometimes casually dimensions this space at the 30-inch minimum, but in reality becomes only 29 or 28½ inches after drywall and shear panel is installed.

Water Closet Control Valves

Occasionally the water closet control valve is placed too close to the floor during the rough plumbing. If this happens, the baseboard needs to be cut around the valve escutcheon cover plate (Figure 17.2). The finish carpenter who is installing the baseboard seldom makes a neat cut around these valve cover plates because he or she does not have the correct tool (i.e., power jig saw, band saw) on hand to create the custom fit. The converging angles and the shape of the two materials makes their joining difficult. Furthermore, because of their different relative contours, caulking afterward is also difficult and unsightly (Figure 17.3). By placing the control valve higher above the floor, the baseboard can run below the valve cover plate without interference.

Check Height of Toilet and Lavatory Top Banjo

The bathroom lavatory top, illustrated in Figure 17.4, narrows in depth as it passes over the toilet. This area of the lavatory top is commonly called the *banjo*. The builder should check during the purchasing or specification of toilets in the plumber's contract that the type and brand of toilet selected will fit underneath the banjo.

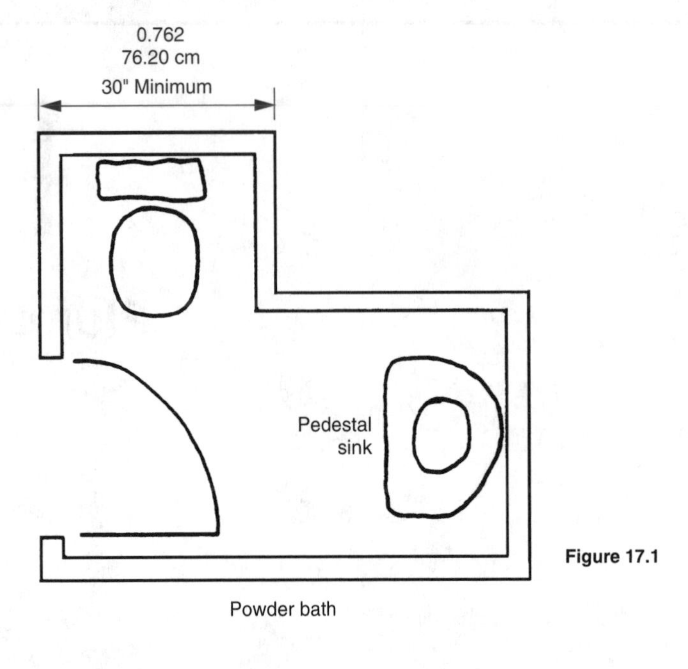

Figure 17.1

Powder bath

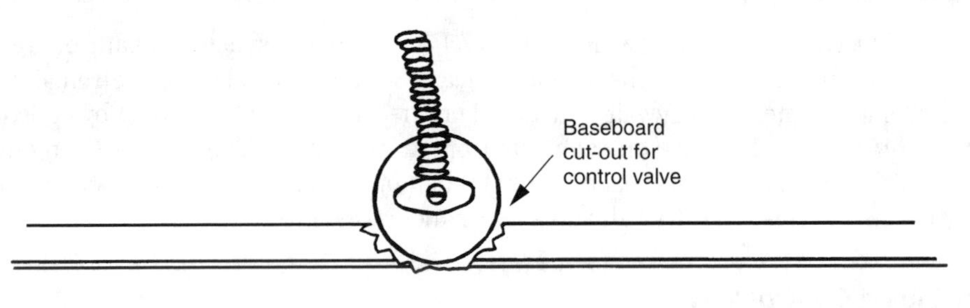

Figure 17.2

On one particular project, the type and brand of toilet was changed midstream in the project and the purchasing agent forgot to look at the dimensions of the new toilet. After being delivered to the job site, unpacked, and readied to be installed by the plumber, it was discovered that the new toilet was 3 inches higher than the previous one and did not fit under the banjos.

Shower Head Height

There are two considerations when deciding on a shower head height during the rough plumbing phase (Figure 17.5). If the shower head nipple is placed at 5 feet, 10 inches above the floor, then the downward bend of the nipple, combined with the length of the shower head, makes the height of the water spraying out of the head at about 5 feet, 4 inches. This height allows most women to take a shower without getting their hair wet. It also places the plumbing nipple below the standard 6-foot-high top edge of a fiberglass tub/shower enclosure.

If it is desired to place the shower head higher, then the nipple has to be placed at about 6 feet, 3 inches or higher. This enables the plumbing nipple, along with its escutcheon cover to miss the top edge of the 6-foot-high tub. This also places the shower head spray at about 5 feet, 9 inches or higher.

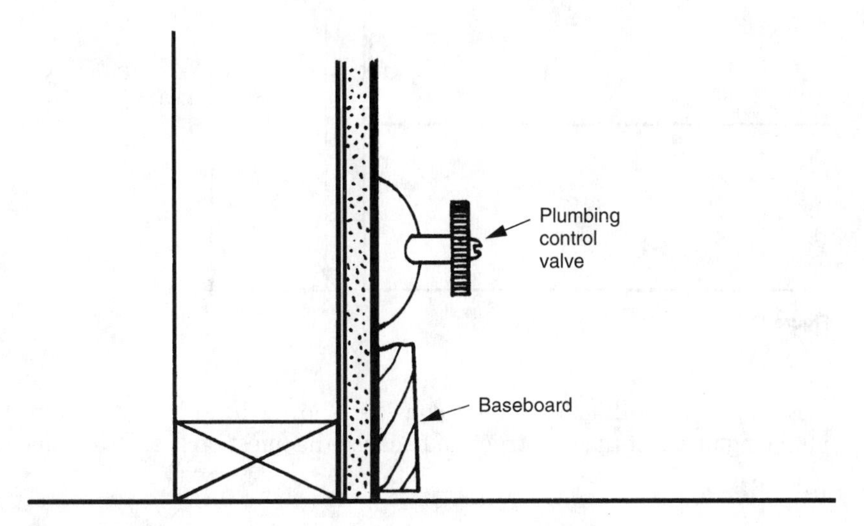

Figure 17.3

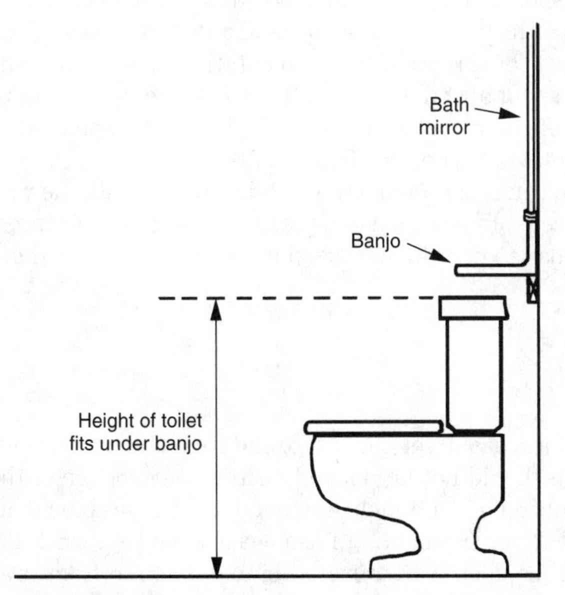

Figure 17.4

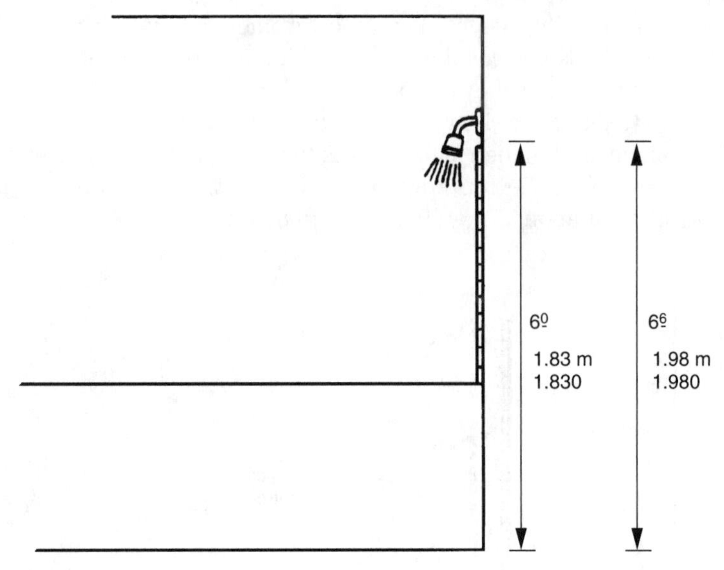

Figure 17.5

Either way, the builder should consider shower head heights before the rough plumbing phase of the construction because the question is sure to come up.

Tub Spouts

For deck mounted tub spouts on Roman tubs, the builder should check whether the spouts have set screws or have to be turned around a threaded valve (Figure 17.6). If the spouts do not have set screws for attachment, the builder and plumber must know the length of the spout ahead of time so that the faucet valves are placed outside its turning radius. If the faucet valves are misplaced, the spout will hit one of them before it can be rotated into place (Figure 17.6).

On one particular project, this problem was not discovered until the valves for the tub spout and faucets were already tiled into place. The plumbing fixtures could not be changed, so the tile had to be torn out, the faucet valves moved farther away, and the tile work redone.

Gas Pipe behind Dryer

The gas pipe stub out for a gas dryer should extend from the wall surface no more than 3 inches. This rule should not be relaxed for the convenience of the plumber who happens to have only 6-inch or 9-inch nipples on hand. A gas pipe that sticks out too far can prevent the dryer from sitting flush against the back wall. In a laundry closet that is 30–36 inches deep with louvered bi-fold doors in front, every inch is needed for the doors to open and close smoothly (Figure 17.7).

Outlets behind Range

The builder should take care to see that all outlets and projections from the wall behind a kitchen range fall within the open well space at the back of the range. Some kitchens are so tightly designed that the clearance between a kitchen cabinet drawer or a dishwasher door opening across and in front of the range oven door handle may be an inch or less. A gas pipe nipple or an electrical outlet cord that interferes with the range being placed flush against the wall may prevent a cabinet drawer or dishwasher door from being fully openable (Figure 17.8).

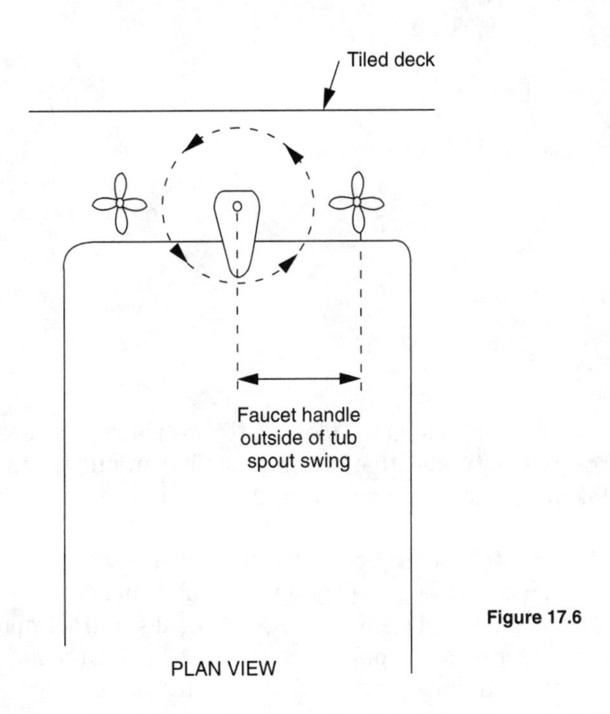

Figure 17.6

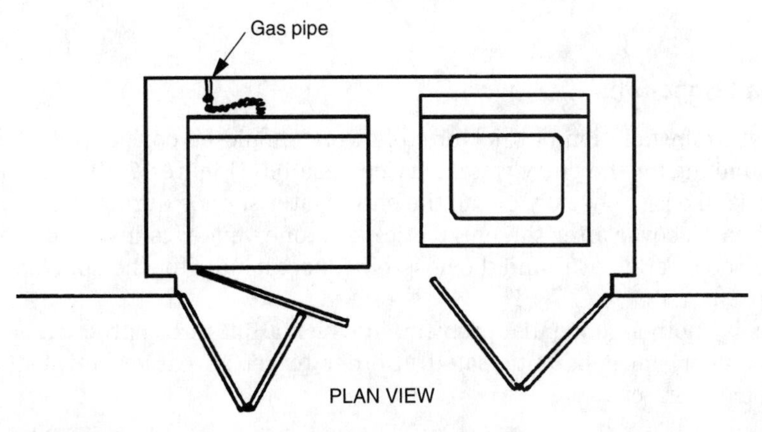

Figure 17.7

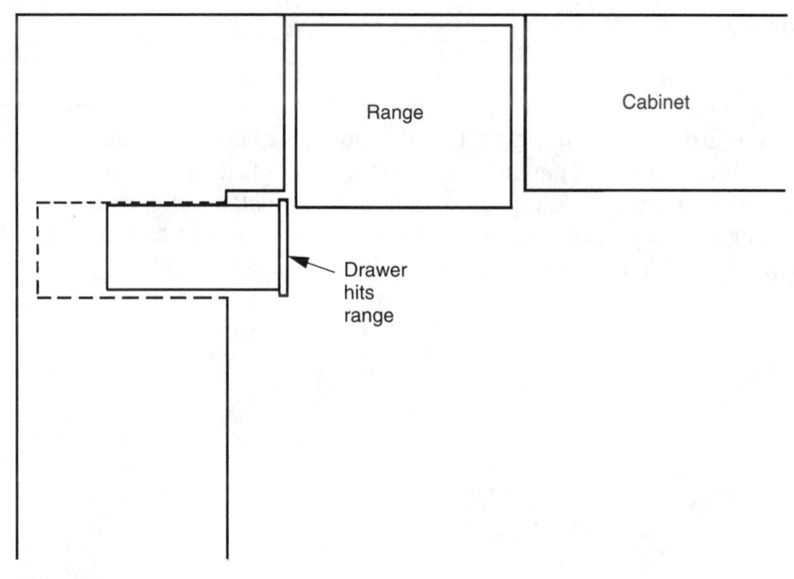

Figure 17.8

Log Lighter Placement

Occasionally the gas log lighter is placed too close to the fireplace with regard to the dimensions of the brick, tile, or marble that is to be installed around the fireplace. If the log lighter is mistakenly placed 12 inches from the firebox edge and two courses of brick are laid edge-to-edge on each side of the fireplace (Figure 17.9), the log lighter will be buried inside the brick. The brick will have to be cut and pieced around the log lighter and an extra long log lighter key will be needed.

A safe dimension for log lighter placement is 20–24 inches in tract homes. This provides enough clearance for most decorative materials, allowing the log lighter to end up outside the face of the fireplace, on the drywall. This also enables the homeowner to hold a lighted match in one hand and turn the log lighter key in the other hand.

Clearance around Hose-Bibs

The future installation of brick or stonework should be considered when placing the plumbing for the entry water service hose-bib (Figure 17.10). An all too common mistake is to casually install the entry water service during the rough plumbing, then discover after the entry brick or stone veneer is installed that there is little or no clearance around the hose-bib faucet handle for the homebuyer to turn on the water.

This is another sequential problem, in which a future construction activity (the brick veneer) must be anticipated in order to get an earlier activity (the rough plumbing) correct.

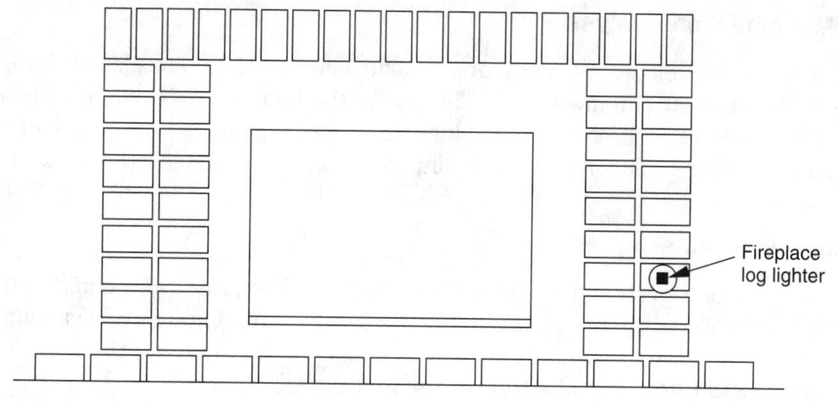

Figure 17.9

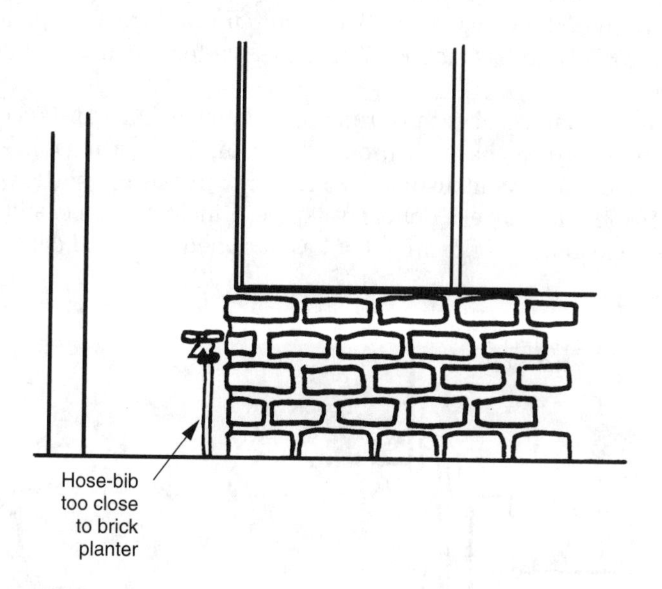

Figure 17.10

Placement of Solar Panels

When possible, solar panels should not be placed on roofs above bedrooms. Solar panels usually have a protective recirculating pump, which circulates the water through the lines when the outside temperature reaches 42 degrees or below. This prevents the water from freezing, expanding in the lines, and breaking the copper piping. The noise generated from the water flowing through the solar unit on the roof, however, can be annoying when directly above a bedroom—especially because temperatures tend to drop at night, when people want to sleep.

Washing Machine Water Valves

The hot and cold water valves for a washing machine should be placed above the height of the machine (Figure 17.11). This allows the hook-up to be made without reaching over the back of the washing machine, or worse yet, having to move the machine away from the back wall far enough to climb behind the machine for the hook-up.

Water Heater Platforms

Several factors should be considered in the placement of and construction around water heaters in garages. Often there is barely or not enough vertical dimension between the garage floor and a dropped ceiling or F.A.U. duct for the water heater platform, water heater, and water heater vent pipe.

Water heater platforms are usually framed 18 inches above the floor, but do not have to be that high. The UMC requires 18 inches from the floor to the water heater burner, which is about 4–6 inches above the legs of the water heater. This 18 inches is to the lowest portion of the garage floor. With a step upward from the garage slab to the water heater/F.A.U. area (Figure 17.12), the platform could be built only 14–16 inches high.

If the vent pipe to the water heater must run horizontally for some distance before turning upward into a vertical chase or through the roof, the vent must have slope and minimum clearances from combustible materials. For instance, a single wall vent pipe must have 6 inches clearance; a double wall pipe, 1 inch clearance; and a triple wall vent pipe, zero clearance. If the architect has designed a vertical chase for the

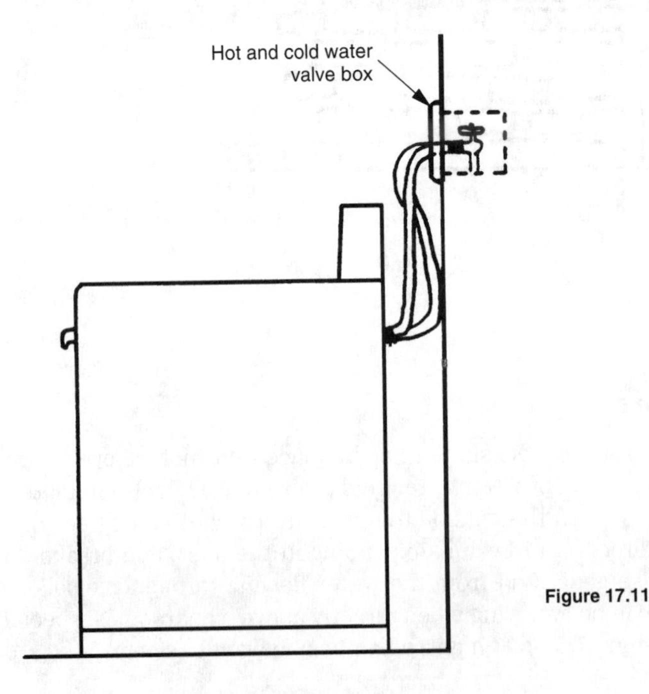

Figure 17.11

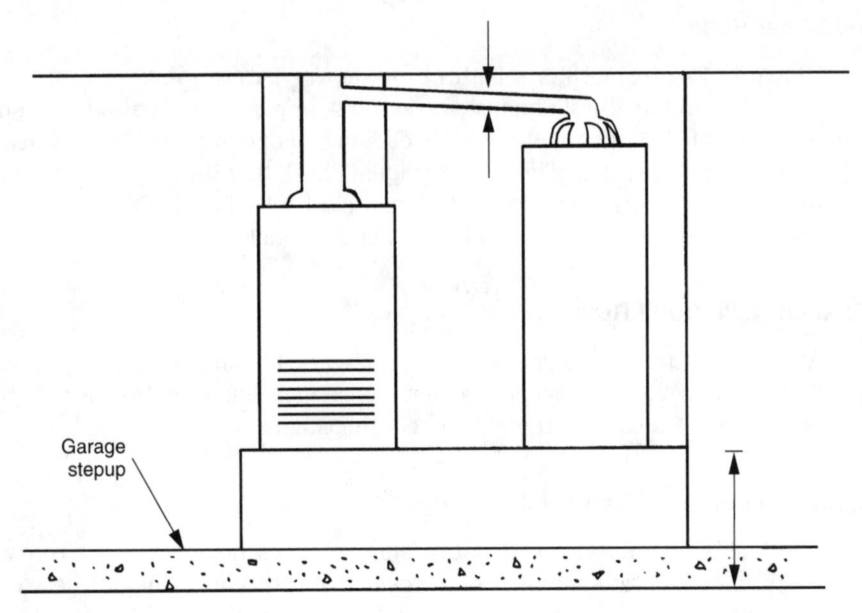

Figure 17.12

water heater vent several feet away from the water heater, sometimes there is not enough space to give the horizontal vent the required ¼-inch per foot slope and the clearance at the same time. Double wall or triple wall vent pipe must then be used instead of single wall.

Bathtub Deliveries

The contract for bathtubs and fiberglass showers should specify whether the delivery is curbside, garage, or spread into the units at the location of the bathrooms. This issue must be worked out with the plumber, and possibly the framer, and spelled out in their contracts as well.

If the tub delivery is curbside or garage and there is no provision in the plumbers contract to spread the tubs from the ground floor up to a second or third floor, the builder in all fairness should give the plumber an extra to spread the tubs or find someone else to do it. Even when the delivery contract is to spread the tubs to a second or third floor, they may not fit up a stairway, and thus may require the use of the framer's forklift to get the tubs up to a second or third floor window or balcony. The builder should not pay extra to have the tubs spread to upper floors, then allow the delivery company to have the use of the framer's forklift and driver as a favor. There should be an hourly charge to the tub company for the use of the forklift. All of this should be considered ahead of time and addressed in each contract.

Tubs and Shear Panel

Drawn in Figure 17.13 is a bathtub or shower bathtub unit surrounded by three walls. The architectural dimensions on the floor plans do not always account for the thickness of shear-panel plywood placed on any one or two of these walls called for on the structural plans. The builder should check that the necessary net opening for the tub, plus the ½ inch extra for getting the tub into the opening, exists between the two side walls even with the addition of shear panel.

Drain Tubs so Nails Don't Rust

After the bathtubs are filled with water for the plumbing inspection, they should be drained soon thereafter. This prevents loose nails amongst any loose debris on the bottom of the tubs from rusting and causing stains.

No Plumbing in Center of Bathroom Ceilings

The builder should check that the plumber does not mistakenly run water, gas, vents, or condensate lines across the center of bathroom ceilings. These could interfere with the future installation of flush lights and bath fans. Because plumbing goes in before rough electrical wiring and sheet metal vents, this is one problem that the plumber must keep in mind and the builder must double-check.

Paint Overspray on Bathtubs

The builder should check for the conflict between the painting of exterior woodwork and paint overspray blowing in through bathroom windows and getting onto bathtubs and showers. Because fiberglass tub and shower units are probably the earliest finish items delivered and installed during the construction, the scheduling of the spray painting of facia board and roof overhang eaves should consider the proximity of bathtubs with regard to paint overspray (Figure 17.14).

Dryer Vents and Concrete Steps

Illustrated in Figure 17.15 is a typical conflict between a dryer vent placed during the rough framing phase of the construction and concrete steps that are formed and poured later. The builder should analyze where the concrete steps leading to entry doors will lay out at the exterior of an apartment or condominium building, for example, and have that information at hand during the rough framing. This will allow all the vents to be placed far enough away from the future concrete steps so that a dryer vent or some other vent does not get partially covered by concrete.

Pedestal Sinks and Flooring

The pedestal sink, illustrated in Figure 17.16, is sometimes used in a small half bath because it takes up little space. These type sinks require more work because they must be attached to the wall and the hot and cold flexible hoses to the shut-off valves

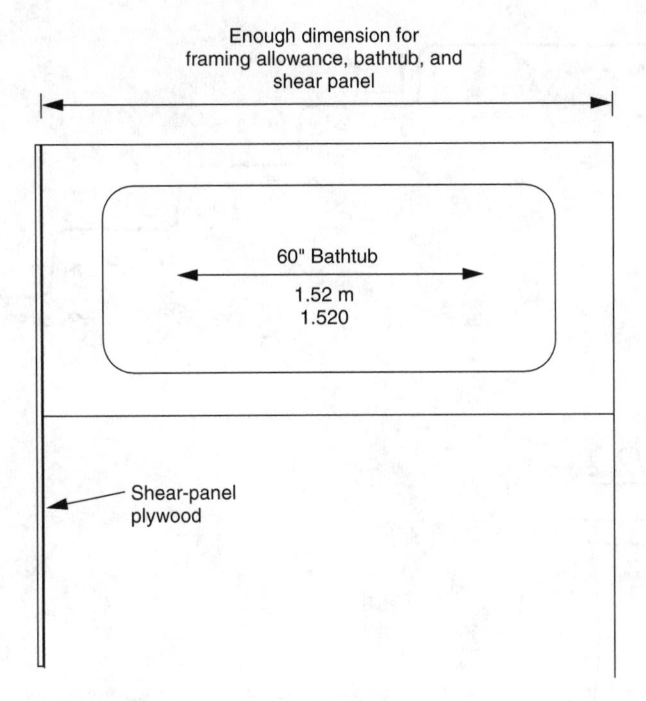

Figure 17.13

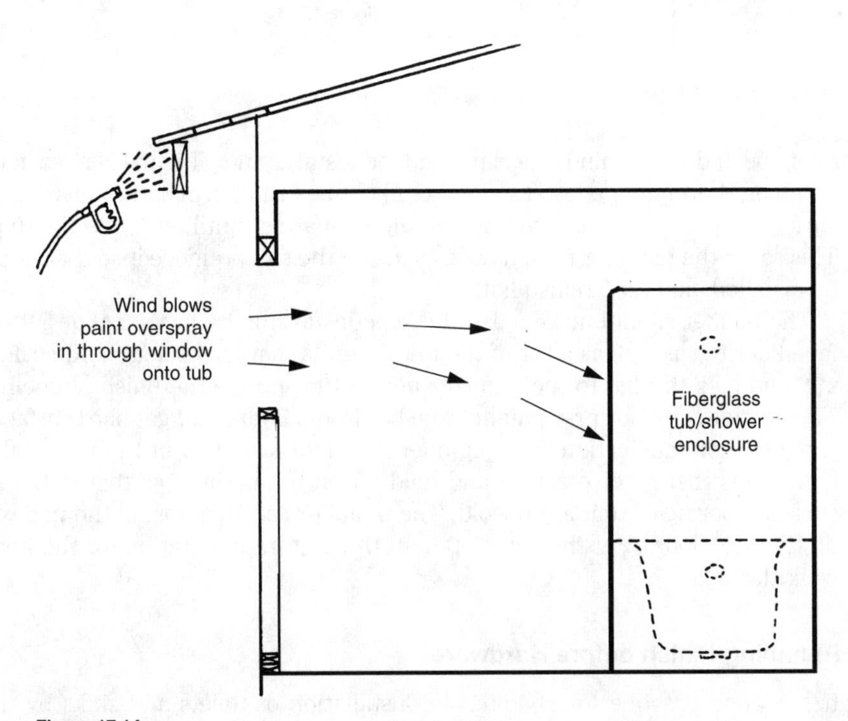

Figure 17.14

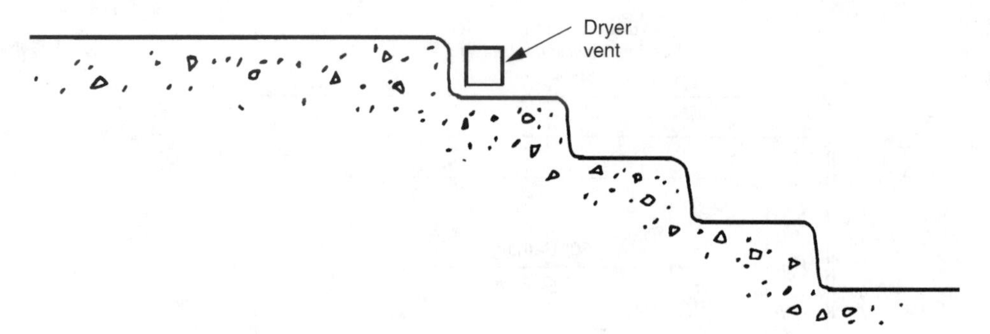

Figure 17.15

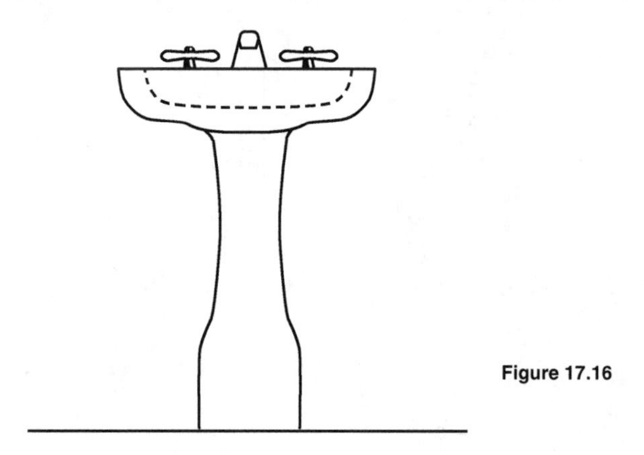

Figure 17.16

must be hidden behind the sink and pedestal stand. To make it easier on the plumber, therefore, pedestal sinks should not be installed in bathrooms that get hard surface flooring such as vinyl, wood, tile or marble, until after the flooring is laid. This saves the plumber from having to install the sink, remove it so that flooring can be installed, and then reinstall it.

The builder should make a list of the pedestal sink bathrooms that get hard surface flooring, as well as a list of the unsold units that may select hard surface flooring, and give this list to the plumber before the start of the finish plumbing phase. The builder can also spray paint on the bathroom floors that get hard surface to help avoid miscommunication. The plumber can then install all of the pedestal sinks in bathrooms that receive carpet and hold off on the bathrooms that either get hard surface flooring or that are unsold. The plumber can then install the pedestal sinks on the new flooring at the completion of the unit in preparation for the homebuyer walk-through.

Schedule Plumbing Finish before Hardware

It is a good practice to schedule the installation of toilets and sinks by the finish plumber prior to the installation of towel bars, towel rings, and toilet paper holders in bathrooms by the hardware installer. The reason is that the plumber may accidentally or on purpose lean up against or grab hold of these towel bars and toilet paper holders while setting the toilets and sinks. Some small bathrooms are so tight that there may not be enough room to install toilets and sinks without hitting these items if they are installed ahead of the finish plumbing. If towel bars, towel rings, and toilet paper holders are not in place during the finish plumbing, then the temptation to use these items as grab bars to lift oneself up from the floor is removed.

Roto-Rooter Drain Lines

It is a good practice to include in the plumbing contract that the plumber will snake out the main house sewer line at the end of the construction. This can be done from a first floor toilet to the main sewer line in the street or courtyard or from one of the clean-outs inside the garage to the main sewer line.

The best time to discover that there is a blockage in the sewer line is just after the finish plumbing fixtures are installed, prior to the laying of carpet and other finish flooring. The worst time to discover a sewer blockage is after the new homebuyer moves in, flushes a second floor toilet, and floods the entire first floor.

Spin Garbage Disposal with Wrench

As part of the final preparation before the homebuyer walk-through, the builder should spin the kitchen garbage disposal with the hex head wrench provided with the appliance to see whether any construction debris such as nails, screws, or tile grout is in the garbage disposal. It is embarrassing when the homebuyer turns on the garbage disposal during the walk-through or after they move in and hears the loud racket created when something is in the disposal.

Run Water through Faucets before Final Cleanup

Water should be run through all of the faucets in each house before the final cleanup stage. All sorts of dirt, sand, and sediment are contained in the new plumbing pipes prior to their initial use. The final cleanup crew does not always run water through all of the faucets in order to clean the sinks, tubs and showers. They only run the water they need for cleaning, which may be the cold or hot side faucet, but not both.

When an unused faucet is turned on after the final cleanup, out comes alternating streams of clean and dirty water until the line has been flushed. Because of initial pockets of air in the lines, the dirty water sometimes comes out of the faucet with great force, splashing the dirt and sand all over the clean sink and countertop.

Tub and Shower Valves Turned Off

One of the scariest things in construction for the builder or superintendent is to discover a house or condominium apartment that is flooded with water. This scenario can be a combination of three occurrences. For example, on one particular condominium project, the bathtubs and showers were located on the third story. The water for each apartment was turned off at the main valves at the street level going into each apartment. The interiors were drywalled and painted, and the exteriors were stuccoed. The stucco plasterer hooked up a hose at one of the apartment's hose-bibs to supply water for a mixer for plastering the adjacent building. When he turned on the main valve, water not only went to the hose-bib, but also to the entire apartment. The problem was that a particular shower valve was in the on position and the shower drain was clogged with debris. The shower ran for several hours without anyone noticing, with water flooding the second floor, running down two flights of stairs and out through the garage. The problem was only discovered after it was noticed that the exterior stucco walls were wet at the plate lines and that water was running out of the garage.

The solution to this potential problem is to prevent at least one of the three events from occurring. The plumber should check that all of the valves are off before turning off the main water valve to each house or apartment. The builder should also check that the tub and shower drains are clear. And a water source should be made available at a nearby fire hydrant or at the building where it is needed.

Keep Appliances the Same

For simplicity during construction, appliances such as garbage disposals, dishwashers, ranges, and microwaves should all be the same for tract houses. For example, offering several models of garbage disposals as options only makes the management of these items more difficult for the builder and the tradesmen. Offering a ⅓-HP, a ⅓-HP quiet, and a ½-HP garbage disposal to homebuyers makes it more difficult to order and install the correct units in the proper house.

The same can be said of offering two levels of dishwashers, ranges, and microwaves. Unless the price differences are large, the builder should pick the best appliance and set the selling price of the houses accordingly. To casually offer a variety of appliance choices to homebuyers in a production tract house setting is to be ignorant or insensitive to the materials management problems of the construction.

Another aspect to consider is the displeasure of homebuyers ordering one type of appliance and getting another. The supposed marketing advantage of offering a wide selection of appliances can quickly be offset by homebuyer dissatisfaction over mistakes in ordering and installing.

Purchasing and Spreading Materials

In an attempt to save on subcontractor's markups, some housing developers prefer to purchase several of the finish materials and appliances themselves. Ranges, dishwashers, light fixtures, interior and exterior hardware, medicine cabinets, range hoods, kitchen sinks, and even wardrobe doors may be purchased by the developer taking advantage of volume discounts. They are then spread to each house by the field staff before installation by tradesmen.

Such a policy works up to a point, but can be counter-productive from a construction standpoint, if not covered adequately in the field with enough people. For example, suppose a typical housing project has one laborer, one assistant superintendent, and one superintendent. If all of the above items must be taken from storage bins or garages and spread to individual houses in blocks of 8 to 10 houses at a time, the supervisory staff becomes, in essence, a moving and delivery team for large blocks of time each day. The time and energy spent spreading and managing materials replaces the time that should be spent supervising and checking the construction. The problem with this type of policy is that the same companies that try to save money purchasing their own materials are also economical in staffing the job site.

Directly purchasing materials can save money, but some consideration should also be given to the management and spreading of these materials in the field. To simply assume that the field people will take care of the problem is yet another of those bottom-line money decisions made in the main office by people who do not understand construction.

Pigtail Appliances

The builder should check that the electrician has installed the electrical pigtail cords on the dishwashers, ranges, and garbage disposals prior to them being installed by the plumber. This is much easier to do while the appliances are sitting in the middle of the kitchen floor rather than in place inside the kitchen cabinets.

The builder should also look into ordering the appliances from the manufacturer with their pigtail extension cords already attached. This eliminates the possibility that the plumber would mistakenly install an appliance without its pigtail.

Check Appliance Model Numbers

When the appliances are delivered to the job site, the builder should not only count the number and types of appliances delivered against the shipping invoice, but also check that the correct model numbers were sent.

This may seem like another of those useless activities, like double-checking that the number on the bags of stucco color coat mix is correct or checking that the accurate color and type of roof tiles have been loaded on the proper roofs. It is easy to assume that these types of mistakes will never happen, and are therefore not worth checking, yet when they do occur, they are often the most difficult to correct.

On one particular project, the wrong type of microwave was delivered to the job site, but both types looked the same. The only difference was that the model numbers varied by only one letter. The wrong type was supposed to be connected to a sheet metal duct vented to the outside, and the correct type was supposed to vent internally through a charcoal filter. The electrician assumed that the microwaves were correct and installed about 75 of them. The discrepancy was discovered only after a few homebuyers complained that air was blowing out the front and sides of the microwave.

In this particular case, the fix was simply to remove each microwave, remove a few screws, flip over some interior parts, and slide in the charcoal filter. This took about 15 to 20 minutes per unit. However, if the purchasing department had ordered the correct model number and the field people had also double-checked the appliance model numbers, all of this could have been avoided. The correct model number would have had this conversion done in the factory, with the microwave ready to install for a self-venting application.

Deliver Garbage Disposals Separately

When appliances are purchased as a package, including garbage disposals, the builder should remember to split up the delivery so that the garbage disposals arrive ahead of the other appliances. This allows the plumber to complete the kitchen sink plumbing along with the other finish plumbing work, irrespective of any other activities affecting the appliance delivery schedule, such as box-out and sweep or kitchen flooring.

Store Microwaves Until Installed

The best way to have appliances delivered to the job site is to have the delivery people spread the appliances to each individual unit. This must be agreed upon, however, in appliance purchase agreement.

Unlike the other appliances, microwave ovens should be separated from the delivery and placed in a locked sales model garage or storage bin until the time they can be installed. Microwaves are smaller than kitchen ranges and dishwashers and can be easily stolen by one person and hidden in the back of a truck or inside a car. Ranges and dishwashers need two people to lift and are too big to conceal.

Appliances should be scheduled to be delivered to houses that remain locked except for workmen. The time span between the delivery of appliances and their installation should be as short as possible. By placing the microwaves in a separate location straight off the delivery truck, the builder helps to keep the tradesmen honest by removing the temptation.

Installing Microwaves

Shown in Figure 17.17 in side view is a microwave oven attached to the underside of a kitchen upper cabinet. There are two important rules when installing a microwave. First, the mounting plate that attaches to both the wall surface and the back of the microwave must be installed. Second, metal washers should be used at the screw heads when attaching the top of the microwave to the upper cabinet shelf. For example, if a 10mm size bolt is run through a 9mm size hole through the cabinet shelf without a metal washer, eventually the screw head will work loose through the wood hole due to the weight of the microwave unit and the microwave will fall down.

On one particular project, the mounting plates were not installed, and the metal washers were not used. Almost like clockwork, according to the customer service repairman, the microwaves would fall down from the upper cabinets at about 10 months after the homebuyers moved in. It took about that long for the screws to work loose from the upper cabinets.

Collect Appliance Booklets

As soon as all the appliances are installed, the builder should start collecting the appliance warranty and instruction booklets from the inside of each appliance. The builder should also collect cookbooks, microwave plates, and any other parts such as knobs, broiler pans, and cooktop grates that the builder thinks may disappear if left

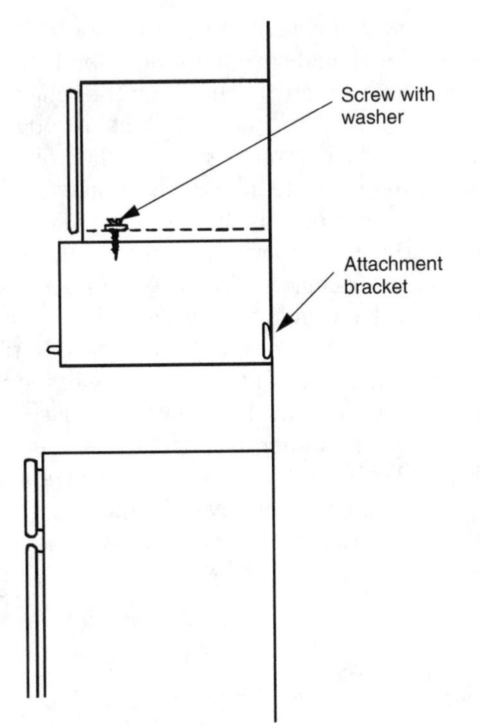

Figure 17.17

in the appliances. These items should be placed in separate boxes or large manila folders and labeled per lot number. They can then be placed in the units when the homebuyers come for the walk-through.

The builder should collect warranty and instruction booklets for the range, dishwasher, microwave oven, A/C thermostat, A/C condenser, A/C furnace, trash compactor, smoke alarms, garbage disposal, garage door opener, solar heating, Jacuzzi tub, central vacuum system, intercom and radio, security system, and any other appliance that was put into the unit.

Get Tradesmen to Collect Appliance Literature

Some tradesmen have the habit of attaching the appliance instruction pamphlets or booklets to the appliances. For example, the electrician will wedge the instruction pamphlet for a smoke alarm between the smoke alarm and the ceiling drywall—8, 9, or 10 feet above the floor. Similarly, the HVAC man will leave the instructions to the exterior A/C condenser units stuffed between the air baffles on the top or sides of the units. The service warranty card to the thermostat is often wedged between the thermostat subbase and the drywall by the A/C man. And in keeping with the routine, the prefabricated fireplace installer will leave the fireplace instruction booklet inside the firebox of the fireplace.

Although leaving the pamphlets or booklets near the appliances seems to be a good idea, they often get lost, rained on, blown away by the wind, or mistakenly thrown away as trash long before the end of construction or before the builder can collect them. For this reason, the builder should have the superintendent collect the appliance literature as a formal construction activity shown on the schedule so that this literature can be given the homebuyer at the time of the walk-through. Another way of handling this issue is to have the subcontractors collect their own literature as part of the installation and then hand over this literature to the superintendent when their work is complete. There is no reason why the A/C man cannot collect the instruction and warranty booklets to the condensers and furnaces at the time each unit is installed. The same can be said of fireplaces, fireplace glass doors, and thermostats. Water heaters usually come with a plastic pouches attached to the sides of the water heaters for the storage of the instructions and warranties. In the case of the smoke detector, if the literature is one page only, the easiest method may be to make photocopies of the one sheet and give it to the homebuyers in place of the originals so that the electrician does not have to bother to collect each sheet.

The builder should look at this problem prior to the start of construction, choose which appliance pamphlets and booklets are more easily collected by the tradesmen themselves during installation, and then write this requirement into the contracts.

Install Microwaves before Ranges

When it comes time to install the appliances, the builder should schedule the installation of the microwave ovens ahead of the ranges for two reasons. First, it is easier for the installers to attach the microwaves to the underside of an upper cabinet if there is not a range already installed and in the way. The installers can get directly below the upper cabinet for measuring, drilling, and placing the microwave without having to lean over the range. Second, the drilling of the holes for the microwave, through the upper cabinet shelf, leaves sawdust all over the top of a range that is already in place below the microwave. This creates extra unnecessary work for the final cleanup people.

Fire Sprinkler Systems

There are several questions and problems to consider when apartment or condominium buildings have monitored fire sprinkler systems. For example, these systems have flow meters that notify the fire department when a sprinkler head goes off. These flow meters are usually located inside the garages of each apartment or condominium unit at the water riser to the unit. If a fire should start inside the unit, activating a sprinkler, water goes through the flow meter, setting off the building fire alarm, and setting off the alarm at the fire department. The problem to avoid is having a flow meter that does not have a delay mechanism, a back flow valve, or a 4–8-gallon-a-minute allowance, before setting off the fire alarm.

On one particular project, the flow meters had a zero tolerance, with no back flow valve or delay mechanism. Once a week a water truck would wash down the streets in the occupied areas of the project, to remove dust and dirt caused by the construction. Each time the water truck would hook up to the fire hydrant in the street to get water, a reverse pressure was created in the main water line to the project, which pulled water backwards through the fire sprinkler flow meters in each building. This would set off the fire alarms and bring out the fire department responding to what they thought was a fire. After several such occurrences, the builder was finally required to go back and change all of the flow meters to a model that had an 8-gallon-a-minute allowance before setting off the fire alarm.

Another problem to avoid is the requirements by the water utility company to prevent fire sprinkler water from contaminating the domestic drinking water. For example, the Irvine Ranch Water District will not allow you to branch off from the domestic water line at each individual condominium unit for fear that the water in the sprinkler pipes may be siphoned into the drinking water. They require the branch separation to occur at the main building meter at the street curb, with separate water lines going to each unit for fire sprinklers and domestic water. They also require a backflow valve at the water meter to prevent the mixing of the two water sources.

You can see, a mistake in either of these two areas can result in costly retrofit work if the fire department or the water utility company will not accept substandard work that is already in place.

Fire Sprinkler Monitoring from the Beginning

When the fire sprinkler system for apartment or condominium buildings is tied to a central monitoring alarm system, the fire alarm system should be adequate and finalized at the time of the first building in the project. What sometimes happens in apartment or condominium projects with several separate buildings is that the fire sprinkler plans are not approved in time for the models building construction or that the fire marshal/inspector has not been consulted ahead of time as to fire alarm monitoring requirements in their jurisdiction. The result is a model building that does not meet the fire alarm requirements at the time of the final building inspection, but which is okayed for temporary occupancy as sales models with the requirement that the building be brought up to the level of the production buildings at the time it is converted to regular units for sale at the close of the project.

This gives the builder two choices at the end of the project. The fire alarm system for the lone building can either be tied to the other buildings, requiring the possible tearing up of streets and landscaping for underground trenching to connect the buildings. Or the model building can be treated as a separate monitoring entity, thereby requiring no connection with the other buildings in the project. This scenario, however, would add an extra monthly monitoring fee to the homeowners association.

The fire sprinkler system issue is not one that can be postponed and resolved after the start of construction. It also cannot easily be integrated into the construction at some point into the project. If a centrally monitored fire alarm system is to be required in a multiunit project, the time to find out is prior to the start of the construction so the underground wiring can be planned along with the other underground utilities off site work. If this issue is not given some forethought, the builder must choose between paying for expensive repair and replace street trenching work or offending the homeowners association with an added expense to their budget.

Fire Sprinklers and Ceiling Fans

When apartments and condominiums have fire sprinklers, the builder should check that the layout and installation of the heads during construction does not result in a head being placed in the center of a bedroom or family room. These are rooms where homebuyers sometimes install ceiling fans after they move in. It's best to avoid the potential conflict between a future ceiling fan that must be centered in a room and the fire sprinkler head that does not have to be centered in the room. If the builder allows the casual placement of heads in the center of rooms merely to simplify the fire sprinkler layout and coverage, the result could be a potentially unhappy homeowner.

Paddle Size for Flow Switches

An important factor in the effectiveness of fire sprinkler alarm system flow meters is the matching of the size of the plastic paddle in the flow meter to the water pressure on the project. Monitored fire alarm systems in multiunit projects have flow meters attached to the fire sprinkler water piping so that if one of the fire sprinkler heads goes off in a fire, the change in water pressure in the piping is noticed in the flow meter switch, and the fire alarm system is activated.

The plumber should avoid selecting the wrong size and shape flow meter paddle from the wide choice of paddles provided with each flow meter and not checking that the amount of water pressure in the piping has the force to actually turn the paddle and activate the alarm. Simply installing the flow meters with paddles in place is not enough. For example, a fire sprinkler water pipe of ¾-inch diameter, with a water pressure of 75 psi, should have a different paddle size and shape than a water pipe with a 1-inch diameter and a water pressure of only 55 psi. If the wrong size and shape paddle is installed, the fire sprinkler head can go off, water can flow through the line, but the water may simply flow around the paddle, without pushing it to turn on the alarm switch.

Figure 17.18 shows a typical paddle card that comes with each flow switch so that the installer can choose the correct paddle size and shape for a particular installation.

Remove Handle to Fire Sprinkler Hose-Bib

In apartment or condominium projects with a monitored fire sprinklers system, it is a good practice to remove the hose-bib handle at the manifold flow meter. This prevents an uninformed homebuyer or an unsuspecting guest from mistakenly using the fire sprinkler hose-bib to wash a car or water plants, which in turn sets off the flow meter switch and triggers the fire alarm. By removing the handle, the user has to look for the "real" hose-bib for outside water use.

If the fire sprinkler system needs to be emptied for service repair by the subcontractor, a handle can simply be borrowed from the regular hose-bib for the duration of the repair.

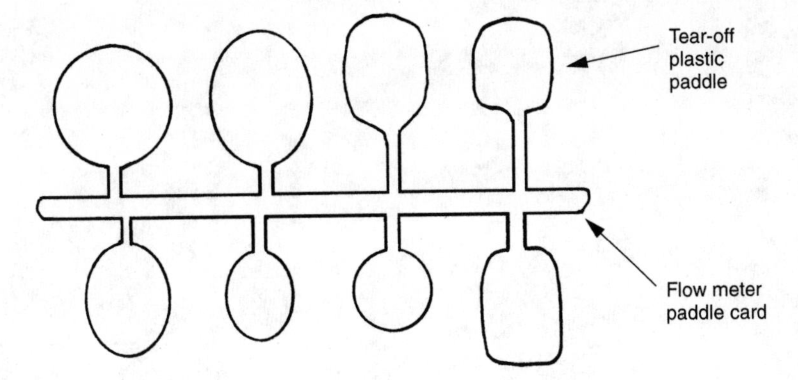

Figure 17.18

Chapter

18

HVAC

Return Air Registers and Baseboard

HVAC return air registers located along the floor should have their rough-framed openings and metal ductwork placed high enough above the floor so that the register cover will clear the baseboard (Figure 18.1). Even when a neat, straight cut is made in the baseboard to match a register, the difference in the thicknesses between the two will still leave the cut edge of the thicker baseboard exposed (Figure 18.2). For this reason, caulking the joint between the register and the baseboard does little to improve or solve this mistake.

Clearance for A/C Return Air Register

Shown in Figures 18.3 and 18.4 are the side view and front view of a typical problem that arises when an HVAC forced-air-unit is placed on top of a platform inside its own

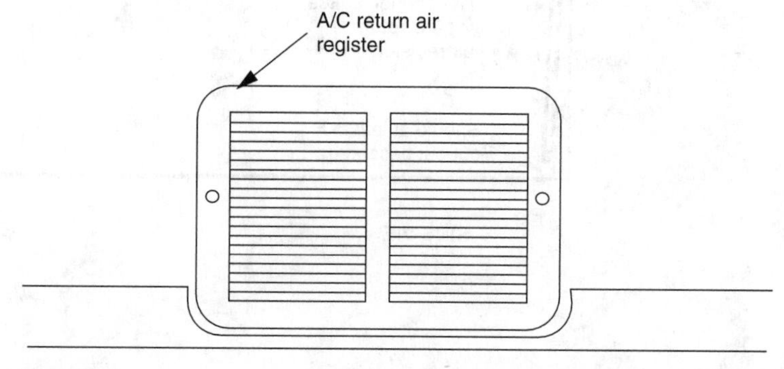

Figure 18.1

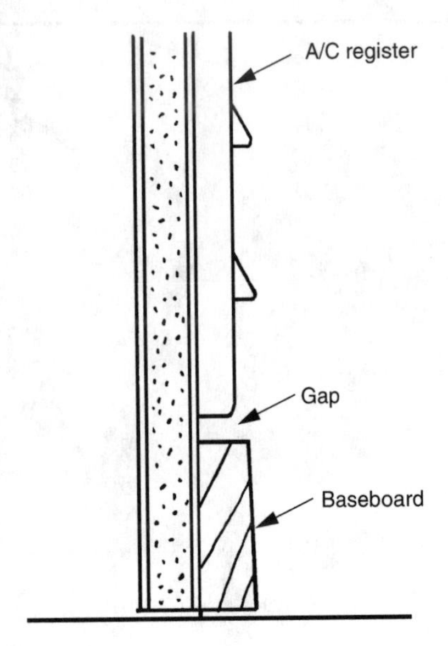

Figure 18.2

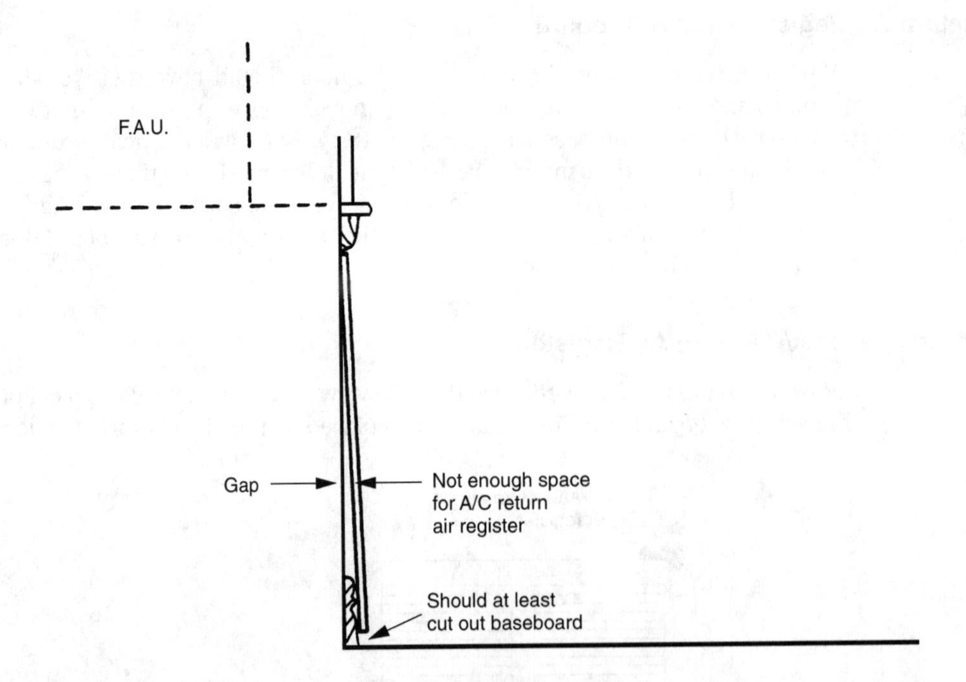

Figure 18.3

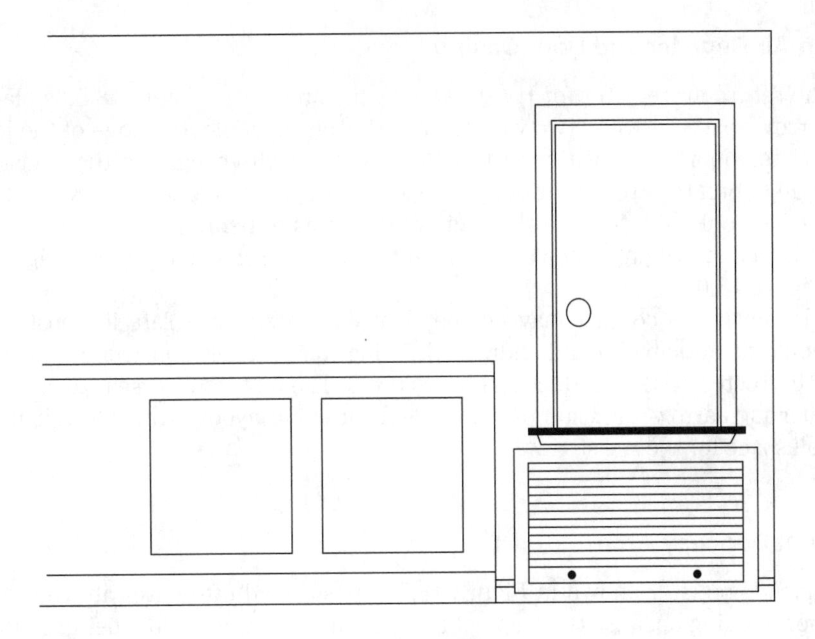

Figure 18.4

closet—the return air register (below the door) is too large for the space between the stool and apron and the baseboard.

This problem results from having to provide a minimum square-inch opening, such as 225-square inches, for sucking in the return air from the interior of the house into the F.A.U. If, as in this example, the F.A.U. closet is next to a door opening on one side, a linen cabinet on the other, and with vertical dimensional restrictions involving the floor to ceiling height in relation to the platform and the height of the F.A.U., there may not be enough space for everything to fit.

In this particular example, in a million dollar spec house, the builder did not consider all of these factors beforehand, and simply had the HVAC finishman slap the return air register over the floor baseboard. This resulted in a gap around the sides of the register. At the very least, the builder should have cut out the baseboard at the register so that the register could fit tightly to the wall.

To avoid this problem, the builder should draw a story pole on a piece of paper during the preconstruction plans review period to determine whether there is enough vertical dimension in the closet for the ductwork, F.A.U., platform framing, stool and apron trim, return air register, and baseboard. The time to discover that they do not fit and changes have to be made is prior to the actual construction.

HVAC Return Air Register and Door Casing

The return air register for HVAC, shown in plan view in Figure 18.5, is placed in a narrow space between two wall corners. In this example, the edge of the louvered register door was positioned too close to the wall corner. As the register door swung open to replace the return air filter, it hit the adjacent bedroom door casing. The only way to open the register door was to bend it awkwardly around the door casing, which placed a permanent kink in the thin sheet metal register door (Figure 18.6).

Because this conflict may not be clearly apparent until late in construction, it should be anticipated and addressed during sales models construction. It can then be corrected on the production houses plans. The fix for this case is to either go to a taller and narrower register or to change the framing layout to provide a slightly wider wall space for the register.

A/C Register at Doorway Arch

The A/C register shown in Figure 18.7 is positioned off-center above the arched door opening because the slope of the ceiling prevented it from being centered. In this example, the register is off-center because the builder forgot that even though the rough HVAC duct fit within the space so as to be centered over the archway, the larger sized register cover did not fit due to the slope of the ceiling. The rough HVAC duct should have been positioned between the doorway arch and the sloping ceiling so that there was enough clearance around the duct for the register to fit.

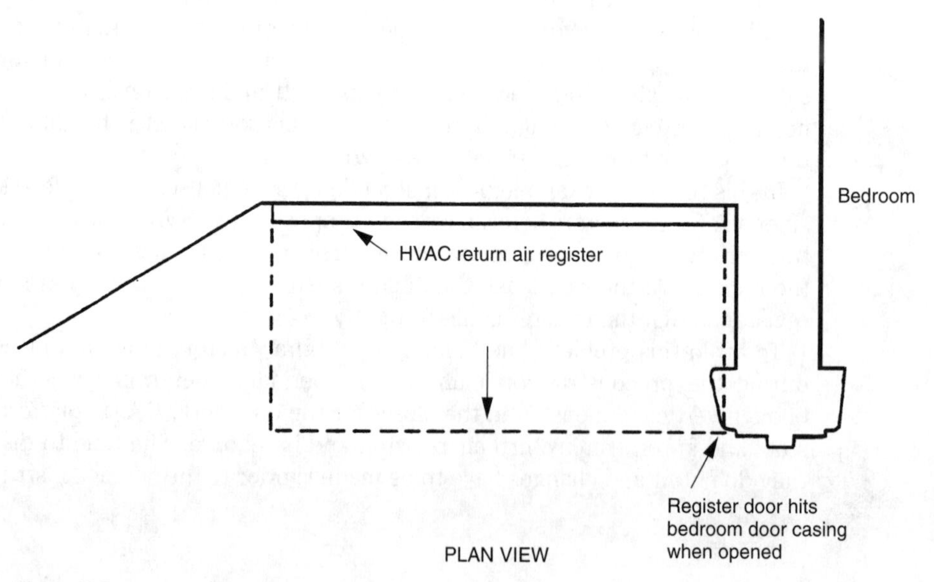

Figure 18.5

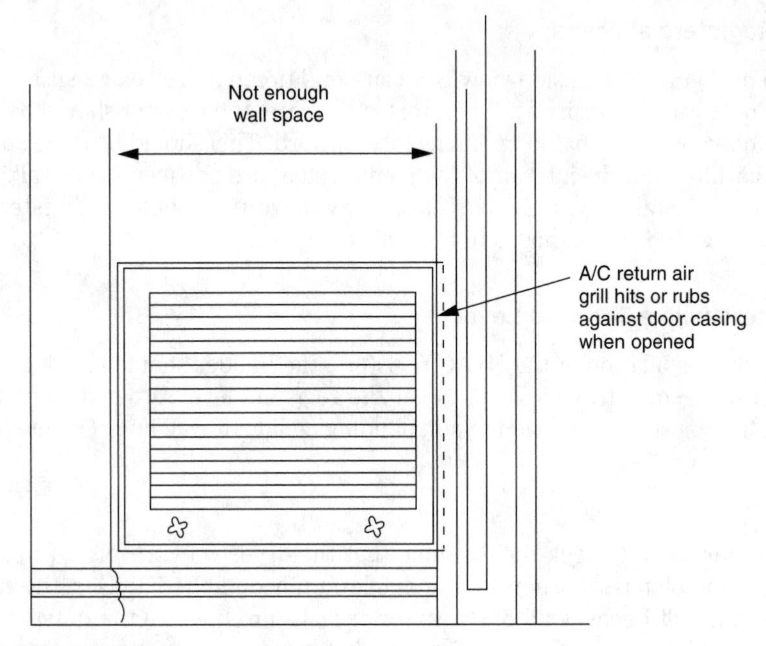

Figure 18.6

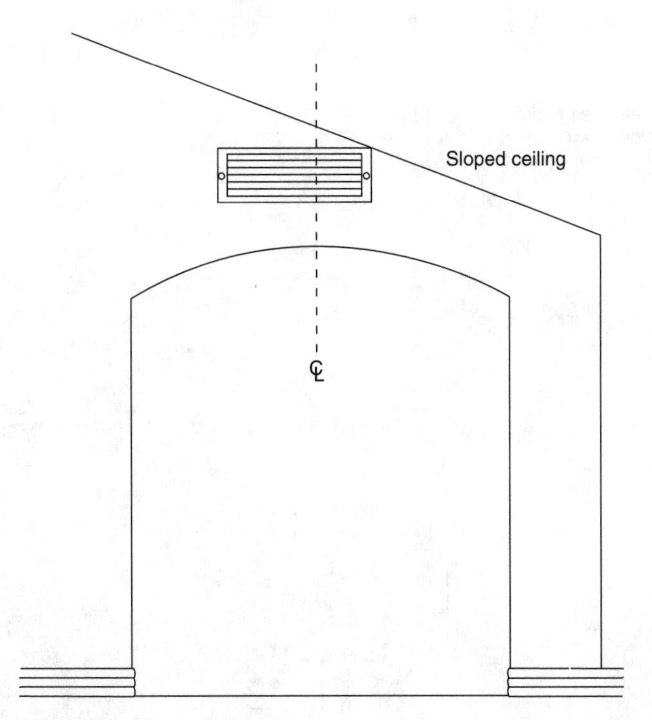

Figure 18.7

A/C Return Air Registers at Stairs

Shown in Figure 18.8 in side view is an air-conditioning return air register that was placed on a wall in a stairway. When this is the case, the register should be installed upside down with the baffle openings face upward. This should be done so that the return air filter and the interior of the plenum shaft are not seen while walking down the stairs. If installed with the baffles slanting downward, the colored filter and the wall framing studs are exposed to the stairway.

Set HVAC Registers with a Torpedo Level

During the installation of the HVAC registers, the builder should check that the installer uses a small torpedo level to get the registers level rather than merely eyeballing the registers in relation to surrounding ceiling or wall lines (Figure 18.9).

Dryer Vents

The builder should check the distance that the dryer vent travels during the preconstruction plan review period. For single-family detached houses, this is usually not too difficult because all of the exterior walls are open to the outside. For duplexes, townhouses, and condominiums, however, getting dryer vents to the ex-

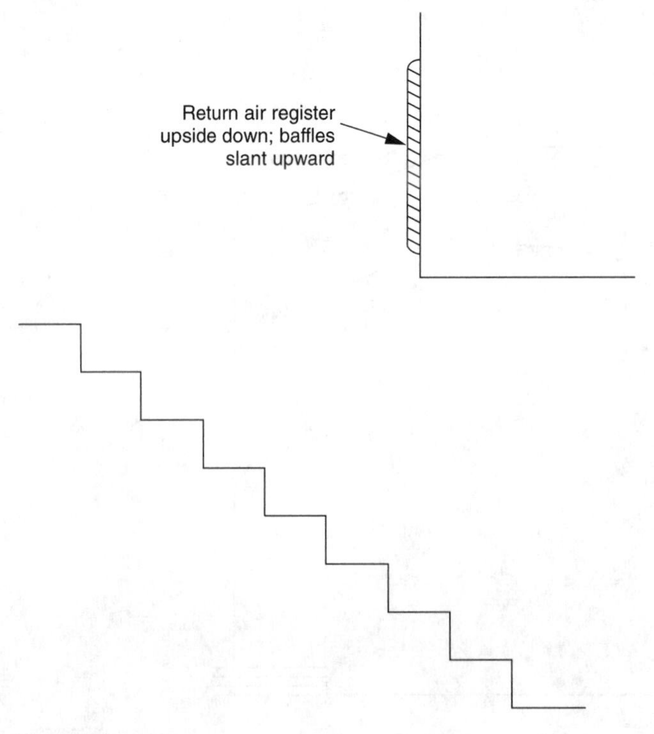

Figure 18.8

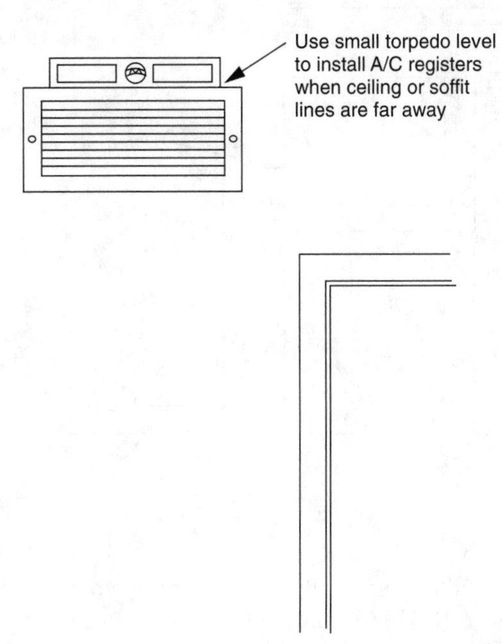

Figure 18.9

terior without exceeding the maximum length vent allowance can be more difficult. This is especially the case when units share common garage walls and the laundry areas are located within the garages. A dryer cannot be vented from one garage into another.

The builder should look for structural beams and posts that would block the passage of dryer vents, thus requiring the addition of soffits or chases not shown on the plans.

If the allowable length is exceeded for the diameter of the vent pipe and there are a number of 90-degree turns in the pipe, the builder should consider using a dryer vent booster fan motor, which increases the allowable length of the dryer vent.

All of these issues are best explored while the structure is still on paper, prior to the start of the construction.

Dryer Vent Duct and Hose Connection

Washing machines and dryers are sometimes placed within closets in a side-by-side arrangement or stacked vertically. In either case, working space to make the necessary hookups around the appliances is often at a minimum (Figure 18.10). To make the hookup of the flexible vent hose from the dryer to the dryer vent easier, the dryer vent duct should have adequate clearance from the adjacent wall corners. A duct that is placed too close to a wall or ceiling corner makes this dryer vent

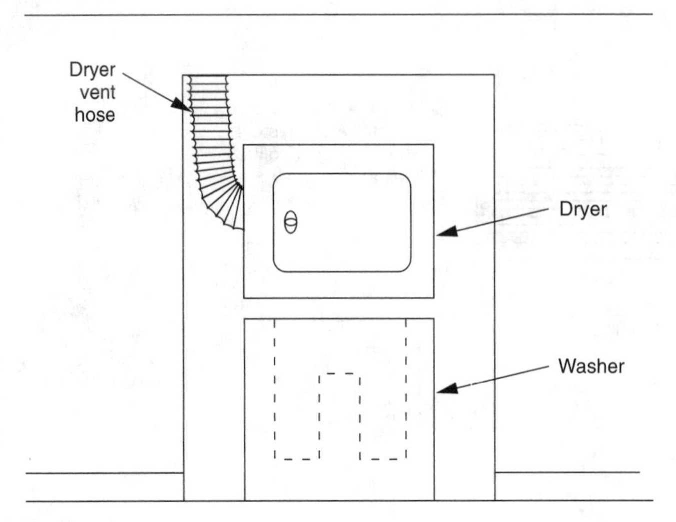

Figure 18.10

hose connection difficult, especially when working space within the closet is tight to begin with (Figure 18.11). A minimum of 3–4 inches from wall corners should be provided so that the installer can get his or her fingers and hands around all sides of the dryer vent to fit the flexible hose over the vent.

F.A.U. Combustion Air in Attic

Illustrated in Figure 18.12 is a combustion air duct for an F.A.U. closet. The duct extends up into the attic roof for its air supply. In this example, the duct should extend 6 inches or more above the top of the ceiling joist, so that the attic insulation will not mistakenly be placed over the duct, blocking off the air supply.

A/C Condensate and Laundry Box

Shown in Figure 18.13 is the front view of a recessed plastic valve box used to house the laundry hot and cold water valves and the washing machine discharge hose. Entering on the left side of the plastic box is a primary or secondary condensate discharge pipe from the air-conditioning unit. A problem arose because there was a circular rim around the connection to the plumbing waste pipe, which was above the level of the bottom of the box. This circular rim acted as a dam so that any water coming from the condensate pipe into the box would leak out the front of the box and down the wall rather than up and over the rim and into the plumbing waste pipe. For this reason, air-conditioning condensate lines should T into plumbing waste pipes directly, as the primary means of discharge. The secondary condensate line should penetrate through an exterior wall to the outside of the structure.

A/C Condenser Placement

Illustrated in Figure 18.14 is the plan view of the sideyard between two houses. The builder should check that sideyard setbacks are wide enough for the placement of A/C condenser units. Some housing tracts have houses placed so close together that minimum sideyard separations exist. The A/C condensers must then be placed on rear walls rather than side walls to avoid encroaching upon the required sideyard clearances.

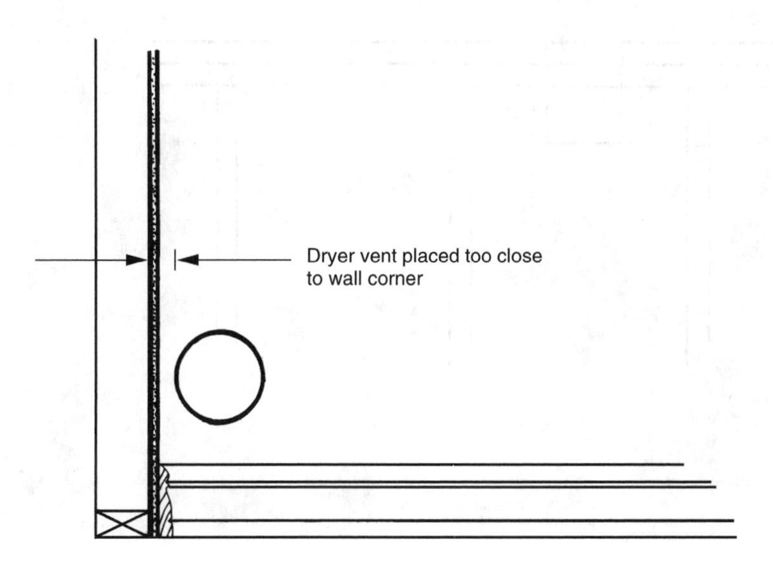

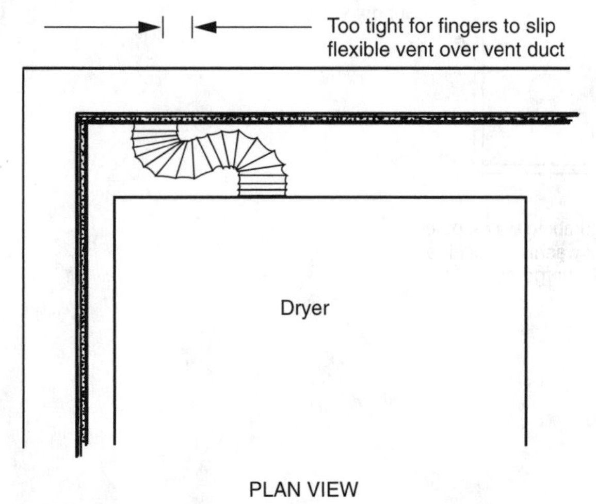

Figure 18.11

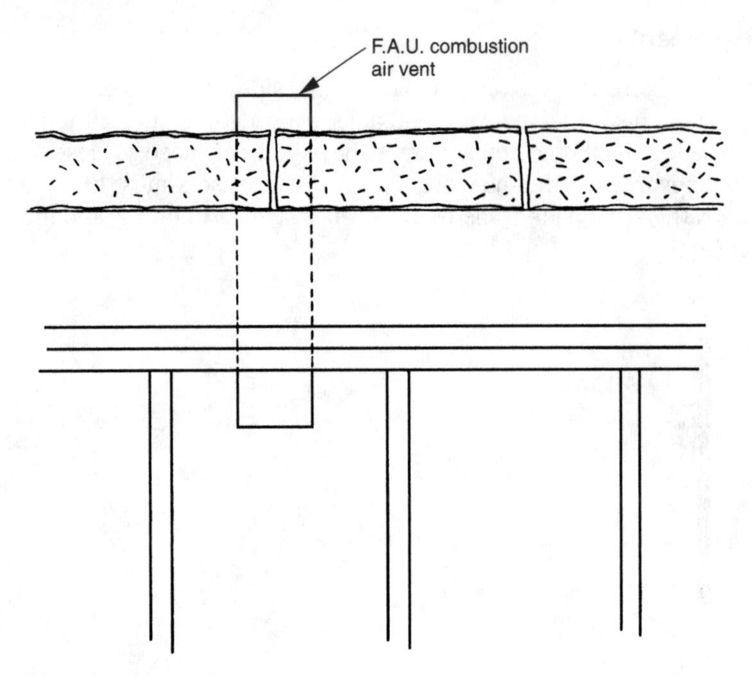

Figure 18.12

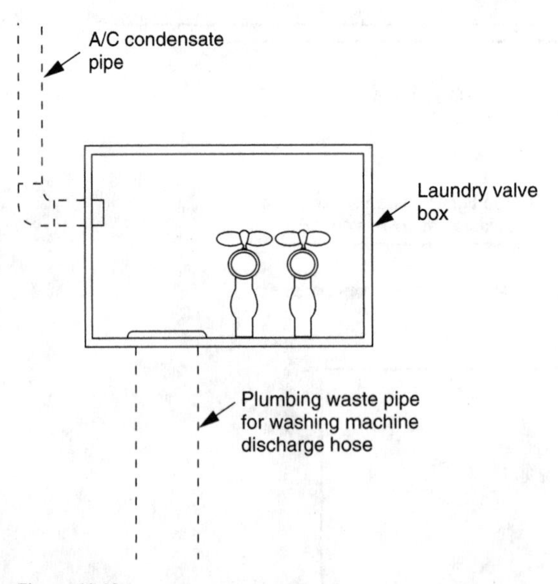

Figure 18.13

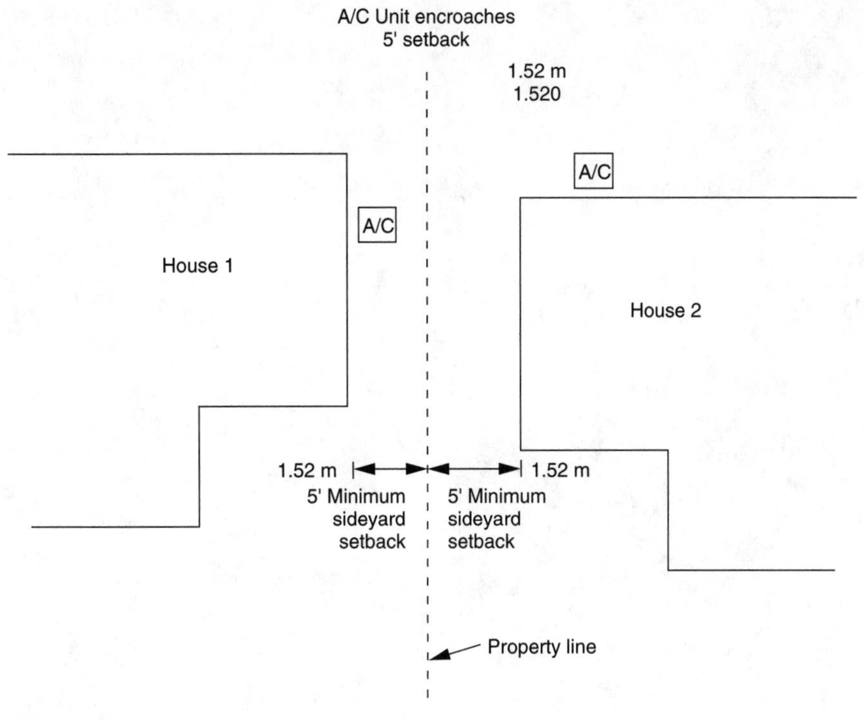
A/C Unit encroaches
5' setback
1.52 m
1.520
A/C
A/C
House 1
House 2
1.52 m
1.52 m
5' Minimum
sideyard
setback
5' Minimum
sideyard
setback
Property line

Figure 18.14

Chapter

19

Electrical

Outlet in Bath Splash

Some bathrooms are designed so that pullman cabinets are bordered by a bathtub on one side and a shower on the other. With a mirror placed on top of the pullman splash, the only remaining location for the placement of an electrical outlet is either in the splash tile or in a special cutout in the mirror.

Figure 19.1 illustrates an outlet placed sideways in a pullman splash. The builder must know the correct centerline height above the floor of the splash before this outlet can be correctly positioned. This involves knowing the height to the top of the cabinet rough top, the thickness of the countertop material, plus half the height of the splash.

Kitchen Cooking Top

Electrical convenience outlets are normally placed 12 inches above the floor. If all of the convenience outlets for kitchen appliances are also placed at this height, the cooking top extension cord will not reach the electrical outlet because a cabinet shelf will block it off (Figure 19.2). By placing this one electrical outlet higher than 12 inches above the floor—somewhere between the cabinet shelf and the countertop height—the cooking top cord can be directly plugged into the outlet without having to drill an ugly and awkward looking hole through the shelf (Figure 19.3).

Outlets under Kitchen Sink

Rough electrical boxes should not be placed too close to plumbing water line stubouts beneath the kitchen sink. Because the electrical work comes after the rough plumbing, it is the electricians responsibility to place the outlet box far enough away

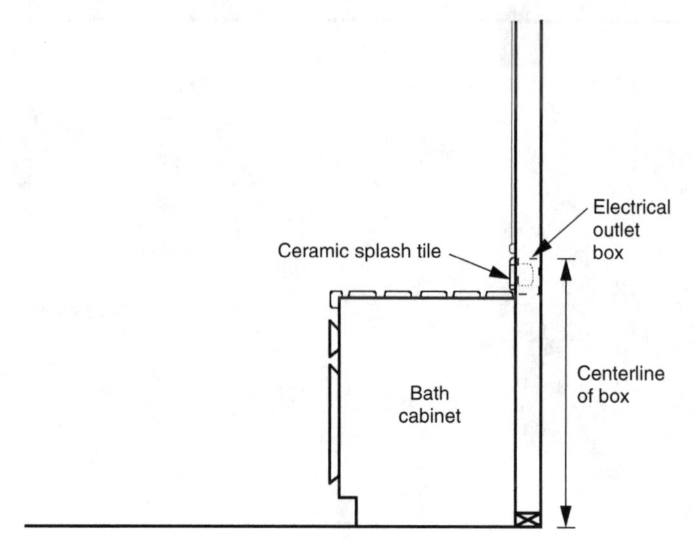

Figure 19.1

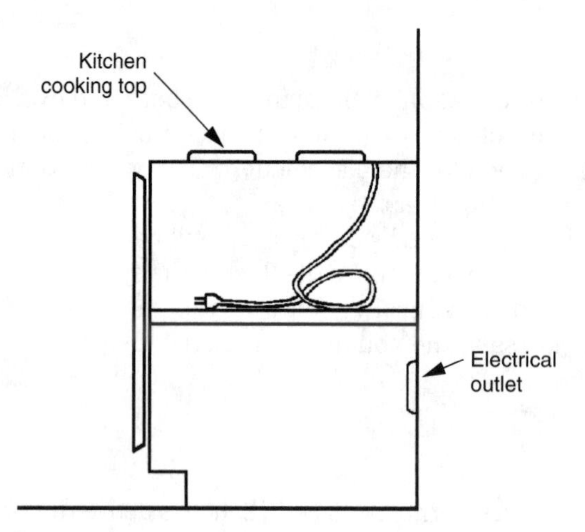

Figure 19.2

from the plumbing to avoid any future interference. For example, if the water valve stub-out and an electrical outlet box are placed too close together, there may not be enough clearance for both the round plumbing escutcheon cover plate and the electrical cover plate to fit (Figure 19.4). During the finish electrical phase of the construction, the electrician will see this problem underneath the kitchen sink, decide that the fault is the plumbers, and not install the electrical cover plate. The builder's customer service crew is therefore sometimes stuck with having to cut out the electrical cover plate to fit around the plumbing escutcheon cover plate.

The builder should check the placement of the plumbing and electrical work under the sink during the rough mechanical construction before this becomes a problem.

Garbage Disposal Outlet

The planning of the placement of the electrical receptacle underneath the kitchen sink for the garbage disposal should be done during the rough electrical wiring phase (Figure 19.5). There are different lengths of appliance cords for disposals, ranging from 3 to 9 feet. The builder should check that each disposal cord reaches the electrical outlets after the disposals are installed.

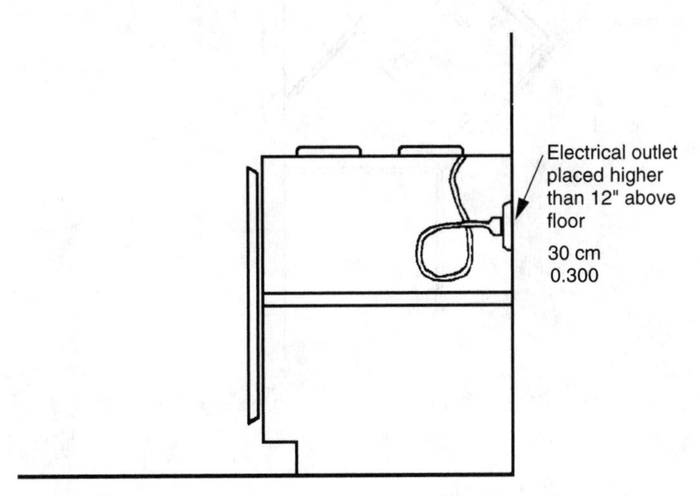

Figure 19.3

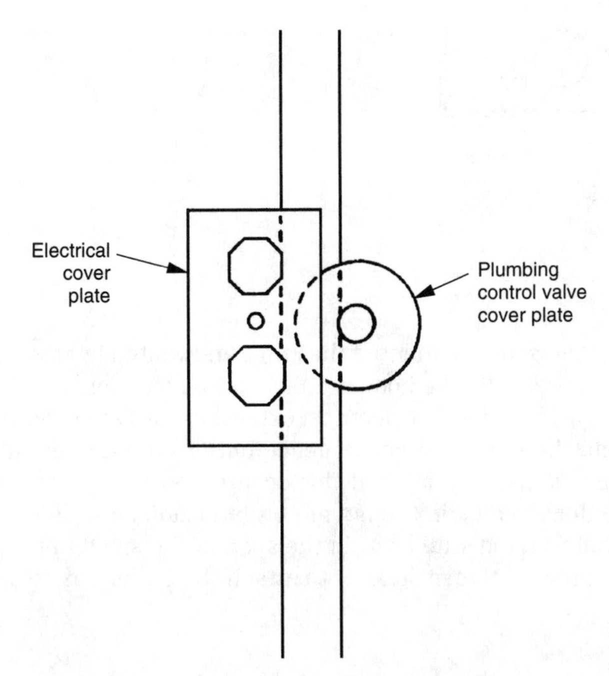

Figure 19.4

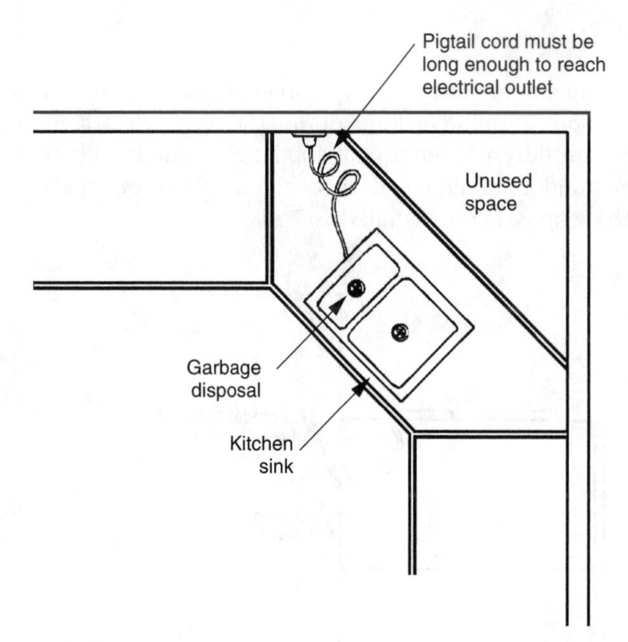

Figure 19.5

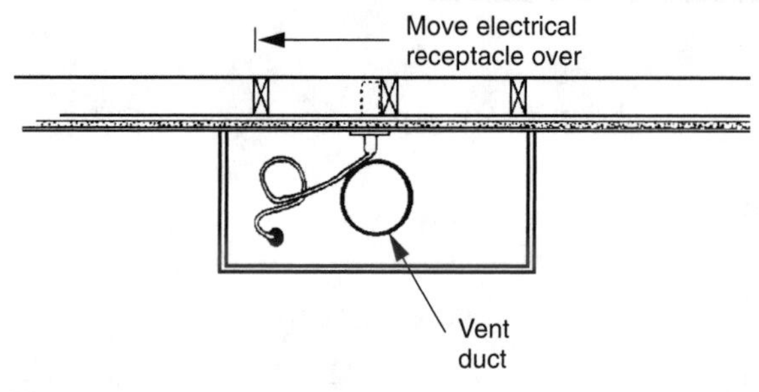

Figure 19.6

Range Hood and Outlet

A kitchen range hood that is independent of the range and vented by means of ductwork is illustrated in Figure 19.6. The hood is attached to the underside of an upper cabinet in which the vent duct and fan electrical extension cord are concealed.

Try to avoid placing the electrical convenience outlet directly behind the exhaust vent duct, which is installed later. If this occurs, there may not be enough clearance behind the duct to plug in the fan motor extension cord. Because there are more than two wall framing studs within the span of the small upper cabinet, all that is required to prevent this mistake is to attach the outlet box to one of the studs that is not centerline with the kitchen range, yet still within the upper cabinet (Figure 19.7).

Cold Water Bond

During the rough construction, the builder should check that the cold water bond connection to a copper water pipe for electrical grounding is not placed directly behind the water heater tank. If the water heater tank is placed too close to the wall, access to the bonding clamp and the grounding screw is cut off (Figure 19.8).

Cabinet Flex Conduit

When an electrical receptacle is required at the end of a kitchen cabinet, as shown in Figure 19.9, the builder should check that the electrical flex conduit serving this outlet actually comes out of the wall at the correct location inside the cabinet.

This is not as easy as it sounds because the electrical flex conduit is laid-out and installed long before the cabinets go in. If, for example, the electrician incorrectly

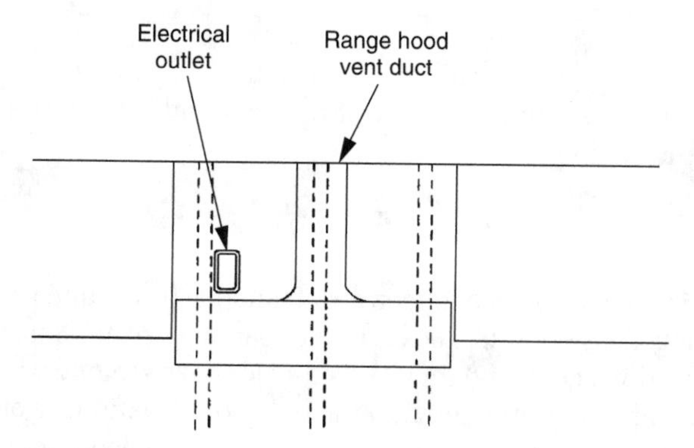

Figure 19.7

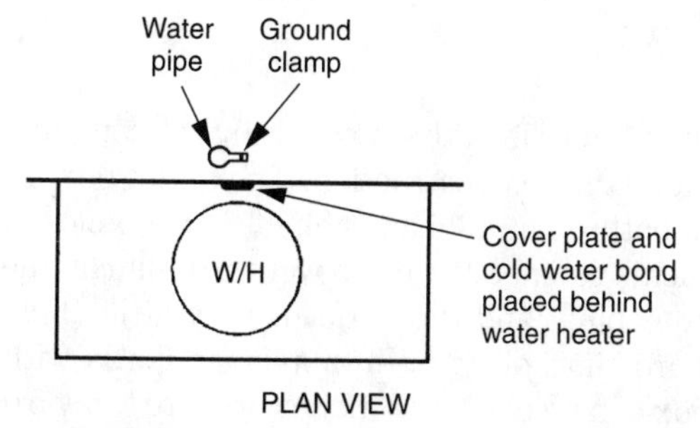

Figure 19.8

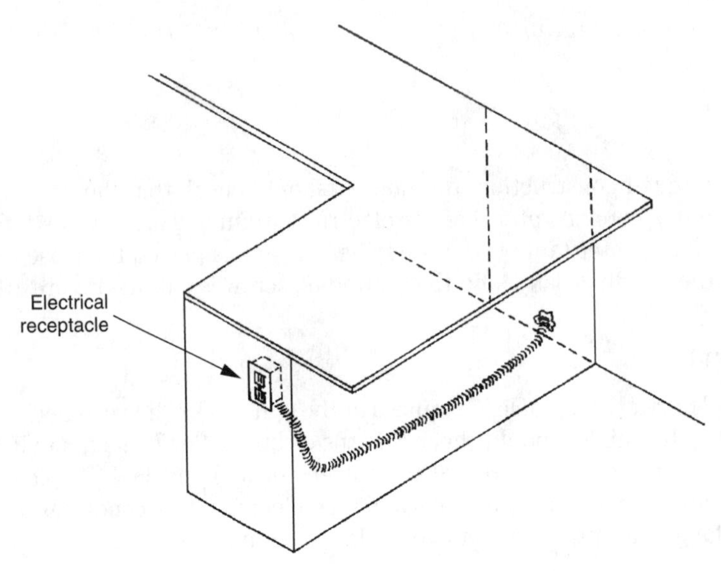

Figure 19.9

reads the line denoting the edge of the pass-through countertop on the floor plans as the line for the cabinet, the flex may come out of the wall underneath the countertop rather than inside the cabinet.

Outlet Clearance for Foam Trim

Shown in Figure 19.10 is the plan view of decorative foam plastic around the exterior side of an entry or sliding glass door. In this case, the foam pieces are 8 inches wide, cemented to the stucco brown coat, and then stucco-color coated along with the rest of the exterior to add an architectural detail to the exterior elevation. Avoid placing exterior electrical outlets, such as waterproof receptacles at balcony decks, entry doorbells, and wall mounted patio light fixtures, closer than 8 inches from the opening of the door so that they miss the foam decorative trim.

Outlet and Cosmetic Box

The placement of electrical outlet boxes in bathrooms during the rough-in phase can become important toward the end of the construction. The outlet box illustrated in Figure 19.11 is placed above a bathroom pullman cabinet, but too close to the wall corner. With the mirrored medicine cabinet centered over the pullman, the space between the projecting medicine cabinet and the outlet is very little (Figure 19.12). The homebuyer will have a hard time plugging in appliance cords with this little space. However, if the outlet box can be placed 12 inches or more away from the wall corner, the interference between the outlet and the medicine cabinet can be eliminated (Figure 19.13).

Switch Covers and Door Casing

Figure 19.14 shows a typical problem that occurs when the builder uses a wider-than-standard door casing and the electrical outlet box for a switch is placed too close to the rough door opening. When this is the case, the cover has to be cut to a narrower width in order to fit. In this example, the rough electrical outlet box was placed next to the door framing king stud, but a 1×4 must have been used as a trimmer instead of a 2×4. When the king stud and the trimmer are both 2×4s, there are 3

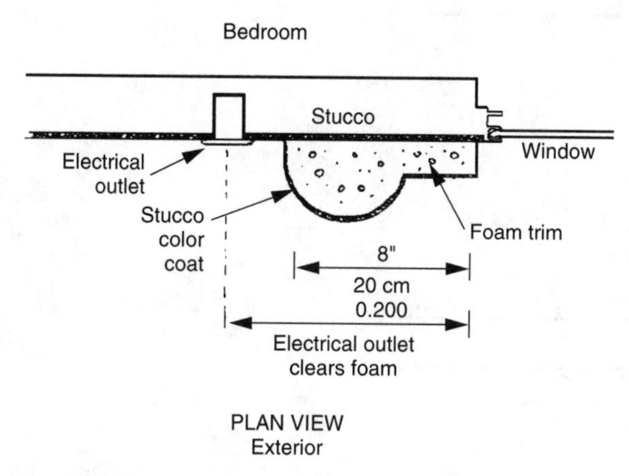

Figure 19.10

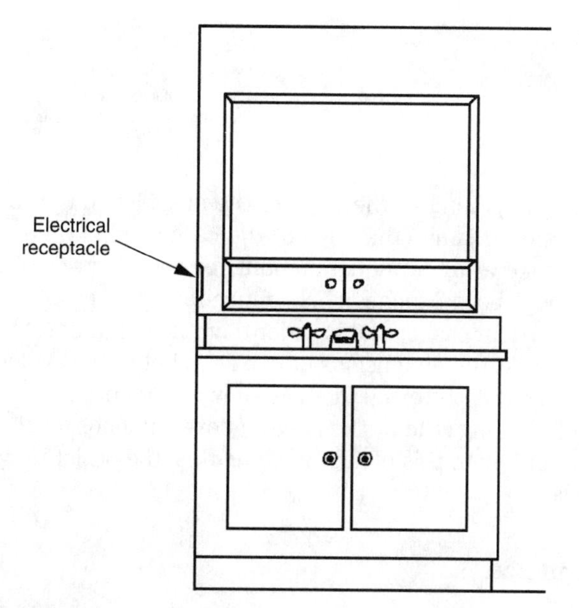

Figure 19.11

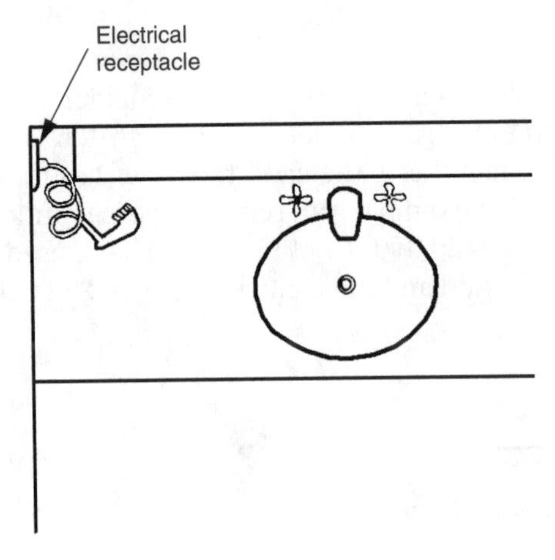

Figure 19.12

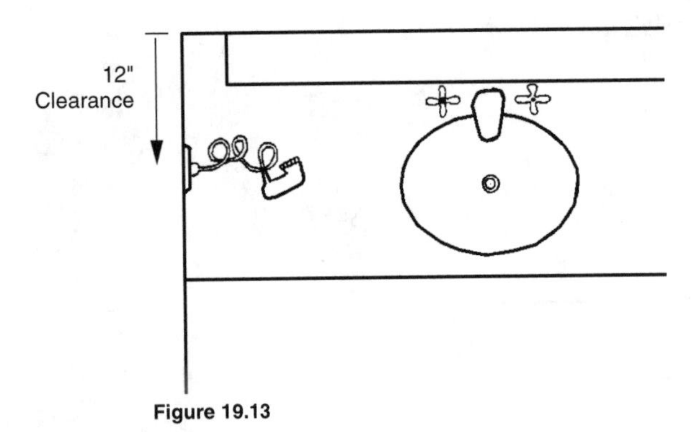

Figure 19.13

full inches from the rough opening to the edge of the electrical outlet box. Add to that the ½ thickness of a door jamb (the edge of the casing splits the jamb in half, leaving a reveal), then take away an equal amount for the electrical outlet cover plate, and you should have about 3 inches of clearance for door casing.

When the builder uses a door casing that is wider than the minimum standard door casing, there must be at least the dimensions of two 2×4s between the door opening and the electrical outlet box. A better method, as shown in Figure 19.15, is to add a 2×4 block to the side of the king stud at the height above the floor for the electrical outlet, then nail the outlet box to this block. By doing this, the outlet is even farther away from the door casing.

Electrical Outlets and Board and Batt

Shown in Figure 19.16 is an electrical outlet on an exterior balcony deck wall that ended up on the edge of a wood siding batt. As with exterior light fixtures, the builder should attempt to lay out the position of electrical outlet boxes during the rough framing construction so that they do not end up in the same location as the wood siding batts. In this example, because the wood batt is a 1×2, it sticks out ¾ of an inch from the siding boards and prevents the weatherproof outlet cover from reaching the surface of the wood siding board. The batt has to be cut out to make room for the weatherproof outlet cover, but with enough clearance for the rubber gasket on the cover to seat against the siding board.

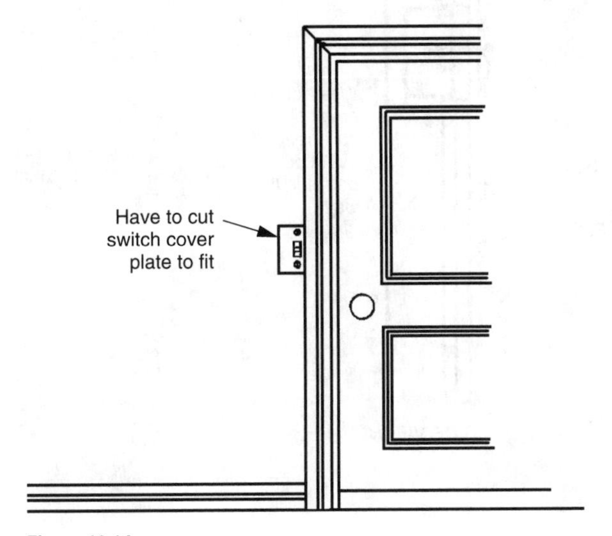

Figure 19.14

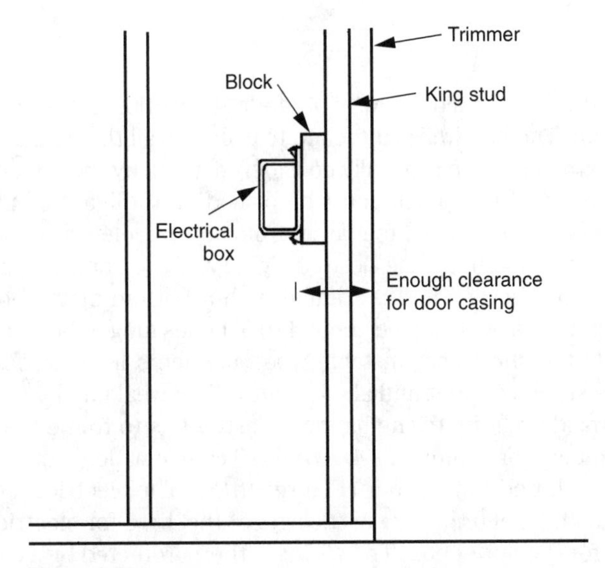

Figure 19.15

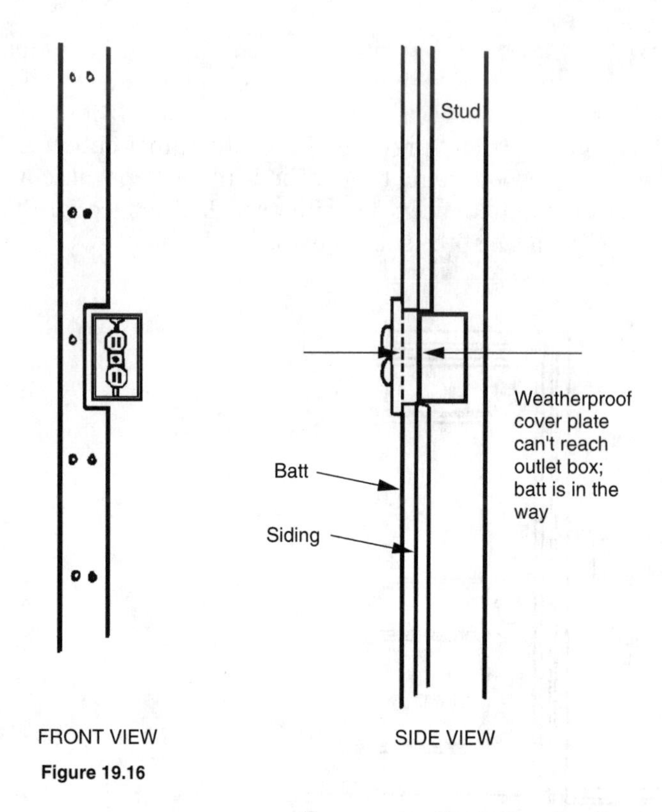

Figure 19.16

If the electrical outlet is originally laid-out to end up in the center of a siding board, between two batts, then this problem is eliminated.

Handrail Partition

Illustrated in Figure 19.17 is the side view of a handrail partition between a kitchen nook and a family room. The handrail is attached to a short wall that stands about 10 inches high. This handrail partition is usually considered a wall by the building codes. For this reason, an electrical receptacle must be placed on the handrail partition if any part of the partition is more than 6 feet away from another electrical outlet.

The problem to solve is obtaining the desired rigidity of the handrail partition. If the short wall supporting the handrail is framed with a bottom plate and two top plates, the short cripple studs will only be about 4 to 5 inches long. If holes are drilled through these already too short and unstable wood members for electrical wiring, some of these cripple studs will split and all of them will be weakened.

A better way to provide rigidity to the handrail partition is to frame the supporting wall with solid lumber. For example, a 4×8 or 4×10 can be nailed to a 2×4 bottom plate mudsill that is anchored to the floor (Figure 19.18). The electrician can use a chain saw to cut out a channel or groove in the side of the 4×10 for electrical wiring and to cut out a hole for the outlet box. The wiring is then protected by a continuous row of metal protector plates that are nailed to the side of the 4×10 to cover the channel. The handrail can then be attached later to a very solid and rigid base.

Outlet Box Depth

Some exterior and interior wall surfaces can have as many as two or three eventual layers of materials. An exterior wall may have, for example, a layer of shear-panel plywood, a second layer of waterproof drywall underlayment, and a final layer of wood siding boards (Figure 19.19).

The builder should check that the electrician knows which walls have more than one layer of materials so that electrical outlet boxes are set at the correct depth. An outlet box that does not project the correct distance from wall framing in order to be

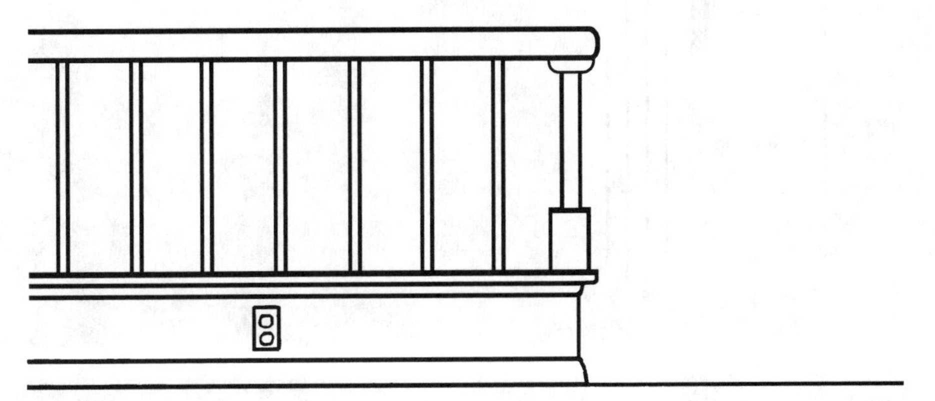

Figure 19.17

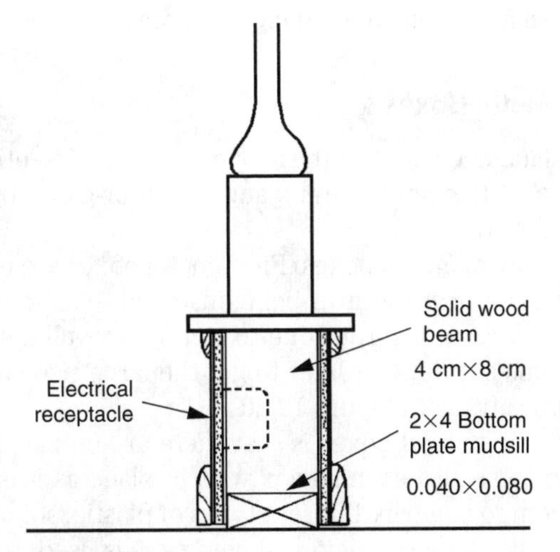

Figure 19.18

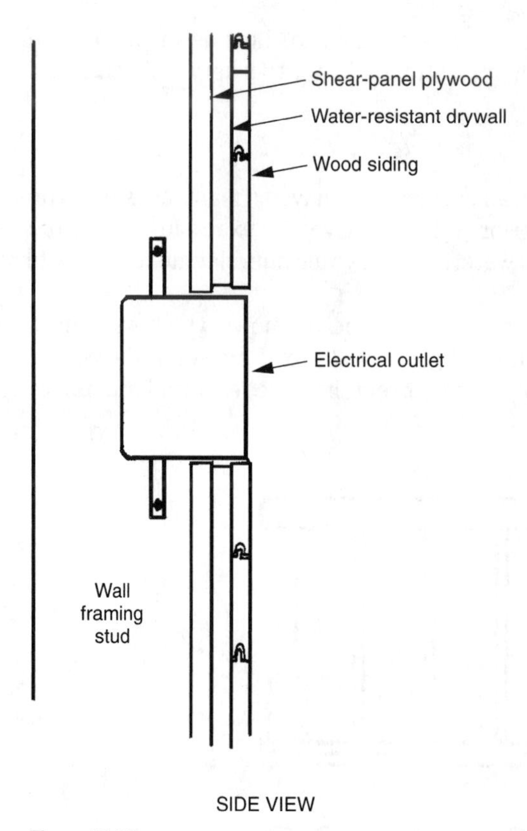

Figure 19.19

flush with the eventual finish wall surface requires longer receptacle and cover plate screws. They also often result in an unsolid and spongy receptacle.

Two-Gang and Three-Gang Electrical Metal Boxes

The builder should consider placing a clause in the scope of the work section in the electrical contract that states that two-gang, three-gang, and four-gang rough electrical outlet boxes be metal rather than plastic.

There are two advantages to having metal boxes. First, metal boxes are more rigid than plastic ones and will not bend or twist after being installed. Plastic boxes will sometimes be pushed up or down at one corner due to sloppy drywall cuts around the box. The undue twisting makes it difficult later to get the large size cover plate to screw into the box and still be straight (Figure 19.20).

The second advantage with using metal boxes is that there are no plastic tabs to break off, which are supposed to secure each romex wire in place as it enters the box. Plastic electrical boxes seem to come in different types of plastic, some soft and flexible and others rigid and hard. When the rigid and hard type is used, these plastic tabs often break off during the make-up of the box wiring. Many electrical inspections are turned down because the inspector sees that a few of these tabs are broken off on two-gang and three-gang boxes.

Television and Telephone Jack

In a master bedroom, electrical outlets for the telephone and television should be placed on opposite walls. The telephone jack should be placed on the wall that is most likely to be nearest the bed headboard. The bedroom telephone can then be placed on a night table next to the bed. The television will most likely be placed on a wall nearest the foot of the bed (Figure 19.21).

The telephone and television outlets should not be casually placed next to each other on the same wall. The telephone and television would then have to be placed next to each other (unless long extension cords are used), and the homebuyer would either have to get out of bed and go across the room to answer the telephone or have the television placed right next to the bed.

Telephone Jacks above Kitchen Tiles

Outlet boxes for telephone jacks and electrical receptacles should be placed above kitchen countertop tile splashes, whenever possible. This eliminates the tile setter from having to cut pieces of tile to fit around the outlet boxes. It also prevents the loosening of small tile pieces or the cracking of tile grout joints by the electrician who is completing the outlet and installing the cover plate (Figure 19.22).

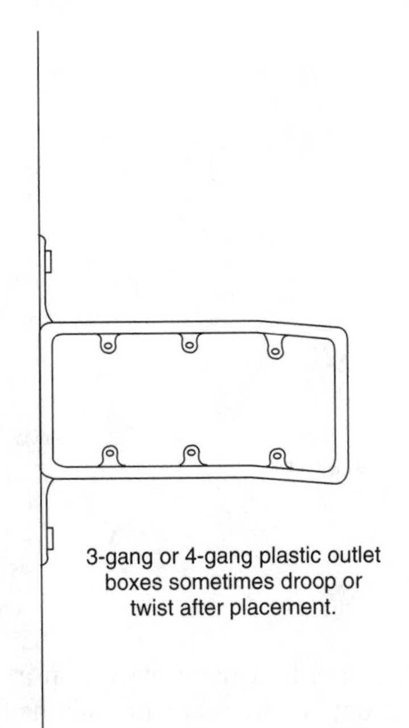

Figure 19.20

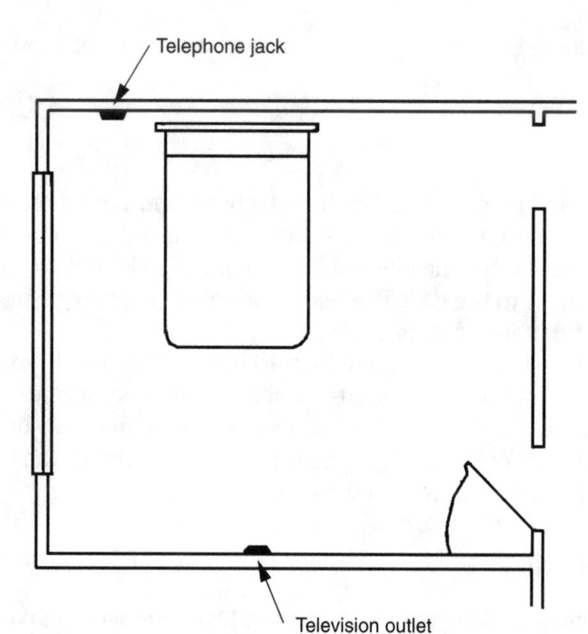

Figure 19.21

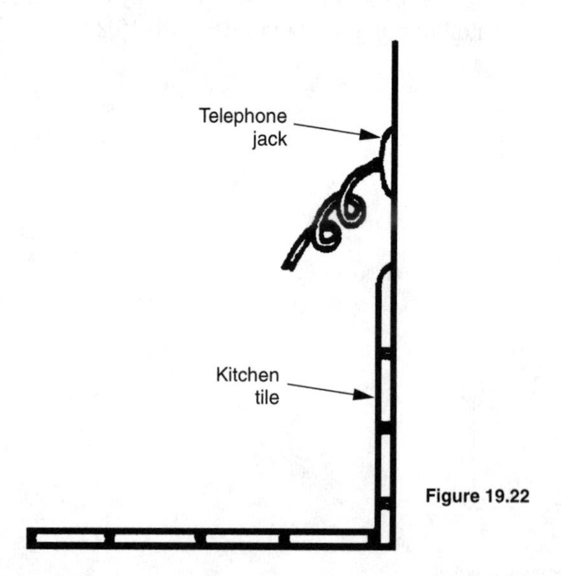

Figure 19.22

Intercom Location

A combination intercom/radio main panel is usually placed on a wall in a central room, such as a living room, dining room, family room, or main hallway. If this main panel is placed on a dining room wall, its location should consider future interference with a dining room hutch furniture cabinet. When possible, the panel should be put either on a narrow wall space between a door opening and a corner or next to door openings. The panel should not be placed 2 or 3 framing bays into the center of a dining room wall, which will effectively eliminate that wall space for the homebuyer's furniture (Figure 19.23).

Electrical Futures

Electrical futures are outlet boxes and wiring that are installed during the rough construction so that they may be in place later, should a homebuyer choose the option. Figure 19.24 illustrates an electrical future placed in the ceiling for an optional light fixture, such as a ceiling fan. The outlet box is nailed to the ceiling joist exactly flush with or slightly recessed above the ceiling joist line. This allows the box to remain buried in the ceiling if the option is not selected or to be exposed if the option is selected. The builder should mark the location of electrical futures on the subfloor with spray paint so that they are easy to find.

Entry Doorbell Tube

The electrician should not be allowed to merely attach the entry doorbell wire to a nail that is driven into a wall stud (Figure 19.25). Instead, a cylindrical tube made of metal or plastic should be nailed to the side of a wall stud at the entry to enclose and

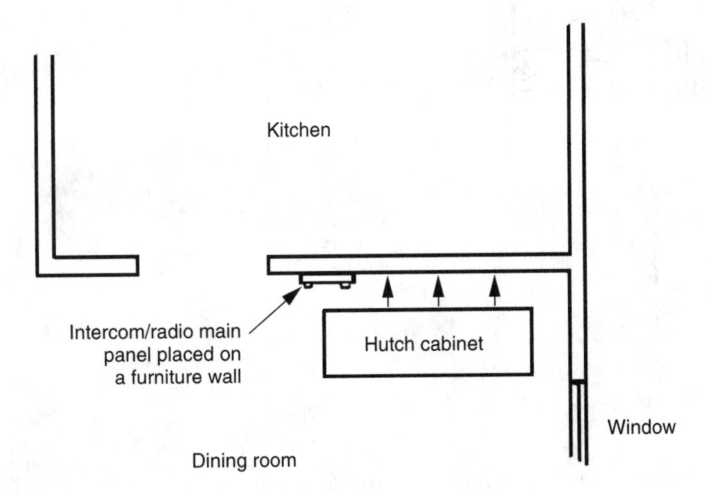

Figure 19.23

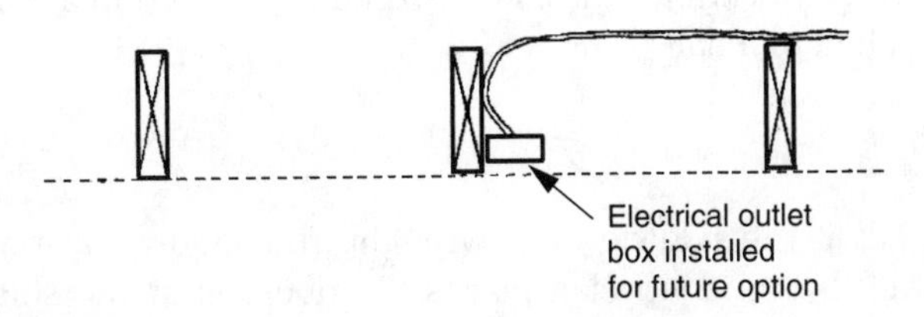

Figure 19.24

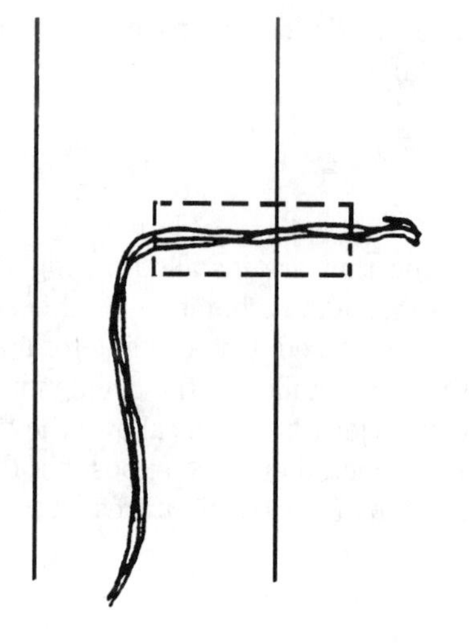

Figure 19.25

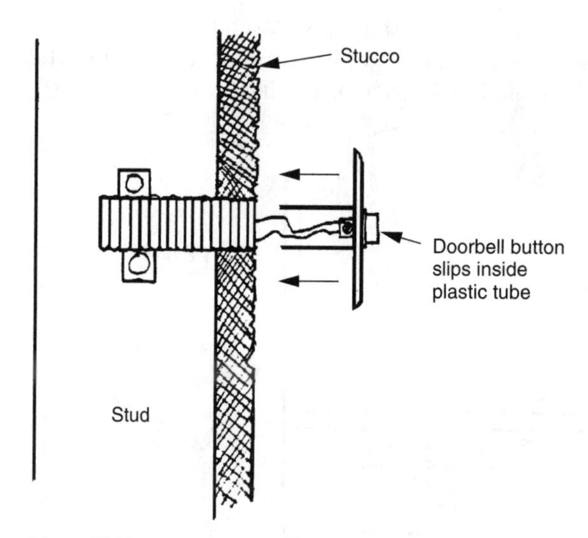

Figure 19.26

protect the doorbell wire through the various stages of the construction until the doorbell is installed (Figure 19.26). The tube is set at a depth projecting from the wall framing to help prevent it from being buried during lathing and plastering. This also helps to signify to the other trades that the doorbell wire is part of a real fixture, not just some wire attached to a nail.

Entry Doorbell and Handrail

Illustrated in Figure 19.27 is the side view of a wrought iron handrail as it meets the exterior entry door wall. If the electrician places the doorbell at the standard 42 inches above the floor, not knowing that a handrail is to be installed later, the handrail may hit part of the round escutcheon cover plate (Figure 19.28).

In this example, it was discovered in one of the model units that the wrought-iron handrail hit the bottom of the doorbell cover. To alleviate the problem in future units, the doorbells were placed at 48 inches above the floor.

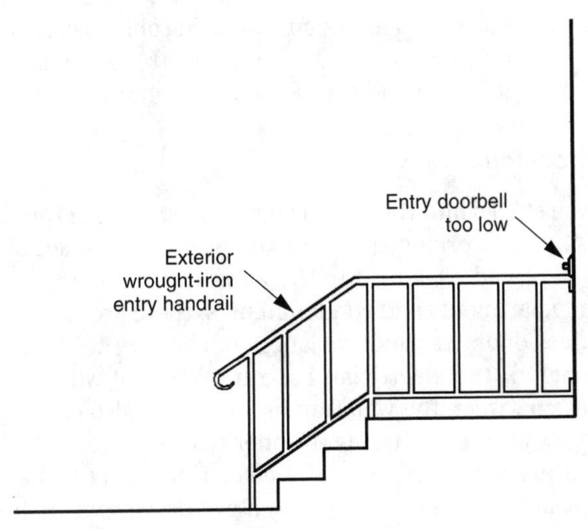

Figure 19.27

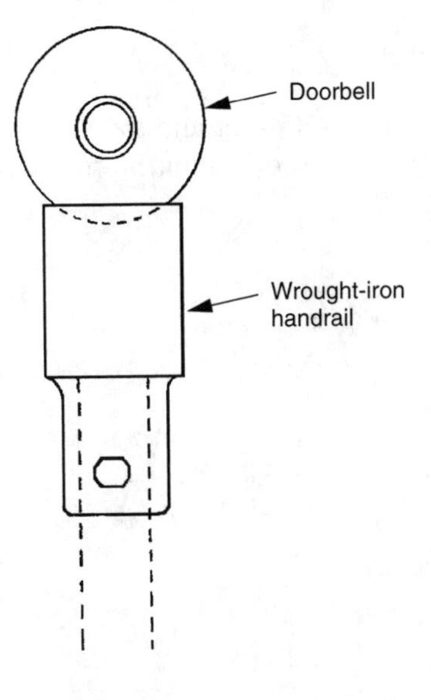

Figure 19.28

FRONT VIEW

Doorbells Audible on All Floors

A common occurrence in housing construction is to place the doorbell in a central hallway or stairway, which is sometimes not audible in the lower or upper floors. For example, if the homeowner is doing the laundry on the lower floor, or talking on the telephone in a bedroom on the upper floor, the ringing of the doorbell sometimes cannot be heard above the sound of the washing machine or the telephone conversation.

Because doorbell wiring and doorbells are not expensive, the builder should look at this issue from a customer satisfaction standpoint. Not being able to hear the doorbell, thereby missing a visit from a friend or neighbor, is a construction mistake that the homebuyer quickly becomes aware of and complains about shortly after moving in.

Electrical Wiring and Drywall Stocking

The placement of drywall stacks into the units is best done after the rough electrical work is complete. This allows working space for the electrician to set outlet boxes without stacks of drywall placed right next to the walls and in the way.

In two and three story apartments and condominiums, there are sometimes no exterior balcony sliding glass doors or windows large enough for the 4×12-foot sheets of drywall to pass through. If the electrician has run electrical wiring through the walls, then the open bays between the wall framing studs are also closed off as possible access areas for drywall stocking to these upper floors.

Figure 19.29 shows a method of running the electrical wiring higher along the walls to be used for drywall stocking so that the sheets of drywall will fit between the studs and below the wiring. This, thus, allows the electrical work to be completed before the drywall stocking, while still leaves an opening for the stocking of the drywall into the building.

A/C Duct and Wiring

Illustrated in Figure 19.30 is the end view of an A/C duct and the electrical conduit in a ceiling. The electrician should avoid running the conduit on top of the flexible

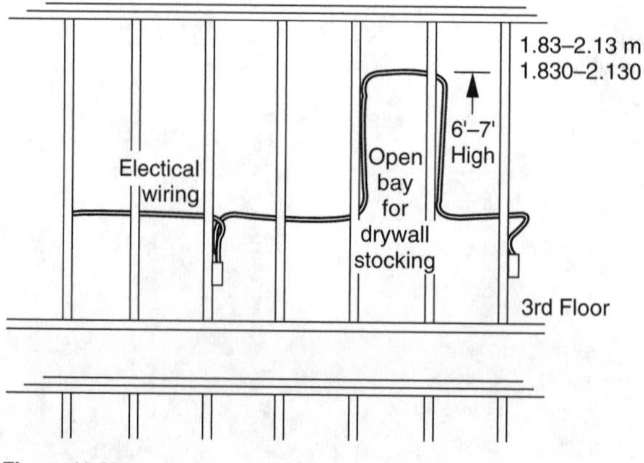

Figure 19.29

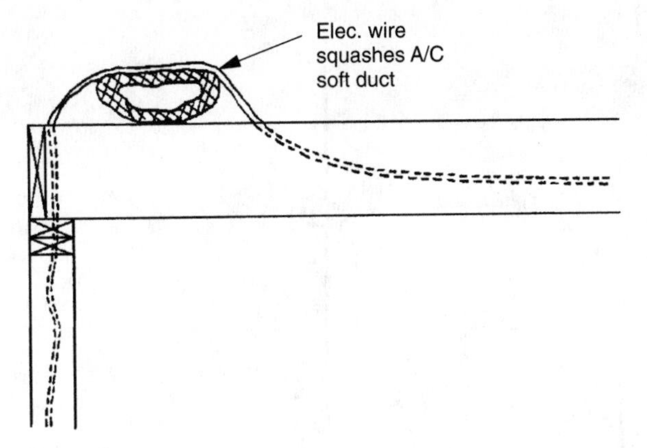

Figure 19.30

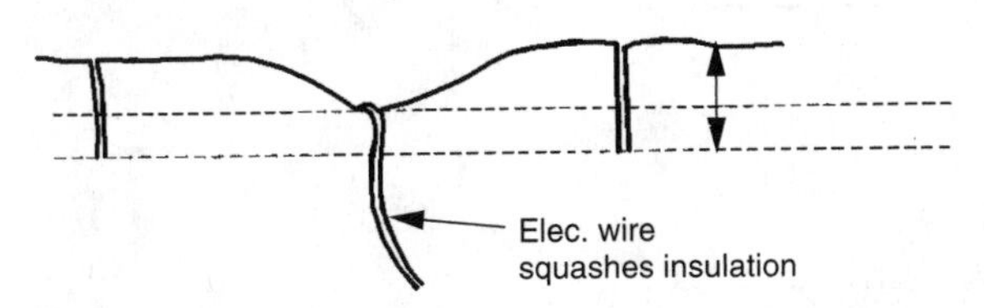

Figure 19.31

hose-duct in order to reach a switch or plug. If pulled tight, the conduit will squash the duct, reducing the flow of air.

Don't Squash Insulation

R-19 insulation batts used in ceilings need the full 9-inch thickness to form the rated heat barrier. The builder should check that pipes, electrical wiring, and A/C ducts do not squash down the insulation above roof trusses that have 2×4 bottom chords. If 9-inch R-19 insulation is squashed to 4 inches because other trades rest their materials on top of the 2×4 truss chord, the R rating is reduced in those areas (Figure 19.31).

Don't Make-Up A/C Disconnect Box

Illustrated in Figure 19.32 is an A/C condenser unit that is placed at the house exterior. If the exterior walls are stucco plastered, the rough wiring in the electrical disconnect box to the A/C should not be made-up—the wires should not have the rubber sheathing stripped off. This leaves the wire ends covered until after the exterior plastering is complete. The electrician can then strip the wire ends during the installation of the disconnect. To strip the wires and make-up the box before plastering is a waste of time because the wire ends will more than likely get coated with plaster, and must therefore be cut off and new ends stripped.

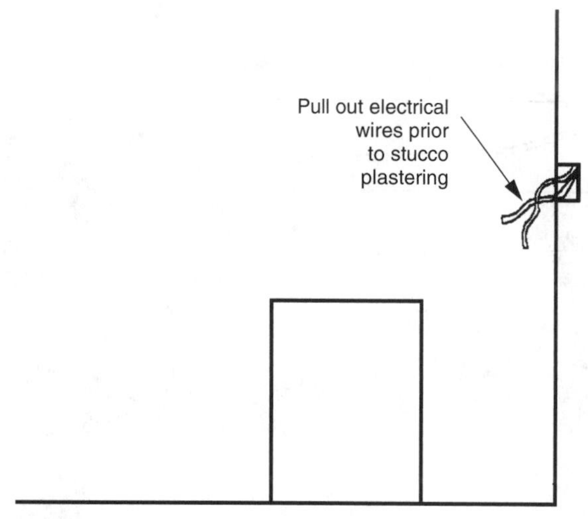

Figure 19.32

Light Bulb in Attic

Whenever equipment, such as an F.A.U. furnace or a water heater, is placed in the attic, a switched light fixture is required to provide light for servicing the equipment. The switch to the light fixture is placed in the attic, within easy reach of the attic access. The light fixture is usually a keyless fixture that is placed on the underside of a roof rafter above the equipment or on a wall close by.

For convenience, the builder should have the electrician place a light bulb in the fixture upon installation during the rough phase of the electrical wiring. Because the attic is not easily accessible without climbing up through the framing, the light bulb will probably be left undisturbed throughout the construction and be in place for the final walk-through. This prevents having to climb around inside a dark attic at the end of construction to install one light bulb that was missed during the finish electrical work.

GFI Outlet for Jacuzzi Tub

Shown in plan view in Figure 19.33 is a Jacuzzi bathtub adjacent a bath cabinet. In this case, the Jacuzzi motor is placed at the corner of the tub next to the cabinet. An access door to service the motor is inside the cabinet.

The electrical circuit to the Jacuzzi motor must be separate and GFI protected. The electrical outlet for the motor must be placed at the same corner of the bathtub, within reach of the motor's electrical extension cord. This outlet should not contain the GFI reset button because the homeowner would have to go through the cabinet access door just to push the reset button. Another outlet, with a GFI reset button that is tied to the outlet at the Jacuzzi motor, should be placed at a more accessible location, such as inside a nearby closet or on one of the bath walls.

A/C Units above Flush-Lights

The A/C furnace pictured in Figure 19.34 is located in the attic. By chance, a ceiling flush-light was positioned directly below the furnace. If the size of the ceiling joists is 2×10 or larger, then the 9½ inches of joist depth is enough clearance for the flush-light to fit. However, most flush-lights require an additional 2 or 3 inches of clearance

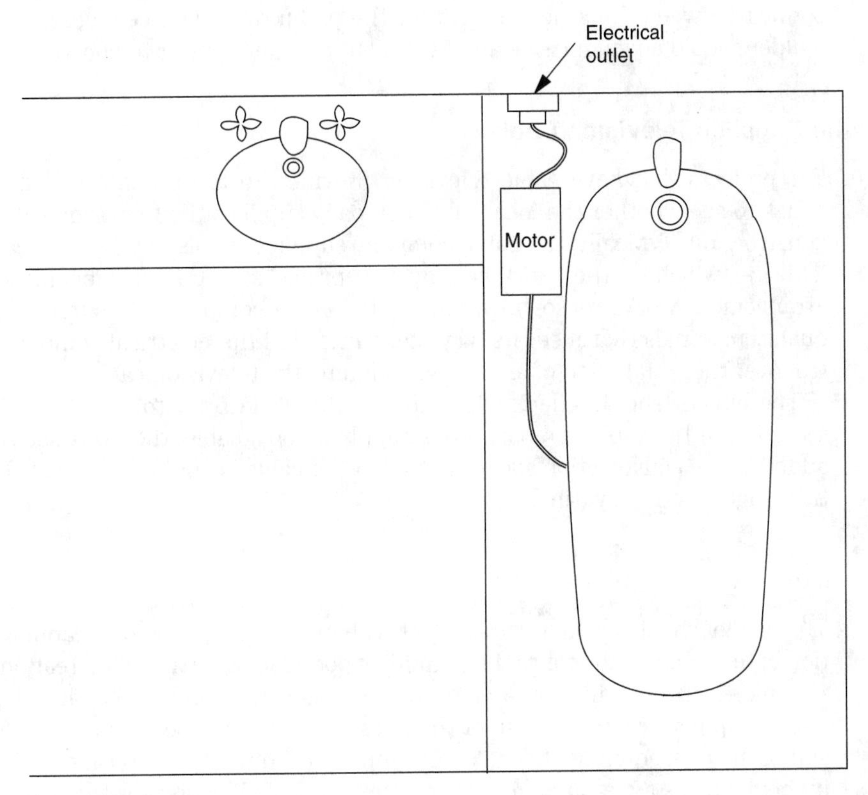

Figure 19.33

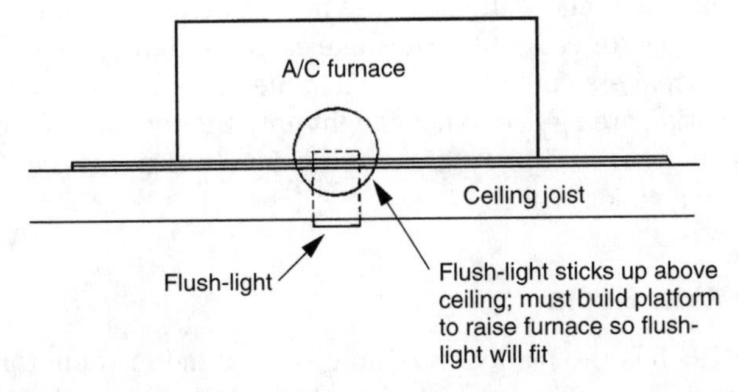

Figure 19.34

above the flush-light in order to dissipate heat. If the plywood platform, built to support the A/C furnace, is directly above the flush-light, the framer should cut a hole in the plywood above the flush-ight to provide the required clearance. If the furnace is directly above the flush-light on the plywood platform, then the builder should consider having the framer build-up the platform on a 2×4 frame on top of the ceiling joists. With the wood on edge, this adds another 3½ inches to the depth of the joists at the platform area, providing the required clearance. In order to avoid the possible conflict between the attic platform for the A/C furnace and ceiling flush-lights, the builder should check furnace and flush-light placement prior to construction.

Check Who Supplies Television Cable

For projects that have cable television provided to the units, the builder should check to see whether the local cable company supplies the television cable for new construction. Typically, the cable company supplies spools of television cable to the job site, which is then installed by the project's electrical subcontractor. This arrangement works reasonably well, as the cable company is insured of providing quality materials that meet its own standards, and the electrical company has control over the installation of all wiring, including the television cable.

The builder should check who supplies the cable prior to the start of the construction. If the cable company does supply its own cable, then this activity can be added to the construction schedule so that the cable is on the job site at the start of the rough electrical wiring.

Electrical Meter in Pool Equipment Room

The builder should check with the electrical utility company before planning the location of the electrical meter to the swimming pool, Jacuzzi spa, and recreation building.

Some electrical utility companies may not allow the meter to be placed inside the pool equipment room because of the possibility of water being on the floor as a result of the pool equipment. The pool equipment room may be considered by the utility company as a wet area, and therefore an unsafe location for the placement and servicing of meters and breaker panels.

In one instance, the breaker panel and meter were mistakenly drawn on the plans by the architect to be placed on a wall inside the pool equipment room. Luckily the electrical utility field inspector caught this mistake early enough in the construction so that the meter and breakers could be placed in a meter pedestal in a landscaped area outside the recreation area fence. If not caught until the recreation building and surrounding landscaping was complete, the retrofit involved in moving the meter panel would have been a major ordeal.

Door Clearance in Electrical Meter Room

Illustrated in Figure 19.35 is the plan view of an electrical meter room for a condominium or apartment building. The builder should check that enough clearance is provided for the meter room door to swing all the way open. If the main switch gear

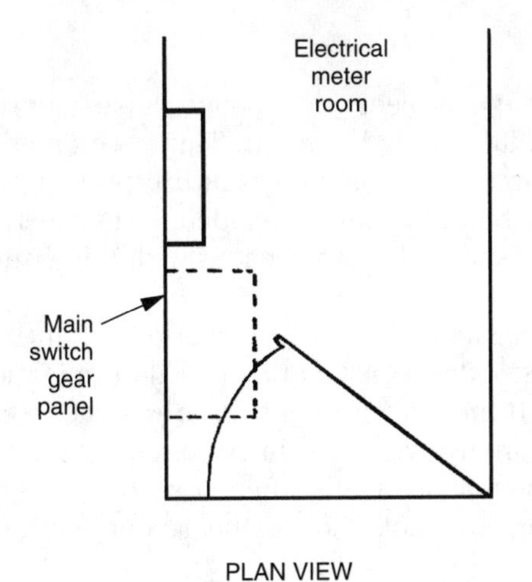

Figure 19.35

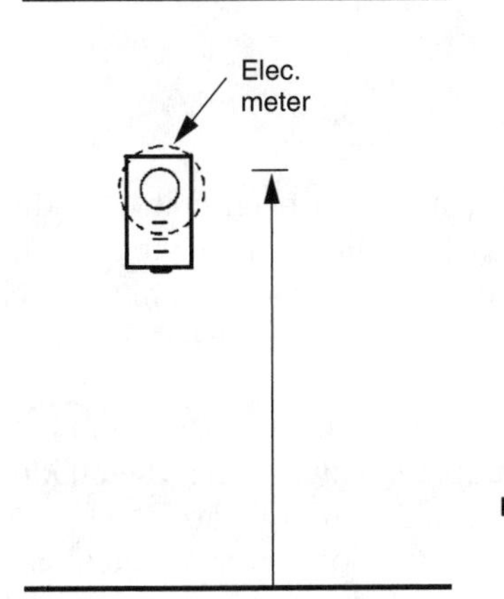

Figure 19.36

panel is placed near this door, as shown in the figure, the gear panel may project too far into the room and into the swing of the door.

Check Electrical Meter Dimensions

When an apartment or condominium building has an electrical meter room, the builder should obtain the specifications for setting electrical meters from the utility company. For example, Southern California Edison requires electrical meters, which are installed by the builder's electrical subcontractor, to be no higher than 6 feet, 3 inches above the floor to the center of the meter, and to have at least a 10-inch radius clearance around each meter glass cover. In a small electrical room that may have a panel for telephone, television cable, and a large switch gear panel, the builder should check that there is enough space to set the electrical meters at the correct dimensions (Figure 19.36).

Lighted Address Plaques

In some apartment and condominium projects, the police department requires the address number plaques to be illuminated at night. This is enforced through the building or the planning department. To alleviate future problems, this requirement should be shown on the building plans, included in the electrical contract, and address number plaques should be coordinated with the exterior light fixtures during construction.

Because this is an obscure requirement that may be overlooked until the final plan check or after the start of the models construction, the determination of the location of the address plaques and the exterior light fixtures sometimes falls to the job site superintendent. This can involve some time-wasting interaction waiting for someone to make the decision at the main office or second guessing from the main office, should the superintendent choose the layout of the address plaques and light fixtures.

To avoid the problem, specific locations and instructions should be included in the plans design phase and covered in the various construction contracts.

Single-Line Drawing for Multiunit Buildings

For multiunit buildings, the builder should remember to obtain electrical single-line drawings showing the individual load calculations for each unit, plus the total load estimates for the entire building. This is usually required by the building inspector or building department at the time of the combination rough framing/mechanical inspection. This allows the building inspector to determine whether the switch gear for the entire building is the correct amperage size.

The problem is that the electrical floor plans shown in the architectural set of plans often does not show the electrical amperage load calcs. The builder often has to hire an electrical engineer to do the calcs or instruct the architect to do so. If this whole issue is a surprise to the builder at the time of the first framing inspection of the sales models building, the inspection can be held up a few days or longer waiting for a single-line drawing to be generated.

House Pack Light Fixture Deliveries

When the builder purchases light fixtures directly from the manufacturer or distributor, they should be labeled and organized per each house when delivered to the job site. The complete set of light fixtures for each of the houses should then be separated, labeled, and stacked in a storage bin or garage so that they may be spread easily to each house by the builder or electrician for installation. This requirement should be included in the light fixture contract.

This allows the builder to be aware of shortages and backorders immediately upon delivery, thus leaving time to obtain the missing fixture for the normal installation period. It also relieves the builder from having to sort and label hundreds of fixtures with unfamiliar model numbers and names.

Have Light Fixtures Preassembled

In an effort to help the finish electrician install the light fixtures more quickly, the builder can write into the light fixture supplier's contract that all light fixtures come preassembled in their boxes.

Having all of the various parts preassembled, along with the wire threaded through the links of the light fixture chain, requires only that the electrician connect the wiring and attach the fixture to the outlet box. This not only cuts the light fixture installation time in half, but also forces the light fixture company to ensure that there are no missing parts for the fixtures.

User-Friendly Light Fixture Numbers

In an attempt to make the ordering and accounting of light fixtures simpler, the builder can insist that the light fixture company translate the confusing array of light fixture numbers into more user-friendly names and numbers.

For example, the following is a list of the locations for light fixtures that are typically supplied by a light fixture company:

1. Dining room
2. Breakfast nook
3. Stair ceiling
4. Stair wall sconce
5. Exterior entry wall or ceiling
6. Exterior balcony deck wall
7. Walk-in closet ceiling
8. Hallway ceiling
9. Bathroom bar-light
10. Bathroom vanity cosmetic strip lights
11. Ceiling track lights
12. Keyless fixtures for garage walls, garage ceilings, and attic spaces

All of the above light fixtures comes with letter and number designations that are probably familiar to the light fixture salesmen who deal with them every day, but they are totally unfamiliar to the builder. Why should the builder be stuck with having to compare and match obscure letter and number designations listed on invoices to the actual light fixtures at the time of delivery, and then spread or hand-out these same light fixtures according to these letters and numbers at the time of installation? All of these fixtures can be renamed to match floor plan types and locations for simpler management and control in the field. By doing so, the builder leaves the translating back and forth from the letter and number designations to the user-friendly field names to the light fixture salesmen.

Light Fixture Installation

To prevent the theft of hanging light fixtures, such as a dining room fixture, the installation of light fixtures can sometimes be divided into two stages. The basics, such as bath bar lights, exterior patio lights, closet ceiling globe lights, entry lights, and hallway sconce lights, can be installed along with the plugs and switches during the finish electrical phase. Light fixtures, such as those in the dining room, entry, and stairways, can be installed later in the construction, just prior to the final cleanup and homebuyer walk-through, when the fixture supplier has its own installers.

This method has some advantages. First, by postponing the installation of hanging fixtures until the latest possible time, the chances that they will be stolen are considerably decreased. Second, by having the supplier install its own fixtures, the problem of broken glass or missing parts no longer belongs to the builder and the electrician. If the installer opens a box and discovers cracked or broken glass, there is no question whether it was broken by the shipper, the builder while receiving and handling the fixture box, or the electrician installing the fixture. Because the builder and electrician are not involved, it is the suppliers responsibility to replace the damaged fixture.

One drawback to this method is that the shipping boxes that the light fixtures come in create excess trash at a time after the box-out and sweep phase. These boxes would have otherwise been thrown away with the plumbing fixture boxes and other trash generated by the interior finish trim work. This problem can be solved, however, by having the light fixture installers remove and throw away their own trash.

Extra Light Fixtures

A typical occurrence during the finish electrical phase of construction is for the electrician to open up the box of a particular light fixture, only to discover that the glass fixture is broken or that a vital part is missing. The electrician then hands the light fixture box back to the builder, who must telephone the light fixture supplier and inform them that a defective fixture has been discovered. The fixture cannot be hung then because it must be sent back in exchange for a new fixture. This process takes the builder sometimes weeks or even a few months to get the new light fixture. The exact fixture must first be ordered by the builder from the light fixture supplier, who in turn orders the unit from the manufacturer or through another middleman wholesaler.

The builder should write into the light fixture contract that at least one or two extra light fixtures of each type are always on hand at the light fixture supplier's warehouse. This may be a small inconvenience for them, but it allows damaged fixtures to be exchanged immediately. This in turn allows the electrician to complete the finish electrical work and eliminates the need for the builder to have to install a light fixture later. This also saves the homebuyer the wait for a light fixture that is on order.

Light Fixture Placement

The placement of light fixtures within the sales models should be planned in relation to windows, future dining room tables, and nearby upper cabinet doors. Once these locations are correctly established, the same exact locations must be used in the following production houses.

The electrician should be given simplified, reduced floor plans, showing exactly where the fixtures are to be placed before the start of the electrical work. These light fixture details accomplish two things. First, they notify the electrician of the exact location of the light fixtures and remove any excuses later over misplaced fixtures. Second, they hopefully ensure that the placement of light fixtures in the production houses will duplicate the sales models. This gives the builder the correct answer when homebuyers later complain about dining room light fixtures that do not center directly over their individually different dining room tables. The answer the builder must be able to give with confidence is: "check the models."

Tie Up Hanging Light Fixtures

The builder should have the electricians tie up hanging light fixtures, such as those in dining rooms and nooks, during the finish electrical work. This can be done with scrap pieces of electrical wire by lifting the light fixtures above head height and tying together two links of the hanging chains.

This raises the fixtures above the traffic of tradesmen, homebuyers, and furniture movers who are walking below and prevents someone from accidentally bumping into the light fixtures and breaking or bending something. After all the furniture is in place and tables are centered underneath hanging fixtures, the homebuyer can untie the electrical wire knots that are holding up the fixtures.

Light Fixture Heights

Illustrated in Figure 19.37 is a ceiling light fixture that is hanging on a long chain above a second floor open hallway. Because these fixtures come with more chain than is needed, the electrician must choose the height of the fixture above the floor by shortening the chain (cutting off a certain number of links).

In one particular tract of house, these hanging light fixtures were placed directly in front of and above the hallway linen closet cabinets. Invariably, some of the fixtures were casually hung too low by the electrician, and were within the swing of the upper linen cabinet door. When the cabinet door was opened, it

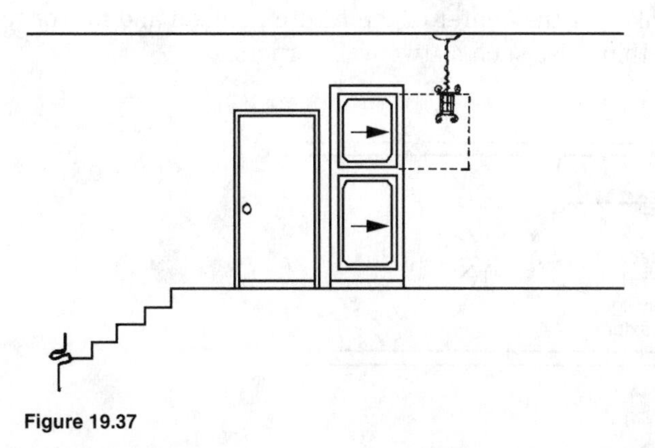

Figure 19.37

would smack the light fixture. The electrician had to go back and remove more links from the chain, thus raising the fixture.

The builder should consider each floor plan prior to the installation of electrical fixtures and check for possible interference with doors and cabinet doors. This will prevent future pickup work and cracked or broken light fixtures.

Hallway Light Fixtures

The three factors that result in a hallway ceiling light fixture obstructing the swing of an upper cabinet door are: the width of the hallway, the size of the cabinet doors, and the placement of the light fixture within the hallway (Figure 19.38).

In many housing and condominium projects, one particular floor plan may combine two or more of these factors, resulting in a cabinet door or some closet door striking or narrowly missing a light fixture or smoke alarm that is placed on the ceiling. When this oversight occurs, the only solution is to move the light or ceiling fixture to another location or to change the light fixture to a flush-light.

This unwanted problem can be prevented by analyzing the three factors for each floor plan before the rough electrical wiring phase of the construction, and by checking the actual location of the ceiling outlet boxes when installed.

Center Bar Light with Pullman

The bathroom mirror that is illustrated in Figure 19.39 goes almost to the ceiling and has a bar light placed over the mirror. In order to attach the light to the rough outlet box in the wall, a hole is cut in the mirror.

The electrician who places the rough outlet box should make sure that the box is centered with the pullman and mirror rather than being centered on the wall. If not, the mirror installer has no choice but to cut out the hole in the mirror to match the outlet box, resulting in a bar light fixture that is off center with the mirror, but centered between the wall corner and the closet door.

To avoid this mistake, the electrician needs to know the length of the pullman cabinet prior to the rough electrical box layout. The builder should also check that the box is actually installed at the center of the future pullman and mirror rather than centered with something else such as two wall corners.

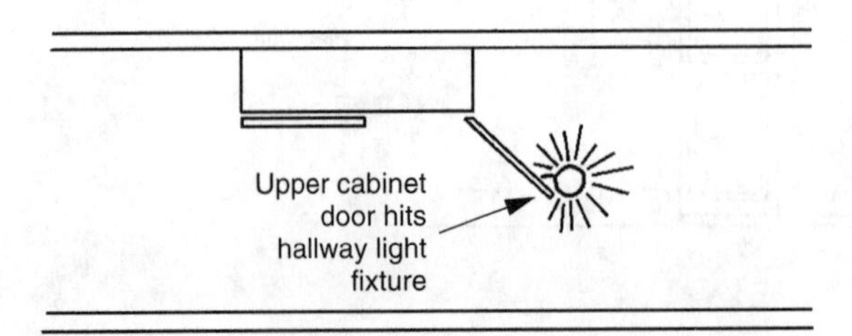

Figure 19.38

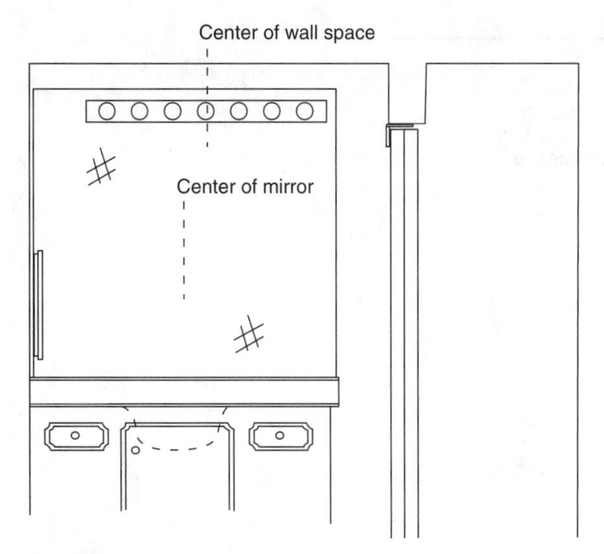

Figure 19.39

Bar Light Clearance

Bar light fixture panels should be installed in bathrooms as far down as possible from the ceiling. The bar light bulbs can become very hot and, if placed too close to the ceiling drywall surface, it will not start a fire, but will burn the paint and leave a brown spot (Figure 19.40). The bar light panels should therefore be placed just above and resting on top of the bath mirror, thus leaving the most clearance between the light bulbs and the ceiling.

Get Height for Undercabinet Light Fixture

Shown in Figure 19.41 is the side view of an upper kitchen cabinet that has a fluorescent light fixture. Because the electrical wiring must be placed within the kitchen wall and covered with drywall long before the upper cabinet is installed, the builder must know the exact location above the floor that the metal flex conduit encased wiring comes out prior to the start of the rough electrical work.

This information can be obtained from the cabinet company. It will show the height above the floor that the upper cabinets are installed. The flex conduit should come out of the wall exactly below the bottom of the cabinet, so the fixture will hide the conduit and the hole in the drywall. If the conduit comes out slightly below the light fixture, and the fixture is not deep enough to cover the hole, then drywall repairs must be done.

Center Light Fixtures Considering Cabinets

The architect has correctly designed the sliding glass door that is pictured in Figure 19.42 to center in this kitchen nook with the addition of a bank of pantry cabinets. A

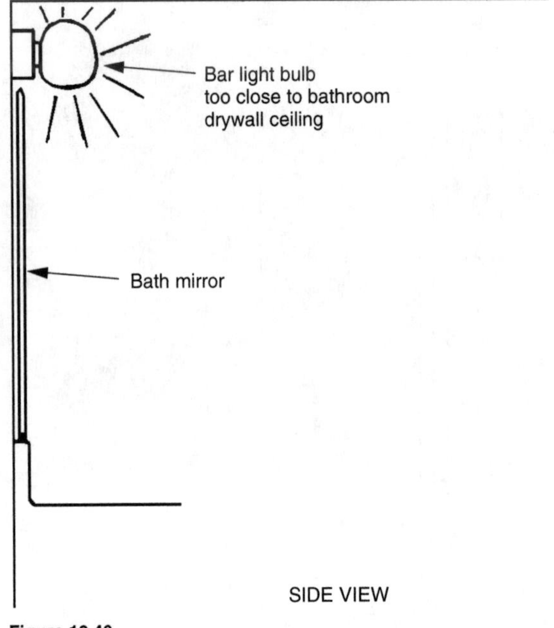

Figure 19.40

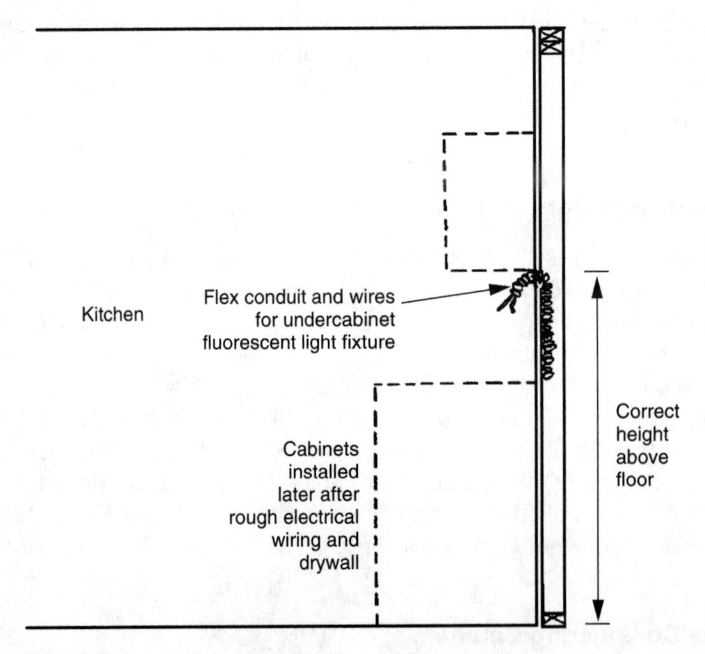

Figure 19.41

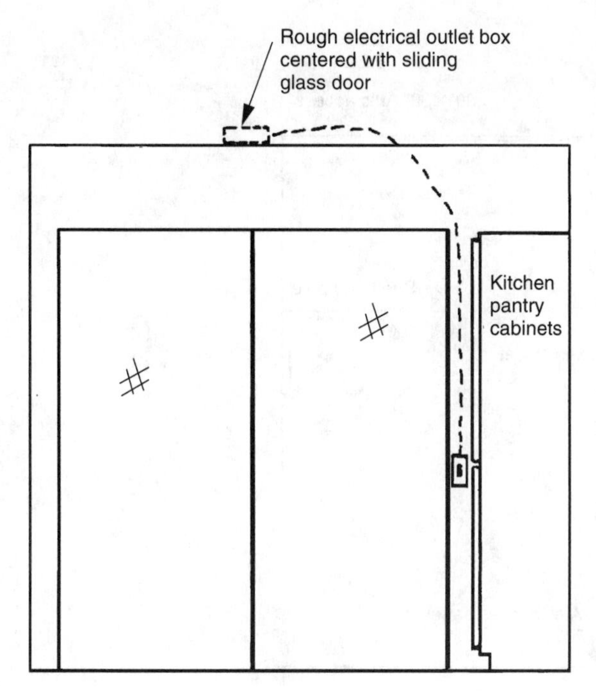

Figure 19.42

mistake that is commonly made is that the electrician may install the outlet box for the future light fixture, centered in the room rather than centered on the sliding glass door. By centering the outlet there, once the cabinets are installed, the fixture will be off-center.

The builder should check the light fixture layout in rooms with cabinets, first on the plans, then on the electrical rough wiring cheat sheets, and finally during the rough electrical wiring.

Closet Light Fixture and Attic Access

Illustrated in Figure 19.43 is a small walk-in closet with a mushroom globe light fixture and an attic access panel. During the preconstruction plans review period, the builder should check whether there is enough room in this type of closet for the attic access, the light fixture, and the closet shelving. The light fixture by code is supposed to be at least 18 inches away from the closet shelving. Add to that a 30-inch×30-inch attic access panel, and some advance planning may be required to get everything to fit.

One way to get around the conflict of space at the closet ceiling is to place the light fixture on the wall above the closet door (Figure 19.44). With an 8-foot-high ceiling, there is plenty of space above the closet door header for the typical size globe type

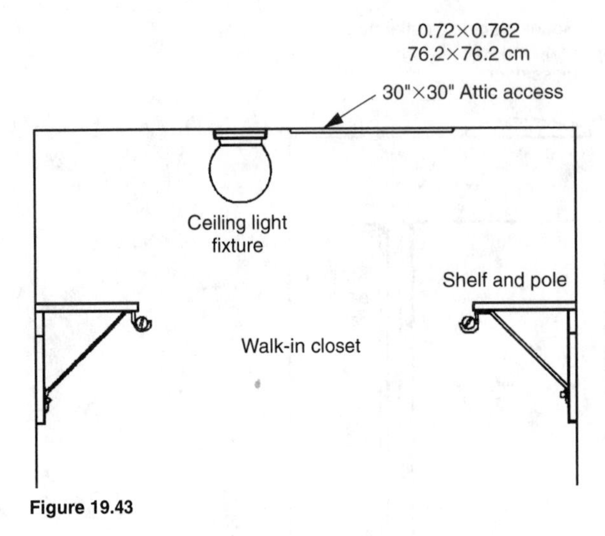

Figure 19.43

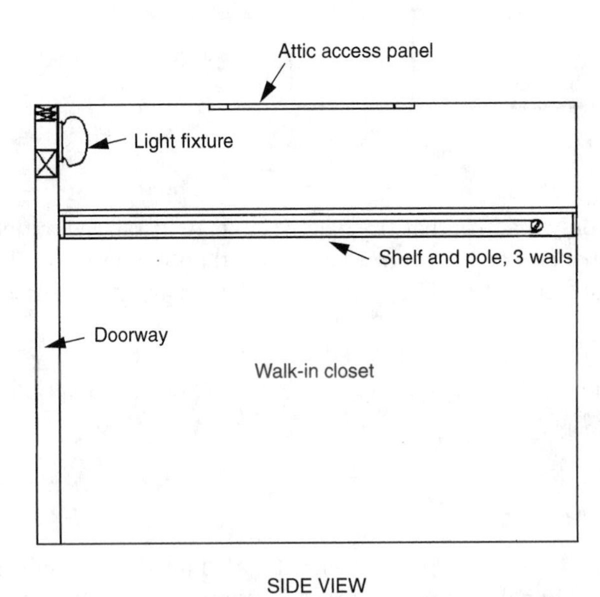

Figure 19.44

light fixture. This also places the light fixture far enough away from the closet shelving to meet code requirements, which might otherwise be difficult if the light fixture has to compete with the attic access for ceiling space.

A/C Register and Sconce Light Fixture

Figure 19.45 shows a design that placed an A/C return air register on the ceiling, directly above a wall sconce light fixture. In this example, the hinge side of the register was placed on the side closest to the light fixture. When the register was opened to periodically replace the return air filter, the register door would swing downward and hit the light fixture glass globe.

To avoid the possibility of breaking the sconce light fixture, the return air register should have been placed somewhere else or the hinge side of the register should have been installed facing another direction.

Doors and Wall Sconce Light Fixtures

As was the case in the last example, the wall sconce light fixture pictured in Figure 19.46 was inadvertently placed behind an exterior door. In this case, the dimension from the wall to the door was not deep enough to allow the door to swing into the wall at an angle that would miss the light fixture. Furthermore, the design and shape of the glass globe on the sconce fixture projected out further from the wall than the length of a standard door stop. In addition, the height of the ceiling did not permit the light fixture to be placed above the elevation of the swinging door, and the section of wall here was not wide enough to move the fixture over sideways to be out of the door swing. The builder's only recourse is to avoid this type of situation in the future by not placing a wall fixture at such a location in subsequent condominium or apartment projects.

Wall Sconce Light Fixture

Shown in Figure 19.47 in side view is a sconce light fixture installed on a stairway wall. In this example, the electrician should use a normal depth outlet box during the

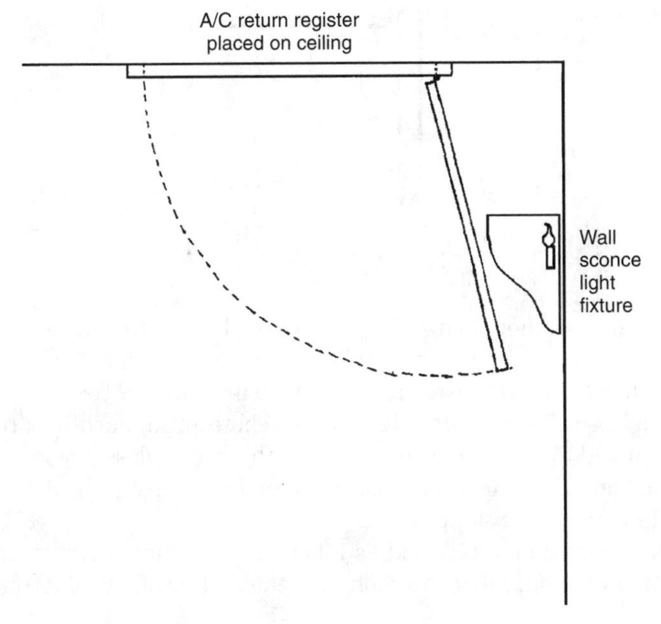

Figure 19.45

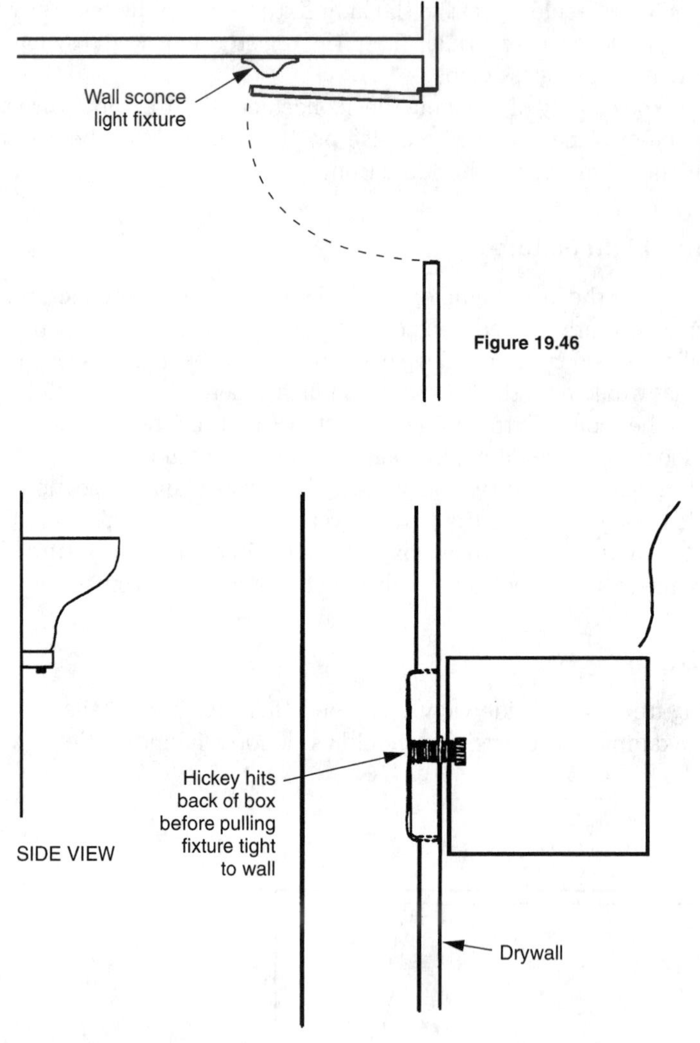

Figure 19.46

Figure 19.47

rough electrical phase rather than a thin "pancake" box. If the shallower pancake box is used, the hickey may hit the rear wall of the pancake box before being able to pull the light fixture tight to the wall during fixture installation. Because of the top-heavy design of most wall sconce light fixtures, the bottom supporting portion of the fixture must be pulled tight to the wall surface or the upper glass portion will have a gap between itself and the wall. If the glass is not pulled tightly at the sides, light will get out at the sides and looks sloppy.

A pancake box should only be used as a last resort, when a structural post or beam is in the way of positioning the sconce in the center of a wall or on a certain layout.

Exit Doors

Exterior light fixtures that illuminate an exit area must be placed high enough above or far enough away from exit door openings in order to be clear of the door swing (Figure 19.48). Figure 19.49 illustrates one type of exterior light fixture in which the glass globe of the fixture extends below the rough electrical outlet box. Even though the outlet box may clear the exit door, the light fixture and globe projected down into the door swing.

An exterior light fixture obstructing the full swing of an exit door may be the result of a mistake on the plans or a simple oversight by the electrician. In any case, the location of exterior outlet boxes can be easily checked by the builder during the

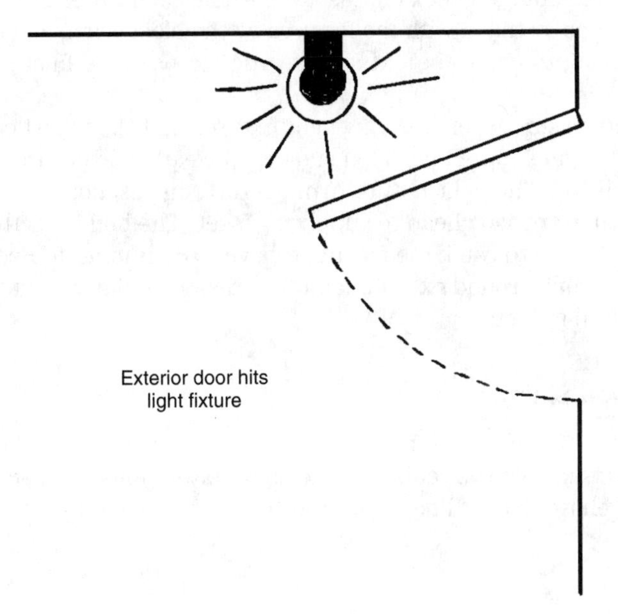

Figure 19.48

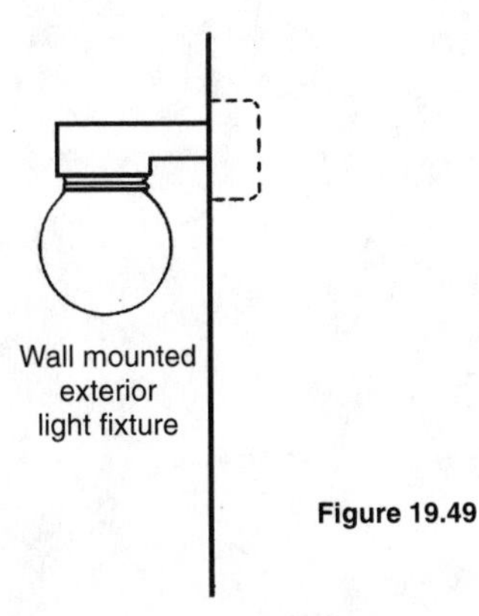

Figure 19.49

rough electrical wiring phase. This will prevent having to add a floor mounted door stop or some other means of protecting the fixture.

Exterior Light Fixture Gaskets

The NEC requires that exterior light fixtures that are subject to the weather must either have a weather seal gasket between the fixture and the exterior wall or be sealed with a weather resistant caulking.

The builder should check during the selection and ordering of light fixtures that rubber foam or some other type gasket accompany the exterior fixtures. These gaskets can be included in each light fixture box or be delivered in bulk in a separate box. The builder should double-check on this issue when the fixtures are delivered. The gaskets can then be installed with the fixtures by the electrician, and thus save the builder from having to caulk the fixtures in order to pass the final inspection (Figure 19.50).

The builder should be sure when ordering light fixtures that they don't come with a small tag attached saying, "when installed subject to weather exposure, this light fixture must be caulked." The light fixture manufacturer in essence is saying that, with these fixtures they are too cheap to supply a gasket. The builder is then stuck with getting the electrician to caulk the fixture or have to do it himself. Because nobody likes to silicone caulk around exterior light fixtures, the builder should look for fixtures that come with gaskets.

Light Fixtures and Bevel Wood Siding

Illustrated in Figure 19.51 is an exterior wall covered in bevel wood siding with a wall-mounted light fixture. The outlet box for the light fixture must be placed during the rough electrical phase to coordinate with the layout of the wood siding, which is

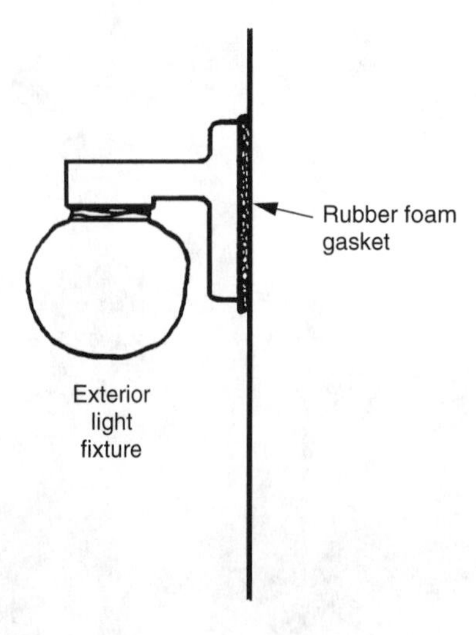

Figure 19.50

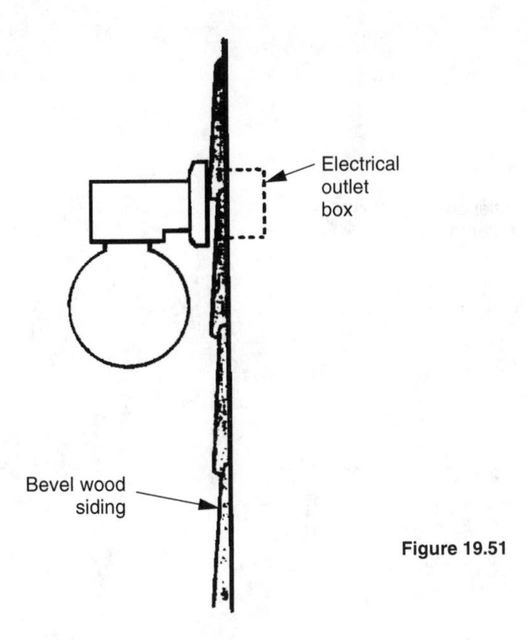

Figure 19.51

installed later. If the box is installed late in construction, after the siding is up, the fixture may be placed on a joint between two boards.

If a light fixture spans the joint between two boards, it can still be installed, but a hole must be cut in the siding so the entire wall portion of the fixture fits inside the siding boards. The joints between the fixture and the edges of the cut-out siding are then caulked. If the siding is not cut out to receive the entire fixture, the fixture may sit crooked against the siding and have uneven gaps at its sides. Having part of the fixture buried inside the siding may not look good either.

By having the electrical outlet box placed so as to fall in the center of one of the siding boards, a smaller hole can be cut at the outlet box for wiring the fixture. The fixture can then lay flat against the surface of the siding board (Figure 19.52).

Light Fixtures at Bevel Siding

Laying out the electrical outlet box so that it falls within the center of a siding board works when dealing with bevel siding that is 12 inches wide or when light fixtures are small. However, problems arise when an exterior light fixture is placed over 1×6 bevel siding. The fixture ends up spanning two or three boards (Figure 19.53).

One method of hiding the gaps between the light fixture and the bevel siding is to cut pieces of wood siding the same width as the light fixture and install them upside down behind the fixture. When two pieces of bevel siding are placed back-to-back, with one piece upside down, the result is a flat surface of uniform thickness. The siding pieces behind the light fixture fill-in the gaps exactly, providing a flat surface for the light fixture (Figure 19.53).

The problem with this method, however, is that the builder must know the dimensions of the light fixture and have the siding installers or the finish carpenter cut and

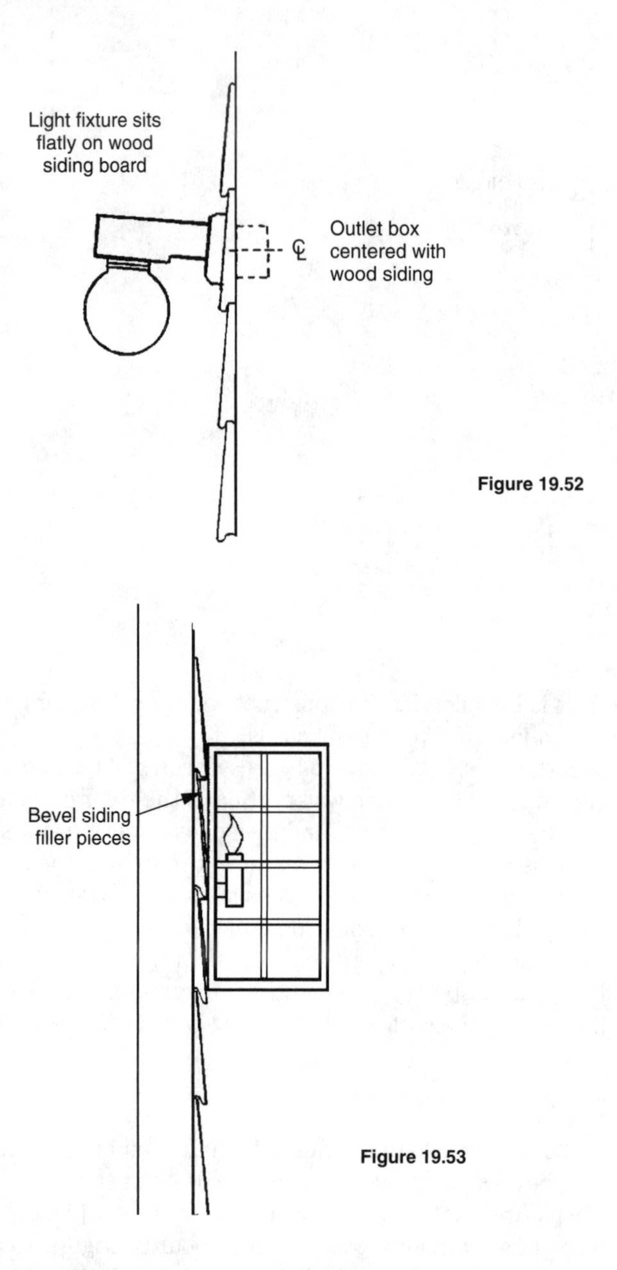

Figure 19.52

Figure 19.53

nail on these filler pieces prior to the finish electrical phase. This involves getting one typical light fixture early so that it can be used as a template for cutting the siding board filler pieces. If all of the exterior light fixtures are different and unique for each house, the electrician may have to cut and nail up these siding pieces as part of their light fixture installation.

Light Fixtures and Exterior Board and Batt

Shown in Figures 19.54, 19.55, and 19.56 are typical mistakes that are made with exterior wood board and batt siding when light fixtures are chosen that are wider than the space between the batts or when the electrical outlet boxes are not placed with the batts in mind.

Figure 19.54 shows a light fixture placed on a wall next to the man-door going into a garage. The electrical outlet box was laid-out and installed during the rough framing and ended up in the middle of a wood siding batt. The batt had to be cut out for the light fixture, which not only looked bad, but was not even centered with the light fixture. Figure 19.55 shows a batt that ended up at the edge of the light fixture and had to be scribe cut to fit around the fixture. Figure 19.56 shows a light fixture that is wider than the space between the two batts. This results in the builder having to cut or notch the batts no matter where the outlet boxes are placed.

If the builder wants exterior light fixtures to be centered within the spaces between wood siding batts or to fall in the center of a batt, then some thought must be given to their spacing at the time of the rough electrical work. For example, if the siding will be installed from one wall corner toward a doorway containing a light fixture and the siding boards are 1×12s with an actual width of 11½ inches, then a series of 11½-inch increments should be laid-out from the wall corner to the position of the outlet box. This box location should blend in correctly with the wood siding that will be installed later.

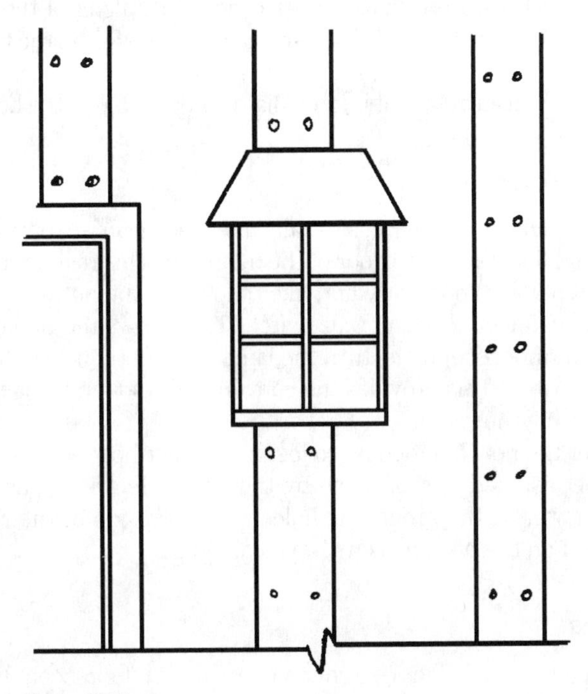

Figure 19.54

Figure 19.55

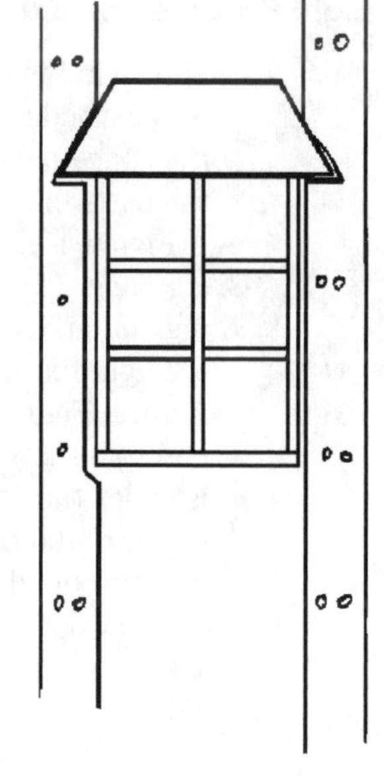

Figure 19.56

Set Electrical Cover Plate Screws Vertically

It is a good quality-control feature to have the electrician install all of the electrical outlet cover plates with their screw head slots turned vertically during the finish electrical phase of construction (Figure 19.57). This gives the impression of precision and uniformity in construction in an area that is immediately noticeable to the homebuyer.

Half-Hot Electrical Receptacle

In bedrooms without ceiling lights, there is usually one electrical receptacle that operates from a wall switch located at the door of the bedroom. This receptacle has one of the two plug-in receptacles constantly hot, like the rest of the outlets in the bedroom, but the other is connected to the wall switch. This allows the homeowner to plug a lamp into the half-hot receptacle, turn the lamp on, and then switch the lamp on and off at the wall switch. This provides the convenience of being able to turn on a lamp light upon entering the room.

To make it easier for the new homebuyer to locate this half-hot receptacle in each bedroom, they can be installed upside down by the finish electrician, as shown in Figure 19.58. In other words, the grounding holes are on the top of the receptacle plug-in holes instead of on the bottom as with typical plugs.

Electrician Installs Light Bulbs

When the units have ceiling flush lights, kitchen undercabinet fluorescent lights, and attic lights, the electrical contract should call for the electrician to provide and install bulbs for these fixtures. Flush lights cannot be found in the local grocery store, and are therefore more easily obtained by the electrical contractor than the homebuyer.

The homebuyer warranty book that lists contractors for repair service, should also include a note telling the homebuyer to call the electrical contractor if light bulbs do not work. This will prevent the new homebuyer from calling the builder for every flush light that doesn't work or every fluorescent light that flickers. The electrical contractor should be responsible for replacing defective light bulbs only once at the time of the homebuyer move-in.

Ring Out Electrical before Wallpaper

After the electrical plugs, switches, and fixtures are installed, the electrician *rings out* each house. This is done by hooking up a low-voltage battery to the main breaker panel and testing all of the outlets using a specially rigged receptacle plug attached to a bell. By doing this, the electrician can test all of the circuits for any cut wires or buried outlet boxes concealed within the walls or ceilings.

Because the electrical ring-out process sometimes turns up electrical problems that require holes to be cut through the walls and ceiling drywall, wallpapering in sales models should not begin until this process is complete. The time to discover that, for example, a master bathroom light switch doesn't work is not after several thousand dollars worth of $100 a roll wallpaper is already installed.

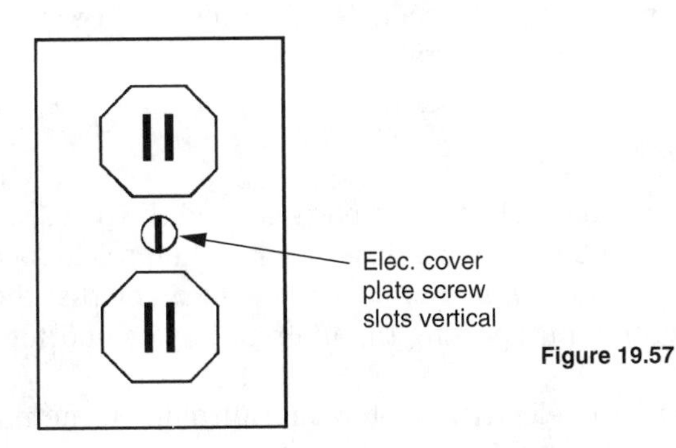

Figure 19.57

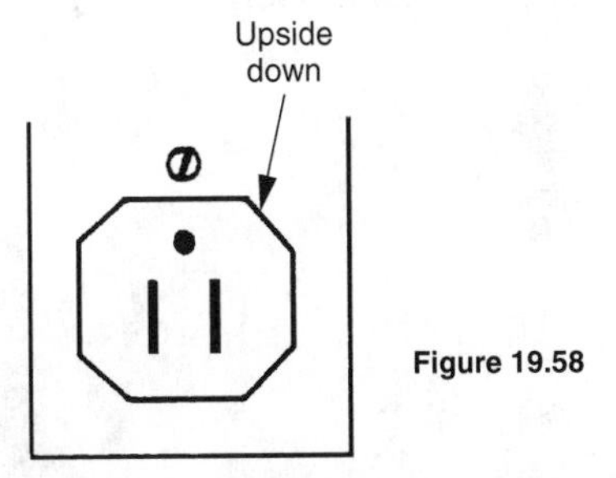

Figure 19.58

Label Breaker Panel

The electrical breaker panel should have each circuit breaker clearly labeled. This should be done by the electrician in such a way that the layman homebuyer can figure it out. Words that must be abbreviated in order to fit within the panel should at least be clear enough to be read by someone who knows nothing about house electrical circuits.

The best method is to have the electrical contractor type up labels for each house. The labels should then be glued or taped over the labels provided in the breaker panel. This eliminates someone scribbling unreadable abbreviations with a ball-point pen or a dull carpenter's pencil.

Electrical Cover Plates for Wallpaper

The installation of wallpaper in sales models should be scheduled after the interior wood trim and doors are enamel painted, but before the installation of mirrors, medicine cabinets, towel bars, toilet paper holders, and toilets. However, in order for electrical receptacles, switch cover plates, and HVAC registers to be wallpapered, they should be installed before papering begins.

Electrical Wiring in Sales Models

In sales models, interior light switches should be wired so that they cannot be turned off and on at the switches. This prevents homebuyers and their children from turning off lights as they walk through the sales models and saves the sales people from having to make sure all lights are on at the start of a day.

Walk Units after Electrical Meters Set

Sometimes the builder elects to have electrical meters set and the power turned on for houses, apartments, and condominium buildings prior to their sale or occupancy. This service is kept at the minimum *show and clean* rate, meaning that the electrical demand is limited to lighting for showing the units to buyers and for periodic cleaning until they are sold.

Immediately after installing the electrical meters and turning on the power, the builder should walk each unit to check for lights, bath fans, appliances, and air conditioning that were installed during the construction with their switches in the *on* position. This will prevent these items from running continuously until someone turns them off.

Chapter

20

Drywall

Clean Yard before Drywall Stocking

Prior to the stocking of drywall in a house, the exterior yard should have all trash and debris removed. This should be a standard, normally scheduled rough cleanup that is called out in the cleanup subcontractor's contract.

This cleanup will provide for safer and faster movement around the perimeter of the house by the forklifts that are stocking drywall and loading roof tile. It will also provide a clean working area for the lathers and scaffold builders.

Stock Drywall Stacks Evenly

The stocking of drywall in stacks on floors above the first floor should be spread out as evenly and equally as is practical. This helps prevent excessive loading on second and third floors and minimizes any possible permanent downward deflection or bowing of floor joists.

Spreading out drywall stacks is something the builder should discuss ahead of time with the drywall stocking company. It also prevents some areas inside the units from being overstocked simply because it's easier and more convenient for the stockers to load the drywall all in one place.

Stockers and Window Frames

Sometimes it is necessary for the drywall stockers to remove the center mullion piece from a second or third floor window frame in order to fit drywall sheets through the opening. This is done when there is no other way to stock the drywall into these upper floors.

The builder should check during the drywall stocking that these pieces are reinstalled by the drywall stockers with the same care and level of quality as existed prior to their removal. The builder should make it clear that the drywall stocking crew has the right to remove pieces and parts from other tradesmen's work, only if they are put back in their original condition.

Drywall Stockers Replace Studs that Are Removed

Sometimes several wall framing studs are removed during drywall stocking so that the sheets can be unloaded directly off the forklift without having to turn each sheet on edge to fit through the wall framing (Figure 20.1).

The builder should check that these studs are nailed back in place by the drywall stockers with the same quality and in the same condition as originally set. This involves carefully removing the studs without splitting or breaking them, hammering in the nails remaining at the top and bottom plates, and then carefully nailing back the studs on their previous layout so as to be vertically straight.

Drywall Stocking and Shear Panel

Illustrated in Figure 20.2 is a condominium or apartment partywall that is plywood shear-paneled on both sides. Because one of the two sides cannot be covered until the mechanicals and insulation are installed and inspected, the second side is left uncovered. Once everything in the partywall is approved, the second side can go up.

Avoid stacking the drywall too close to the wall that has not been shear-paneled. If there is not enough clearance between the drywall stack and the wall for a hammer

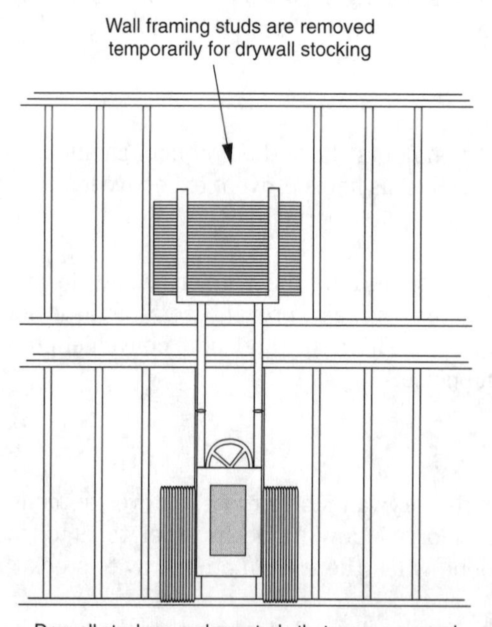

Figure 20.1

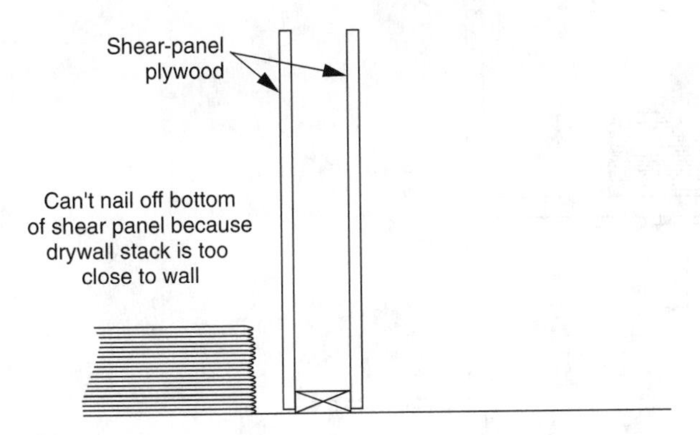

Figure 20.2

or a nail gun, the bottom portion of the plywood cannot be nailed off without moving the drywall. Moving 20 or 30 sheets of drywall 12 to 18 inches inside a small room is not easy, yet the drywall cannot be hung, and thereby gotten rid of, until the plywood is nailed off and inspected.

In cases like this, the builder should spray paint the location of drywall stacks on the floor and tell the drywall stockers to leave at least 2-feet clearance from these two-sided shear-paneled walls.

Support under Drywall Stacks

The overstocked stack of drywall illustrated in Figure 20.3 is supported from underneath by two 2×4 studs. This support helps prevent a large stack of drywall from causing permanent downward deflection of the floor joists as a result of the excess weight in that location for several weeks until the drywall is installed.

The builder should discuss drywall stocking issues with the drywall contractor prior to the start of construction. It should be determined whether the drywall can be spread around the upper floors of a project in small enough stacks to prevent structural damage to the units or whether temporary bracing needs to be added under larger drywall stacks. The question of who supplies, installs, and removes these extra 2×4 support studs should also be addressed.

Preliminary Drywall for Showers

The fiberglass shower enclosure pictured in Figure 20.4 has drywall between the unit and the wall framing at the apartment partywall. The drywall is required for sound and fire separation between the two apartments.

The builder should get the drywall installation dimensions from the shower specification sheets and field installation representative. The correct height and side-wall depth is required so that all drywall requirements are met. The drywall is then installed during the preliminary drywall phase, after the plumbing is complete and the partywalls at the shower areas are insulated, but before the shower units are installed (Figure 20.5).

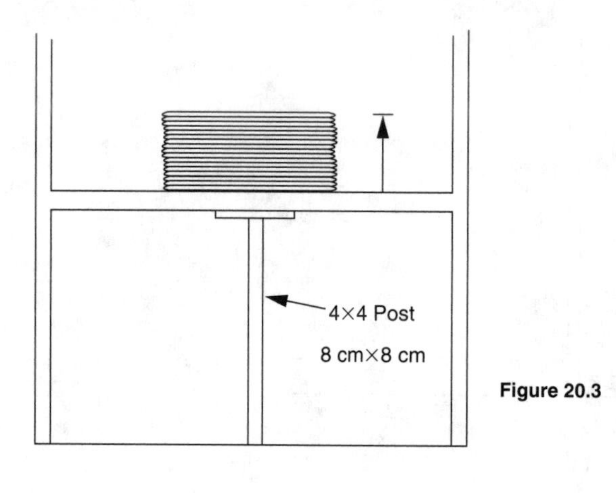

Figure 20.3

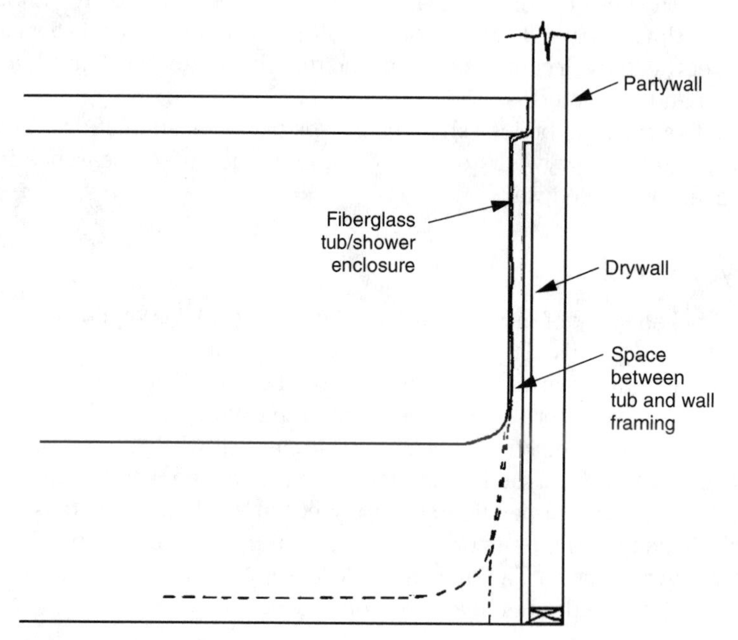

Figure 20.4

Preliminary Drywall on Partywall Joists

Figure 20.6 shows an architectural detail of a partywall between two condominiums. In this example, the architect wants the floor joists framed over the double top plates so that drywall can extend up to the bottom of the floor sheathing. This then gives a one- or two-hour fire protection to the entire wall assembly, from bottom plate through the floor joists.

When this is the case, the builder should schedule the installation of the preliminary drywall covering the floor joists, prior to the start of the rough mechanicals, such as plumbing or electrical. This allows the drywallers to cut and nail the 11½-inch×12-foot long pieces of drywall in place (for 2-inch × 12-inch joists, for example) without having plumbing pipes, air-conditioning ducts, and electrical wiring in their way. An unobstructed 14½-inch open bay between the wall and the next floor joist also helps to make the installation and nailing of the preliminary drywall easier.

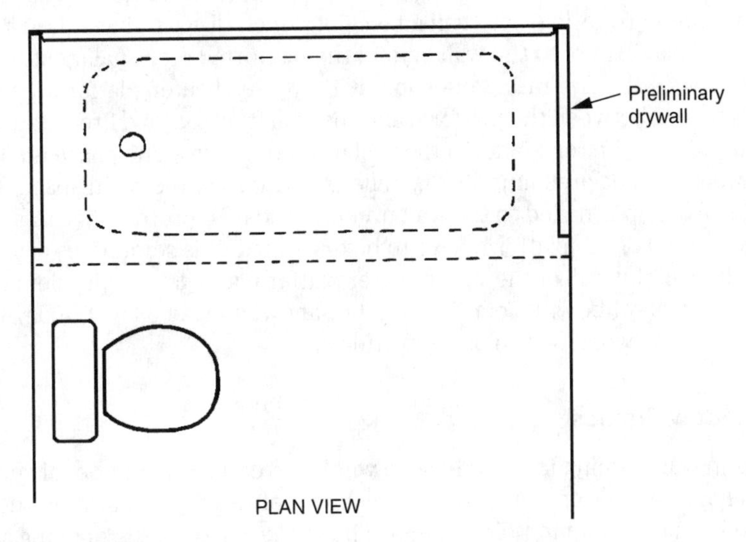

Figure 20.5

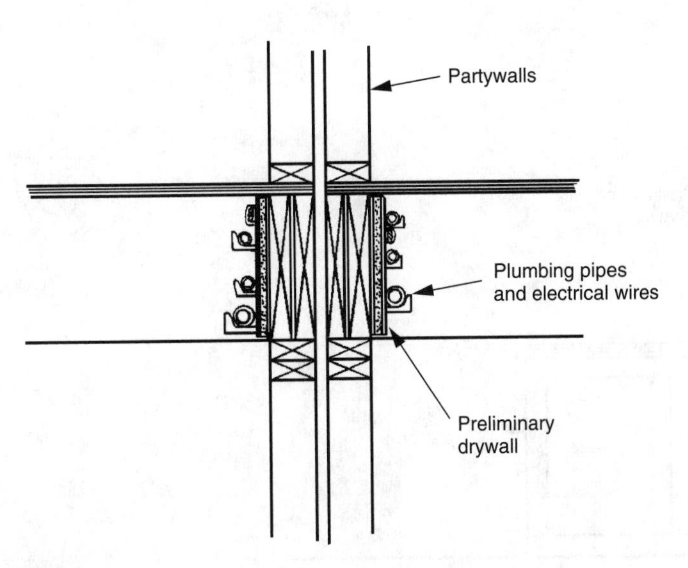

Figure 20.6

Preliminary Drywall behind Water Heater Platforms

The water heater platform illustrated in Figure 20.7 is located on a garage wall, which is a partywall between two condominium units. In this case, the partywall is a one-hour rated firewall, requiring one layer of ⅝-inch thick, type-X drywall on each side of the wall.

One method to simplify the building of the water heater platform while maintaining the required fire rating, is to nail a piece of ⅝-inch drywall that is larger than the water heater platform onto the wall behind the platform before nailing it to the wall. This then solves the fire-rating question at the water heater platform because the drywall is placed between the platform and the wall framing. The fire-rated drywall is then continued on the remainder of the wall when the drywalling phase is completed.

If the walls are not preliminarily drywalled, an unprotected open space is created where the platform is nailed to the wall framing studs. To protect this area from fire, the sides and top of the platform have to be drywalled. It is sometimes easier to preliminary drywall the wall at the water heater platform and get the problem out of the way rather than bother with coordinating the sandwiching of a layer of drywall along with a sheet of plywood on top of the platform.

Drywall around Window Frames

Openings in wall framing for metal-frame windows require an additional vertical and horizontal allowance to provide for the thickness of the drywall around the interior sides of the window opening. The framing allowance added to the opening should be such that when the drywall is installed, it butts into the window frame, leaving an

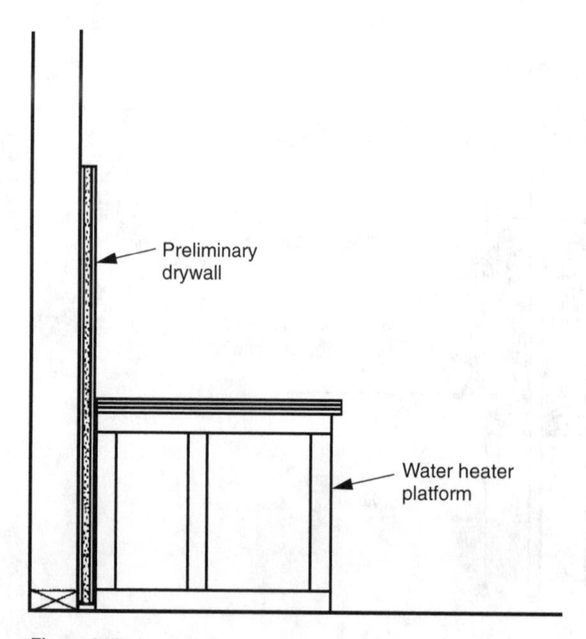

Figure 20.7

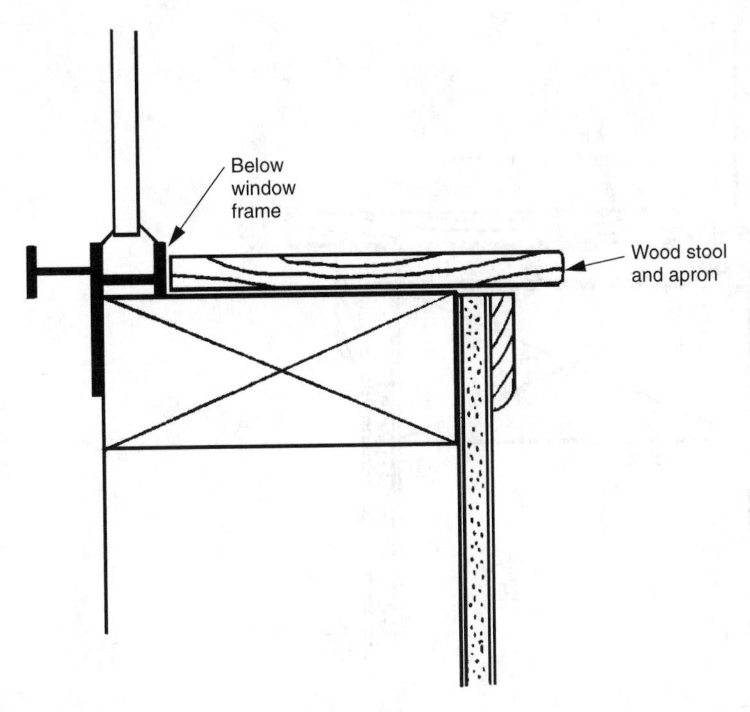

Figure 20.8

equal reveal of the frame exposed on all four sides. This provides an easier drywall joint to finish and lessens the appearance of framing inaccuracies around the window frame.

If the window sill is to be finished with wood stool and apron, the same framing allowances are added to the opening, but drywall is not installed on the window sill. The wood stool, being nearly the same thickness, merely takes the place of the drywall (Figure 20.8). If both drywall and stool are added to the window sill, an additional allowance must be added vertically to the window opening to allow for the thickness of the stool. If this allowance is mistakenly omitted, the wood stool may extend above the lip of the window frame, making it difficult or impossible to remove the sliding half of the window from the inside (Figure 20.9).

Drywall in Doorways

Drywall should not be allowed to extend into rough-framed door openings. The drywallers should trim or saw off this excess material during the installation phase (Figure 20.10).

In production housing, because of framing inaccuracies, the finish carpenters who installs the interior prefit door units often has to trim off the door openings. Because he or she doesn't have the proper tools (drywall saw) or skill to quickly accomplish this task, interior door installation is then slowed and unnecessary debris is added to the floor as a result.

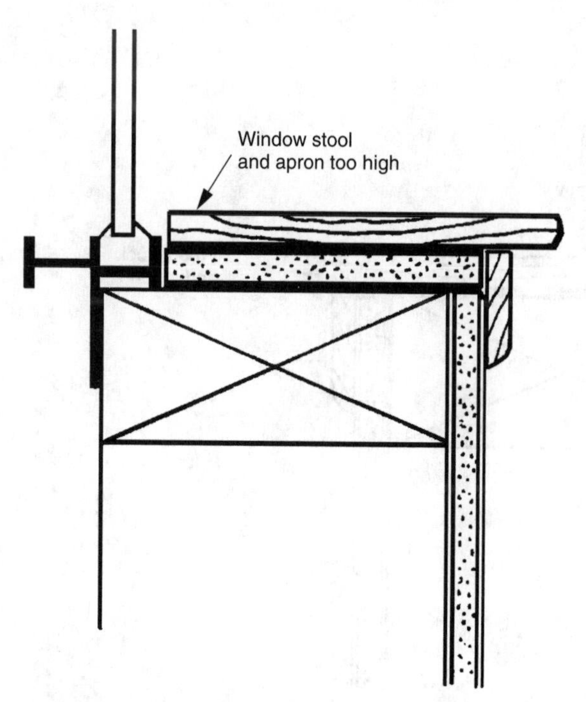

Figure 20.9

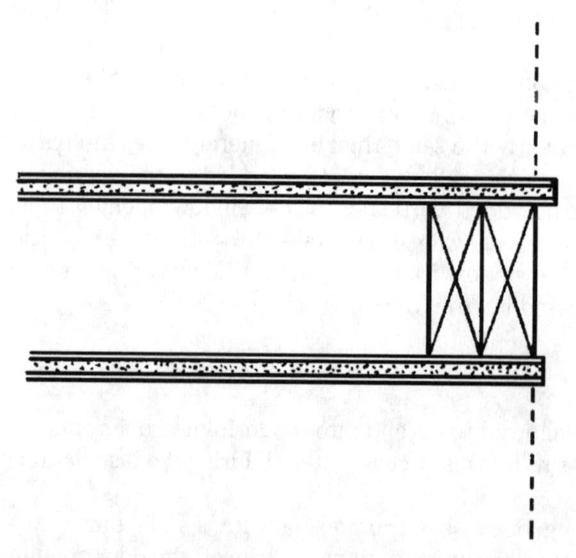

Figure 20.10

Drywall Cutout around Outlet Boxes

Drywall should be cut out closely around at least the top and bottom of a wall electrical outlet box. This ensures that when the receptacle is installed, the two metal tabs on the top and bottom of the receptacle will have solid drywall to rest against.

If the drywall is cut loosely around the outlet box and the outlet box is not flush with the drywall surface, the receptacle has nothing to rest against to keep it flush with the wall. The plastic wall cover plate screw will pull the receptacle toward it and hold it in place, but the cover plate is not designed to resist the force of an electrical cord being plugged into a receptacle. The result is a loose feeling receptacle or a cracked or broken wall cover plate (Figure 20.11).

Drywall over Doorways

Drywall installation around doorways should be *horned-over* the opening, as shown in Figure 20.12. Having the drywall extend past each side of the opening to a cripple above the header adds solidity to the door opening and prevents cracks from forming above the striker or latch side of the door when it is slammed shut.

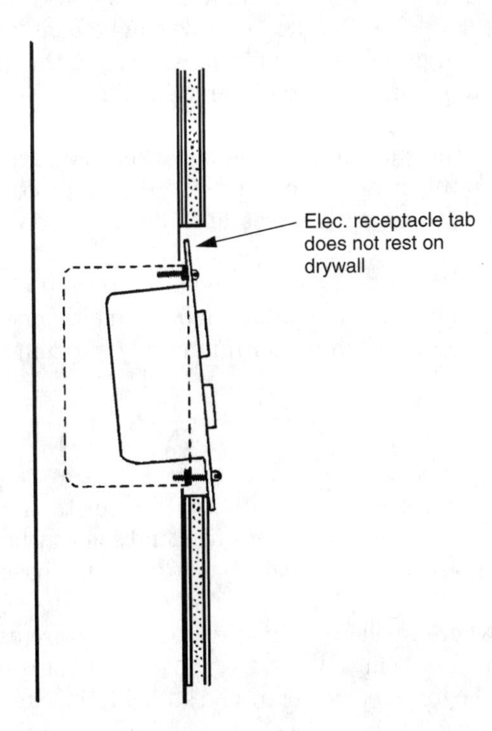

Figure 20.11

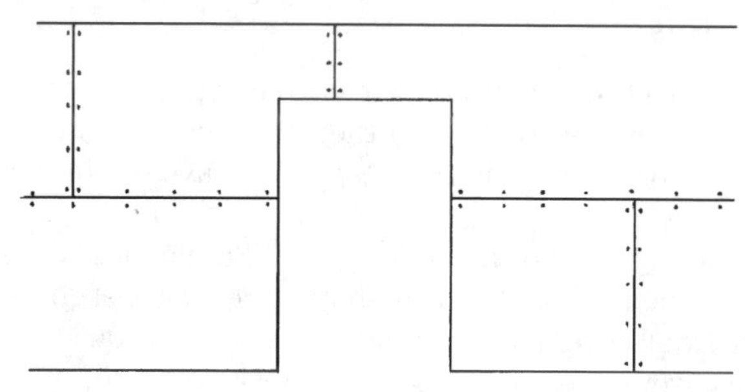

Figure 20.12

Drywall on Fireplace

Some prefabricated metal fireplaces are made for ⅝-inch thick drywall so they can be used on apartments and condominiums as well as houses. The metal tabs on each side of the fireplace, which are nailed to the wall framing, are set back from the front edge of the fireplace by ⅝ of an inch. This allows the required one-hour ⅝-inch drywall to come out flush with the fireplace face. When ½-inch drywall is used in single detached houses, the fireplace projects past the drywall ⅛ of an inch.

The builder should try to get both the fireplace and the drywall flush. If the fireplace side tabs for ⅝-inch drywall are flexible, the fireplace can be bent inward the ⅛ inch to be flush with the ½ inch drywall. If the nailing tabs are rigid, the drywall on the fireplace wall can be changed to ⅝ inch to make everything flush.

Decorative materials, such as marble, tile, or wood, are often added over the joint between the metal fireplace and the surrounding drywall. When the fireplace projects past the drywall, the decorative material that rests on the fireplace front will also project past the drywall. This will leave a gap that must then be filled with caulking or grout.

The mistake to avoid is letting the initial ⅛-inch problem turn into a ¼-inch to ⅜-inch problem due to poor workmanship. Few things look worse than a fireplace with a ⅜-inch wide caulked gap between the beautiful marble surround and the wall.

Gap between Drywall and Floor

The gap between the drywall and the floor should be no greater than ½ inch to ¾ inch. A gap of this size or less gives baseboard trim a firm and solid surface to nail against (Figure 20.13). When the gap is greater, nailing may cause the baseboard to tilt (Figure 20.14).

This problem usually occurs where small drywall scraps and pieces are used to fill in around columns, posts, and door openings. When the gap at the floor is too large, even though the baseboard may be installed correctly and straight, the later installation of carpet or shoebase trim may tilt the baseboard away from the wall (Figures 20.15 and 20.16).

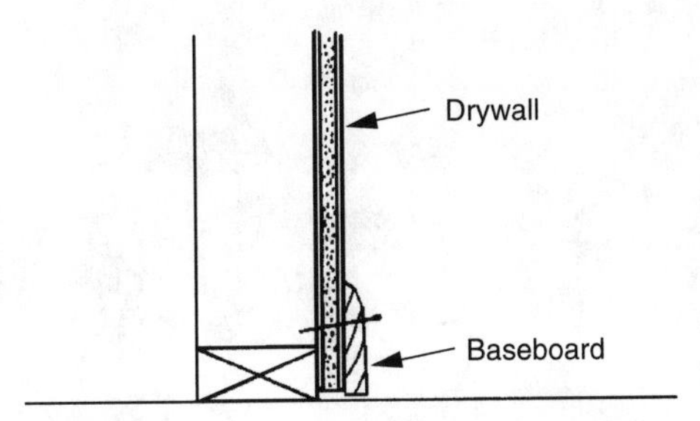

Figure 20.13

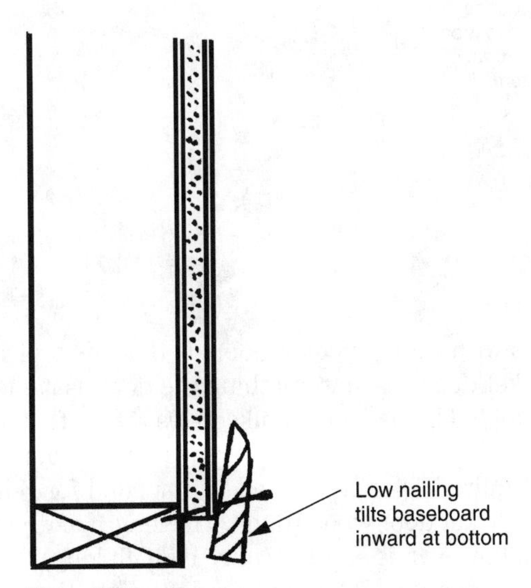

Figure 20.14

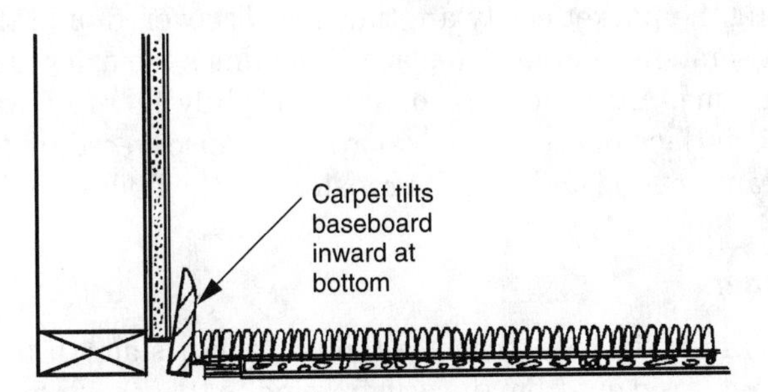

Figure 20.15

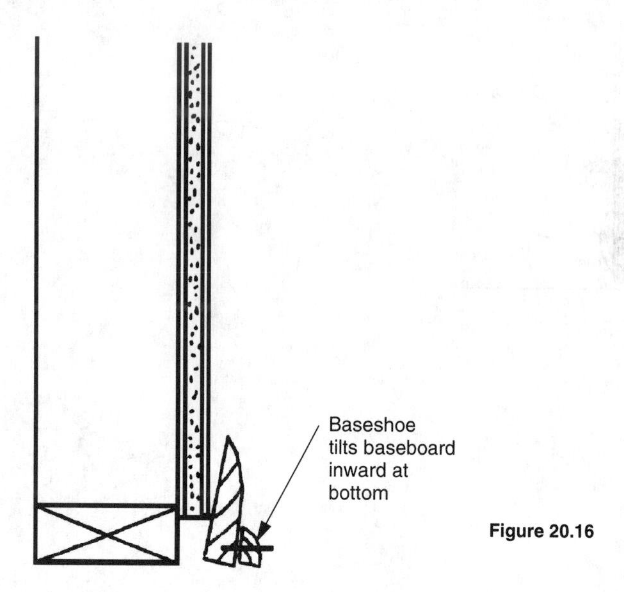

Figure 20.16

Drywall and Pocket Doors

A common problem that occurs with sliding pocket doors is that the typical drywall nail will penetrate into the pocket door space, scratching the door as it slides in the pocket. This problem can be avoided by using drywall screws rather than nails. The required length of drywall screws is shorter than that of nails for the same thickness of drywall. For example, the required 1⅜-inch drywall nail needed for ⅝-inch thick drywall, may be too long in the pocket door wood frame. The length of drywall screw for the same thickness drywall, however, is shorter (only 1⅛-inch long). This length will not penetrate the drywall and the ¾-inch thick wood frame into the pocket cavity space, scratching the door (Figure 20.17).

Pocket doors are installed by carpenters after the drywall has been taped and textured. Nails penetrating the pocket cavity are thus not discovered until the pocket door is installed and is scratched or binds in place. Sometimes the nail will not touch the door until a later time, after the door or frame slightly bows or warps. The scratch on the painted surface of the pocket door is then discovered by the homebuyer during the warranty period and must be repaired by the builder.

Cut Drywall Close to Stair Corner

Illustrated in Figure 20.18 is the side view of a run of interior stairs. If framed correctly, the stair jacks are furred out from the wall framing so that the drywall can slip in between the stair steps and the wall.

The builder should check, however, that the drywall ends up below the inside corner of each stair step. Because the carpet must be tucked downward tightly into the wall, thinner carpet, such as Berber, will not always cover a gap at the edges of stair steps. A small hole at a stair step inside corner will be difficult to repair if the stairs were incorrectly framed.

Fire Resistant Fiberboard Next to Fireplaces

The two-sided fireplace pictured in Figure 20.19 has an adjacent wall that is closer to the firebox than is allowed by the building codes. When adjacent walls, pop-outs, and mantles are closer than 8 inches to the opening of the firebox, the builder should discuss with the drywall contractor the possibility of using one of the several fire resistant products in these areas in place of the drywall. These products are the same thickness as drywall and can be taped, textured, and painted just as the other wall surfaces.

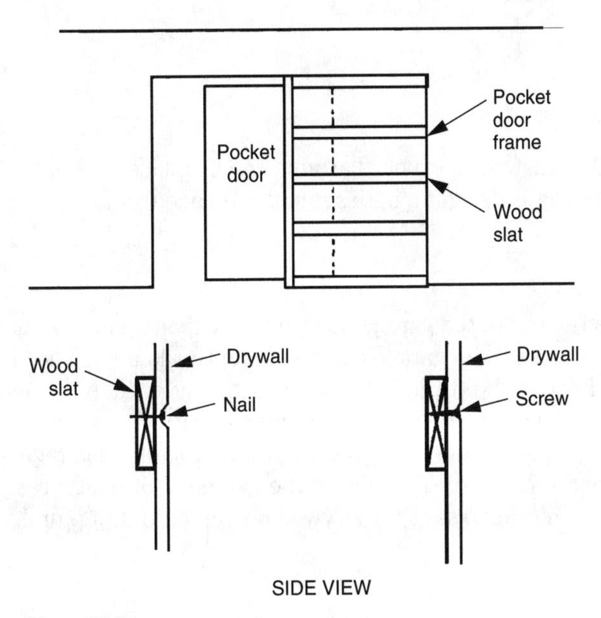

Figure 20.17

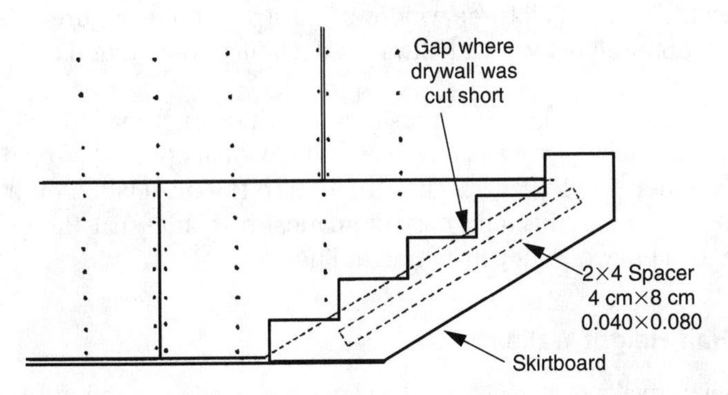

Figure 20.18

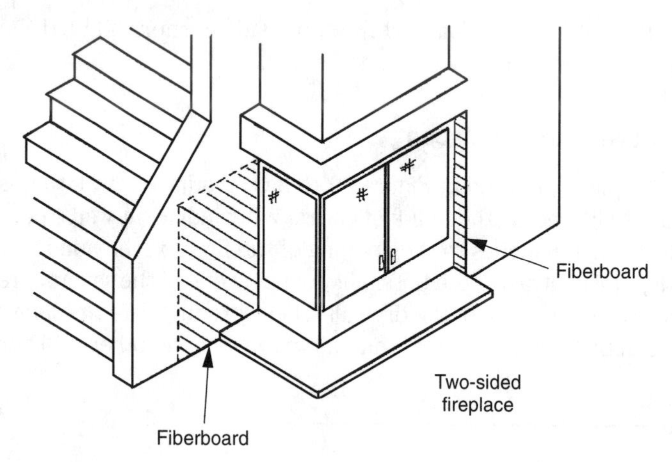

Figure 20.19

Using approved fire resistant fiberboard around the fireplace fireboxes allows the architect and builder greater freedom in designing fireplace fronts.

Garage Door Drywall

A 4×12 header spans the top of the garage door opening that is pictured in side view in Figure 20.20. When closed, the garage door rests on this header. The sides of the opening are trimmed with a 2×6 to hide the gaps at the side when it is closed. The tricky part is coordinating the placement of the drywall with the garage door.

One method of doing so is to cut the drywall above the top of the garage door, then trim the edge with metal mill-core. This allows the garage door to still rest upon the 4×12 header, without the thickness of the drywall pushing it out of plumb.

Align Window Cornerbead

Illustrated in Figure 20.21 is a simplified representation of three window openings from the front view. This illustration shows the importance of aligning the cornerbead on the window sills of all three windows in a straight line. Figure 20.22 depicts a pot shelf partition wall between a hallway and a family room that has three similar openings.

A common mistake in production housing is for the cornerbead installer to place the cornerbead on each opening as if it were an individual opening instead of part of a group. The builder should pay special attention to the installation of drywall cornerbead in situations like this, and not automatically assume that the installer will take the time to align everything in a straight line.

Level Cornerbead on Half-Height Walls

For all half-height walls, especially thicker ones like that depicted in Figure 20.23, the drywall cornerbead should be installed using a level. The half-height wall can be framed perfectly level, but if the drywall cornerbead is nailed on crooked, the front of the wall will end up crooked after the entire thing is taped, textured, and painted.

Cut Out Drywall at Plate Straps

Metal framing straps and protector plates are nailed on the sides of wall framing bottom and top plates by the framer and mechanical tradesmen to create structural ties and to prevent plumbing pipes and mechanical vents from nail punctures.

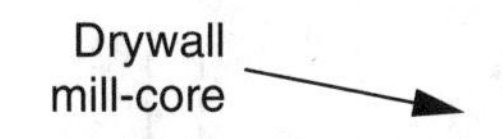

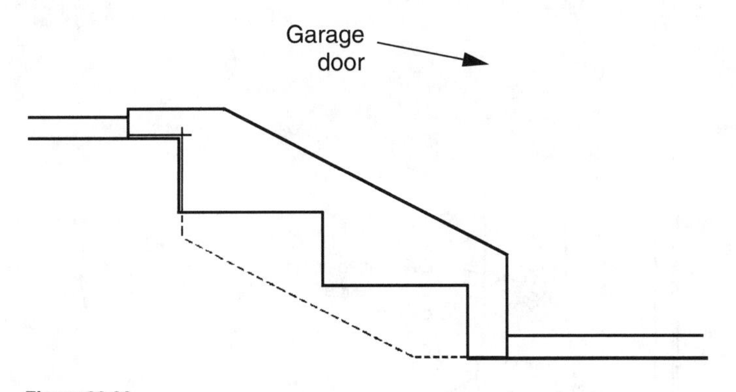

Figure 20.20

Figure 20.21

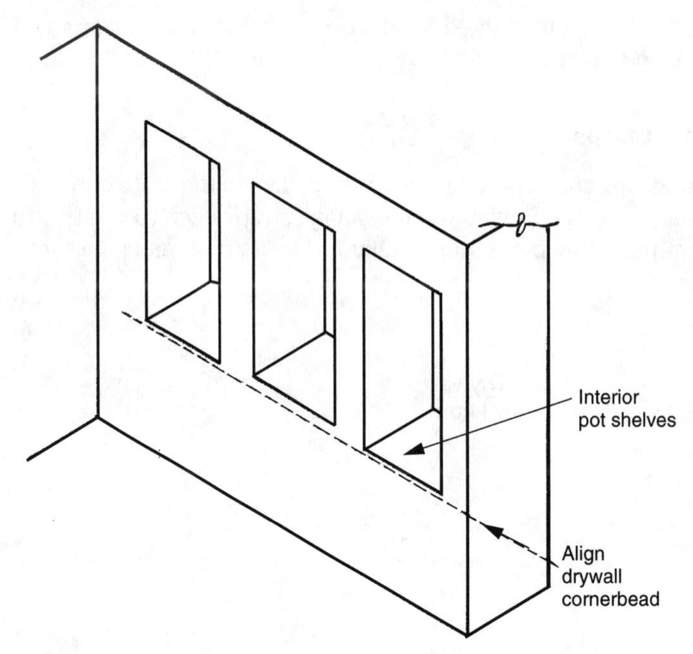

Figure 20.22

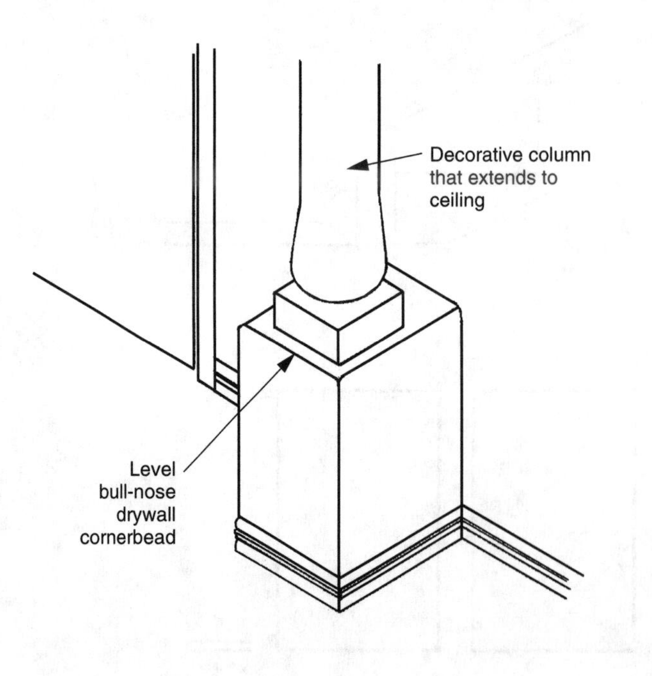

Figure 20.23

To prevent these straps and their projecting nail heads from causing unsightly bulges on the wall surfaces, the drywall installers should cut out the drywall at these members (Figure 20.24). The drywall can then fit tightly against the wood framing around the metal straps, with the voids filled-in later by the drywall tapers. This practice greatly improves the straightness of walls at the floor and ceiling lines.

Wall Drywall Bulges

The builder should check for wall and ceiling line bulges after the drywall is hung, before it is taped. If the framing straightedging was thorough, the wall and ceiling lines should be fairly straight.

Small bulges, due to plumbing pipes, metal protector plates, and crooked framing that was missed during the straightedging, become immediately noticeable after the drywall is hung and nailed off. If the bulge is large enough, the piece of drywall should be taken off and the problem itself should be fixed or a new piece of drywall back-cut to fit around the problem. If the bulge or bow is slight, these areas can often be floated out by the taper.

If wall bulges can be identified prior to the start of the drywall taping, the problems can be fixed and the taping repairs worked into the normal drywall taping without becoming an expensive extra. If not caught until after taping and texturing, then repairs involving tearing out and replacing drywall result in extra costs added to the contract.

Waterproof Decks

When a second floor balcony deck is designed above a garage, the plywood sheathing covering the deck should be waterproofed before the drywall is installed in the garage (Figure 20.25). This will prevent water damage to the drywall ceiling under-

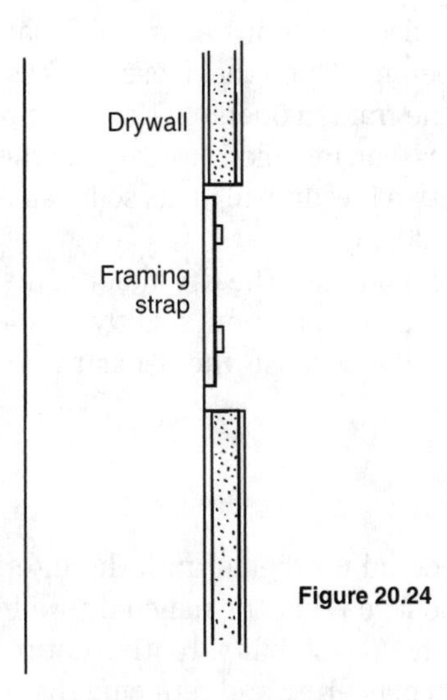

Figure 20.24

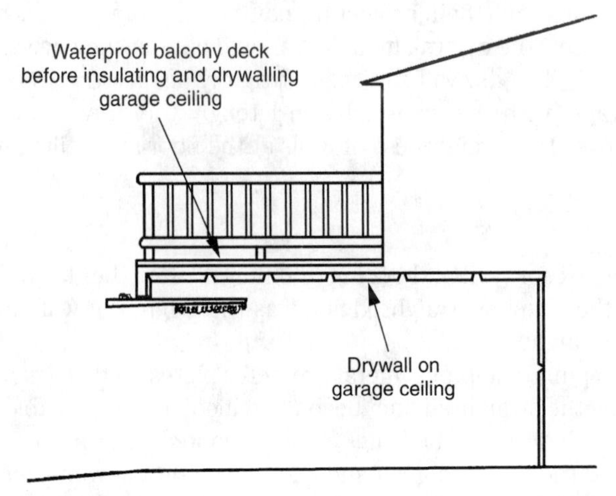

Figure 20.25

neath the deck, should it rain and leak through the unprotected plywood joints. The same holds true for decks with stucco lathing beneath them.

Nailing at Beams

Because solid wood beams may shrink or twist slightly in the first year, extra nailing should be added to the metal cornerbead of the drywall ceiling beams to prevent cracks at the edges (Figure 20.26).

Trim Drywall at Attic Access

One way to trim out the edges of an attic access is to install metal T-bar track with mitered corners at all four sides of the opening. Once the T-bar track is in place, a drywall attic access panel is cut to match the framed opening, then dropped into the track (Figure 20.27). Because the track is not installed until later, the drywaller should cut the panel and tack it in place with a few drywall nails so it can be textured and painted along with the ceiling (Figure 20.28).

When this method is used, the builder should have the drywaller come back after the T-bar is in place to cut the panel to fit the new and now slightly smaller opening. The drywaller is the best choice for this task because the T-bar track installer is probably not familiar with cutting drywall.

Screw Drywall Instead of Nailing

It used to be that the exterior plastering could not begin until the interior drywall was fully nailed off and inspected. It was believed that the nailed drywall in the interior of the building added shear strength and stability to the entire structure, thereby providing a rigid surface for stucco plastering and reducing the potential for cracking. Later it was found, through laboratory testing, that the gypsum around each nail crushes slightly when shear forces are applied, thus reducing the value of drywall shear nailing.

The builder can now schedule the installation of interior drywall to coincide with the exterior plastering, as long as drywall screws are used instead of nailing. This prevents the hammering of nails on the interior side of a wall from loosening the wet stucco that was just applied to the other side of the wall. Being able to drywall and plaster concurrently saves time on the construction schedule.

When the builder chooses drywall screwing instead of nailing so that the tasks can coincide, the builder should discuss with the building inspector the number of nails that will be allowed to tack up each sheet of drywall. Most drywallers prefer to use a few nails to tack up each sheet of drywall, then come back later and screw off the entire house or unit at one time. Tacking up each sheet is easier than trying to hold up

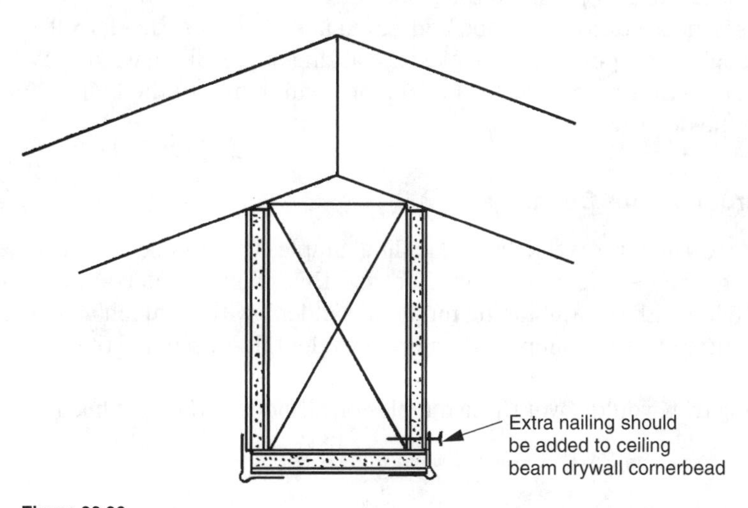

Figure 20.26

Figure 20.27

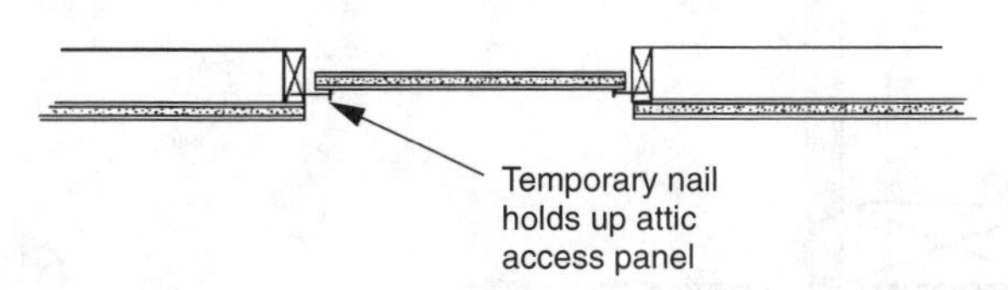

Figure 20.28

a 12-foot sheet of drywall while trying to get a screw in the right place. If using nails to tack up the drywall is cleared in advance with the building inspector, then the presence of a few nails during the inspection will not be a problem.

Coordinate Stucco Window Masking with Scrap-Out

When the builder decides to schedule the exterior plastering at the same time as the interior drywalling, there may be a conflict between the plastic masking of the windows for stucco plastering and the throwing of drywall scraps out of windows for drywall cleanup.

The best solution is to reach an agreement between the builder, drywaller, and plasterer, by designating one window per house or apartment building, for drywall scrap disposal. This limits the repair of window masking to one window per unit only, which should be acceptable to the plasterer.

If this issue is not foreseen and addressed ahead of time, the drywall cleanup men may thoughtlessly pull down the plastic masking out of all the windows. This may result in the plastering contractor tacking on a surcharge to the builder for having to replace the destroyed masking.

Drywall Taping around Door Openings

Figure 20.29 illustrates the drywall nailing dimples that can be seen around the edge of a rough-framed door opening. Because this nailing is not always consistently taped (filled with drywall taping mud and sanded) with as much care as the nailing in the center of a wall, dimples are more prevalent. It is assumed that the finish wood casing will cover up the edges around the door openings, hiding these dimples. Finish casing trim would cover these dimples on all sides of the opening if it were not for

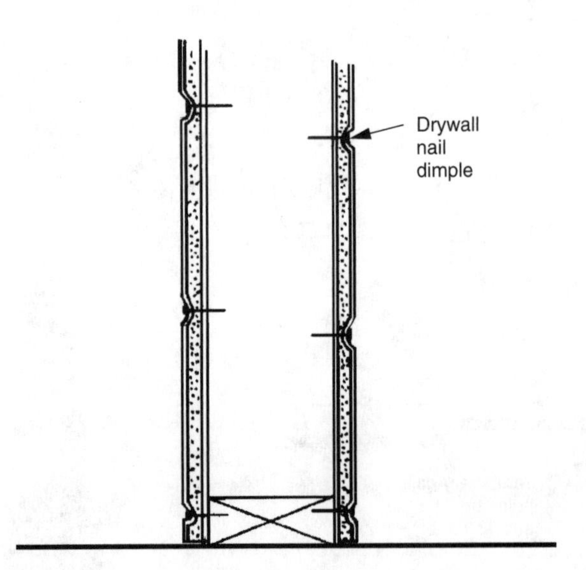

Figure 20.29

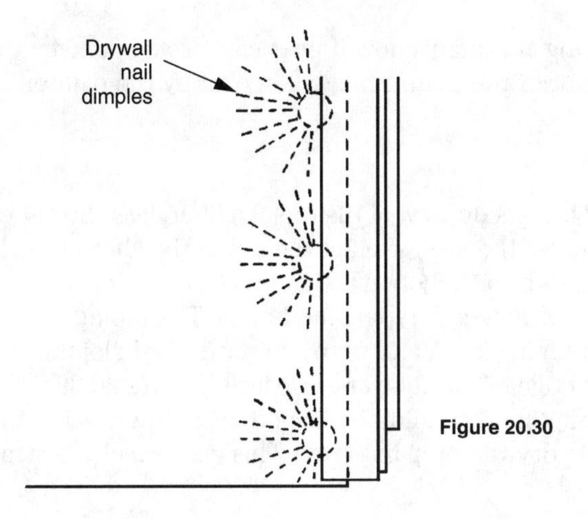

Figure 20.30

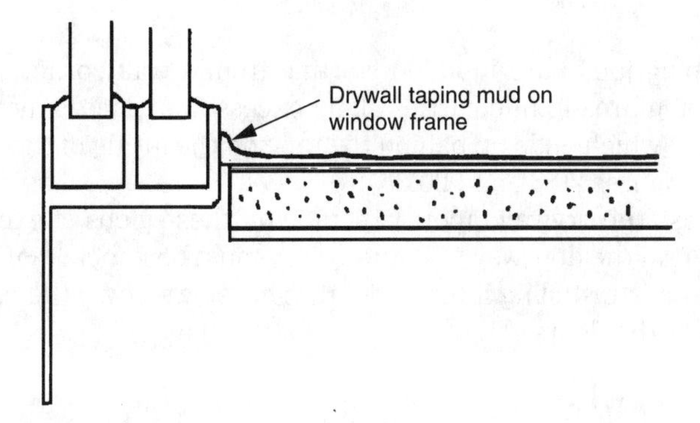

Figure 20.31

the framing allowance, which is added to the width and height of the door opening. The framing allowance provides the finish carpenter with an amount of play or tolerance with which to install the prefit door unit. This often results in the door unit being installed within the opening favoring one side or the other (Figure 20.30).

Drywall nailing dimples that are partially exposed alongside door casing, if not adequately filled and finished, are quite noticeable—especially when the walls are painted with enamel.

Drywall Taping around Windows

When drywall is taped around metal frame windows, some of the taping mud gets onto the edges of the window frame (Figure 20.31). This excess drywall taping mud should be cleaned off the window frames by the drywall cut and scrape crew. This then leaves a clean, straight edge for the painter and eliminates any later disputes as to whose fault it is for uneven paint lines. This is especially crucial for bronze colored window frames, as the joint line between the brown window frame and the surrounding lighter colored wall surfaces is very noticeable.

This cleaning and scraping around window frames can also be done by the drywall prepaint pickup crew just prior the units being taken over by the painter.

Drywall Mud above Shower

Illustrated in Figure 20.32 is the side view of the top of a fiberglass shower enclosure. Drywall and taping mud meets the level shelf at the top of the shower enclosure.

A typical problem in housing construction is that this top edge or shelf around the shower enclosure ends up with excess taping mud on it. The taping mud oozes out around the bottom of the taping knife and onto the downward-sloping shelf as the knife is dragged along the edge of the shower enclosure (Figure 20.33).

This excess drywall taping mud should be cleaned off by the tapers, the drywall cut and scrape crew, or the drywall prepaint crew. This leaves a clean, straight edge for the painter.

Taping Mud in Doorways

Horizontal drywall taping joints are mudded starting from a wall corner toward a door opening. The taping broad knife carries along excess mud as it is moved over the taping joint, some of which ends up behind the tape on the inside of the doorway at the end of the run (Figures 20.34 and 20.35).

The builder should ask the drywall tapers to scrape off these globs of excess mud during the taping. If left in the doorway to harden, they must be knocked off later by the finish carpenter who is installing interior doors. This slows down the carpenter and adds more debris to the floors.

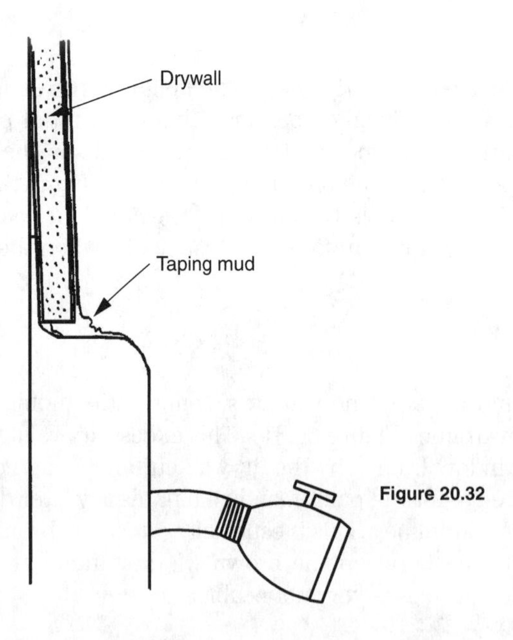

Figure 20.32

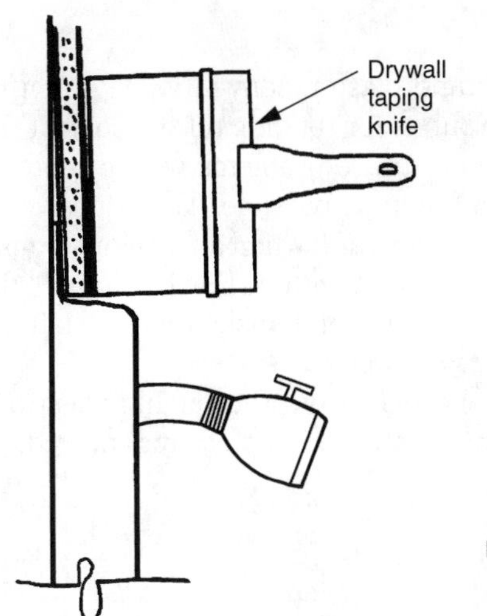

Figure 20.33

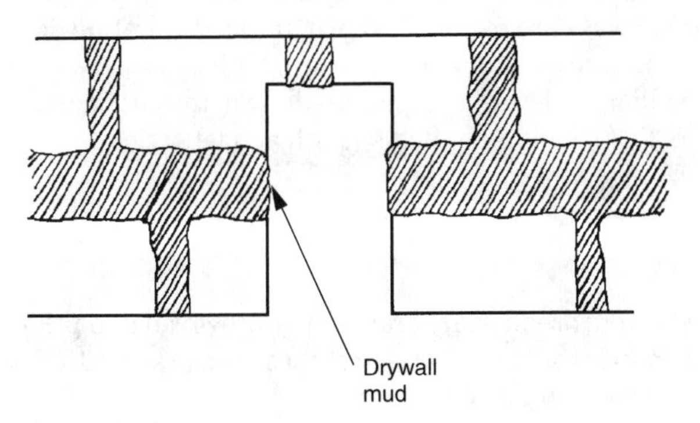

Figure 20.34

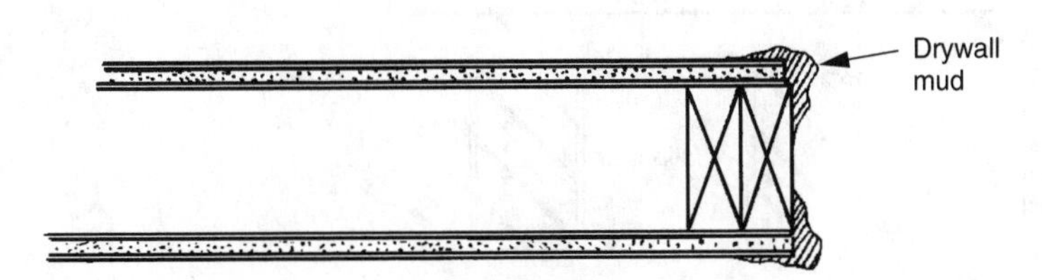

Figure 20.35

Plaster on Wood Jambs

For interior wall surfaces that are lath and plaster, conventionally hung wood door and window jambs act as the plaster grounds for rodding off the correct thickness of plaster around the openings. To alleviate future problems with dried-on plaster, it should be scraped off the woodwork as soon as possible. The adhesive qualities of the plaster cause it to adhere to the raw wood, and when an attempt is made later to scrape off the plaster, some of the wood comes with it. Furthermore, being plaster, it is not easily sanded off, yet is often too rough and unsightly to be left on the jamb and partially exposed after door casing is installed.

It is important, therefore, to have the plasterers thoroughly wash off all wood jambs during plastering while the plaster is still wet and easily removed.

Texture or Wallpaper

When it can be known beforehand which walls are to be wallpapered, those walls should not receive drywall texturing. Drywall texturing can add an unwanted and unattractive feature to some wallpapers. In some cases, such texture can be more dominant than the wallpaper design pattern. Sales models, custom homes, and condominium lobbies and hallways are examples of where the decorating design and selections can be made before the drywall texturing phase of the construction. Wallpaper areas that cannot be determined before drywall texturing, such as the example shown in Figure 20.36, can be sanded smooth at a later time.

Mask Off Tile Scratch

When the bathtub or showers are mortar scratched or brown coated prior to drywall texturing, the drywall contractor should mask off these tile areas with plastic masking before spraying the drywall texture.

Drywall texture over tile mortar forms a barrier between the coats of mortar. Because the mortar needs to be clean and free of any drywall texturing for everything to adhere correctly, it is important to mask these areas.

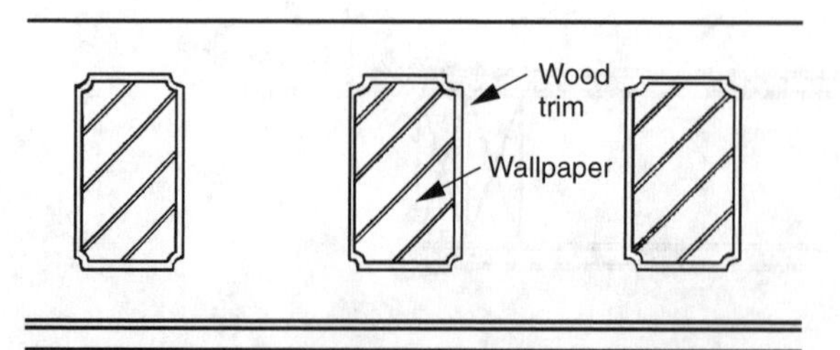

Figure 20.36

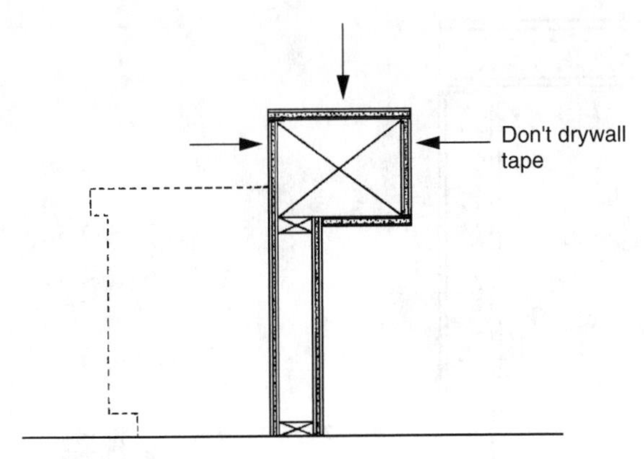

Figure 20.37

No Drywall Mud in Outlet Boxes

For the sake of the electrician installing switches and plugs during the finish electrical phase, the drywall taper should be informed that no excess drywall mud can be left in electrical outlet boxes. If not policed by the builder, some tapers will almost entirely fill the inside of outlet boxes while taping the outside of the boxes. The electrician must then chip out this hardened drywall mud weeks later, in order to install the switches and plugs.

Glue-On Tile and Drywall Mud

Figure 20.38 shows the plan view of an inside wall corner of a kitchen where a ceramic tile splash is to be glued directly over the drywall. In this situation, the inside corners of the drywall should not be taped and mudded where the tile will be installed later. Drywall taping at inside corners does not result in a perfectly square corner, but rather leaves a slight build-up of drywall mud in the corner edge. When the tiles are glued to this, they will kick out at the inside corners. A conscientious tile setter will then have to take the extra time to tear out the drywall taping at the inside corner to obtain a true corner for tile setting.

Just prior to drywall taping, the builder should mark with a carpenter's pencil the correct height above the floor that drywall taping should stop for glue-on tile. This will ensure that tape is only where is needed.

The conflict between drywall taping and glue-on tile becomes more crucial when kitchen pass-through countertops and splashes are entirely ceramic tiled (see Figure 20.37). In this type of design and application, none of the inside and outside corners of the drywall should be cornerbeaded or taped, as that throws all of the corners slightly out of square for the tile setter. If left untaped, these corners are then open and raw for the tile setter to work with, making it easier for the tradesman to line up the vertical corners with the horizontal countertop corner.

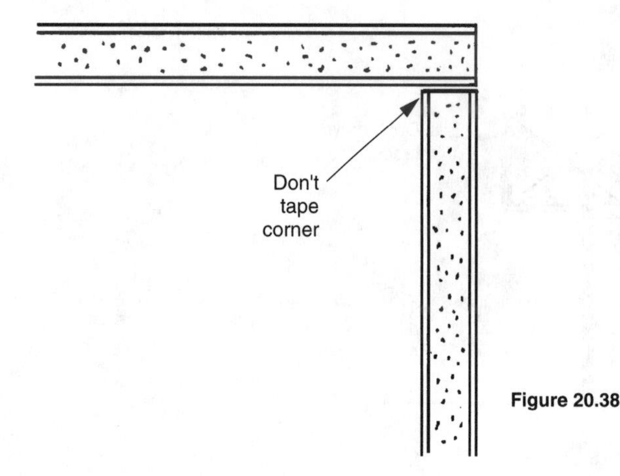

Figure 20.38

Scrape around Sliders

As part of the drywall scrape, cut, and ding operation after texturing, the perimeter edges around sliding glass door frames should also be scraped clean of excess drywall mud (Figure 20.39). The joint between the slider frame and the drywall should be scraped with a putty knife, and the frame wiped off with a wet sponge or rag (Figure 20.40).

Texturing in Units with Stairwells

The builder does not want to be guilty of telling other tradesmen how to do their jobs, but he or she can and should make suggestions on how and where the drywall texture sprayer moves around the units.

For example, in a two- or three-story condominium project, if the sprayer starts on the bottom floor and works his or her way upward, texturing as he or she goes, the spray-gun hose will rub or brush up against the walls that have just been textured, ruining them in some spots. This creates extra work for the drywall company's own drywall pickup crew.

If the builder has a policy of going through each unit and circling the dents and dings for repair for the drywall prepaints and finals, the drywall pickup crew may be at their limits of patience and tolerance for the builder anyway. Whatever the builder can do, therefore, including getting the texture sprayer to change the operation to reduce the number of drywall pickup repairs, works in the builder's favor.

In units with stairwells, especially ones with curved walls, the texture sprayer should either work from the top floor downward or work each floor so that the stairs are textured last. With this scenario, it is less likely that the walls will be damaged by the texture spray hose.

Texture at Stair Steps

When carpeted stairs do not have skirtboard, the drywall fills up all of the space surrounding the stair step treads and risers (Figure 20.41). The builder should check that the drywall texture is sprayed uniformly around all inside and outside corners of the stair steps during the initial drywall texturing or later during one of the drywall pickup phases. The inside corners of the stair steps are especially easy for the texture spray-gun to miss because the projecting nosing at the front of each step is in the way of the spray trajectory as the spray-gun moves up or down the stairway.

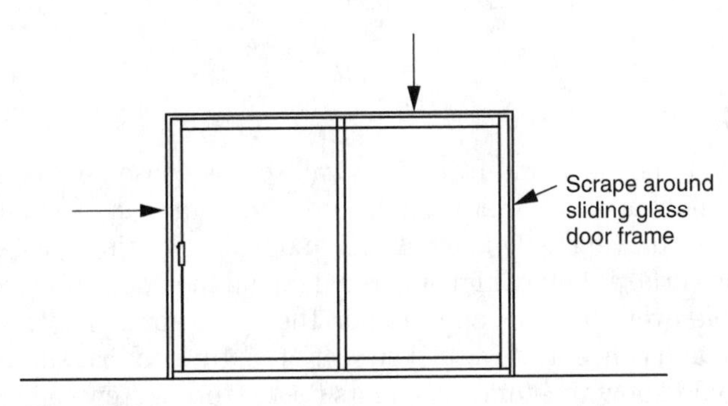

Figure 20.39

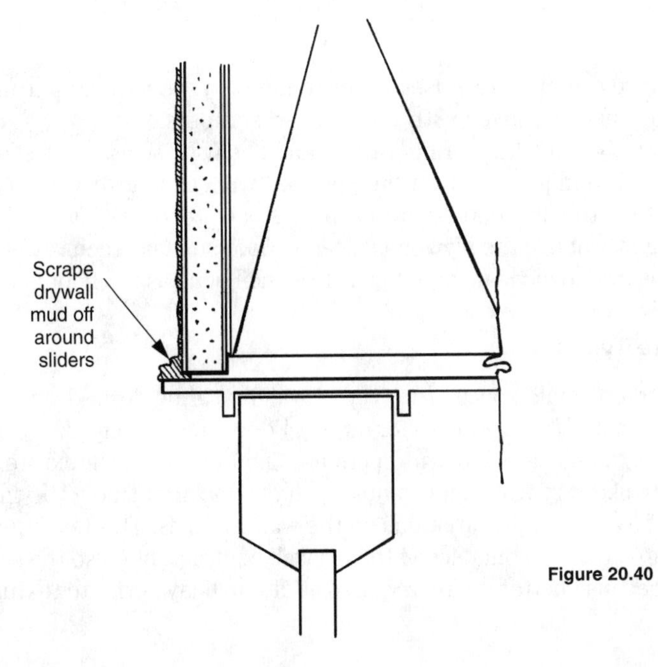

Figure 20.40

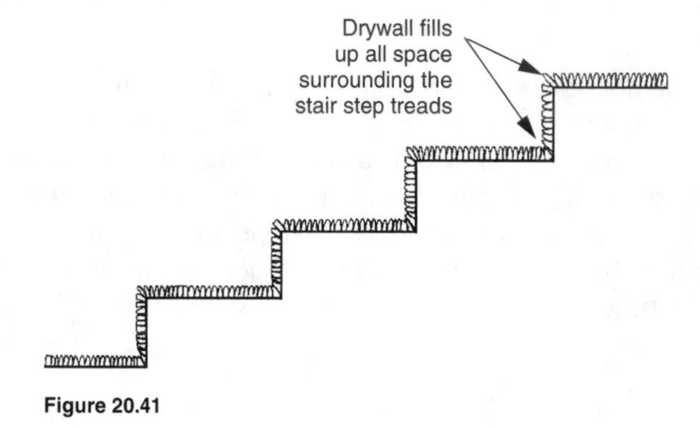

Figure 20.41

Floors Clean for Texturing

All floor areas should be clean and free of drywall scraps, loose boards, and debris prior to drywall ceiling and wall texturing. In order to obtain an even and uniform texture, the worker spraying ceiling acoustical and wall texturing must be able to concentrate on the surfaces before him or her and not on the floors. If the worker has to step and stumble over obstacles on the floor, the spray nozzle may momentarily pause in one place and concentrate a thicker coat of texture on that surface area.

The builder should know the number of units the texturing crew can complete in one day (about five houses a day—ceilings and walls) and check that at least that number has been thoroughly cleaned out.

Texture Cleanup

The drywall texture cleanup crew uses water to moisten the leftover texture on the floors so it can be easily removed with wide bladed scrapers.

On second floor plywood, little or no water should be used so as not to saturate the plywood subfloor. Too much water on the plywood will cause it to swell, forcing up the nail heads. When the plywood dries and shrinks back to its original thickness, the nail heads are slightly above the plywood surface. Floor squeaks then result from the plywood, which is free to move up and down at the nail head as the floor is walked on.

Drywall Taping for Texture Types

The drywall contract should specify the type of wall texturing desired, as this affects the degree of quality and time required for taping. For example, a knock-down texture does not require as much care and time spent as a light orange-peel texture. Orange-peel textures can also vary from light to heavy, all of which demand a varying degree of taping quality to cover slight variations on the wall surfaces. This texturing specification not only affects the economics of the drywall contract, but also the scheduling of construction because better taping requires additional days prior to texturing.

Drywall Prepaints

Drywall prepaints are the patching and texturing of dents, dings, and scratches on drywall surfaces due to the installation of cabinets, interior doors, wood trim, ceramic tile countertops, wardrobe doors, stair handrail, bathroom pullman tops, and other items as they are installed. Drywall prepaints merely refers to the drywall repair that is done prior to the painting of the walls.

The quality and extent of the drywall prepaints is a major factor in the final quality of the house interior, and the smoothness and morale at the end of the job. If most of the dents, dings, and holes can be identified and repaired before the walls are painted, this reduces the number of final repairs that must be made prior to the final paint touchup. If large numbers of prepaint dings are missed the first time around, they roll over into the final drywall repair, and often require large sections of walls to be repainted corner-to-corner instead of merely touched up. If most of the dings are identified and repaired during prepaints, then the only trades remaining to further ding up the walls are the last finish trades such as finish electrical, plumbing, shower doors, appliance delivery, wardrobe mirror doors, bathroom mirrors, and so forth.

It is in the interest of the builder to ensure that the drywall prepaints are thorough. This can be done by going through all of the units before the start of the prepaints and lightly circling the dents and dings with a carpenter's pencil (Figure 20.42). This takes the quality-control value judgment away from the drywall repairman and places it with the builder where it belongs. If the builder is tough and thorough at the prepaint stage, this actually can save both the drywall and painting contractors money in the long run. By circling each dent and ding, the drywall repairman doesn't have to spend time looking for problems or guess at the level of quality that is expected. If the builder is thorough, neither the drywaller nor the painter have to come back four or five times to repair and touch up areas that were missed the first time around.

To reach ceilings and high wall areas, a carpenters pencil can be taped to the end of a piece of baseboard or other wood trim.

Drywall Mud around Electrical Boxes for Head-Start

One way to get a head-start on drywall taping before the drywall nailing inspection is to mud around the electrical outlet boxes on walls and around flush-light cans and ventilation fan boxes on ceilings (Figure 20.43). Because none of these items have anything to do with the drywall nailing, this part of the taping operation can be done without affecting the drywall inspection. By starting this preinspection taping on the same day or even a day before the drywall nailing inspection, time can be saved on the construction schedule.

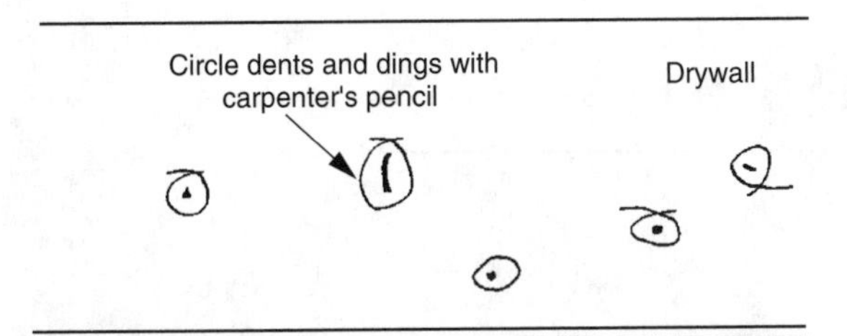

Figure 20.42

Clean Out Skirtboard Channel

Shown in Figure 20.44 is the end view of the gap between the stair treads and risers and the stair wall. In this case, the stairs are framed away from the wall 1½ inches to allow for the thickness of drywall and skirtboard trim to slip into this channel without having to be cut saw-tooth fashion to match the stair steps.

During drywall installation, taping, and texturing, debris gets into these stair channels. The builder should have the drywall texture cleanup crew clean out these channels as part of the texture cleanup. This saves the finish carpenter who is installing the skirtboard from having to clean out the dry-wallers taping and texturing debris in order for the skirtboard to fit within the channel correctly.

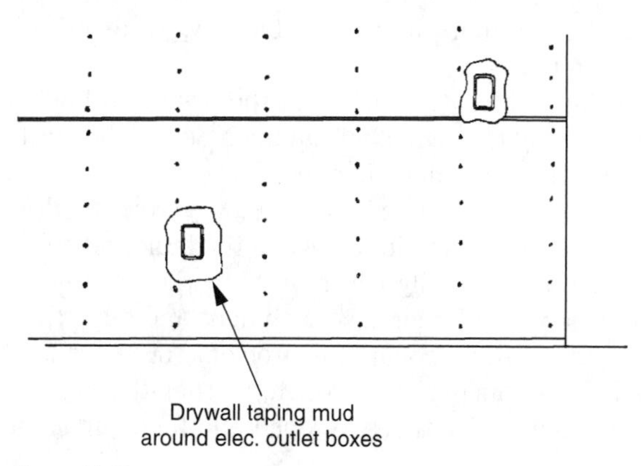

Figure 20.43

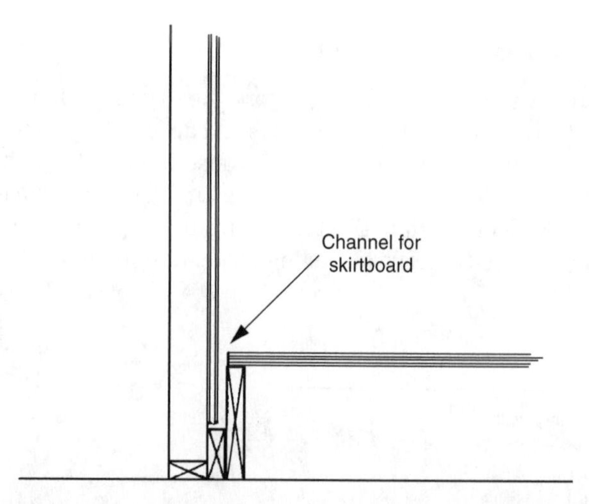

Figure 20.44

Chapter 21

Cabinets

Deliver Cabinets into Units

Cabinets should be delivered to the job site only when they can be placed directly inside the houses. Any drywall texture cleanup or wall painting should be completed beforehand.

Placing cabinets inside garages temporarily because the previous work activity is not complete is not good practice. Moving cabinets a second time from the garages into the houses by the job site laborers, the builder, or the cabinet installers can damage the cabinets, dent and scrape interior walls, and needlessly tire everyone out.

If the schedule of work is behind, the cabinet delivery should be put off until the houses are ready.

Cabinet Quality Sign-Off

Some cabinet manufacturing companies have a policy of walking the units with the job site superintendent, immediately following the cabinet installation, to check for any defects or flaws. If the cabinets are okay or if there are only a few items to fix, the superintendent then signs-off that the cabinet installation is acceptable.

Although this approach sounds good and has the appearance of promoting quality by the cabinet company, the builder should reject it for two reasons. First, a walk-through with the cabinet company's field foreman, looking at 5 or 10 houses worth of cabinets, quickly changes from a walk-through to a run-through. The field foreman does not want to spend 30 minutes per house looking closely at the cabinets. He or she merely wants to get the walk-through over and obtain the sign-off sheets. The job site superintendent is made to feel that he or she has to rush through the inspection and that spending more time examining the cabinets amounts to being too particular. The cabinet company field foreman is certainly not pointing flaws out to the superintendent, so the foreman in this situation is not an ally in the effort to improve quality, but rather an adversary. The superintendent can usually perform a better quality check on his or her own, without being distracted by having to listen to someone else talking at the same time.

Second, the builder or superintendent should not be inconvenienced by having to check the quality of their work with a subcontractor. The cabinet field foreman is free to walk the cabinet installation anytime shortly after the work is complete. He or she should then compile a list of missing pieces and things to fix and correct the problems long before the superintendent does his check. If the cabinet company does their own walk-through and corrects any defects or missing parts, this actually saves time for the superintendent, as there are less defects to find and fewer things to write down on a punch-list. The superintendent can then check the cabinets at his or her convenience and at a comfortable pace.

Wide Face Frame at Doors

Illustrated in Figure 21.1 is a bathroom cabinet with a drawer that hits the door casing. Because the cabinets are usually installed ahead of the interior doors and casing, this problem does not become apparent until after the cabinet drawer slides out and hits the door casing.

The builder should check the width of the cabinet face frames on the spec sheets prior to the start of the cabinet work. The amount of face frame that is scribe-cut off to match the drywall should also be checked during installation.

Anchoring Island Cabinets

A problem with kitchen island cabinets is obtaining solidity and firmness because they are not attached to any walls. To help anchor an island cabinet, at least the two side panels (not containing doors or drawers) should extend all the way down to the kitchen floor (Figure 21.2). Each side can then be nailed to wood cleats that are attached to the floor.

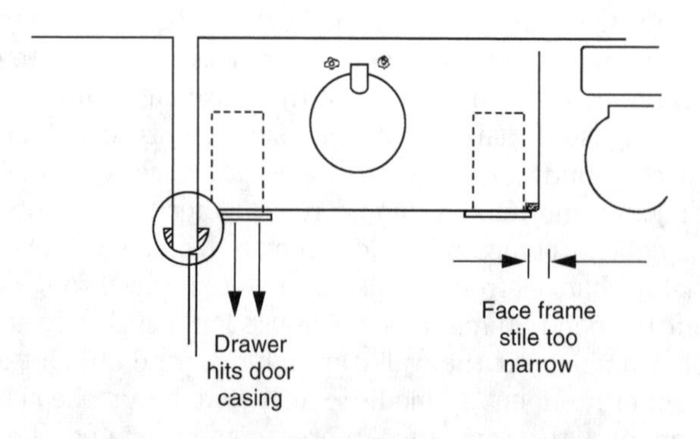

Figure 21.1

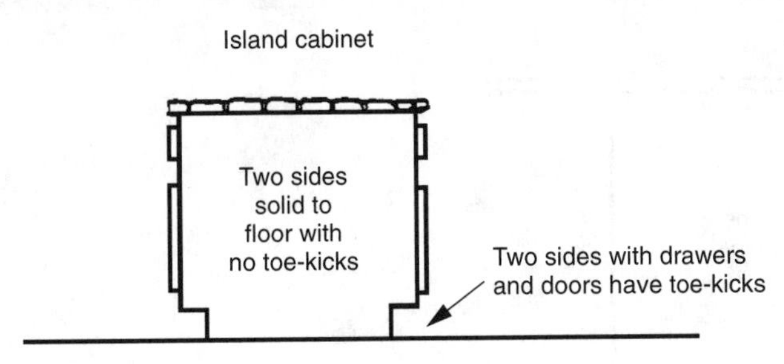

Figure 21.2

An island cabinet with toe-kicks on all four sides has too many wood connectors and joints to obtain the solid feel that any cabinet should have.

Countertop Overhang

The overhang for a kitchen cabinet pass-through countertop should be at least 12 to 16 inches deep (Figure 21.3). This allows enough room to pull up to the countertop in a chair without hitting your knees.

A token overhang of only 6 to 8 inches, with wood corbels to help make the overhang look authentic, will bring criticism from the homebuyers either before or after they move in.

Protect Wood Bar Tops

Occasionally pass-through bar tops between kitchens and dining rooms are made of stained and lacquered wood rather than ceramic tile. Because they are generally finished far before the end of construction, they should be covered to protect them from accidental damage. Cardboard or some other protective material should be placed on them so that the tradesmen, who use the bar top as a convenient counter to lay tools, materials, coffee cups, and so on, do not scratch, dent, gouge, or stain the surface (Figure 21.4).

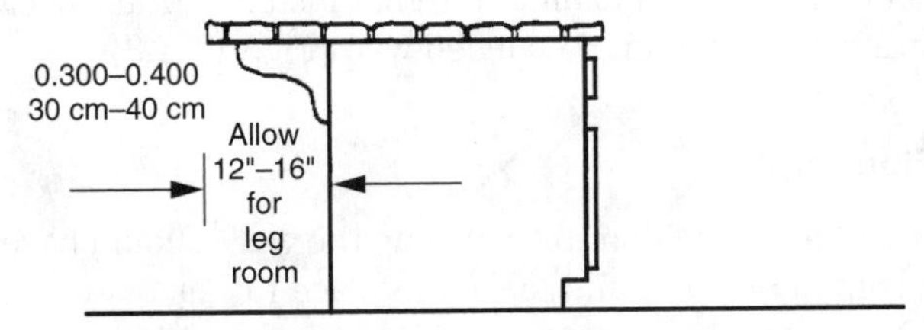

Figure 21.3

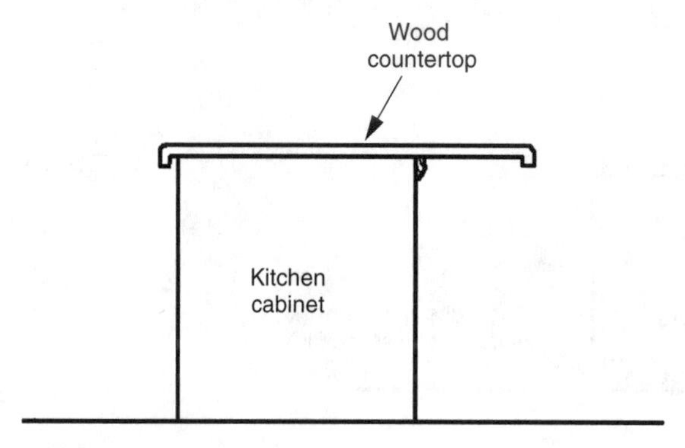

Figure 21.4

Cabinet Cutouts for Outlets

When upper kitchen cabinets have a back veneer, cutouts to be made around electrical outlets, such as the outlet above the range hood, should be as neat as cutouts made in drywall around electrical outlet boxes.

The same size wall cover plate must cover the cut around the electrical outlet within the cabinet as with other outlets throughout the unit. If the cabinet installer gets sloppy, oversized cover plates must be used or special filler pieces made (Figure 21.5).

Cabinet Cutouts for Pipes

Illustrated in Figure 21.6 is the side view of a bathroom cabinet that has a cutout through its back panel for plumbing pipes. This cutout should be clean, square, neat, and precise, without saw kerfs extending past the hole.

Homebuyers can easily see these cabinet cutouts when opening the cabinet doors. They are therefore an aesthetic and visible part of construction. They should be done with the same amount of care as any other part of the exposed construction. The builder should not accept sloppy cutouts around plumbing pipes simply because these holes are underneath and inside a cabinet.

Scribe Molding and Angled Walls

Cabinets that join to angled walls should have scribe molding beveled to match the angle of the wall (Figure 21.7). If the scribe molding joins the angled wall at a 90-degree angle, the leftover gap between the front edge of the scribe molding and the wall may look worse than if the molding were not installed or if the cabinet face frame was merely scribe-cut to match the angled wall (Figure 21.8).

Scribe Molding and Coved Flooring

Kitchens and baths that have vinyl flooring coved up the walls should have the lower cabinet scribe mold trim merely cut and tacked in place rather than fully nailed or stapled (Figure 21.9). The reason is that this cabinet trim must be removed and re-

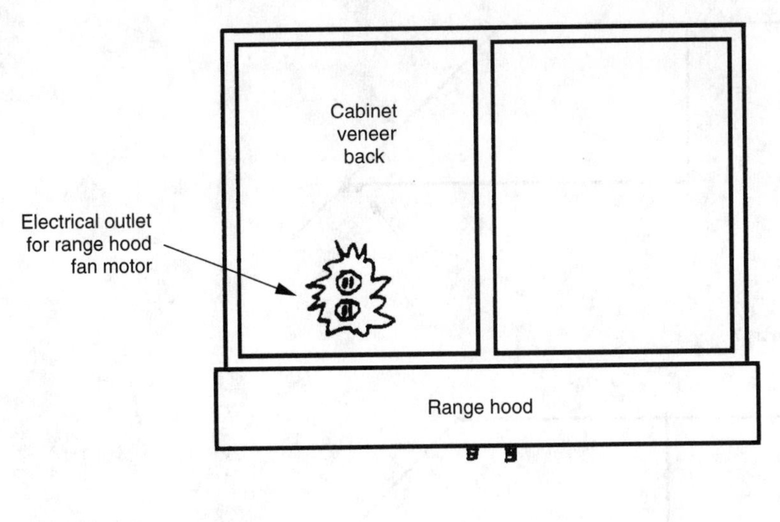

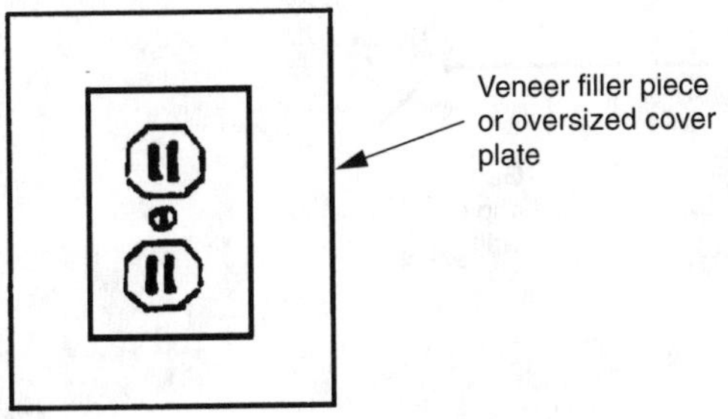

Figure 21.5

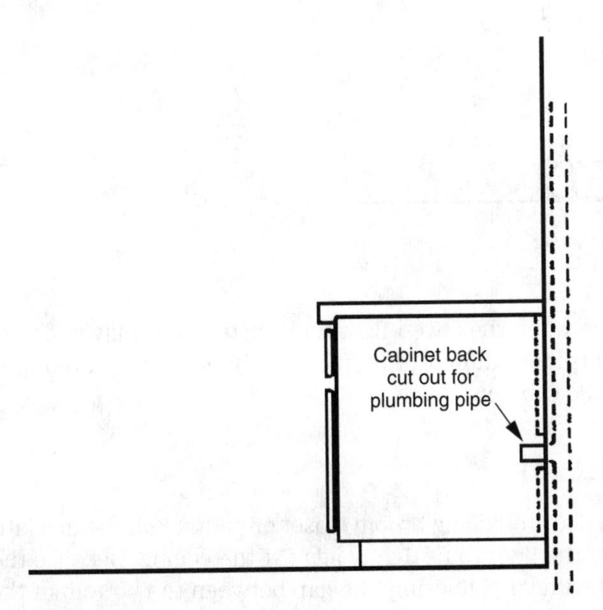

Figure 21.6

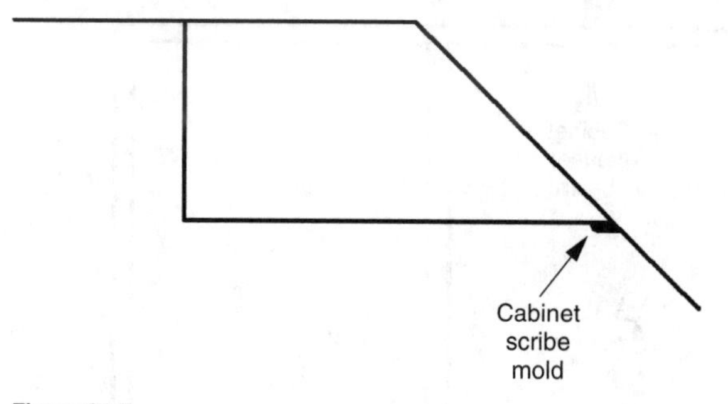

Figure 21.7

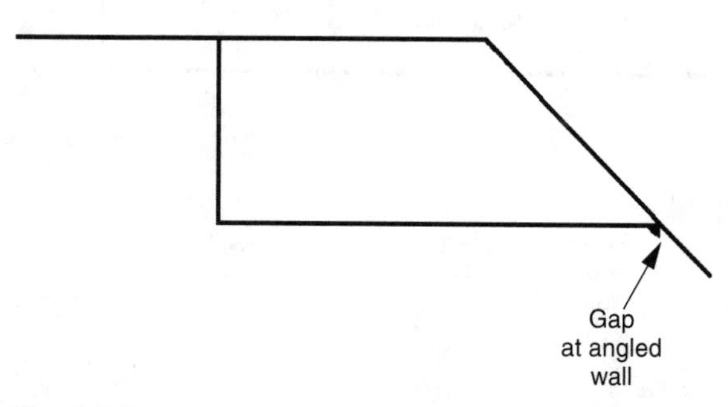

Figure 21.8

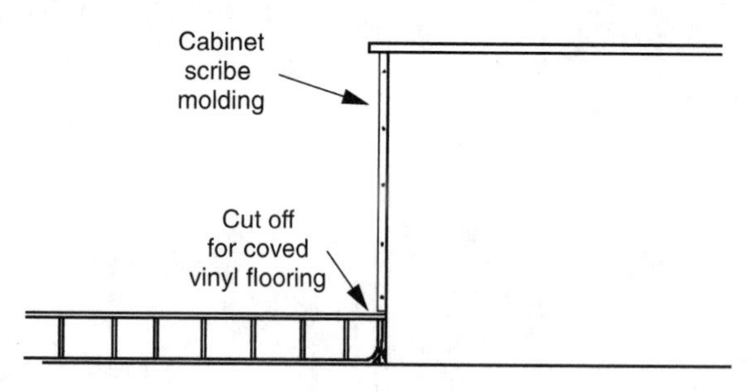

Figure 21.9

cut to a shorter length when the coved flooring is installed. If fully nailed or stapled in place, the molding is more difficult to remove without being broken or damaging the finish surface of the cabinet.

Cabinets and Baseshoe

Figure 21.10 shows a floor-to-ceiling broom closet or pantry cabinet in a kitchen that has been shimmed off the floor to fit tightly against the ceiling. Because the kitchen floor may be covered with vinyl flooring, the gap between the bottom of the cabinet and the floor should not be greater than the size of the wood baseshoe molding. If this gap is too large, the baseshoe cannot be nailed to the bottom of the cabinet and still cover the gap.

When cabinets must be raised off the floor, the builder should check that the installer cut and tacked in place a piece of wood trim that is wide enough to cover the gap. This prevents the finish carpenter who is installing the baseshoe from having to hunt all over the job site for a wider piece of wood trim after discovering that the normal baseshoe will not work.

Cleats for Range Hoods

Cabinets should have built-in cleats or nailer strips at the rear of the upper cabinets, placed at the same depth as the lower rail of the front face frame. When installing a range hood, like the one in Figure 21.11, the cabinet installer should add such a cleat. This allows the range hood to be installed without tilting upward in the back, causing it to be out of level (Figure 21.12).

The builder should check the cabinet installation while the cabinet setters are still on the job site and not wait until the finish electrician has to inform you that range hoods cannot be installed because the rear cleats are not in.

Cut Out Hole for Range Hood Cord

Shown in Figure 21.13 is a kitchen upper cabinet with a range hood below. The builder should check during the cabinet installation that the cabinet setters have cut out holes in the bottoms of upper cabinets for the range hood electrical cords.

Check Openings for Microwaves

The standard opening for a microwave oven, placed between two upper cabinets above the range, is 30 inches (Figure 21.14). The cabinet manufacturer usually

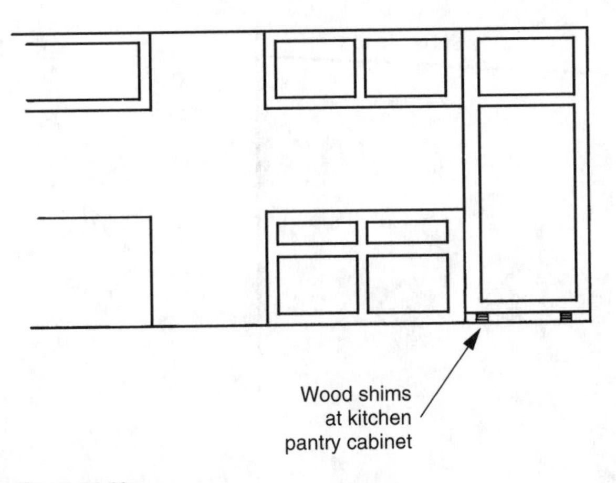

Figure 21.10

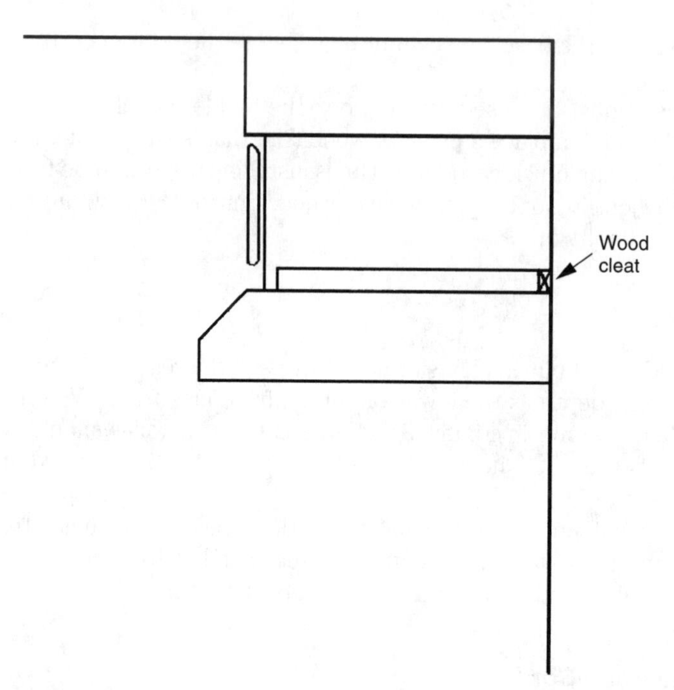

Figure 21.11

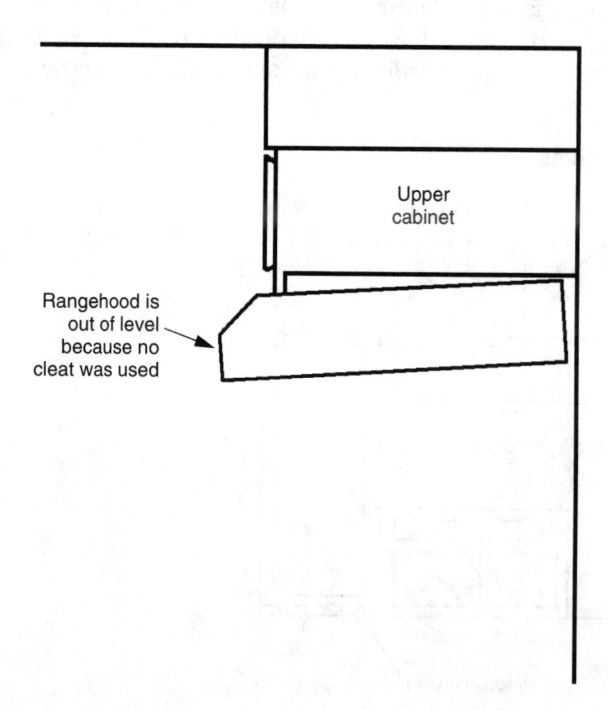

Figure 21.12

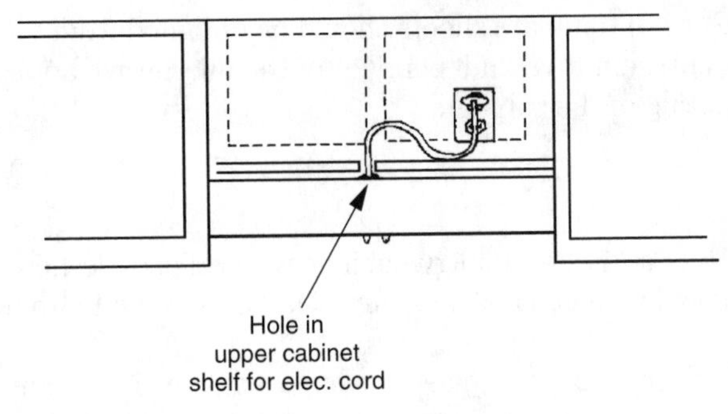

Figure 21.13

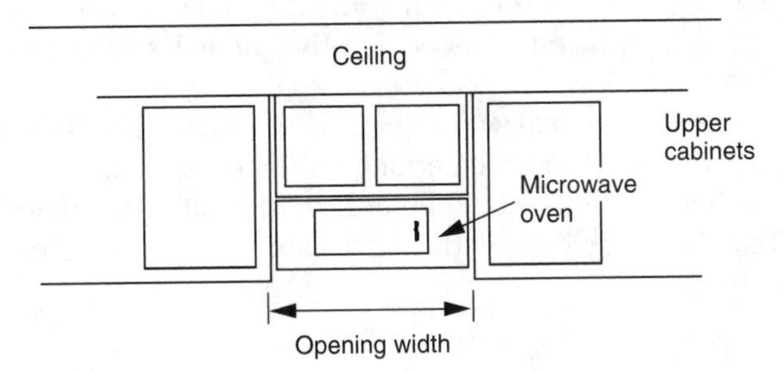

Figure 21.14

builds the cabinets with an opening for the microwave of about 30⅛ to 30¼ inches wide, so that the microwave can fit into place, yet not have a wide gap on either side.

The builder should check the cabinet openings for the correct width using a wood test stick that is 30 inches long. The time to discover that a particular opening is 29 ⅞ inches wide, and will therefore be too tight for the electrician to install the microwave, is while the cabinet installers are still on the job site and can make the repair. Otherwise, the builder must telephone the cabinet company, wait for someone to come out to the job site and fix the cabinet, and then get the electrician to come back out and install the microwave.

Get Microwave Templates

As with range hood fans, the cabinet installer usually drills a hole in the underside shelf of an upper cabinet for a microwave extension cord. Because the microwaves are not delivered to the job site until late in construction, the cabinet setter has no way of knowing where to correctly drill the hole for the microwave appliance cord at the time of the cabinet installation. The builder, therefore, needs to obtain an installation template, which is a sheet of paper that shows the locations for all of the holes that need to be drilled through the underside of the cabinet for the microwave installation. This template accompanies the installation instruction sheet that comes with the microwave. Several template sheets should be obtained from the appliance manufacturer's sales representative, independent of the microwave appliance delivery and prior to the setting of the cabinets.

Pantry Cabinet Sides

Shown in Figure 21.15 is a kitchen pantry cabinet with adjustable height shelves. The shelves can be moved by means of shelf clips that are inserted into holes in the sides of the cabinet.

The problem to avoid is not fixing in place the exposed side of the pantry cabinet, which in this case ran from floor to ceiling. If someone bumps into the side of the pantry cabinet, it may move because it is only being held in place at the wall by the friction created by contact between the cabinet and the wall only. When the cabinet does move, unpainted wall surface that is behind the cabinet side edge and scribe molding is exposed.

The solution in this case was to nail wood cleats on the back wall inside the pantry cabinet at 16-inch intervals to provide something for the exposed side of the pantry cabinet to nail into. Other solutions would be to have the cabinets built with an intermediate height fixed shelf or built with rear cleats as part of the cabinet construction (Figure 21.16).

Fixed Pantry Shelves

From a construction standpoint, kitchen pantry cabinets with fixed, nailed-in-place shelves are preferable to loose, adjustable shelving. Adjustable shelves usually come bundled in packs of four or five, with the shelf clips delivered to the builder. The painter takes the shelf bundles apart and stains and lacquers them along with the

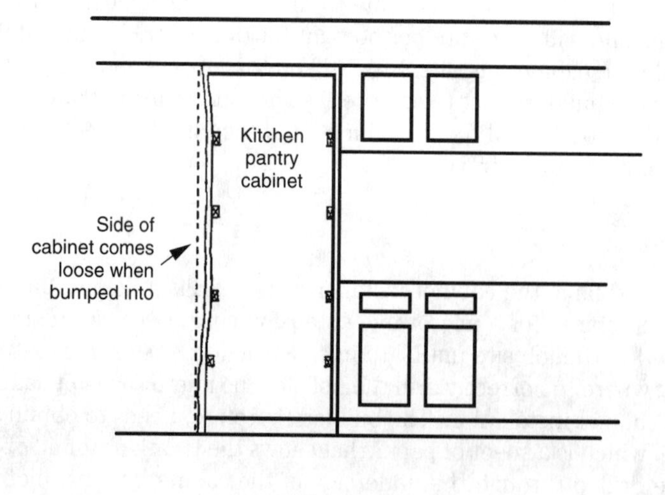

Figure 21.15

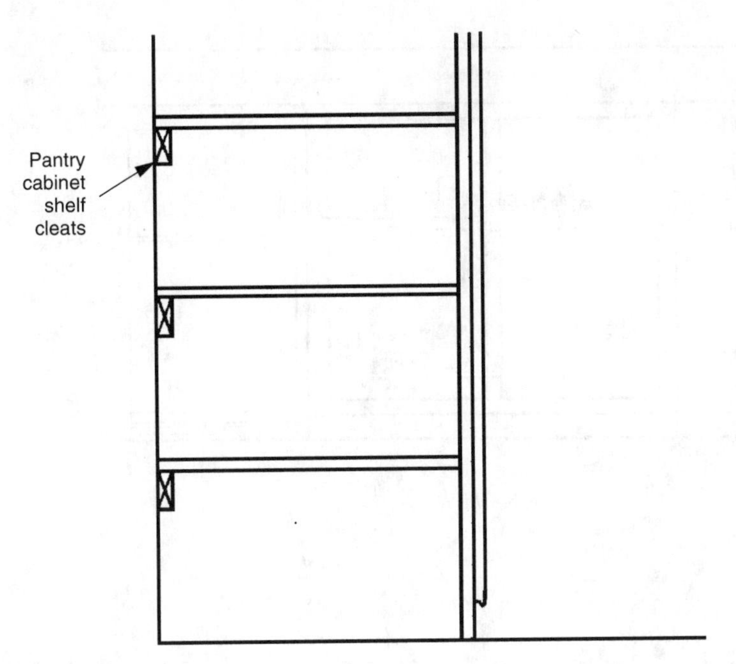

Figure 21.16

cabinet drawers. The problem with this is that they are not always put back inside the pantry cabinet by the painter when finished and are sometimes thrown away as trash by the construction cleanup company workers. If the shelves are fixed, as part of the pantry cabinet, they cannot be removed by the painter and subsequently damaged or thrown away.

Having fixed shelves especially simplifies the builder's supervision task when kitchen cabinets have four or five stain colors. Getting missing or stolen shelves ordered, delivered, stained the correct colors, and installed in time to complete construction can be both difficult and time-consuming.

Center Support Pantry Shelves

Kitchen pantry shelves that are unattached and adjustable to different heights within the cabinet, in some cases depending on the length of the shelves, should have some means of support at the front edge of each shelf.

Kitchen pantry shelves are sometimes stocked full of canned foods. This weight can cause shelves to bow downwards that are even reinforced with front edge wood rails. Figure 21.17 shows the front view of a pantry cabinet shelf in which the homebuyer complained that the shelves were sagging downward. The shelves were about 4 feet long, reinforced at their front edge with a wood rail, but not attached to the center face frame stile. The builder had to drill holes on the inside face of the stile and insert shelf clips to help support the center of each shelf (Figure 21.18).

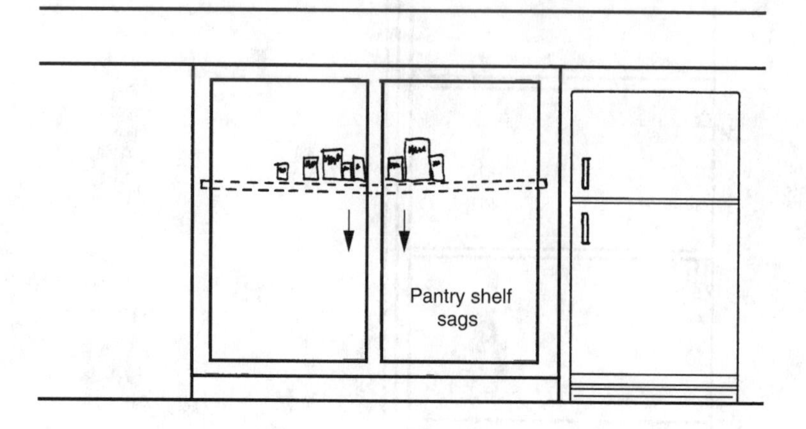

Figure 21.17

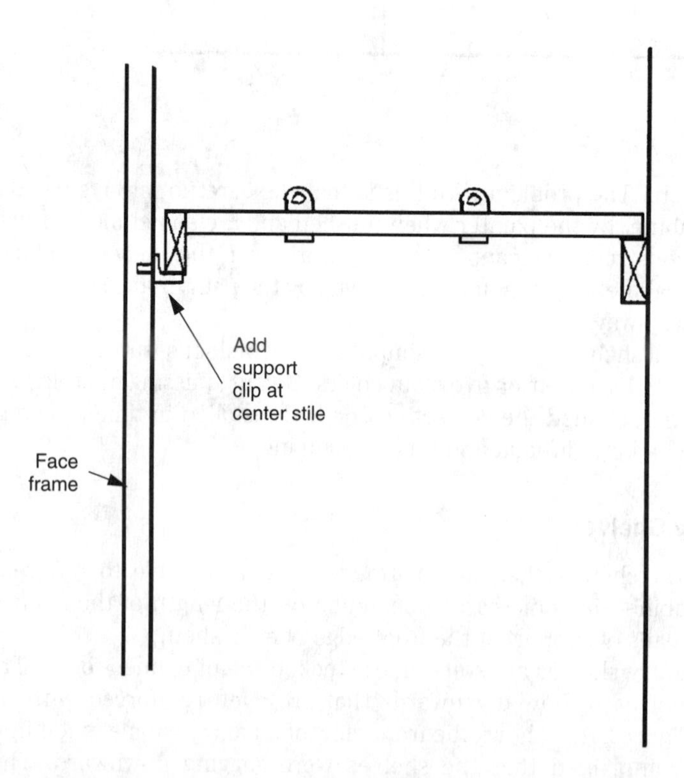

Figure 21.18

Cabinet Door Silencers

Adhesive backed felt pads should be installed on the inside corners of cabinets doors in cabinets that are field stained and lacquered (Figure 21.19). Prefinished cabinets usually come with either small plastic bumpers or felt pads already in place at door corners. Felt pads or plastic bumpers soften or eliminate the noise of the cabinet door closing against the cabinet. They also increase the overall sense of cabinet quality.

The problem arises when the felt pads cannot be installed on field stained cabinets until after they are stained. If this is the case, either the cabinet company should provide the builder with felt pads for installation or the cabinet company should return to add the pads.

Breadboards

Kitchen cabinet breadboards are often delivered to the job site separate from the cabinets and installed as part of the move-in kit just prior to the homebuyer walk-through.

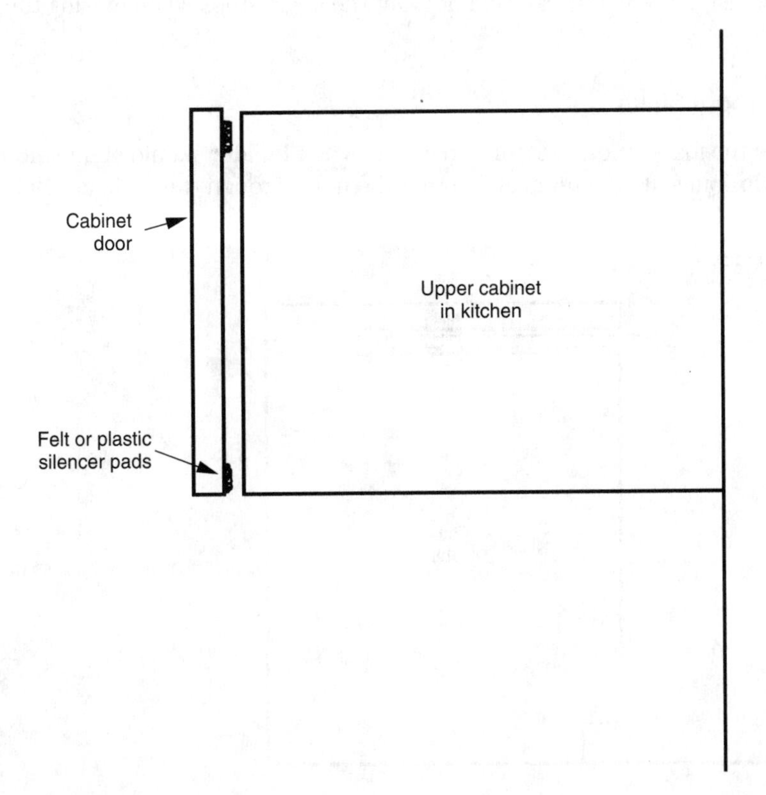

Figure 21.19

The painter stains or paints the breadboards as a group and the superintendent or customer service person installs them (Figure 21.20). By installing them so late in construction, this prevents them from disappearing from the houses. One problem with this approach, though, is that the breadboards can bind or stick in cabinet slots that are too tight. This is typically not discovered until they are installed at the end of the job.

One way to check the breadboard slots while the cabinet installers are still on the job site is to use one typical breadboard as a test piece. The builder should check each kitchen breadboard slot, and then have the cabinet installers fix any slots that are too tight. This saves pickup work at the end of the job and possibly one item on the homebuyer walk-through sheet.

Angle Stops before Cabinets

The builder should have the plumber install the angle stops before the cabinets are put in (Figure 21.21). The advantage of doing so is two-fold. First, it is easier for the plumber to install the angle stops without having to reach inside lower cabinets and bathroom pullmans. Second, water spilled during the installation falls on the floor rather than the base shelf of the cabinets. The one drawback, however, is that the cabinet installer must be careful not to hit the angle stops when moving the cabinets into place.

Cutouts for Bathroom Sinks

Prior to the installation of bathroom cabinets, the builder should obtain either a sample bathroom sink or a template so that the hole through the pullman cabinet rough

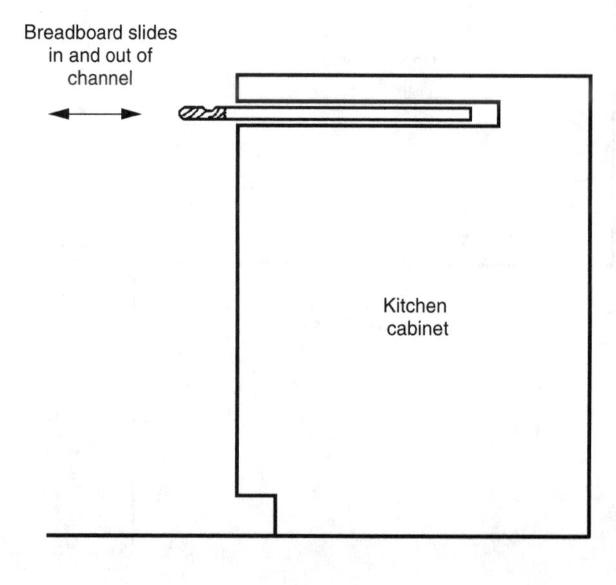

Figure 21.20

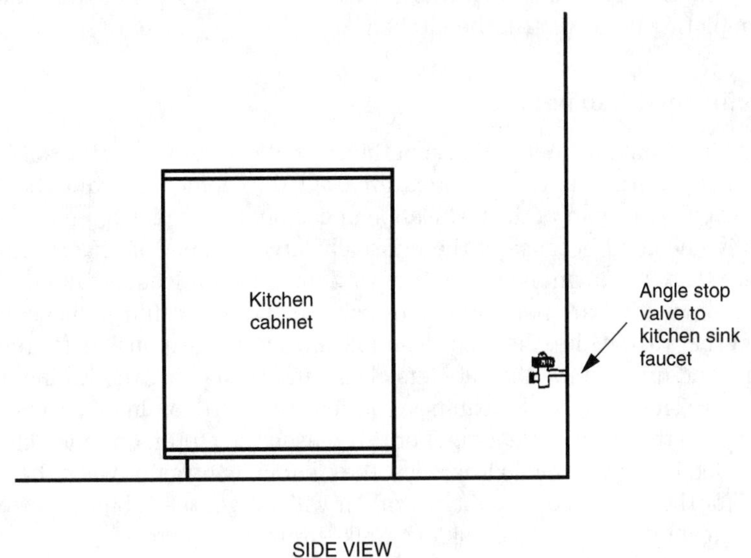

Figure 21.21

top can be cut (Figure 21.22). This hole is cut out by the cabinet setter in preparation for the tile setter.

It is important that the hole be cut out accurately so that the tile setter can cut and lay tile with the hole as a guide. This results in a bathroom sink that drops neatly into place.

Start Cabinets in Bathrooms First

The builder should have the cabinet setters install the bathroom pullmans first so that the tile setter can start work. The tile setter needs to have the bath cabinets in-

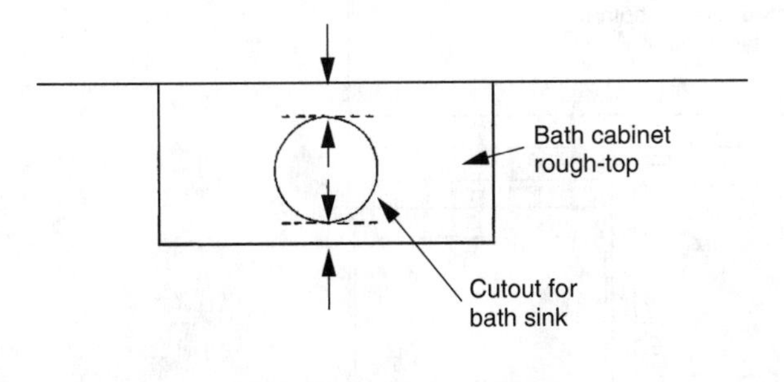

Figure 21.22

stalled to establish the splash height for adjacent bathtub or shower tile (Figure 21.23). The tile setter can then begin work in the bathrooms as the cabinet setters are completing their work in the kitchen.

Smudges on Prefinished Cabinets

The builder should include a clause in the cabinet contract that states: "When white, prefinished, plastic laminated cabinets are used, the cabinet company should clean off all dirt smudges as part of the installation or cabinet repair pickup phase." This clause should be included because of the unusually large number of fingerprints, dirt, and smudges that end up on the cabinets during shipping, off-loading, and installation.

For cabinets that are field stained or painted and for prefinished cabinets stained a color other than white, the final cleanup contractor is responsible for removing any sawdust and dirt, wiping the cabinets clean, and then oiling the cabinets with cabinet polish. The additional cleaning required for white laminated cabinets usually goes beyond this normal cleaning. For this reason the aforementioned clause should be included in the contract. Otherwise, the cleanup contractor will either ask for extra pay for the increased amount of work or will not get the cabinets perfectly clean, leaving cleanup work for the builder's walk-through prep crew.

Cabinet Knob Screws

Illustrated in Figure 21.24 is the side view of a cabinet drawer with a typical cabinet knob and screw. During installation, the builder should not allow the cabinet setters to use a motor screw-gun on the cabinet knobs. The screw-gun will tighten the screw into the knob to the point at which the knob is secure, but then continues to spin inside the screw head a few extra revolutions. This usually creates metal burrs at the screw head slots, which are sometimes very sharp.

If the cabinet knobs are installed by hand, the installer will know by feel when the knob and screw are tightly secure and will not mar the head. It remains smooth and safe for the homebuyer.

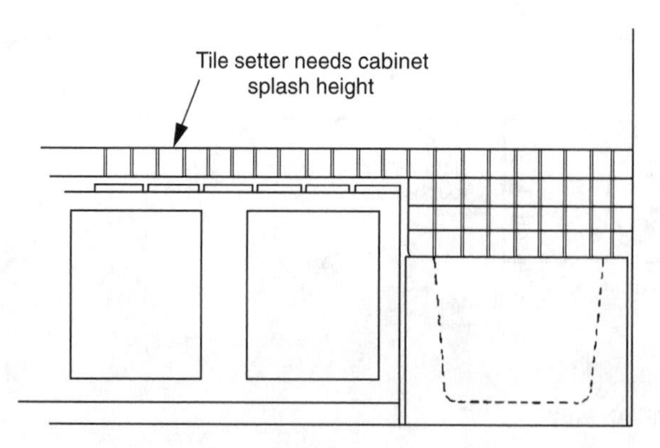

Figure 21.23

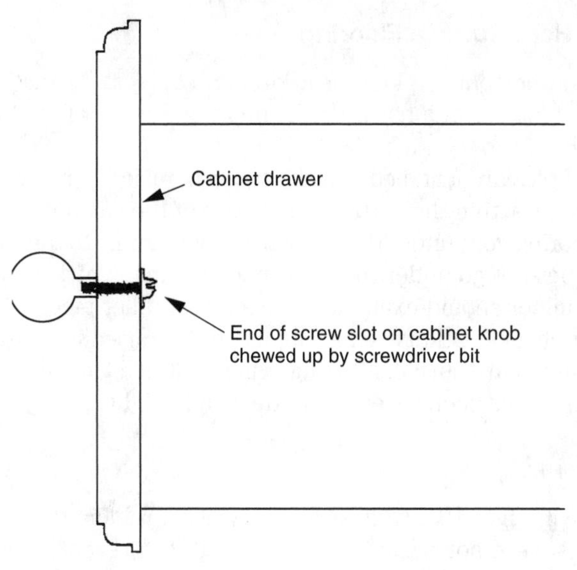

Figure 21.24

Deepen Upper Cabinet Valance for Fluorescent

The upper kitchen cabinet pictured in Figure 21.25 has an undercabinet fluorescent light fixture to meet energy saving requirements. When this is the case, the builder should discuss with the cabinet manufacturer the possibility of using a deeper valance rail to hide the light fixture.

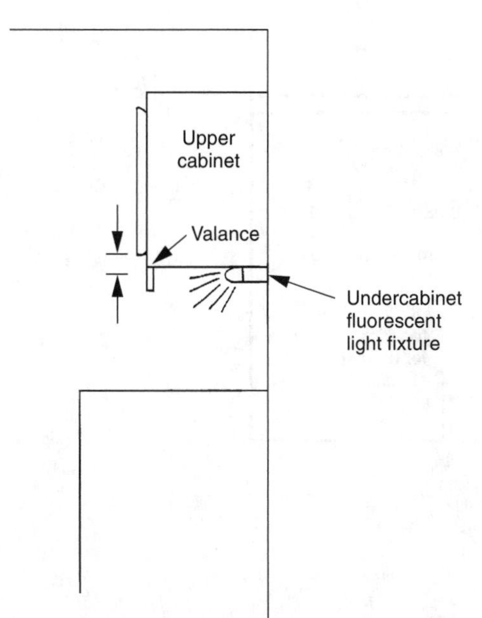

Figure 21.25

Deepen Cabinet Toe-Kicks for Hard Surface Flooring

Figure 21.26 shows the bottom toe-kick of a kitchen cabinet. So that there was enough space at the toe-kick after a hard surface flooring was installed, ¾ of an inch was added to the height of the toe-kick.

When marble or clay tiles are installed and the toe-kick raised, appliances such as the dishwasher will still fit within the vertical dimension of the cabinet opening. This is assuming that the flooring goes under the appliances, not merely to their front edge.

When the flooring does not go under the appliance or is made of a thinner material such as vinyl, the builder should examine the need for a filler piece of cabinet to be placed above the appliance. Otherwise, a cabinet that is raised up, with a deeper toe-kick, in anticipation of hard surface flooring, will result in a gap above the appliance even when the legs have been extended to their full height.

Cabinet Drawers and Knobs

The cabinet face frame in Figure 21.27 illustrates a typical problem that occurs with cabinetry. The cabinets were not designed wide enough to prevent a drawer from hitting the knob on an adjacent cabinet door as the drawer is fully opened. In this example, the drawer could have been opened another 2 or 3 inches if it were not for the projecting cabinet knob. This was not a problem for the bank of drawers immediately below (Figure 21.28).

This type of mistake always makes the builder look bad. There are only two possible solutions: one, ask the homebuyer to live with a cabinet drawer as is, or two, tear apart the cabinetry to add a wider face frame. The second solution may not be possible, however, if the layout of the kitchen is already tight.

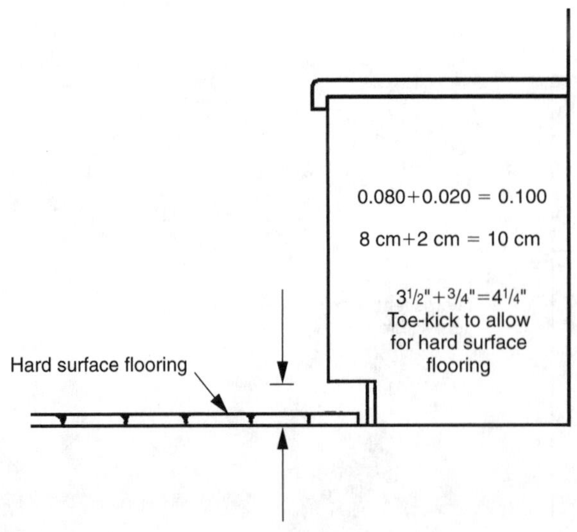

Figure 21.26

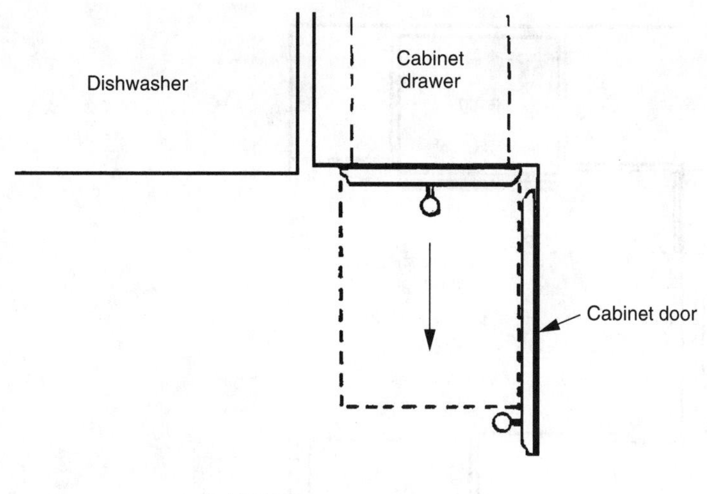

Figure 21.27

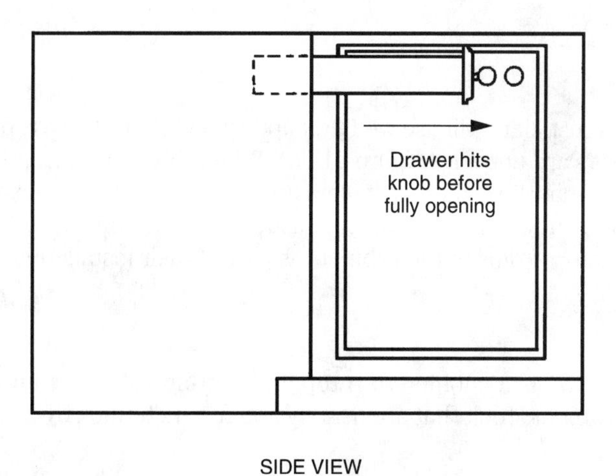

Figure 21.28

Dishwasher and Refrigerator

Illustrated in Figure 21.29 is a plan view of a kitchen. The dishwasher in the illustration opens toward the refrigerator. The dimension of the cabinet that separates the dishwasher from the edge of the refrigerator should be at least 26 inches—the length of the dishwasher door. In this case the cabinet was less than 26 inches deep. If the dishwasher door hits the refrigerator before being fully opened, the bottom tray of the dishwasher will not slide out for loading dishes.

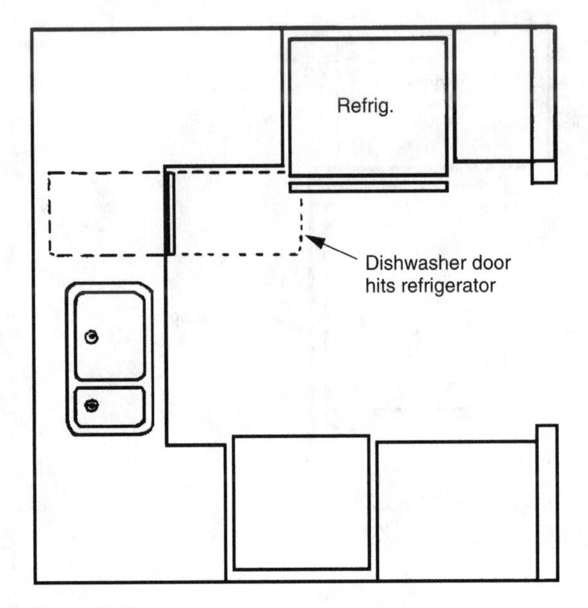

Figure 21.29

Cut Cabinet Shims Flush

Sometimes the cabinet installers will use wedge shaped wood shims to raise portions of the cabinets when floors are uneven (Figure 21.30). When this is the case, the builder should insist and check that the cabinet installers not leave these shims sticking out for the flooring installers to cut off. The cabinet installers should cut off these shims flush with the toe-kicks or the bottoms of the cabinets as part of their installation.

One-Piece Cabinet Valances

Illustrated in Figure 21.31 is a cabinet with top and bottom valance trim. The trim consists of pieces of cabinet stock that are usually about 1-inch thick by 2 or 3 inches

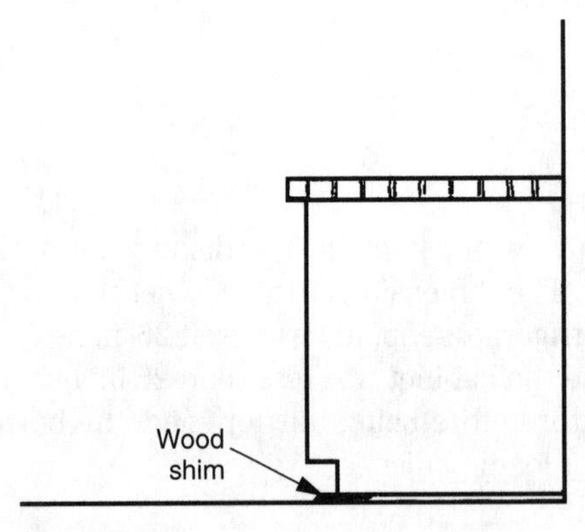

Figure 21.30

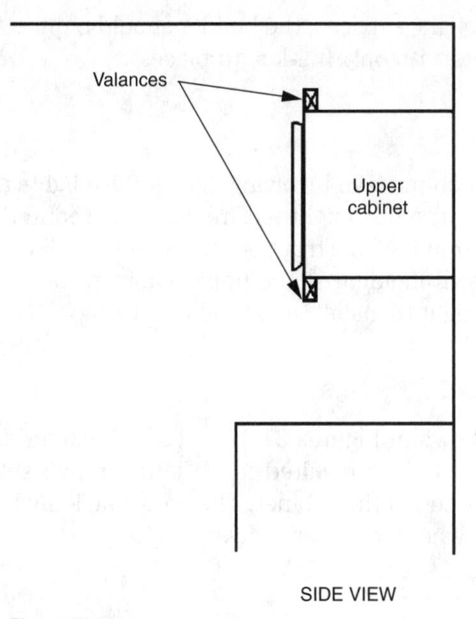

Figure 21.31

wide. They are attached to the cabinet face frame to hide undercabinet or overcabinet fluorescent light fixtures.

The builder should specify in the cabinet contract that valance trim be continuous pieces without splices (Figure 21.32). For example, a particular cabinet in a kitchen may extend 10 or 12 feet along a wall. The valance trim should be long enough to span this length, without requiring any splices or seams.

The reason for this is that homebuyers nearly always notice splices in cabinet valances that are not 100 percent invisible. Because all other seams in the cabinets are perfectly matched, or should be, a splice in a valance appears to the homebuyer to be a cabinet defect. To avoid this complaint during the customer service period,

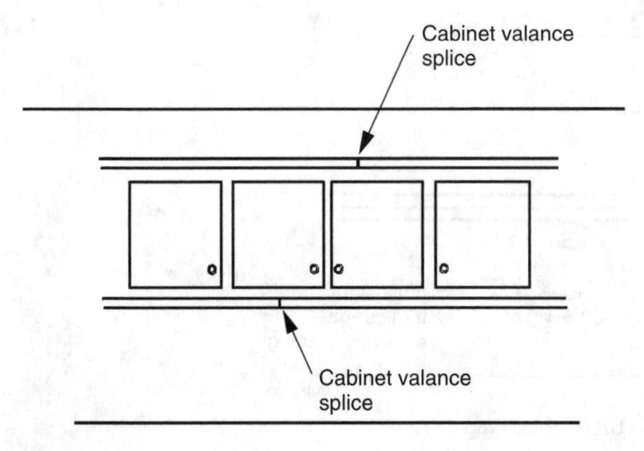

Figure 21.32

when problems arise with the valance splices, the builder should require the cabinet company to replace the valances with only full-length pieces.

Crown Molding and Flush Lights

Figure 21.33 illustrates a common mistake involving ceiling flush lights and cabinet crown molding. In this example, the flush light fixture can is placed in the dropped ceiling soffit during the rough framing construction. If the builder did not consider the thickness of a detailed crown molding at the upper cabinet, there may not be enough room around the flush light to install the round cover ring.

Crown Molding at Bull-Nose Cornerbead

The upper kitchen cabinets, shown in Figures 21.34 and 21.35, extend too close to the drywall bull-nose cornerbead. This resulted in no leftover wall space for the crown molding to die into at the top of the cabinet. The crown molding hits the bull-nose cornerbead past the edge where it curves away.

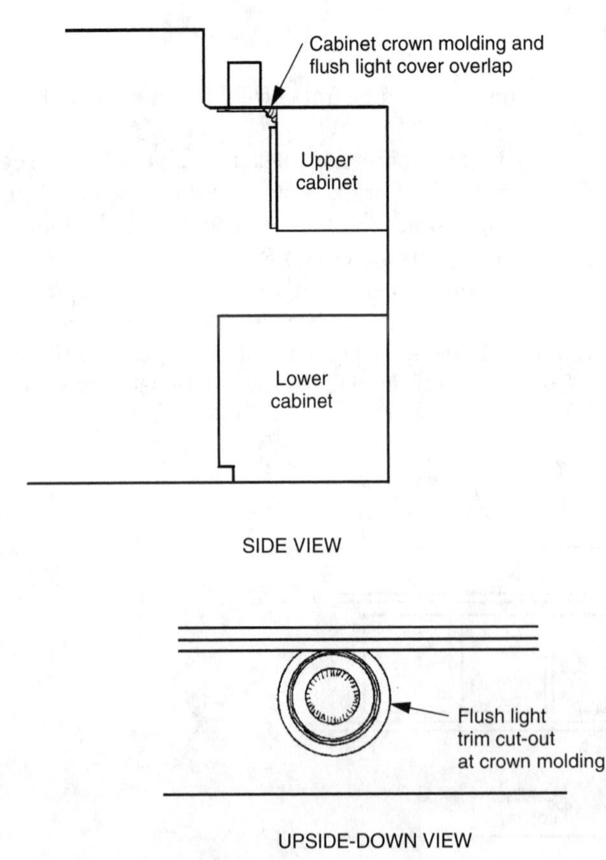

Figure 21.33

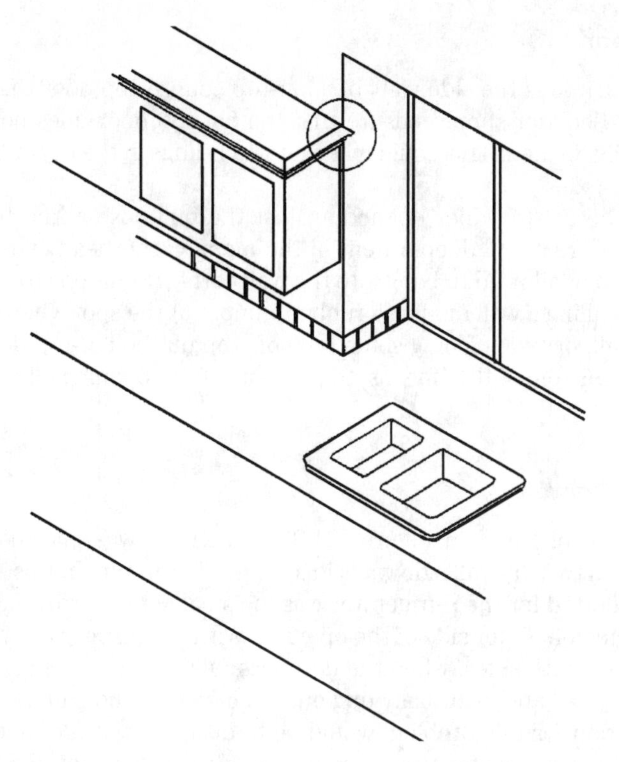

Figure 21.34

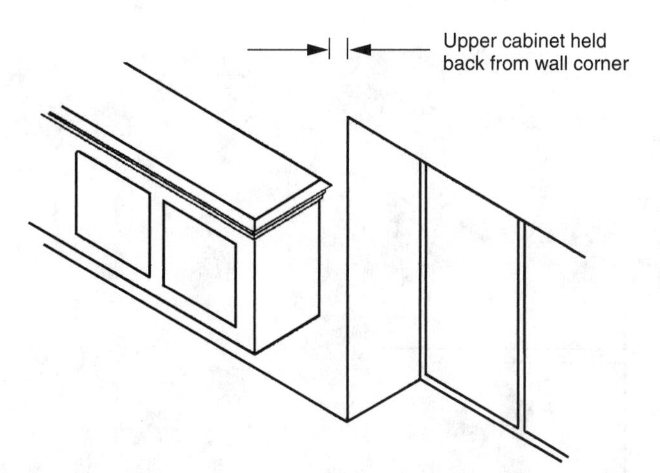

Figure 21.35

When crown molding is used at the top of kitchen cabinets, the molding thickness should be figured into the overall cabinet dimensions for construction purposes. The upper cabinet in this case should have been an inch narrower so the crown molding could have returned into a flat wall surface.

Door Stop for Countertop Door

Illustrated in Figure 21.36 is the side view of a bar-top countertop door that swings upward to open. The builder should ensure that the bar-top door does not hit the drywall when it is fully opened. This could make dents or dings in the drywall surface or mar the door.

In one instance, the bar-top door opened against the bull-nose cornerbead of a wall corner. The door created a deeper dent in the metal cornerbead every time it was carelessly opened or allowed to bang into the wall corner. Solutions to this problem include: one, installing a wall mounted rubber bumper at the spot where the bar-top door hits the wall, or two, placing some type of stop on the bar-top door itself, such as a brass chain or some other means, to prevent it from opening all the way to the wall.

Small Refrigerator Door and Drywall

The small refrigerator, pictured in Figures 21.37 and 21.38, was placed within a lower cabinet between two drywall side walls in a master bathroom. In this example, because the space allotted for the refrigerator was too small, only narrow face frame stiles could be installed on either side of the opening. With the refrigerator placed so close to the drywall on both sides, when the door was fully opened, the refrigerator door hinge hit the drywall and eventually dug out a ⅜-inch deep hole in the wall.

The ultimate solution for this problem would be to design a width of opening for the refrigerator that accounted for the swing of the refrigerator door. Because the

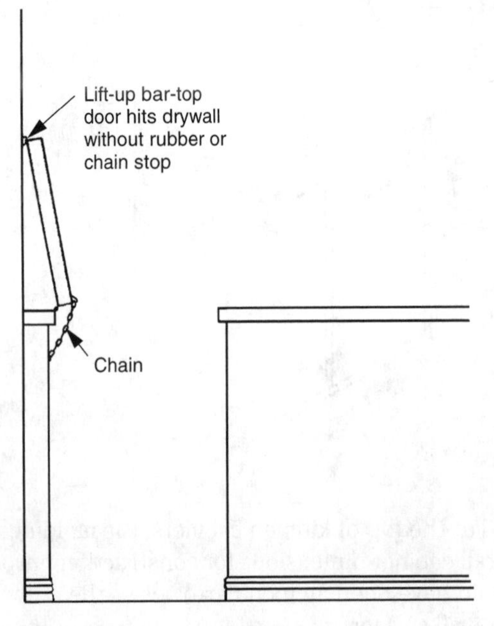

Figure 21.36

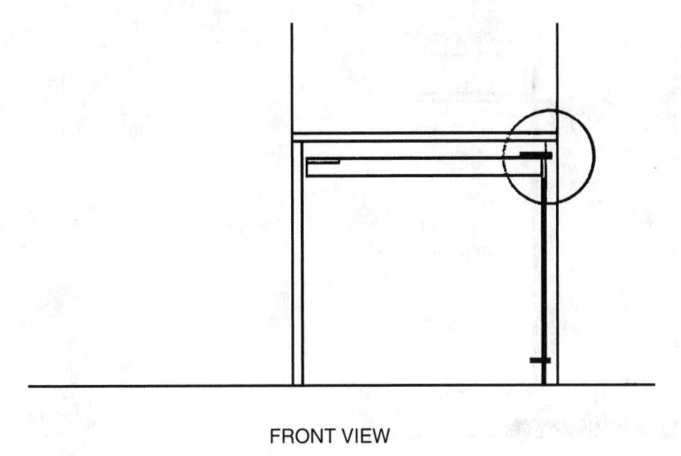

Figure 21.37

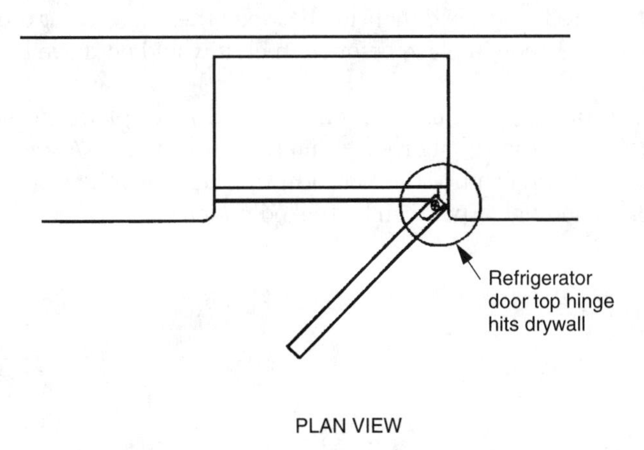

Figure 21.38

cabinet was already made in this case, the only solution was to remove the cabinet face frame at the handle side of the refrigerator door, move the refrigerator over closer to the wall, and replace the other face frame with a wider piece of wood. This left the refrigerator off center with the cabinet, but at least the door missed the drywall as it was swung open.

Small Built-In Refrigerator and Carpet

Figure 21.39 shows the side view of a small refrigerator that was placed directly upon the rough subfloor, surrounded by bathroom cabinetry in the master bath.

In this example, the bathroom carpet was installed up to the refrigerator, which at that time did not have the lower plastic panel cover in place. After the carpet was

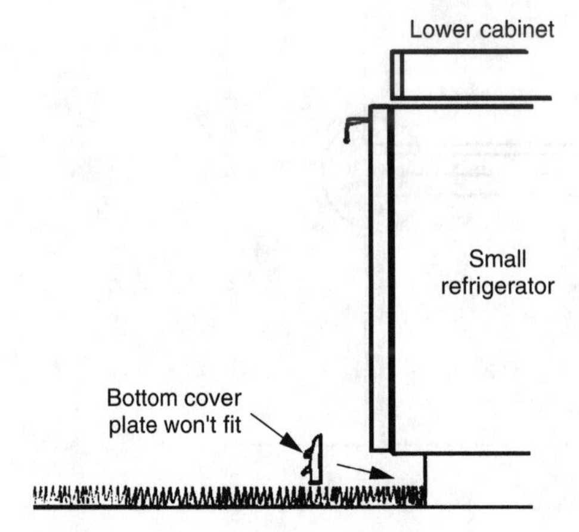

Figure 21.39

complete, it was discovered that the lower panel cover would not fit because the knap of the carpet had reduced the dimension between the refrigerator toe-kick and the floor. If the panel had been in place before carpet, it would not have been able to be removed.

A solution to this problem is to raise the refrigerator up on a platform shelf, at the same elevation as the other lower shelves within the bath cabinet. A toe-kick board can then be provided at the front of the platform so it matches the other cabinetry and so the flooring, including carpet, can be tucked under it.

Chapter

22

Finish Carpentry

Margin between Doors and Jambs

The gap around any door and its surrounding jamb, should be wide enough to allow for the thickness of paint, stain, or lacquer, and slight settling and shifting following construction. This gap should be between ⅛ and 3⁄16 of an inch. A door that is hung with a perfect 1⁄16-inch gap before finishing will end up with so little tolerance that any settling, shifting, or swelling during or after construction will cause the door to stick or bind in place. The binding edge must then be shaved, sanded, and refinished.

Interior Doors and Vinyl Flooring

Interior prefit door units that are commonly used in production housing are assembled within each door opening. Either side of the unit can be lifted to accurately fit the door within the rough framed opening. Likewise, the top or bottom can be moved sideways for a proper fit. The margin between door casing and adjacent side walls or ceiling should be equal when the door is finished (Figure 22.1).

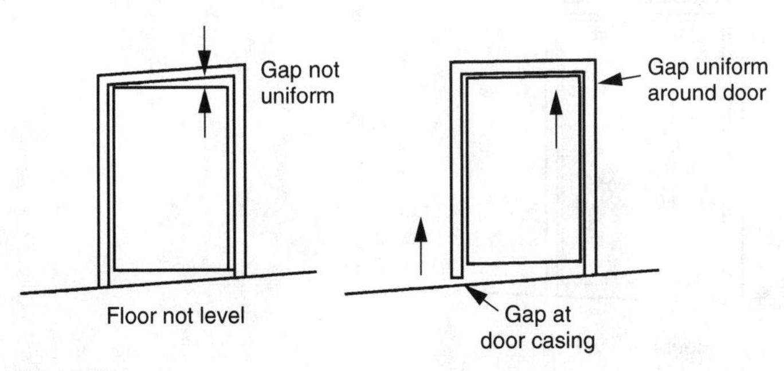

Figure 22.1

When one side of the door unit must be lifted as part of the plumbing and leveling adjustment, a gap is created between the bottom of the raised door jamb and the floor. In carpeted floor areas, the thickness of the carpet padding and carpet usually covers or hides this gap. However, in kitchens and bathrooms where vinyl flooring is installed, instead of raising up one side of the door unit, the opposite side is cut off and lowered. This enables both sides of the door unit to be near or touching the floor (Figure 22.2).

Doors and Wind

Wind can cause damage to interior doors and create additional and unwanted pickup repair work. For example, when windows are left open by tradesmen or get broken, air blows through the interior of the house. These gusts can open and then violently slam doors closed repeatedly before corrected. This can cause a lot of problems with interior doors that do not have hardware installed yet. Slamming doors can knock the jamb stops loose, shake the jambs loose from the casing, or even worked their hinge screws free, releasing the door to fall to the ground.

The builder should check the exterior of the buildings for open windows at the end of each day. Units with broken windows should have their interior doors wedged tightly in the open position each night until the broken window is replaced or the interior hardware is installed.

Doors Fit Stops

Doors should close evenly and solidly against door jamb stops. This allows the doorknob latch to engage at the same time the door touches the stop. If the door and stop do not match and the door hits the jamb stop at some point before the latch can engage, then extra force is required to pull the door shut (Figure 22.3). When extra effort is noticed, the homebuyer generally requests that the mismatched doors and jamb stops be corrected.

The builder can avoid this problem by checking the fit between doors and jamb stops before the painting phase. Once the jambs have been painted, the raw unpainted wood behind the stop is exposed when it is moved to fit the door. This leaves

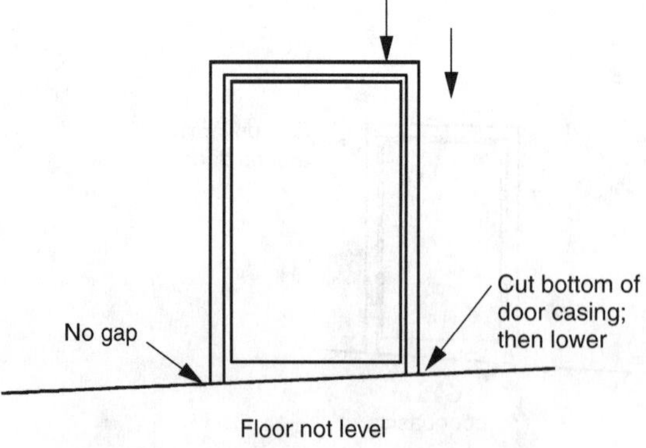

Figure 22.2

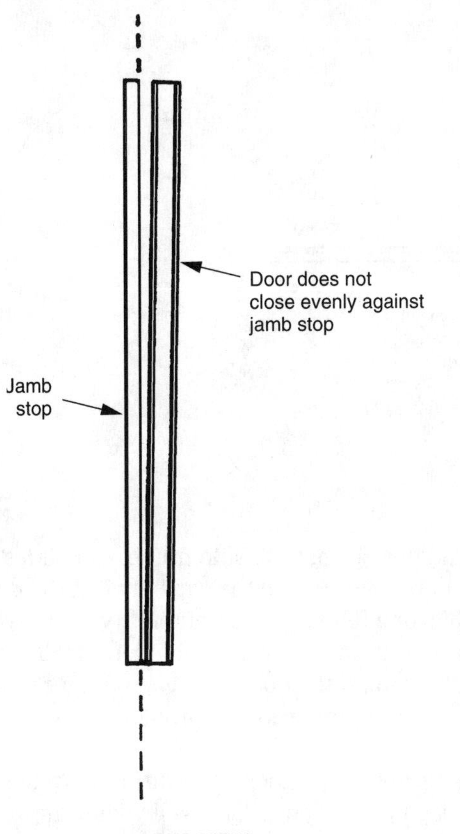

Figure 22.3

an edge where caulking must be scraped off and a line where the paint meets the raw wood. This line must be sanded smooth and the jamb repainted. If the painter used enamel paint, it will be difficult to sand, and sometimes peels instead. The repair then becomes more difficult to complete and expands into a lot of unwanted extra work.

Pocket Door Clearance above Floor

The builder should not only check the fit between a pocket door and its vertical side jamb, but also the adjustment of the pocket door above the floor (Figure 22.4). Interior doors in production housing are manufactured to have about 1½ to 1¾ inches clearance between the bottom of the doors and the floor. This raises the door high enough so that it does not rub on standard height carpet. If a pocket door is installed to match the side jamb without considering the height above the floor, it may rub on the carpet after it is installed. The builder then has to either get the finish carpenter back out to raise and adjust the door or adjust it him- or herself.

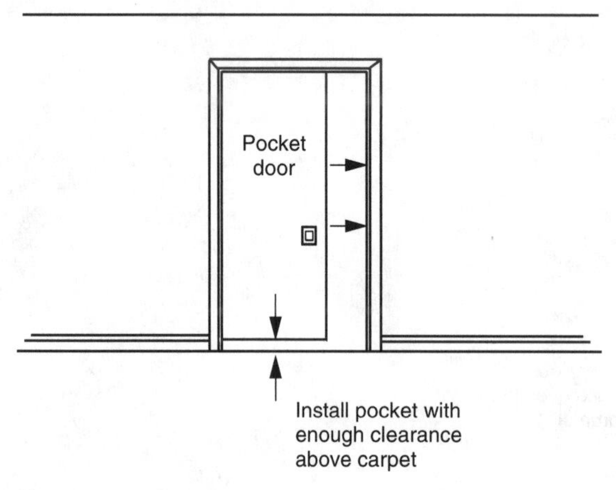

Figure 22.4

Shim Sides of Pocket Doors

One way to fix a pocket door that rubs against the side frame as it slides in and out of the pocket is to wedge a piece of baseboard into the gap between the door and the frame (Figure 22.5). If left like this for a few days, the temporary wedge will bow out the metal portions of the pocket door frame away from the door itself. The result is a slightly wider gap between the frame and the door. The gap remains that way even after the wedges are removed, eliminating the rubbing problem.

A similar preventive measure the builder can take with pocket doors is to ask the finish carpenter who is installing the pocket doors to routinely wedge a piece of baseboard around both sides of each pocket door. This will widen the gap on every door, alleviating the problem before it happens. When it comes time to paint the

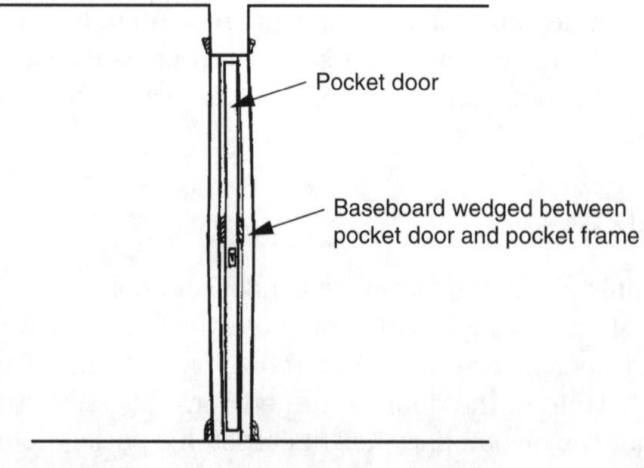

Figure 22.5

pocket door and surrounding jambs and casing, the painter merely pulls out the baseboard wedges and does the job. The result is alot fewer pocket doors that rub.

Take Jamb Off Pocket Door

Illustrated in Figure 22.6 is the side view of a typical pocket door rough frame. In production housing, the frame, track, and jamb are installed as a one-piece unit by the framing carpenter. The pocket door itself is installed by the finish carpenter later in construction, after the surrounding walls have been drywalled.

A better method is to have the framing carpenter remove the pocket door jamb from the rest of the unit prior to installation. The finish carpenter then installs the jamb later along with the pocket door so that the jamb can be shimmed and adjusted to match the striker side of the pocket door. Because the pocket door unit is not a prefit—meaning that the door casing does not come preattached, the finish carpenter must cut and install door casing around the pocket door opening anyway. By being able to shim and adjust the jamb to fit the door, the finish carpenter avoids having to work with a pocket door frame and jamb that is out of plumb or square. It also allows the finish carpenter to get a true fit between the jamb and the door, which is important with pocket doors.

Pocket Door Jamb Stop

An end view of a pocket door header track is illustrated in Figure 22.7. The two pieces of flat wood trim on each side of the header track act as jamb stops to hide the track and rollers at the top of the pocket door.

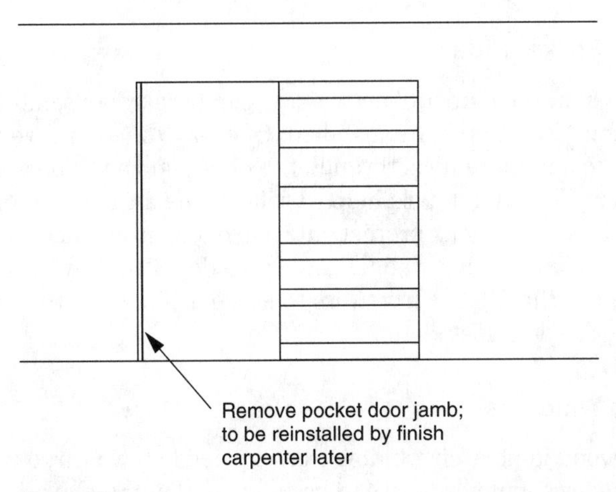

Figure 22.6

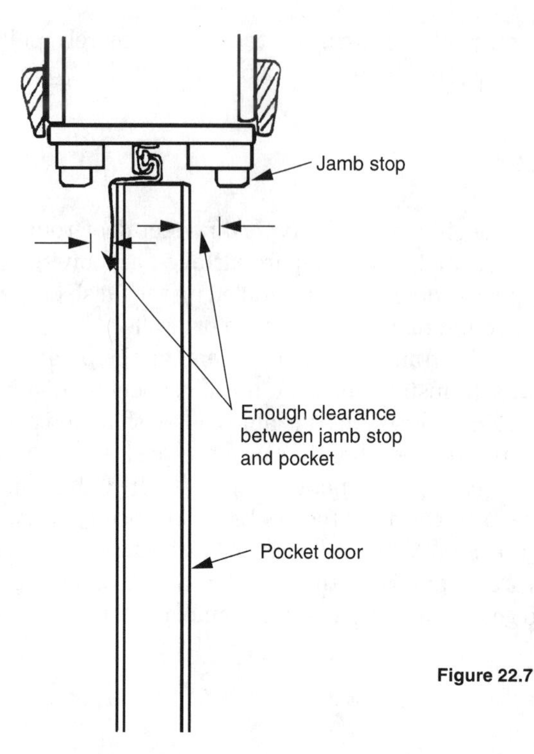

Figure 22.7

These two pieces of wood jamb stop should be installed with enough clearance away from the track so that it can be more easily removed should a problem arise. Some finish carpenters install this jamb stop trim too close to the metal header track, making it difficult to get the pocket door roller wheels free from the track without actually removing the trim.

Add More Nails to Door Jamb Striker Side

Wind whistling through an open or broken window can break door jambs and even blow over doors without their hardware installed. One way to help prevent damage to doors and jambs is to add more nails through the door jamb and door stop on the striker side of the door. The added nails help to solidify the striker side of the door opening. Other precautions to help protect interior doors from the wind include keeping windows closed and placing wedges under the doors so they cannot move.

The builder should ask the finish carpenters to add extra nails to the doors prior to the start of the interior door installation.

Curved Casing to Match Arch Windows

Figure 22.8 shows a wood jamb arch window that is placed above three windows in a living room. The builder should order wood casing trim that matches the radius of the window arch from the window manufacturer or supplier. The casing can then be installed in one piece, with an even and equal reveal around the jamb.

In a million dollar spec house, the casing from the manufacturer did not match the quality of the window and door casing used throughout the house. The finish carpenter then trimmed out the arch window using six short pieces of detailed casing that did not match the radius of the window arch. The result was irregular reveals at the jamb and wavy curves at each casing splice. Apparently the builder and the finish carpenter thought that by cutting a casing that had the wrong radius into small enough pieces, the curvature of the window arch could be closely approximated and would not be noticeable. In this case, the pieced together window casing idea didn't work.

Replace Cracked Finish Trim

Door casing, baseboard, and other wood trim that can split or crack when nailed should be replaced immediately. This should be communicated to the finish carpenters at the start of their work. Split or cracked wood trim seldom goes unnoticed by the homebuyers and must typically be replaced.

Don't putty, caulk, sand, and paint cracked or split wood in an effort to cover up a defect. In production housing, the painter only has enough time to repair the slightest cracks in wood trim. If a cracked piece of wood is going to have to be removed sometime in the future anyway, the finish carpenter might as well replace it when it cracks.

Sand Cut Ends

As a general rule, the builder should insist that the finish carpenter sand smooth all exposed cuts made in wood pieces. This, of course, should be discussed before work begins. Areas to be sanded include the corners of stair skirtboard, exposed end corners of window sill stool and apron, and exposed ends of baseboard at sliding glass doors (Figure 22.9).

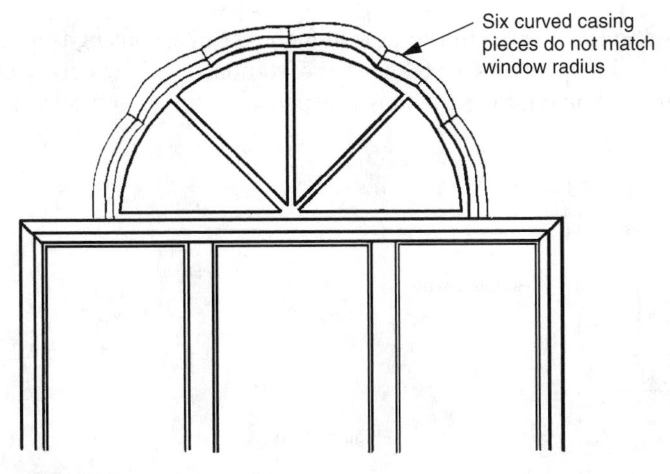

Figure 22.8

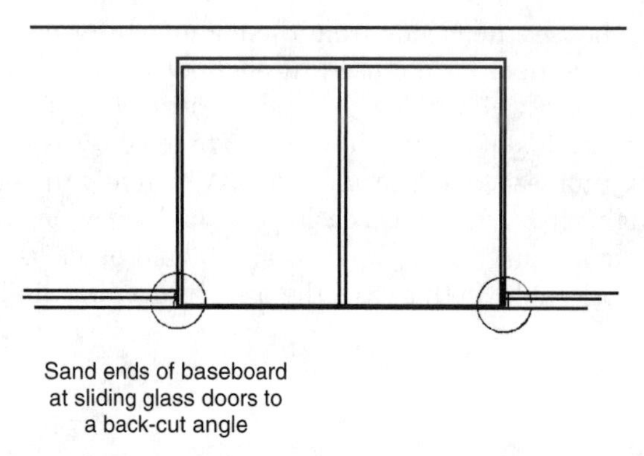

Figure 22.9

Few things look worse and reflect more poorly upon the builder than to see rough-cut, unsanded ends of baseboard, trimmed back at a 22½-degree or 45-degree angle, and then merely painted over by the painter. The builder should insist that all of the finish carpenters, including the baseboard installer, have and use sandpaper.

Rough-Sawn Wood and Trim

Rough-sawn lumber and accompanying trim should be ordered in the exact size in which it will be used. For example, two 2×6-inch rough-sawn plantons used as vertical trim on an interior decorative fireplace front should not be cut from a single 2×12-inch board because the smooth cut edges will contrast with the remaining rough-sawn surfaces (Figure 22.10). Likewise, rough-sawn trim should not be cut or ripped from larger pieces unless the cut edge can be hidden against a piece of wood or wall that is being trimmed out.

Mirror Door Options

If one of the options available to the new homebuyer in a housing tract is to select mirror wardrobe closet doors instead of the standard hollow-core wood veneer doors, the builder should inform the finish carpenter of the option selections prior to

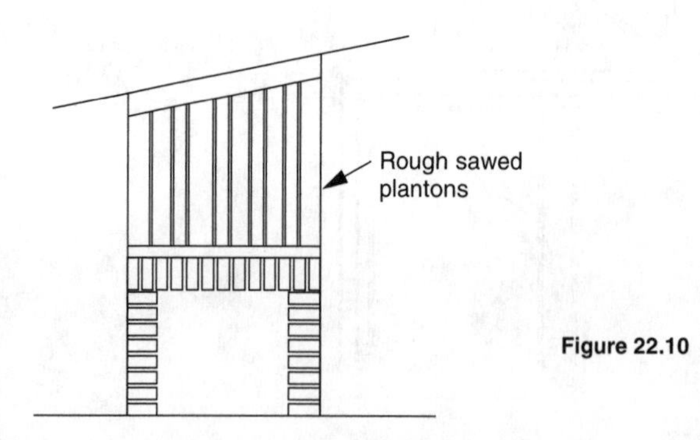

Figure 22.10

materials delivery. This eliminates delivering, spreading, and possibly installing unnecessary closet doors that will be replaced by mirror doors later.

If spread to the units and installed, the standard closet doors will then have to be taken out and returned to storage until they can be used. Sometimes they become trash that must be thrown away. By getting accurate information from the sales office or design center and coordinating this with the finish carpenter, this extra handling of unneeded materials can be prevented.

Another problem with mistakenly installing standard wardrobe closet doors is that valuable time will be spent on a task that did not need to be done. For example, the painter has no way of knowing that some closet doors do not need to be painted along with the rest. If the closet doors are hung, the painter must assume they need to be painted. Few things are more stupid than removing newly painted closet doors, with hardware finger cups and top track and bottom roller guide installed, simply because the builder did not get the information out in time.

Wardrobe Doors

Wardrobe closet bi-pass doors should be adjusted to match the vertical wall or bumper jamb against which it will close. It should also be raised high enough above the floor to clear carpeting that is installed later. Usually the front or foremost closet door will drag on the carpet if not adjusted to the same height above the rough floor as other interior doors (Figure 22.11).

Wardrobe Track

Illustrated in Figure 22.12 is a wardrobe door header track that is cut by the installer to extend past the bumper jambs on each side instead of being trimmed to the jamb. By extending the header track, the air gap that would show if the track was not cut perfectly tight against the jamb is eliminated (Figure 22.13).

Shop Prime Exterior Shutters

False wood shutters should have a primer coat applied to their backsides before being nailed to the exterior of a building. The frontside of the shutter will be spray painted or stained prior to the stucco color coat. By treating both sides of the shutter, they will both be equally sealed and protected from moisture.

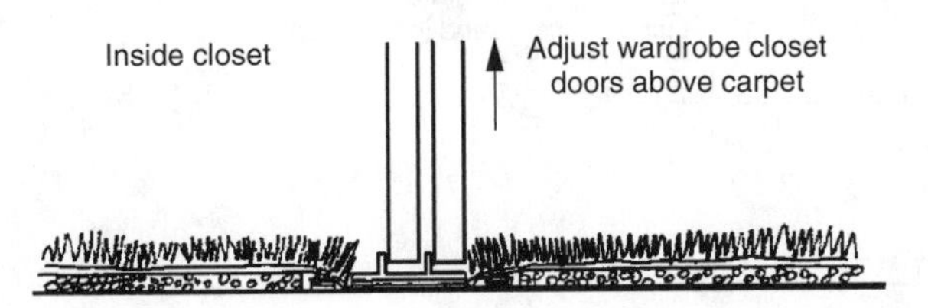

Figure 22.11

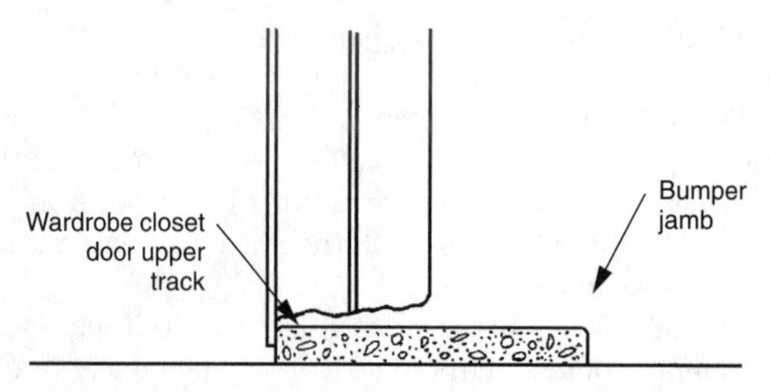

Figure 22.12

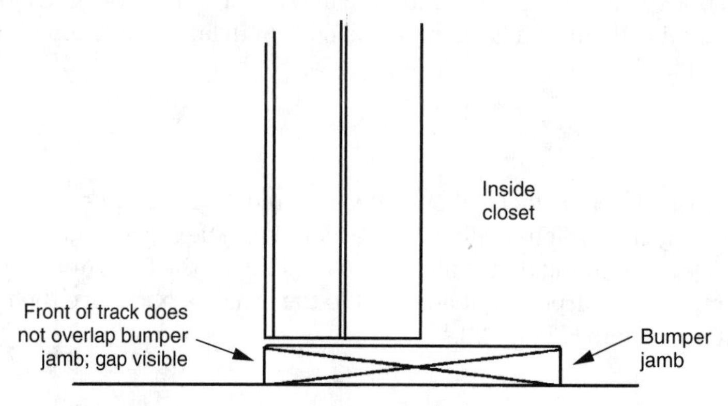

Figure 22.13

The builder should include in the supplier's contract that the shutters are to be back-primed before delivery. The painter can back-prime shutters on the job site, but this means spreading them out somewhere, letting them dry, and then stacking and storing them until installation. If, for example, the finish carpenter is the supplier, he or she should back-prime the shutters in the finish carpenter's shop and deliver the shutter ready for installation (Figure 22.14).

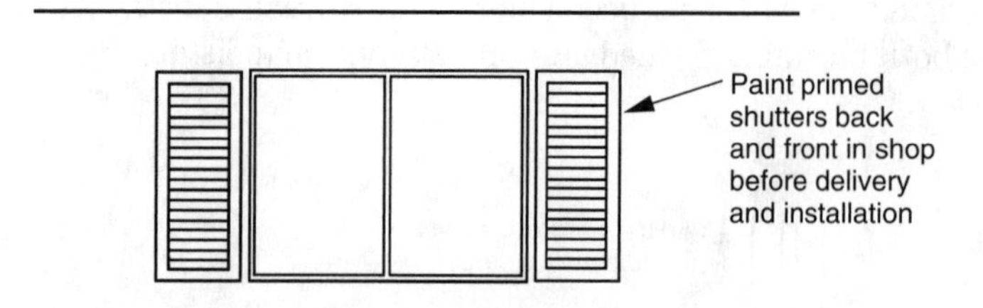

Figure 22.14

Shelf Supports on Drywall

When metal supports or brackets are used for a wardrobe closet shelf and pole, they should not be placed directly on the surface of the drywall (Figure 22.15). Drywall has very little compressive strength. When the weight of clothes and storage are added to the shelf and pole, the metal bracket is forced into the drywall. Eventually, the drywall paper surface will break and the gypsum will crush, turning to white powder.

To alleviate this problem, a short piece of 1×3-inch or 1×4-inch hook strip should be placed vertically on the wall where support brackets are to be placed. The wood hook strip will then distribute the force of the clothing to a larger surface of the drywall (Figure 22.16).

Bumper Jambs

Illustrated in Figure 22.17 are the sides of a wardrobe closet with bumper jambs that extend all the way up to the closet header. This ensures that if the wardrobe door header tract is not cut tight to either side, the bumper jamb will cover the gap from behind (Figure 22.18).

Nailing Particleboard

In production finish carpentry work, woodwork is nailed both by hand and with an air-compressed nail gun. Interior door units, wardrobe shelves and poles, and other components are set and tacked in place by hand nailing. This is followed by a second complete nailing phase with a nail gun.

A problem that may affect quality is that the compressed air for the nail gun is set at around 85 to 90 psi. This pressure will drive and countersink nail gun tee-nails through pine or other soft wood trim, but will not always countersink nails into particleboard. When nailing wardrobe closet shelving and other members that are particleboard, the compressed air must be adjusted to a higher psi or the carpenter must manually countersink the nails.

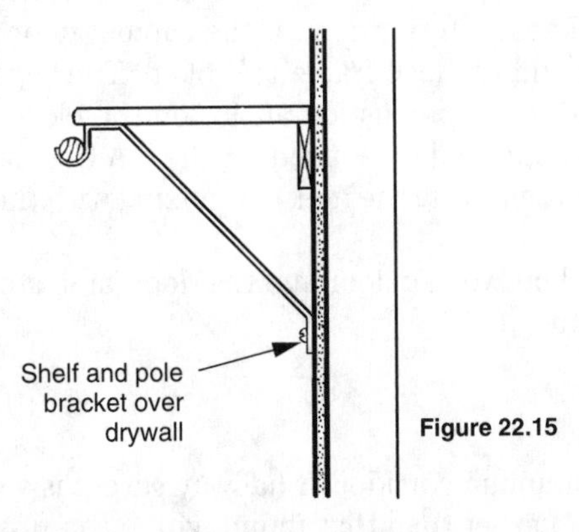

Figure 22.15

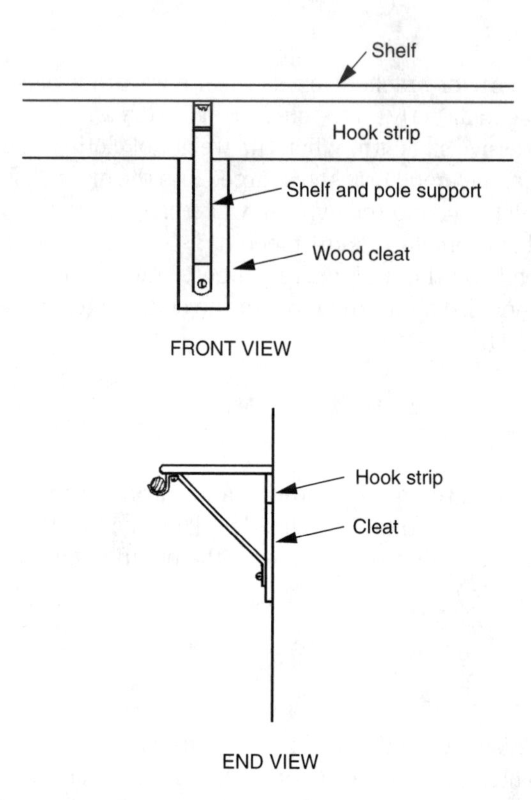

Figure 22.16

Nail gun tee-nail heads that project above wood surfaces look bad and have sharp corners. The builder should check particleboard areas to make sure that these nail heads are countersunk.

F.A.U. Closet Door

F.A.U. closets within the house interior should have a standard door jamb and door rather than a cabinet face frame and cabinet door. A cabinet door is more likely to warp or bow than a standard thickness interior door. If the cabinet door warps, the air seal between the cabinet door and the face frame is broken. The separation between return air outside the closet and combustion air inside the closet is broken (Figure 22.19). Only outside air is supposed to be used for the F.A.U. combustion. It is then vented outside again without entering the house or mixing with the house return air.

A better fit is usually maintained between a door and the door jamb stop than the cabinet door and face frame system.

Corridor Carpet and Skirtboard

For maintenance purposes, condominium corridor or hallway carpet is sometimes a much thinner and tighter grade of carpet than that found within the private apartments. Figure 22.20 illustrates a small flight of corridor steps with skirtboard trimming out both sides of the stairway. Cuts in the skirtboard around the landing or at the top step must be accurate and without saw-kerfs that extend past the intended cut because the corridor carpet is so thin. If unsightly saw marks are not discovered

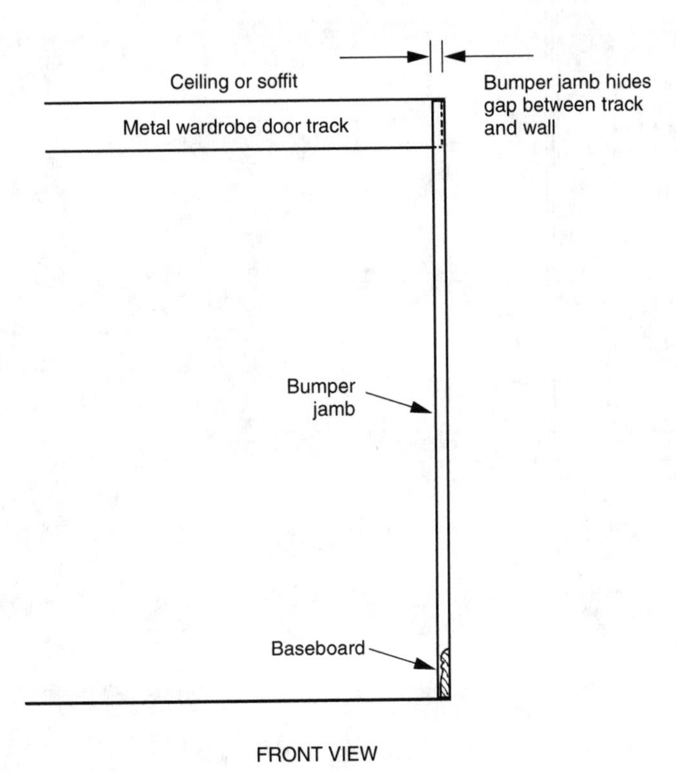

Figure 22.17

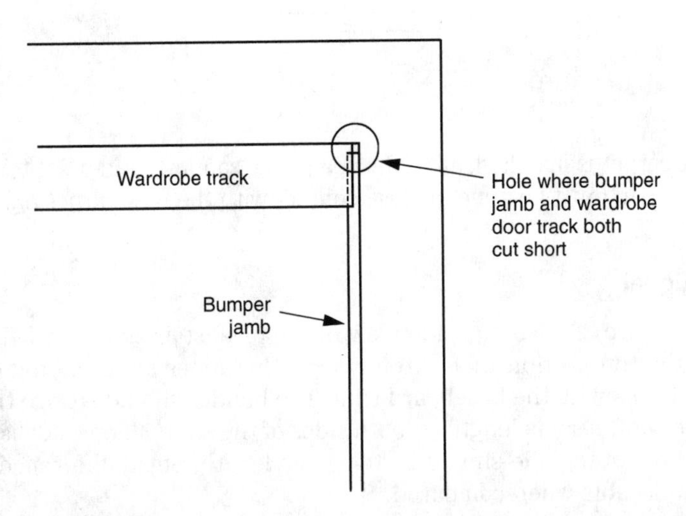

Figure 22.18

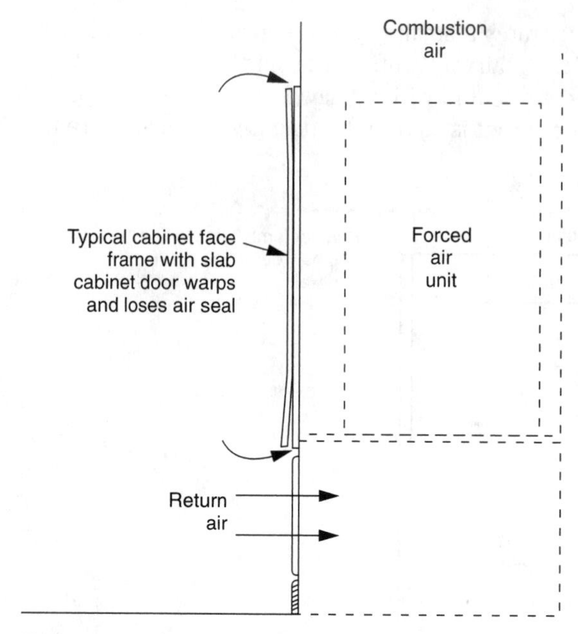

Figure 22.19

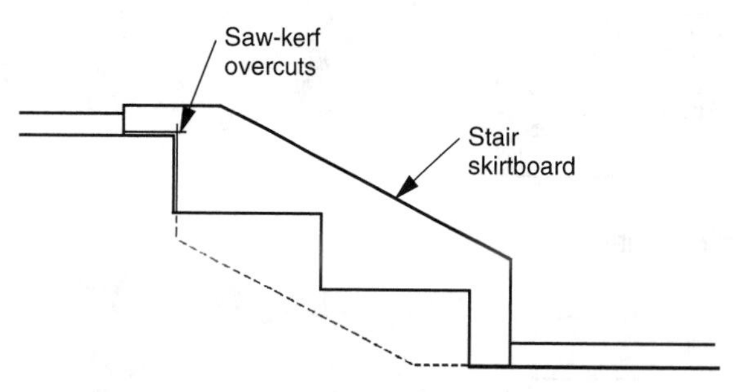

Figure 22.20

until after the carpet is installed, all of the spackling, sanding, and paint touch-up to fill the exposed gap must be done in close contact with the new carpeting.

Stair Skirtboard Ears Equal

Illustrated in Figure 22.21 is the side view of a typical stair skirtboard. The skirtboard ear is the top portion that extends over the upper stair landing or upper floor level and joins with the baseboard trim. The builder should ensure that skirtboard ears are cut the same length on both sides of the stairs. If one side is cut with a 2½-inch ear, the other side should be the same length. Small differences are actually quite noticeable when compared.

Check Skirtboard

The stair skirtboard trim, pictured in Figure 22.22, goes on each side of a stairway. The builder should make sure that the trim measurement is equal from the edge of the skirtboard to the pointed corner at the bottom step and the top step. This shows that the skirtboard was installed to match the slope of the stairs.

Some skirtboard installers are in too much of a hurry or get lazy, and will not pull the skirtboard out to make that last ½-inch cut at the top or bottom to even it out prior to nailing it in place. They think that no one will notice a difference of ½ inch between the slope of the skirtboard and the slope of the stair steps.

Check Particleboard Skirtboard Edges

Sometimes the finished factory edges of manufactured particleboard are uneven, pointed, or wavy. When particleboard is used for stair skirtboard, the builder should check the quality of the material while it is still on the delivery truck. If the factory edges of these 1×12-inch boards are not up to standard, the builder can have the finish carpentry subcontractor send the material back before being off-

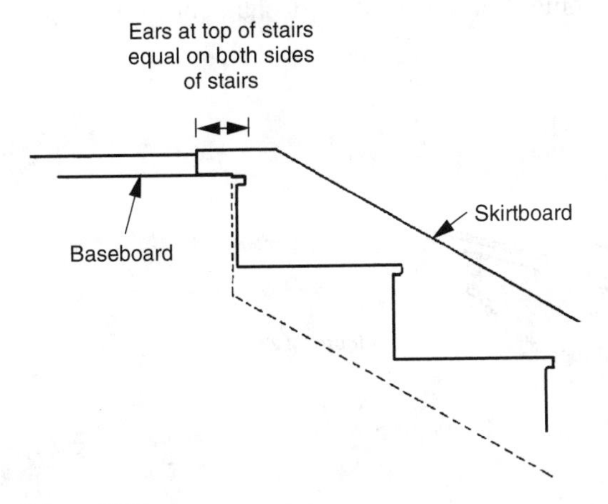

Figure 22.21

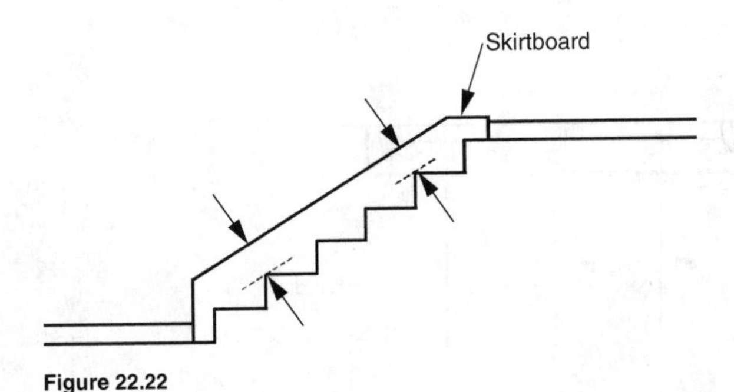

Figure 22.22

loaded from the truck. Otherwise, the builder is placed in the position of trying to get the finish carpenters to sand these defective edges smooth after the skirt-boards have been installed. This sanding will only be done half-heartedly because of the difficulty of fixing rock-hard particleboard edges that are already nailed tightly against drywall.

Two Brackets Minimum per Handrail

The builder should insist that at least two metal brackets be installed per each piece of stair handrail and that they be spaced an equal distance from each end (Figure 22.23). Even for handrail pieces that are only a few feet long, nothing looks worse than to see a short piece of handrail with only one bracket placed somewhere in the center of its span (Figure 22.24).

Vertical Wood Trim in Hallways

A mistake that sometimes can happen in apartments and condominiums is to install the floor baseboard on the walls in hallway corridors during construction, and then later decide to add vertical wood trim in the corridor along with wallpaper as part of the interior decorating scheme. For this type of design, shown in Figure 22.25 the

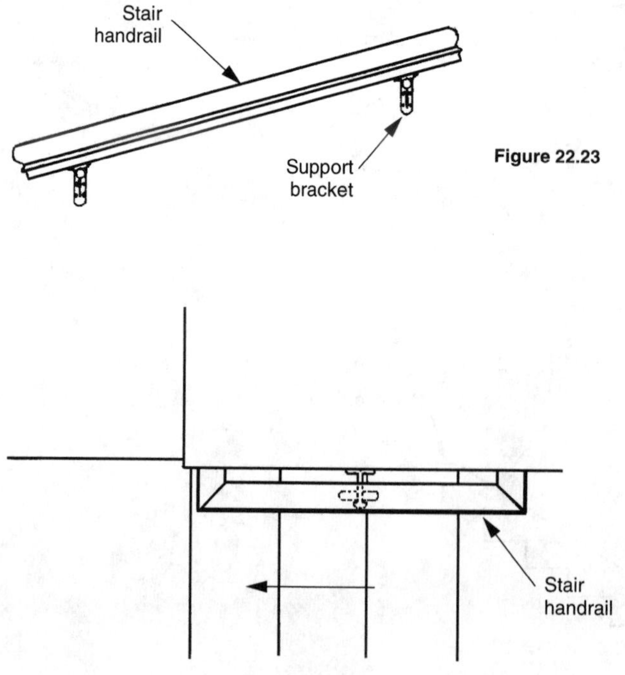

Figure 22.23

Figure 22.24

baseboard should be installed after the vertical wood trim pieces are in place so that the baseboard can be cut to fit the trim.

Baseboard Between Casing and Wall

Figure 22.26 shows a piece of interior door casing adjacent a side wall. This illustrates the correct method of baseboard installation. If the gap between the casing and the wall is roughly equal to the thickness of the baseboard, the temptation of the finish carpenter is to slip the baseboard in between, assuming the painter will caulk the remaining gap or that nobody will see it in a corner behind a door (Figure 22.27).

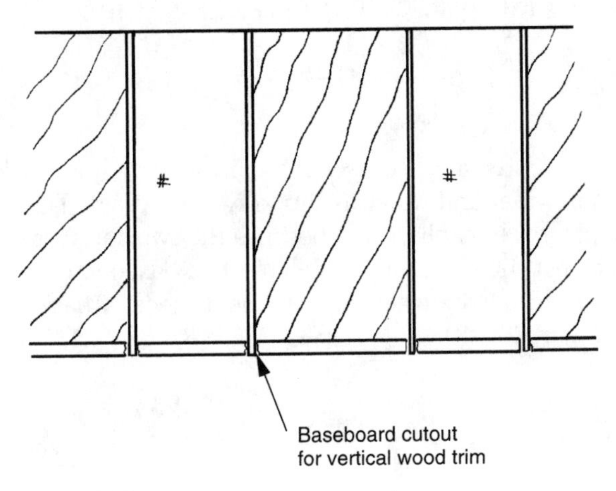

Figure 22.25

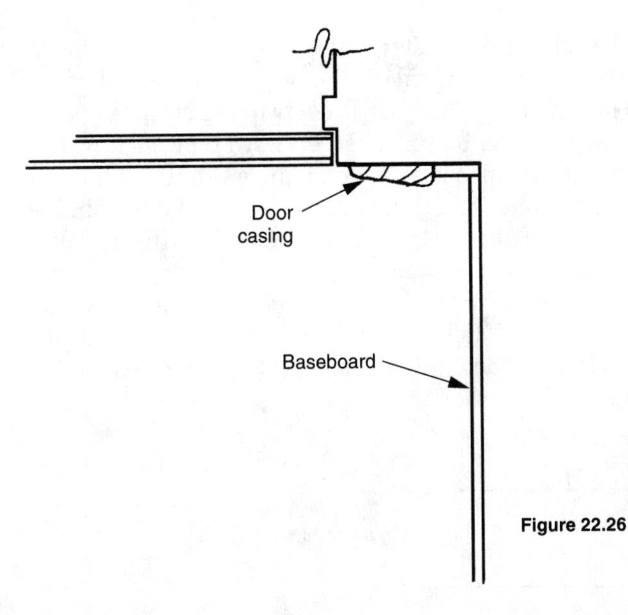

Figure 22.26

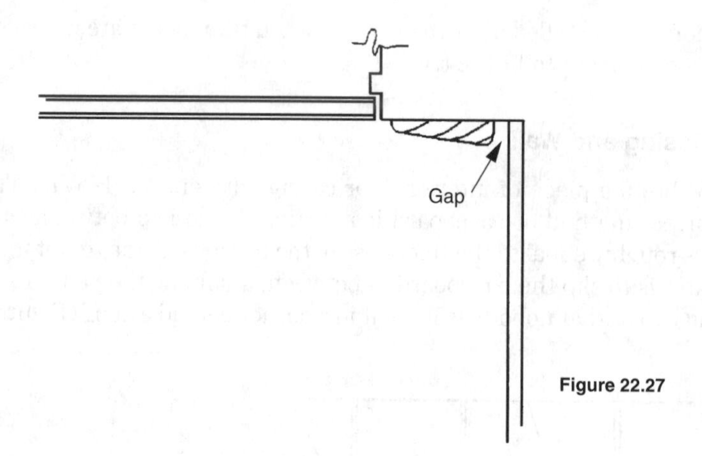

Figure 22.27

This procedure, once accepted at the outset of the finish carpentry work, often results in gaps that become wider and wider as the work progresses. This type of gap is also difficult for the painter to caulk neatly because the two surfaces curve in different directions. To correct the situation, a small piece of baseboard should first be placed between the door casing and the wall. The next piece of baseboard should then be coved into it (Figure 22.28).

Baseboards and Doors

Baseboard installers in production housing seldom or never use a tape measure to find the length of each piece. They usually rough cut pieces slightly longer than they need for each section of wall, then cut off small pieces from one end until the baseboard fits.

Occasionally, the baseboard installer, rather than cut a final ⅛ inch or ¼ inch off the end, will jam the piece in and smash the bow in with a nail or two to make it tight to the wall. This practice is okay between two wall corners of drywall because the drywall at the corners will be squashed outward only slightly. However, when baseboard is forced into a wall section containing a prefit interior door, the baseboard pushes over the door casing that is stapled to the door jamb. The margin between the door jamb and the door is then reduced at the bottom of the door (Figure 22.29).

When several interior doors rub or stick at the bottom after painting, the baseboard installer is generally at fault. The builder should check the door margins during the baseboard installation and before painting in order to prevent finish carpentry and painting pickup work later.

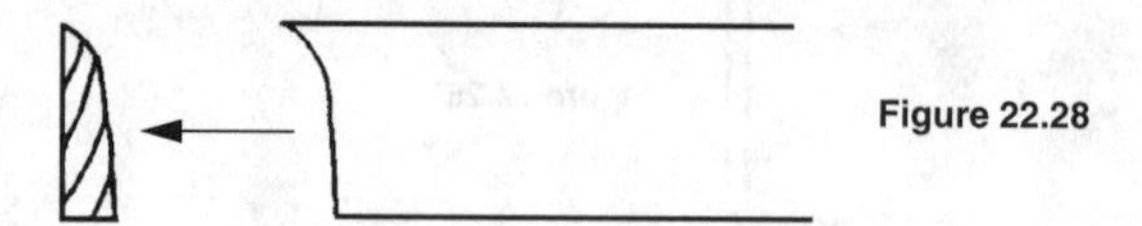

Figure 22.28

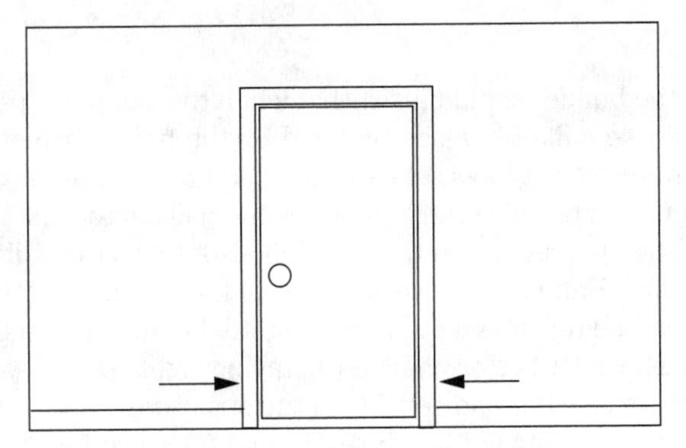

Figure 22.29

Wrap Baseboard around Stair Steps

Illustrated in Figure 22.30 is the side view of stair treads with floor baseboard wrapping around the treads and risers. This is one method used to trim out around stair curved steps that both looks good and ensures that no drywall gaps at the treads or risers are left exposed (Figure 22.31).

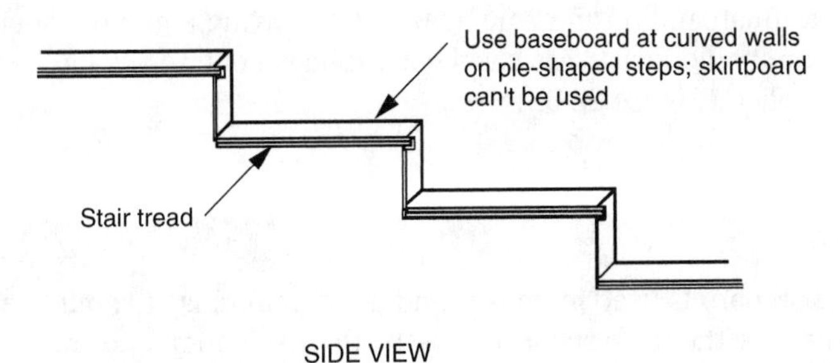

Figure 22.30

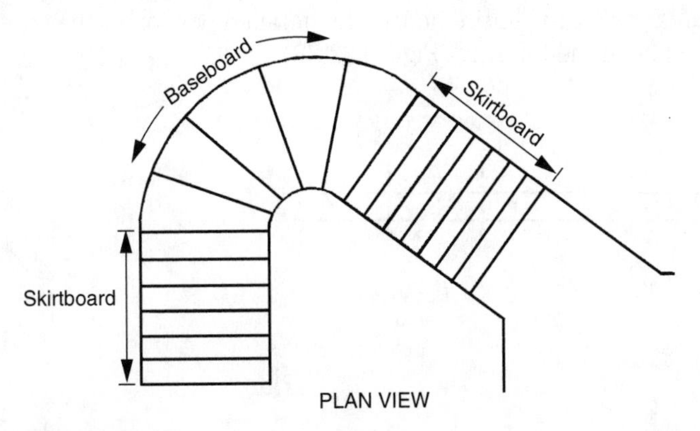

Figure 22.31

Hand Nail Baseboard

In production housing, the builder should insist that when the thinner types of baseboard are used, it should be nailed in place by hand by the finish carpenter rather than with a nail-gun. Production style baseboard, or *speed base*, is often too thin to effectively use a nail-gun with because the gun drives the nails in too fast to pull the baseboard to the wall. The air pressure in the nail-gun also drives the nails half-way through the thin baseboard. Hand nailing will slowly pull the baseboard into the wall surface, with the nail head being hammer-set only slightly below the baseboard surface. This gives the baseboard a better chance of staying tight to the walls during carpet laying and normal drying out and settling of the structure.

For thicker types of curved, detailed baseboard, nail-guns work fine because the baseboard is thick enough so the nail is not shot half way through the piece and it is stiffer and will not bend into curved or bowed portions of walls as easily as thinner baseboard. Hand nailing offers no appreciable advantage over gun nailing with thicker detailed baseboard.

No Butt-Joint Splices in Baseboard

Butt-joint splices in floor baseboard should not be allowed by the builder. Butt-joint splices are seldom flush and often come apart later, leaving a gap between the two pieces. The builder should insist that baseboard splices be mitered and be nailed on both sides of the splice (Figure 22.32).

Raise Detailed Baseboard

When detailed baseboard is used in houses and condominiums, it should be installed ½ to ¾ of an inch above the floor according to the floor selection chosen. The thicker the flooring, the greater the gap. This prevents an installation of marble or tile on top of a composition board underlayment or an installation of thick carpet padding and carpet from burying some of the baseboard detail. By raising the baseboard off the floor in these areas or for the entire house, the detailed portion of the baseboard can remain exposed above the flooring (Figure 22.33).

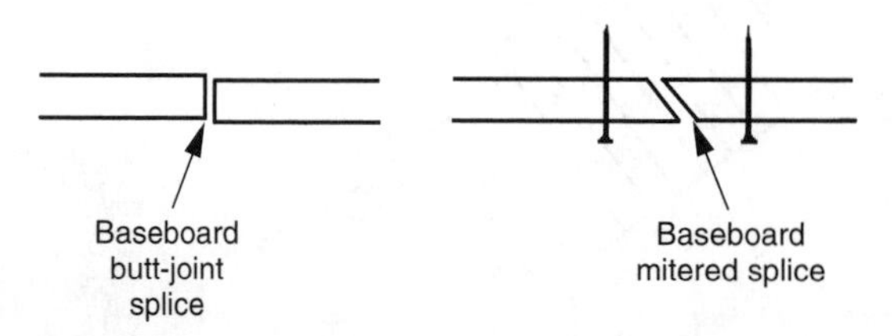

Figure 22.32

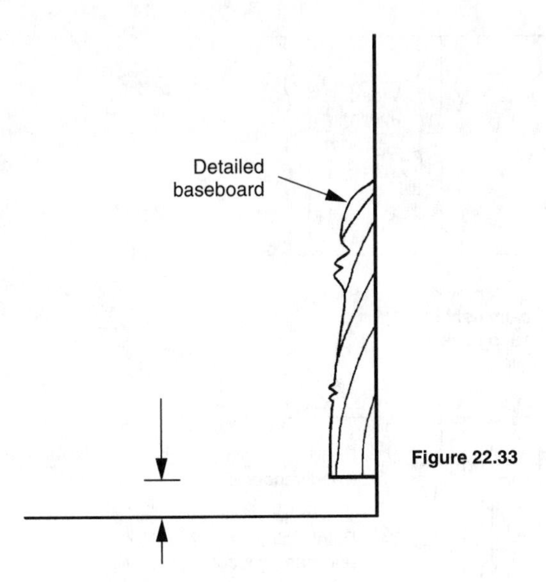

Figure 22.33

Paint behind Paneling Joints

Vertical factory edges of wood paneling sheets are not always perfect. Because paneling edges are not tongue-and-groove or shiplay, any deviation will produce a gap between the adjoining edges, exposing the wall surface underneath the paneling. Few things look worse than dark colored wood paneling with a white wall surface showing through the vertical joints.

This occurrence can be prevented by painting vertical dark brown or black stripes on the wall at 4-foot intervals prior to paneling. A dark colored wall surface behind each paneling joint eliminates any contrast between the wall and the paneling, should gaps occur.

The same idea can be used for thin veneer wood that has square edges butted together. The entire wall to be covered should be painted beforehand with a color that will blend well with the wood veneer (Figure 22.34).

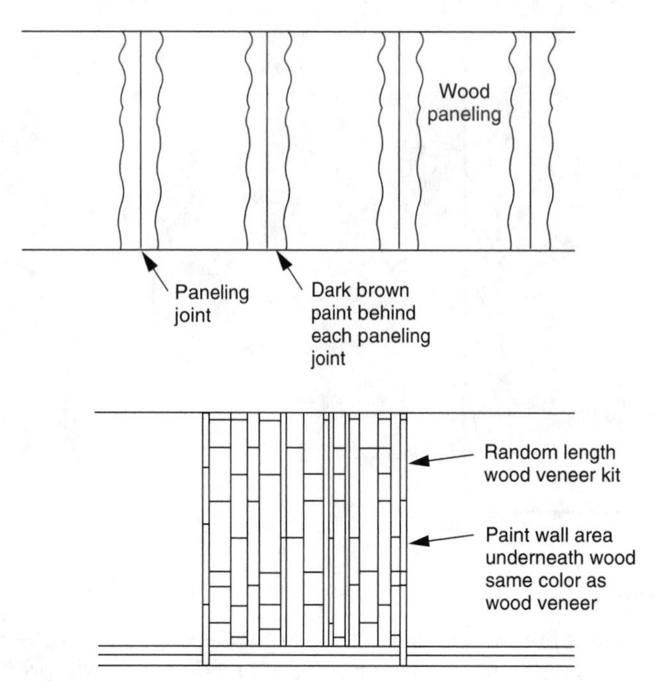
Wood paneling
Paneling joint
Dark brown paint behind each paneling joint
Random length wood veneer kit
Paint wall area underneath wood same color as wood veneer

Figure 22.34

Chapter

23

Tile and Countertops

Avoid Small Tile Pieces

Illustrated in Figure 23.1 are two ways to lay out ceramic tile. The result with these layouts is either small slivers of tiles (top) or larger size pieces of tile (bottom) at each end. The layout displayed on top in the figure starts with a tile grout joint at the centerline of the tile, with full pieces of tile going in each direction. If the span is 1-inch greater than an even number of full size tiles, then the two end pieces are ½-inch wide.

The second layout design starts with a full size tile installed at the centerline of the tile work. In this layout, the two end pieces of tile are ½-inch wide, plus half the width

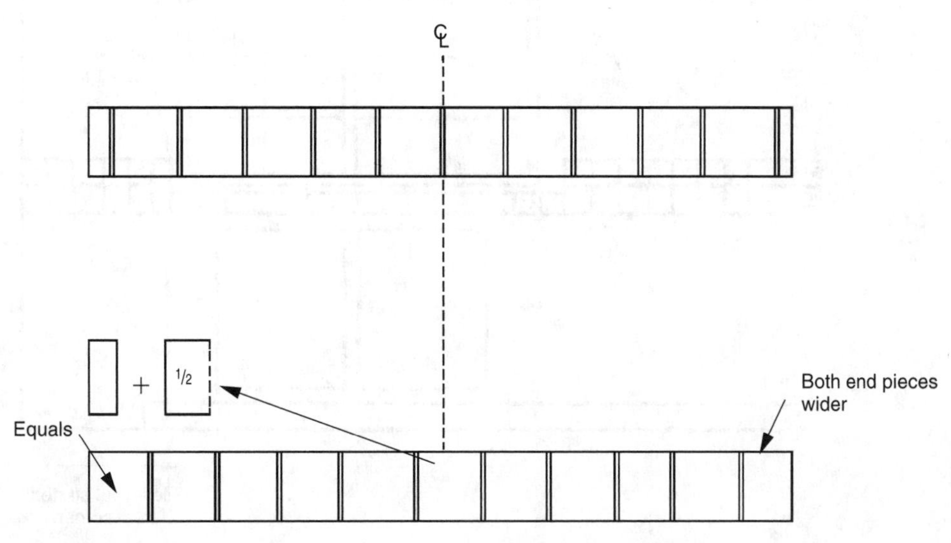

Figure 23.1

of a full tile. If 4-inch tile is used, the end pieces are now 2½ inches wide rather than the previous ½-inch wide.

Center Splash Tile at Kitchen Window

The kitchen front view pictured in Figure 23.2 has the ceramic splash tile that started from the corner of the kitchen wall rather than centered with the sink. This resulted in pieces of tile at the two bottom corners of the window sill not being equal in width.

If the ceramic tile splash layout is started by centering the first piece with the kitchen window and the kitchen sink, then the tile splash will end up with equal size pieces at each end of the window sill. The tile can then continue in both directions with full size pieces (Figure 23.3). This is something the builder should discuss with the tile contractor prior to the start of the tile work. Some builders and homebuyers are not as concerned with the centering of some items within a house. A difference in widths between two pieces of tile at a kitchen window sill may not bother some people. However, the builder and tiler should be aware of the potential problem and arrive at a decision after examining the job.

Install Window Casing before Tile

Illustrated in Figure 23.4 is a typical mistake in upper priced houses. The wood sash window at the kitchen sink has been tiled part way up the window sill. When the wood casing trim is installed later, it is difficult to obtain a good-looking joint between the casing and the curvature of the top row of bull-nose tile.

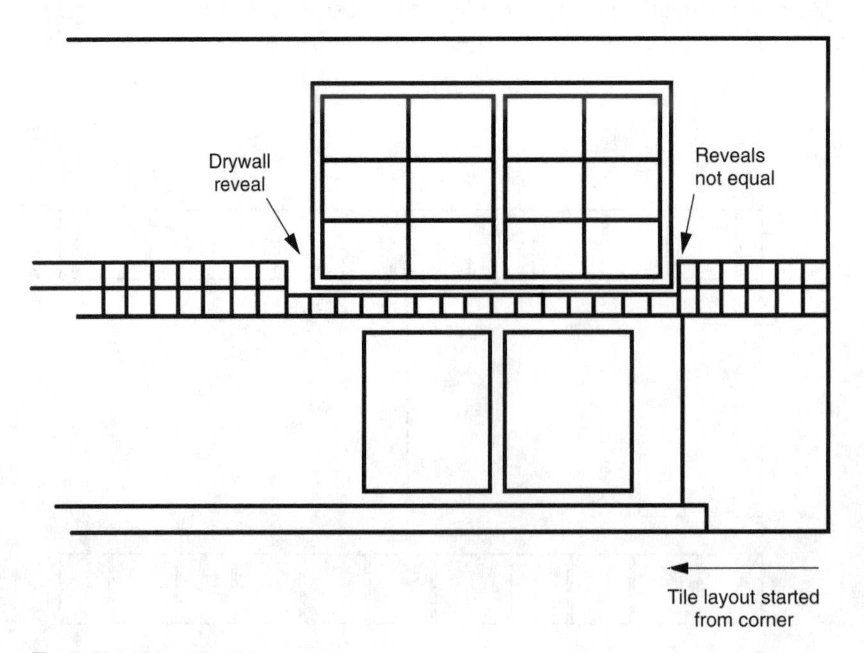

Figure 23.2

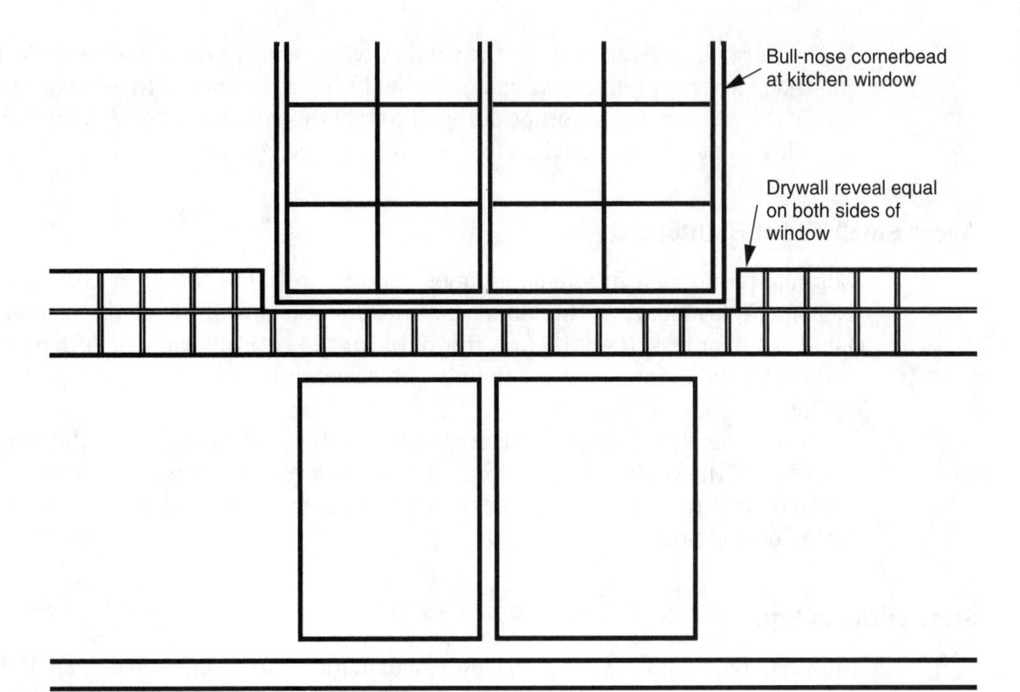

Figure 23.3

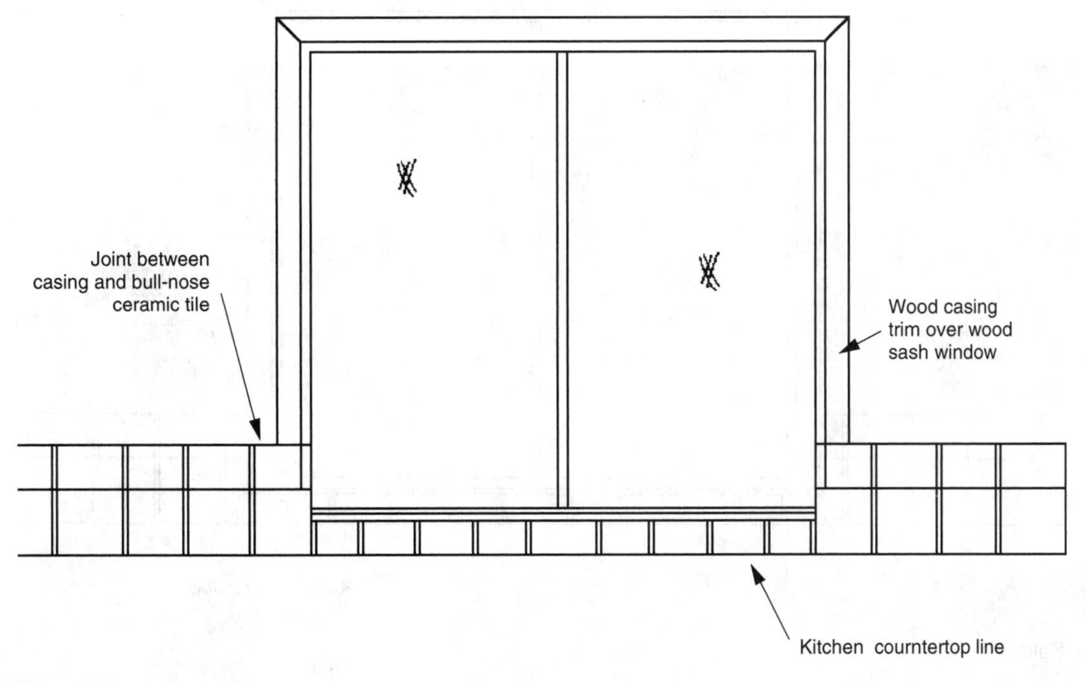

Figure 23.4

A better method is to trim out the window with casing prior to the tile work. The tile can then either die into the casing or be held back a few inches to leave a drywall reveal. If the window gets stool and apron, the stool should be installed without the apron. The tile can then be installed up to the bottom of the window stool (Figure 23.5).

Avoid Small Tile at Bathtubs

Ceramic tile in a bathroom sometimes extends from the top of the bathtub to the ceiling (Figure 23.6). If this dimension results in a row of tile at the ceiling that is 1-inch wide or less, then the first row of tile at the bathtub can be cut to a narrower width and the amount cut off the bottom row can be added to the top row at the ceiling.

The builder should discuss this issue with the tile contractor before the start of the tile work. If the vertical layout from tub to ceiling requires the first row of tiles to be cut to benefit the top row, this fact can affect the economics of the tile contract on a large housing project.

Shower Dams Square

Shown in Figure 23.7 is the plan view of a dam for a bathroom shower stall. If the dam is built to be exactly wide enough for one row of field tiles, the two or three sided dam

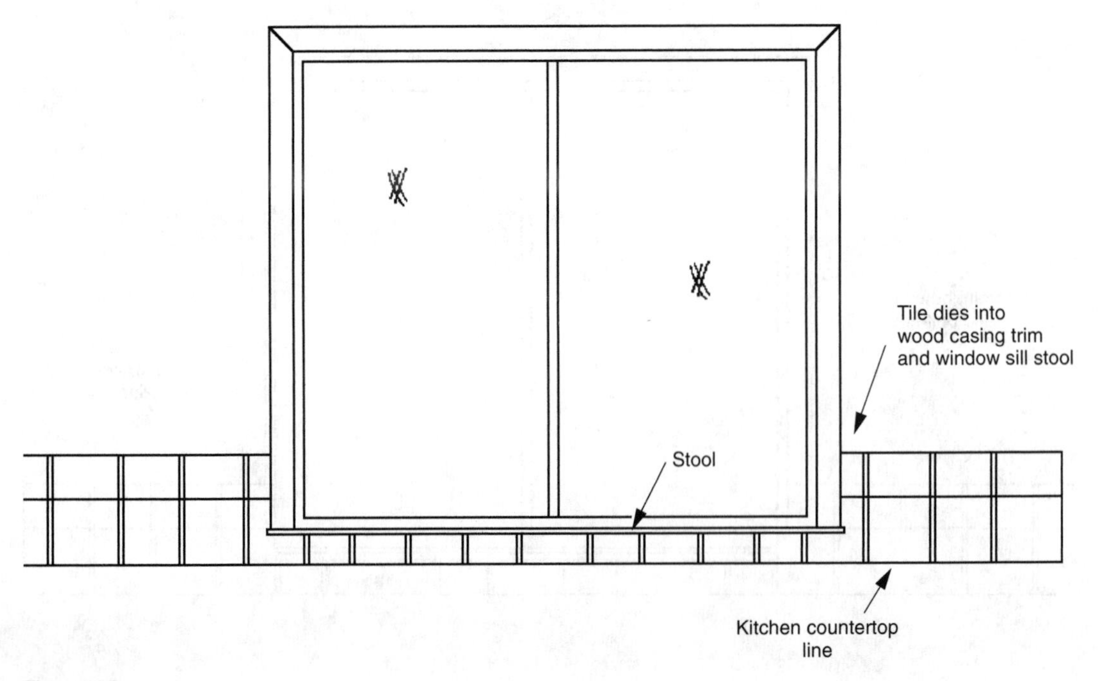

Figure 23.5

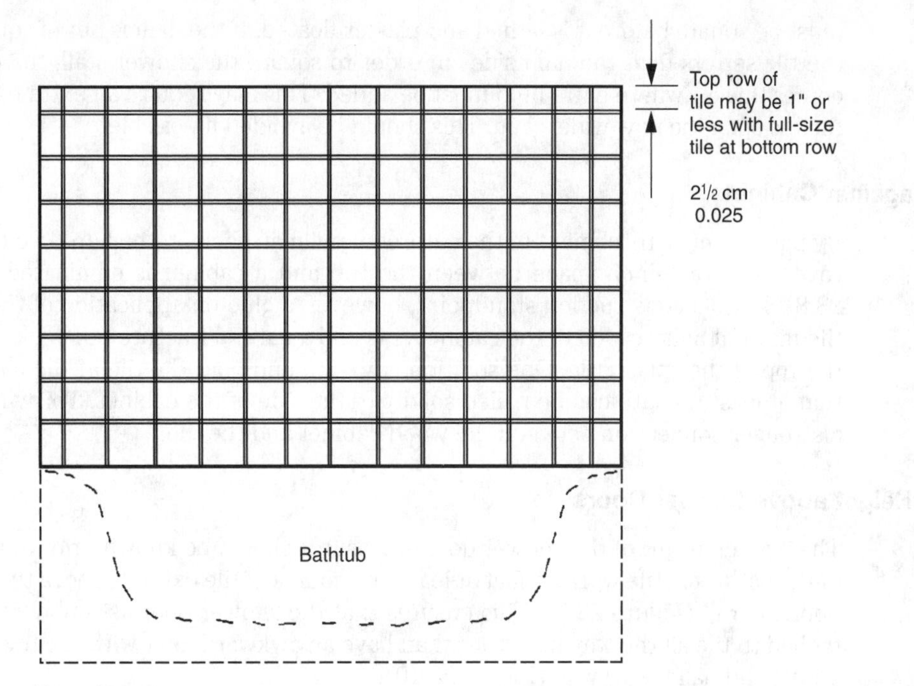

Figure 23.6

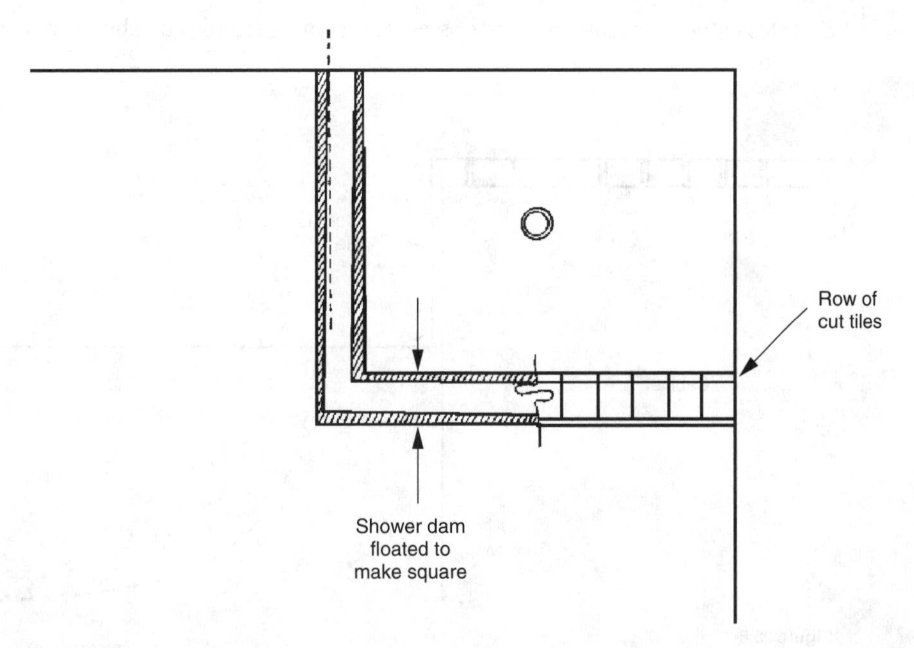

Figure 23.7

must be square before it is lathed and plaster floated. If the dam is out of square and the tile setters float the dam sides in order to square the shower stall, the dam becomes thicker where extra mud must be added. This may require an extra row of cut tiles to span the now wider dam. This should be avoided if possible.

Tile against Cabinets

By placing the bathtub next to the bathroom pullman cabinet, the hard to clean, narrow 2-inch or 3-inch space between the tub and a cabinet is eliminated (Figure 23.8). From a construction standpoint, however, a glue-on application of the splash tile may not adhere well to the cabinet side and a natural fracture point is created at the top of the splash tile. One solution is to lath and float the tile at the cabinet so that at least the lath can be nailed solidly to the side of the cabinet. This will create a stronger connection between the wood cabinet and the tile.

Tile Height above Shower Doors

The overall height of the shower door enclosures should be known prior to the start of the bathroom tile work so that at least one course of tile extends above the shower door top rail (Figure 23.9). This ensures that the shower door assembly will be attached to tile all the way up rather than have an awkward joint with the top rail half on tile and half on drywall (Figure 23.10).

Bar Sink Tile

Stainless steel bar sink countertops covered with ceramic tile should not be floated with mortar that is thicker than ¾ of an inch. Shown in Figure 23.11 is the side view

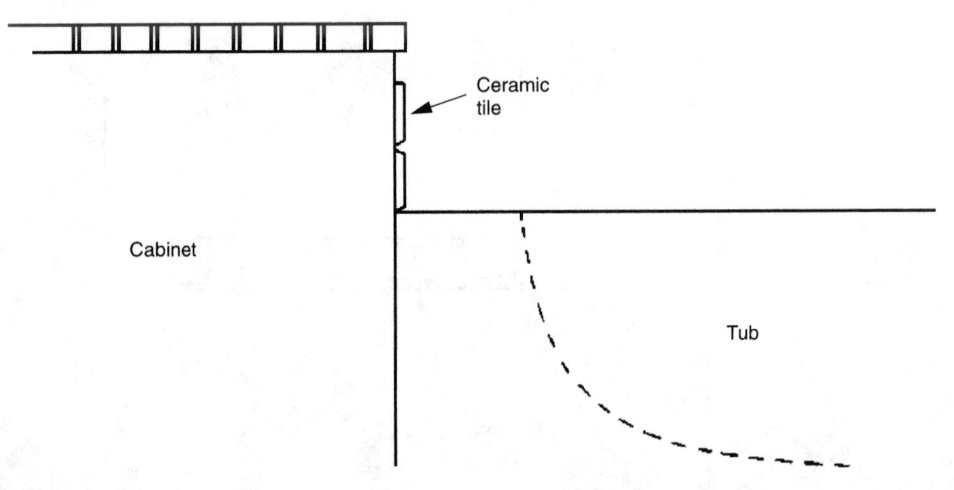

Figure 23.8

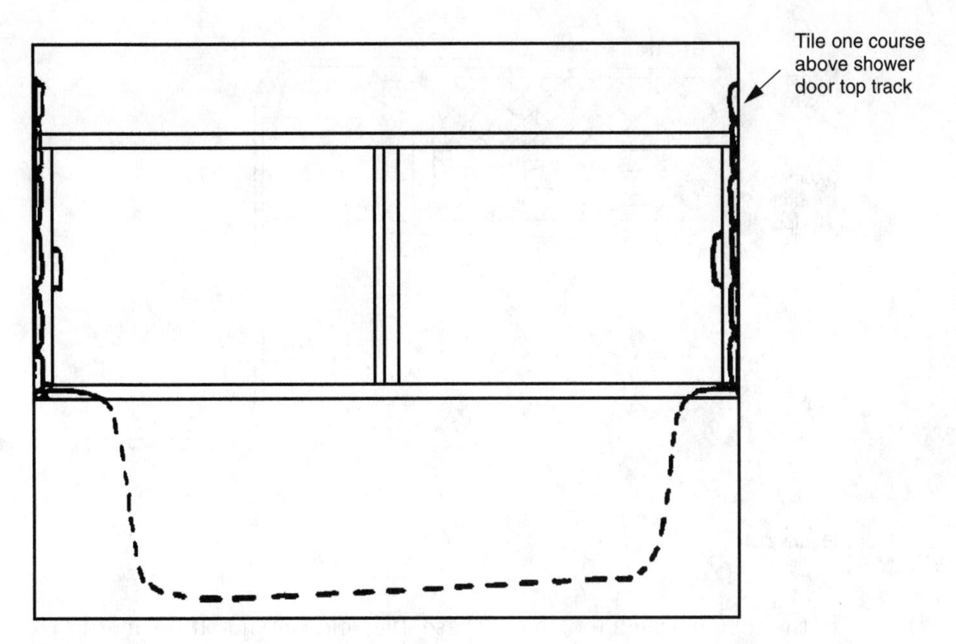

Figure 23.9

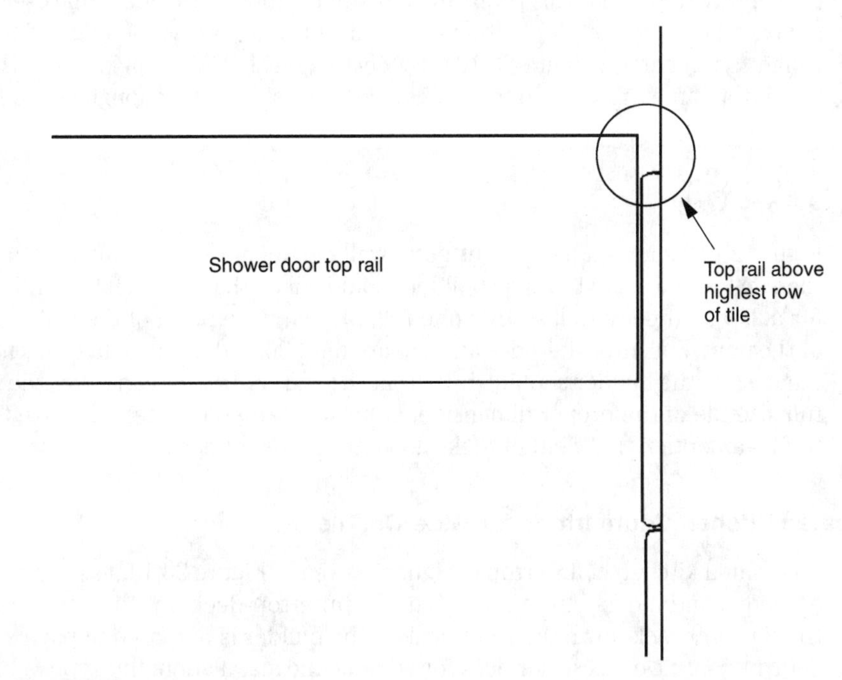

Figure 23.10

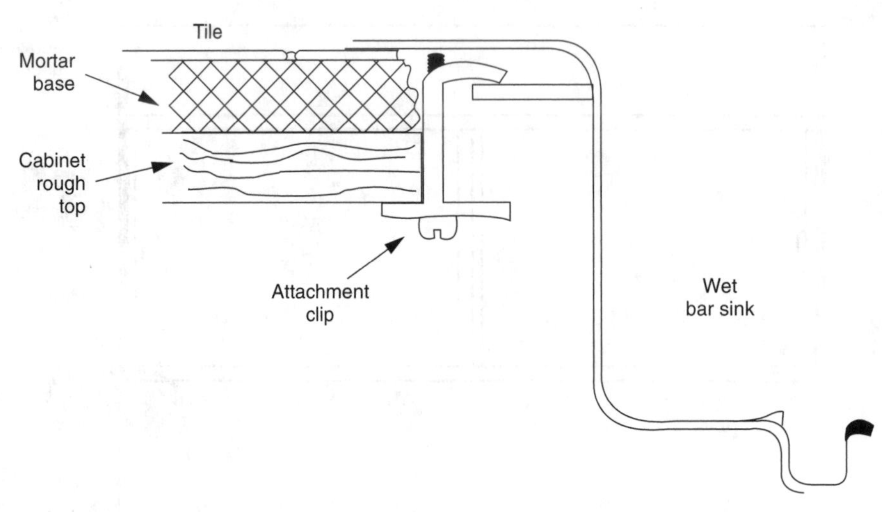

Figure 23.11

of the cabinet rough top, mortar base, tile, and bar sink. If the mortar base is over ¾ of an inch thick, the sink attachment clips may not be long enough to reach below the cabinet rough top wood.

One way to prevent this problem, and still be able to provide a thick enough mortar base beneath the tile, is to cut or rout-out a groove at the edge of the cabinet rough top opening (Figure 23.12). By cutting out half the thickness of the cabinet rough top, an extra ⅜ of an inch is gained. This should be enough space for the attachment clip.

Tiling a 2×4 Pony Wall

Figure 23.13 shows a half-height pony wall between a shower and a bathtub that is tiled on the sides and top. The builder should check that the architectural detail calls for a 2×4 stud pony wall so that one full 4¼-inch wide piece of tile will span the top of the pony wall after the sides and top are mud floated. If the detail calls for another dimension, the builder should check that the tile contractor can cover the top with a full-size tile of the proper dimension. Otherwise, the tile setter will have to cut tiles to fit—an option that neither looks good nor is cost effective.

Prefabricated Kitchen Countertops for Glue-On Tile

A common kitchen countertop design, pictured in Figure 23.14, has ceramic tile that extends continuously from the cabinet countertop deck, up the vertical side, over the top, and back down the front. Unless the builder is prepared to pay extra for the cement floating of these surfaces for ceramic tile installation, this type of kitchen design should be avoided.

Conventionally framed and drywalled countertops are too difficult to get square, straight, and dimensionally accurate for a glue-on tile application in production hous-

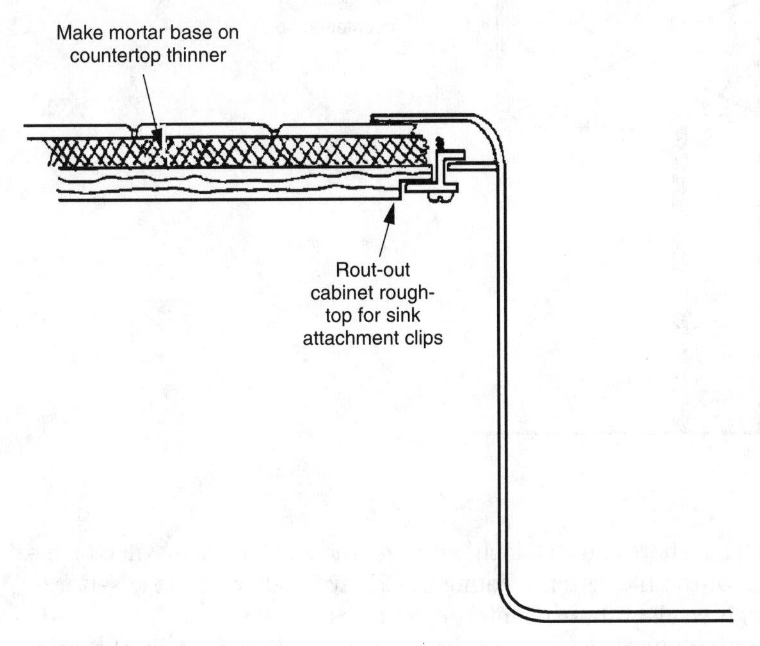

Figure 23.12

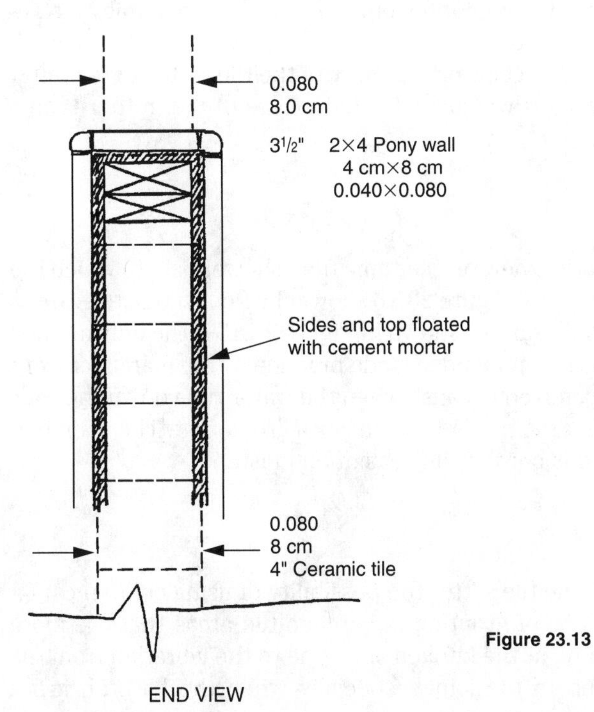

Figure 23.13

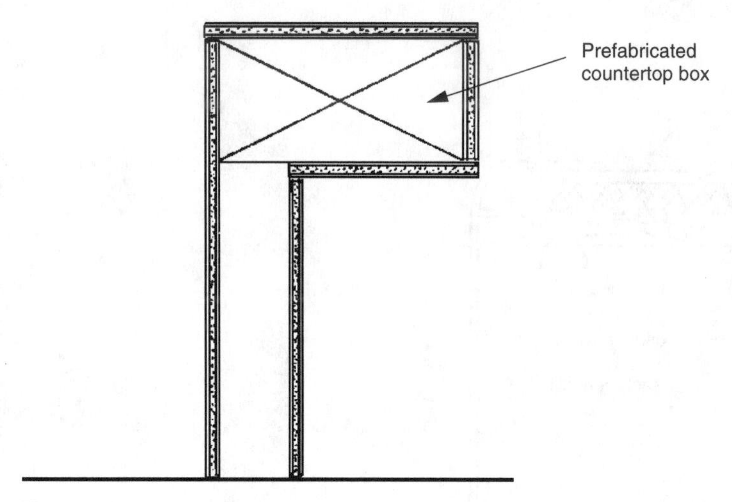

Figure 23.14

ing. All it takes is a ¼ inch here and an ⅛ inch there and the glue-on tile installation becomes a nightmare for the tile setter. Floating the tile work allows the tile setter to square up and straighten all of these countertop surfaces.

If the tile application must be glue-on because of project economics, the pass-through countertop should be prefabricated by the cabinet manufacturer or the finish carpentry subcontractor. This will result in a one-piece box-like structure that can be nailed or screwed to the cabinet countertop. It also has the accuracy advantage of being shop built rather than being framed out of 2×4s at the job site. The dimensions of the prefabricated box can be worked out with the tile setter so that after the drywall is added, the dimensions will allow for full pieces of tile, rather than a long row of cut pieces.

Shower Pan Tile

Illustrated in Figure 23.15 is the side view of a ceramic tiled shower pan, in which the floor tiles go underneath the wall tiles. Figure 23.16 shows the floor tile butting up to the wall tile with a grout joint. The example shown in Figure 23.15 is the only method that is accepted and approved as recommended trade practice by tile manufacturer's organizations. This method is preferred because when the water runs down the wall, it will strike the surface of another tile rather than hitting a grout joint. The main reason for using the second method is because it is easier and faster.

Use Caulk at Tile Joints

The builder should discuss with the tile setter the possibility of using colored caulking, with or without sand, in place of grouting at ceramic tile areas that are more prone to cracking. These areas are: at the kitchen sink, due to the vibration from the garbage disposal; at the bathtub to tile joint, especially when the bathtub has a jacuzzi motor; at the fireplace front face to tile joint; and anywhere else where cracking may occur due to movement or heat (Figures 23.17 through 23.19).

Ceramic tile grout is a rigid material that will crack rather than stretch or move. Caulking is a flexible and adhesive material that will stretch and bend some before coming loose or cracking. Since many tile grout colors have an equivalent colored caulking to match, the appearance is the same but the durability at these vulnerable areas is better.

Factors Affecting Tile Grout Cracks

There are two factors that affects the degree at which ceramic tile grout cracks occur in kitchens and baths at the joints between deck and splash tile.

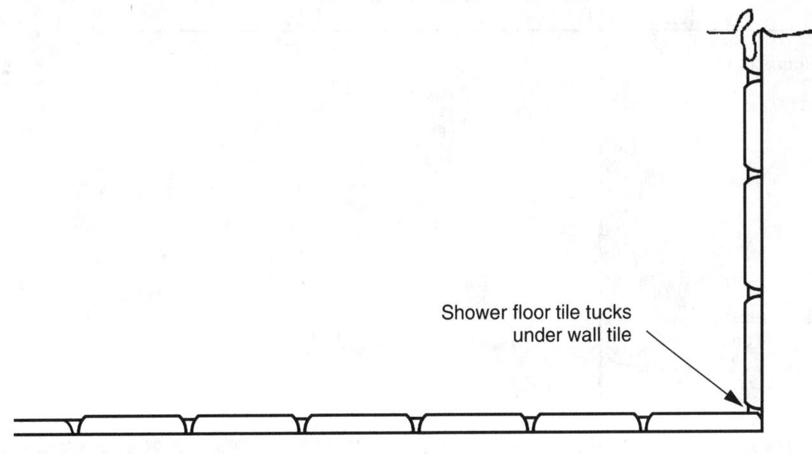

Figure 23.15

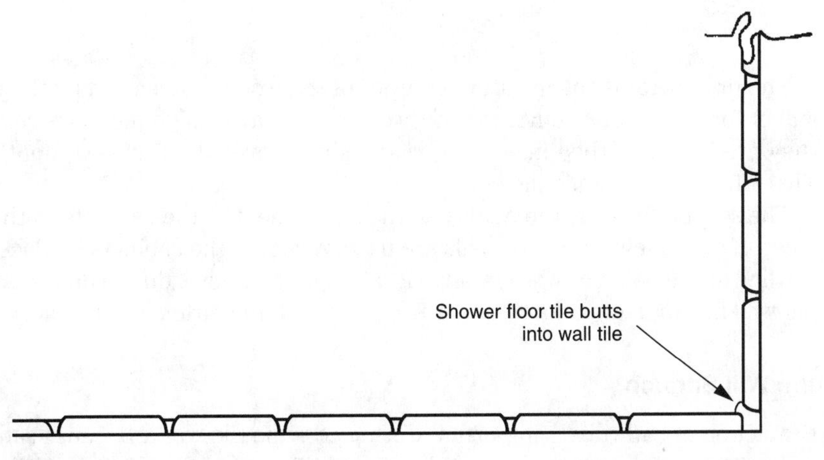

Figure 23.16

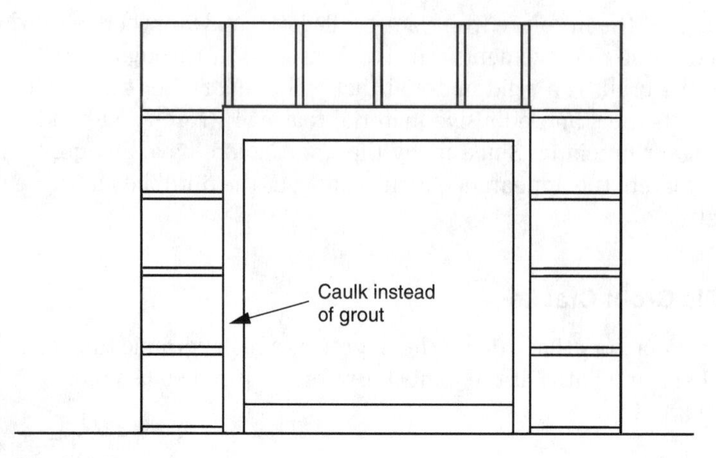

Figure 23.17

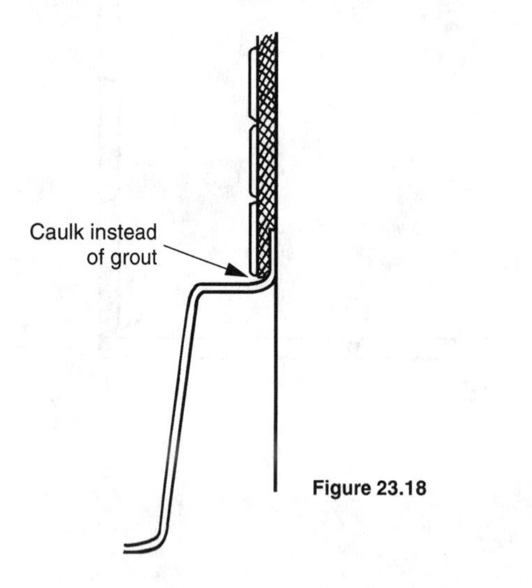

Figure 23.18

The first factor is the moisture content of the wood that is used by the cabinet installer for the cabinet rough-top. If these boards have a high moisture content when they are installed, they may dry out and shrink over the first few months or year. This affects the ceramic tile countertop.

The second factor is the quality of the attachment of the cabinets to the wall surfaces. If only a few screws or nails are used to secure the cabinets in place, then any settling of the floors or walls or any vibration in the house, due to doors closing, people walking, or garage door opener use can result in hairline grout cracks.

Colored Tubs Will Scratch

Protection of bathtubs is important during ceramic tile work. Because sand or grout will spill onto the bottom of the tubs during tiling, scratches can result in the bathtub as the tile setter stands inside the tub. This is especially true of bathtubs that are colored—the worst or most vulnerably color being black or dark blue.

Molded plastic tub covers should be taped in place in the bathtubs as soon as they are installed. If the tub covers must be removed so that the tubs can be filled up with water for the rough plumbing inspection, they should be cleaned off and reinstalled after the inspection. There are also several types of temporary jell coatings that can be applied to bathtubs to protect their surfaces during construction.

The point is that ceramic tile work around bathtubs will scratch the tubs if no protective covers or coatings are used. No matter how careful and clean the tile setter is, sand and grout will fall into the tub and scratch the surface as it is stepped on.

Check Tubs for Scratches before Tile Work

It is good practice to walk the bathtubs with the tile contractor prior to start of the tile work. The condition of the tubs regarding scratches, chips, or holes should be noted by both parties. This relieves the tile contractor of any responsibility for damage caused to tubs before the tile work, while at the same time putting the tile contractor on notice that any additional damage done during the tile work is the tile contractor's responsibility.

Although this is a little more work for the builder and the tile contractor, it is a more fair way of addressing the cost of bathtub repairs required at the completion of each house.

Bathtubs Installed Level

From the standpoint of the ceramic tile setter, bathtubs should be installed level. When bathtubs are mistakenly set out of level, the tile grout joint around the tub or the bottom row of cut tiles at the tub become uneven in width.

The builder should not accept bathtubs that were installed out of level to provide more downward slope to the bathtub drain. All bathtubs are manufactured with a slope toward the drain so that the bathtubs can be installed level. Any unlevelness is due solely to sloppy craftsmanship, not out of a desire to improve the drainage of the tub.

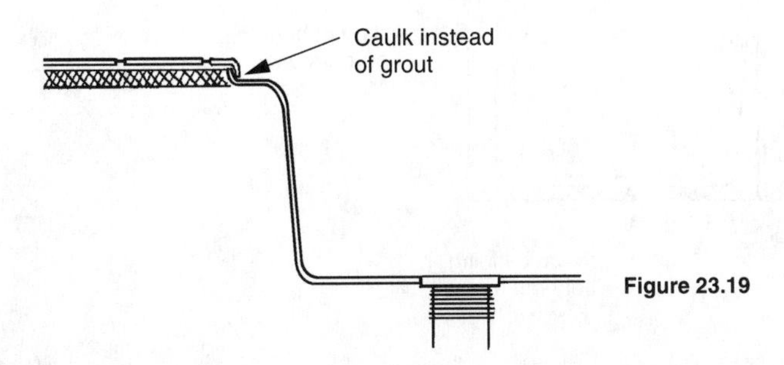

Figure 23.19

Jacuzzi Tub Secure

Because of the vibration from the Jacuzzi motor, Jacuzzi tubs are more prone to tile grout cracks than normal tubs. A Jacuzzi tub should be installed on top of some sort of bedding material—spray foam or mortar—and be supported at the perimeter with wood ribbon board ledger strips (Figure 23.20). This helps prevent grout cracks at the joints between the tub and the tile work.

Shower Shampoo Shelf

A shampoo shelf in a shower or bathtub is a recessed opening in the wall, in which bottles of shampoo can be placed. Some builders will require the ceramic tile setter to have full size tiles to, through, and above the shampoo shelf without any rows of cut tiles.

One way to do this is to frame the opening in the wall for the shelf oversized by a few inches, then lath, scratch, float, and tile the rest of the shower, but not the shampoo shelf. After all of the tile is in, the tile setter merely floats the recessed shampoo shelf with whatever amount of extra mud is required to have the opening match the surrounding full-size tile layout. For example, if the bottom of the opening must come up 1 inch to make a full-size row of tiles, then the tile setter floats the bottom of the opening with 1-inch thick of mud. If the top must come down ½ inch, then ½-inch thick of mud can be added.

By framing the shampoo shelf opening slightly larger than is called for on the plans, the exact opening size and position does not have to be as precise, and the tile setter has some flexibility to achieve the correct tile layout (Figure 23.21).

Shower Valve Covers in Place for Tile

The builder should check that the cylindrically shaped plastic covers are on the shower plumbing valves prior to the start of the tile work. The builder should also

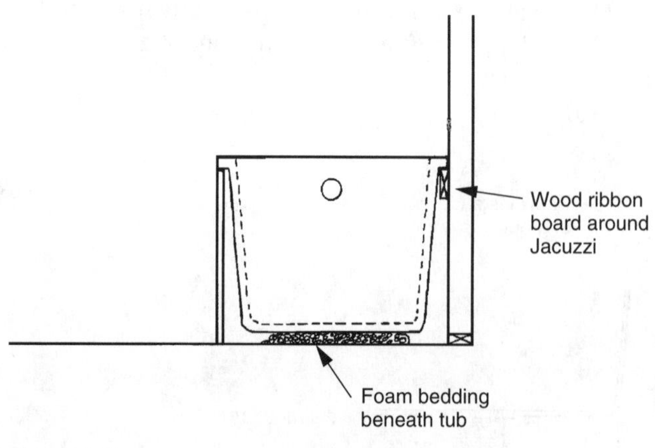

Figure 23.20

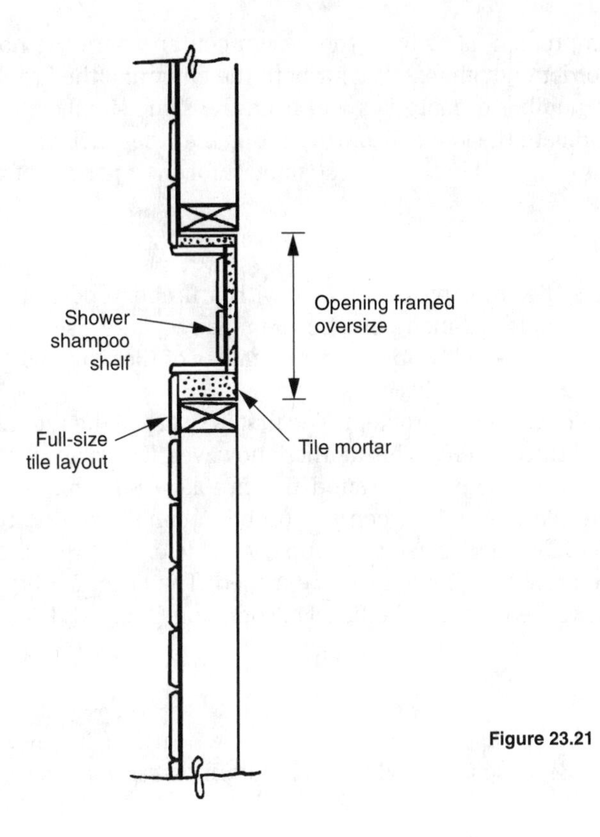

Figure 23.21

ensure that the tile setter does not install tile around shower valves without these covers.

The plastic covers that come with the valves provide the adequate clearance around the valves so that finish trim can be installed later through the hole left in the surrounding tile. If the tile is installed around the plumbing valves without these covers in place, the tile setter must guess how big of a hole to leave. If the hole is too tight, the plumber must chip out tile to install the finish trim, creating the potential for an entire tile to crack or break.

When Entry and Fireplace Tile are Supposed to Match

Sometimes homebuyers will want the fireplace hearth and face tile to be made of the same material as the entry floor tile. The problem with this is that the installers are two different subcontractors. The contractor who installs tile on fireplace hearths and faces is usually the same tile contractor who does the kitchen countertop and bathroom tub and shower tile work. The contractor who installs tile at the entry floors is usually a flooring contractor. If the two separate contractors are allowed to pick out the same brand name and color of tile, marble, clay tile, or slate at different stores or at different times, the materials may not be exactly the same. The homebuyer will complain later that the fireplace does not match the entry.

The solution to this problem is to have the design center where the homebuyer makes the selections order enough material for both the entry and the fireplace from the same batch or lot number of materials and from the same supplier. The design center must then coordinate the two subcontractors so that they pick up the materials from the correct supplier and in the correct amount for their portion of the work.

Fireplace Face and Hearth Tile

Figure 23.22 illustrates the front of a fireplace, with a firebox opening that is 38 inches wide. In this example, 12-inch ceramic tile squares were used. A problem arose when the spacing did not allow for an even number of tiles to reach each corner of the opening and travel up each side.

There are two solutions to such a problem. The first is to lay all the tiles as 12-inch pieces, then cut around the opening. This method, however, leaves an L-cut at each corner and cut edges of tile exposed around the fireplace opening. The second choice is to cut all the tile pieces between the firebox opening to a narrower but equal width so that full 12-inch tiles can travel up each side of the firebox. All of the cut edges can then be grouted so as not to be exposed. The clean factory edges of the full tiles are then exposed around the fireplace opening (Figure 23.23).

Fireplace Tile and Vinyl Flooring

Shown in Figure 23.24 is the edge view of a fireplace tile hearth as it meets the family room vinyl flooring. Because the vinyl flooring, for all practical purposes, can be

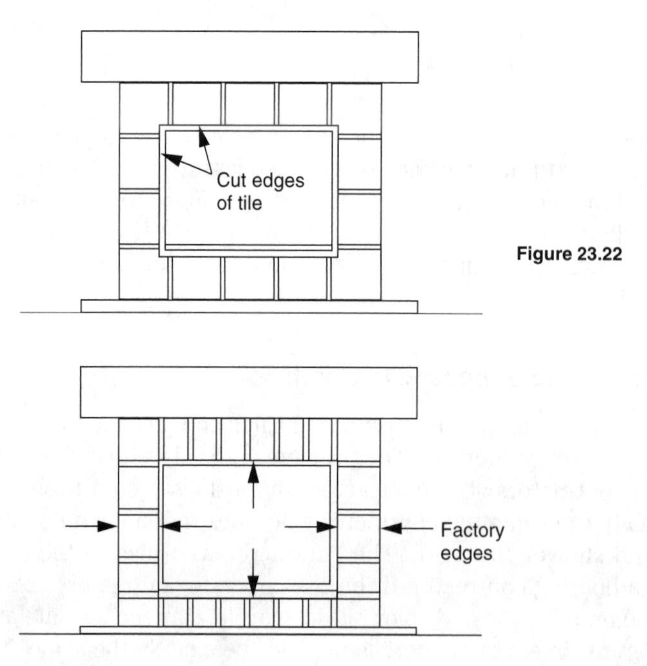

Figure 23.22

Figure 23.23

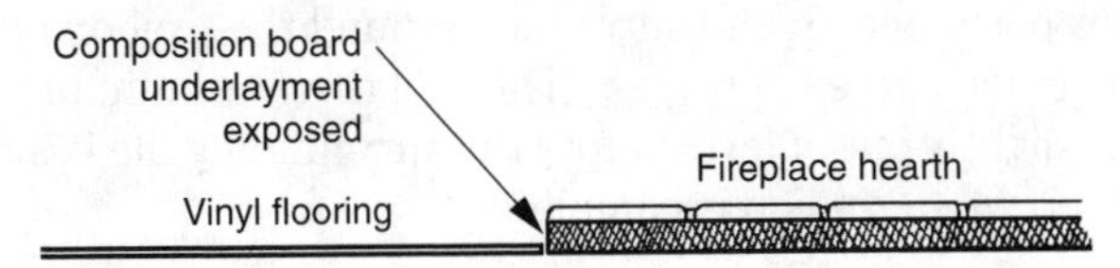

Figure 23.24

considered to have zero thickness, the edge of the underlayment material underneath the tile hearth is exposed. In this case, the underlayment should be held back from the tile edge ¼ inch and the edge filled with grout down to the vinyl flooring (Figure 23.25). The edge of the vinyl flooring and the tile underlayment is then covered by the grout.

Miter and Grout Tile Edges

One method of hiding exposed edges of tile at fireplaces and other locations is to cut them back at a 45-degree angle, then fill it with the same grout used at the tile joints (Figure 23.26). Depending on the thickness of the tile, this can look better than the exposed factory edge.

Fireplaces Installed Level

The builder should check during the installation of prefabricated metal fireplaces that the top of the firebox opening is level. If the fireplace is not level, the tile setter

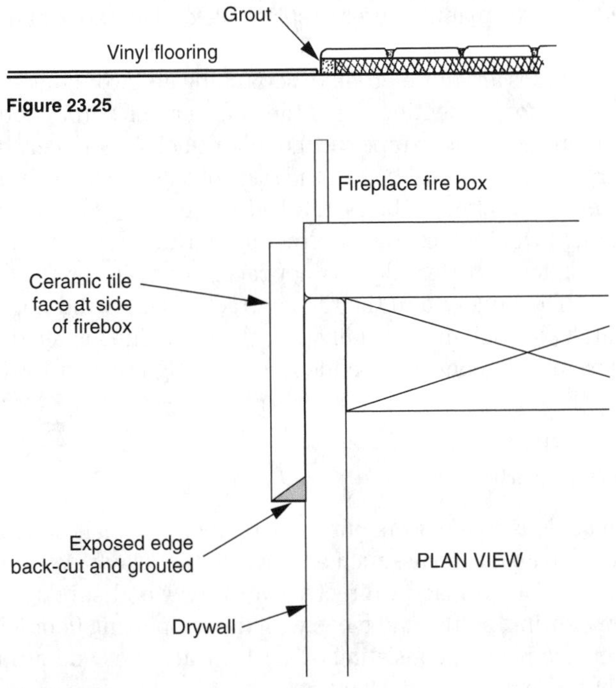

Figure 23.25

Figure 23.26

who is installing the fireplace hearth and surrounds around the firebox opening will not be able to maintain equal and even reveals. This will then result in uneven tile as well. A fireplace that is slightly out of level before the surrounding tile is installed becomes noticeably uneven when the job is complete.

Clearances for Fireplace Glass Doors

The builder should check the clearance around the firebox openings of prefabricated fireplaces before construction to make sure there is enough room for the fireplace glass doors. These clearances are given on the fireplace manufacturer's specification sheet. For example, some fireplaces require no clearance from face tile because the glass doors are installed on the inside edge of the firebox. Other types of glass doors are attached to the front face around the firebox and can require ¾ to 1½ inches of clearance from the edge of the opening to the face tile. These measurements can vary with different models of the same manufacturer.

The idea is to answer this question of clearance upfront rather than discovering late in construction that the fireplace glass doors do not fit and the expensive tile work needs to be removed and redone.

Protect Dark Colored Tile

Some types and colors of ceramic countertop tile scratch more easily than others. Black and dark blue are tile colors that not only show their scratches more readily, but are sometimes selected by homebuyers for countertop tile even though they are expressly recommended by the manufacturer as splash tile or decorative pattern tile only.

When kitchen and bath ceramic tile is a homebuyer's option, the builder should find out which (if any) tiles are more prone to scratching and work out some method with the tile contractor for protecting these tile countertops throughout construction. One method of protection is to tape cardboard or thick paper over the tile countertops after the grouting is dry and before the start of the other finish trades.

In one case, an entire kitchen of black tile had to be replaced due to excessive scratches, even though the builder, tile contractor, and the homebuyer were made aware by the tile supplier that this tile was not recommended for countertops and would scratch easily. The builder and the tile contractor simply did not believe that this tile would scratch so much more readily than the tile being installed in the other houses. The builder and tile contractor ended up splitting the cost for the removal and retiling.

Don't Lay Hearth Tile over Plywood

This may seem too obvious to mention, but some builders have allowed tile contractors to install hard surface materials, such as ceramic tile, clay paver tile, and even marble tile, as hearths for fireplaces, directly over a plywood subfloor. A plywood subfloor will dry out, shrink, settle, and move with the supporting floor joists, beams, and framing. A fireplace hearth, consisting of hard surface tiles and grout, is a rigid system that will not move without cracking.

At the very least, a fireplace tile hearth should be installed over a paper and wire lath base as a buffer between the subfloor and the tile. There are several underlayment products that are as thin as ¼ of an inch so the builder does not have to be concerned with the build-up of the hearth with relation to adjacent flooring. These underlayments will provide the necessary break between the subfloor and the tile hearth.

Trim Building Paper

Building paper beneath cabinet tile work should be trimmed off flush with the grout cement by the tile setters (Figure 23.27). For example, few things look worse in a kitchen than to see the frayed and uneven edges of black paper hanging down a half inch below new ceramic tile.

The paper should be trimmed neatly with a utility knife above the level of the tile so that the cutting line does not show on the cabinet. The builder might also discuss the possibility of grouting or caulking the gap between the tile corner-round piece and the cabinet with the tile contractor.

Tile Floating on Pocket Doors

When bathroom shower tile is plaster floated over a pocket door on one end of the shower, the plaster should be lathed and floated over waterproof drywall greenboard. Figure 23.28 shows a pocket door frame with horizontal wood slats. If this side of the bathtub shower is lathed with paper and wire as a base for the plaster scratch coat, some of the plaster can ooze through the lath perimeter edges into the interior of the door frame. To prevent this, all sides of the pocket door frame must be sealed off. If space permits, it is much easier to install drywall greenboard over the pocket door frame, then attach the lath and wire over the greenboard. This provides a solid surface to lath, scratch, and float against without the danger of getting plaster in the pocket door interior space.

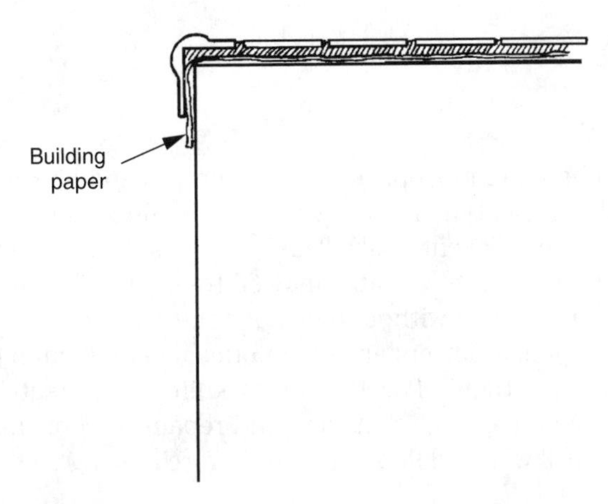

Figure 23.27

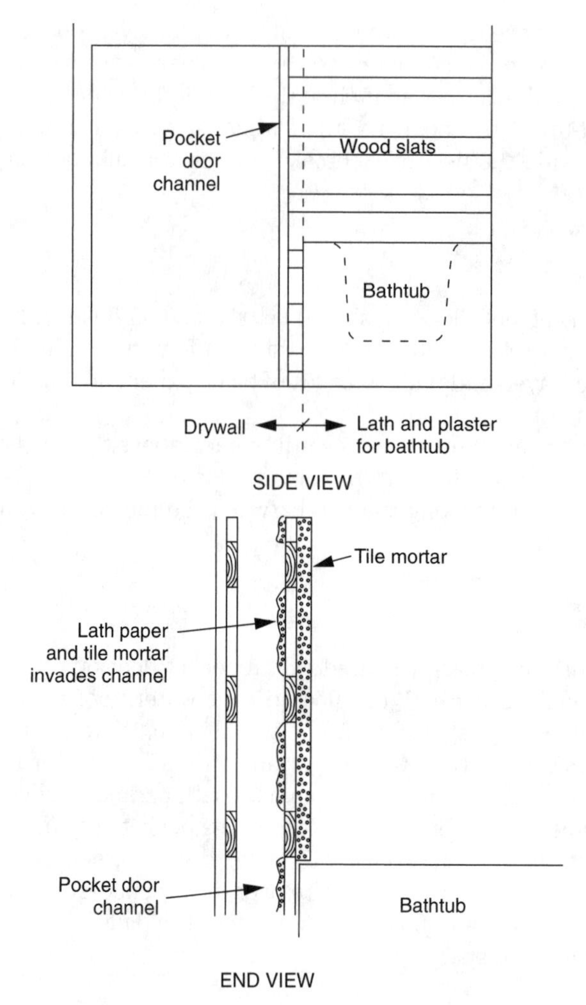

Figure 23.28

Cement Splatter

When ceramic tile for kitchen countertops and bathroom tub and shower walls is laid over a cement base that is floated onto the walls by the tile setter, some of the cement may splatter onto the adjacent walls. This cement splatter is easy to remove while still wet, but if left on the wall and allowed to dry, it will adhere to the drywall and often cannot be removed without damaging the surface. When painted over with enamel, these cement splatters are very noticeable and can only be removed by scraping or chipping them off with a putty knife. This creates drywall pickup work that must then be patched, textured, and repainted. For this reason, the tile setters should clean all wall and floor surfaces before moving on to the next unit.

Shower Tile and Texture

Illustrated in Figure 23.29 is the joint between the top of a bathroom shower, which has been lathed and scratched for tile, and the drywall above. There are two important issues to consider in this instance.

First, the drywall texture should not be sprayed over the plaster scratch coat, as this prevents a bond between the primary scratch coat and the secondary plaster float application. Second, the painting of the bath walls prior to plaster floating for tile, should extend down to the scratch coat and cover all of the drywall texturing. This seals the drywall texture and prevents it from being wiped off as the tile is grouted and cleaned with a wet sponge.

Storage Bin for Tile

Kitchen and bath ceramic tile and fireplace marble and tile squares should be stored in a bin rather than locked in a garage. Not only does the locked garage present problems with access for other workmen and keys for the superintendents, but it also is a less secure place for storing valuable materials. If the house is not relocked up, materials can be stolen over the weekend or by workmen before or after the normal workday hours.

The small added expense of renting a bin is far less than the cost of custom color tile that takes weeks to reorder or expensive tile cutting saws and other tools that can be stolen if only locked in a garage.

Scored Ceramic Tile

Scored ceramic tiles are larger pieces of tile that are scored so that after grouting they look like smaller tile pieces. For example, 4-inch by 4-inch bathroom shower tiles, which are scored each direction down the middle, look like separate 2-inch tile

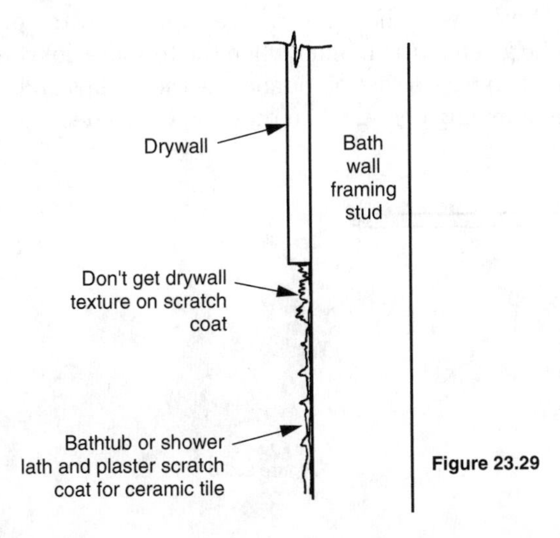

Figure 23.29

pieces after the perimeter edge joints and the center scored grooves are all grouted. Because it is easier and faster to install 4-inch tile compared to 2-inch tile, the look of the smaller tile is achieved without the installation costs.

A possible problem to watch for with scored tile is that when colored grout is used, because of the difference in depth between the grout around the perimeter edges of the tiles and the grout inside the scored grooves, the colors of the two grouted areas may not match. The grout in the deeper joints between the individual tiles may be darker than the grout in the shallower scored grooves in the tile, especially when the grout color is dark brown over white colored scored tile.

Rough Pullman Top Edges

Figure 23.30 shows an end view of the front edge of a manufactured imitation marble top for a bathroom cabinet. These types of tops sometimes come out of the mold with rough and jagged underside edges. The builder should check this and have the pullman top installer file the edges smooth as part of the installation.

Pullman Top Ends

Imitation marble pullman tops are sometimes delivered with exposed ends that are rough. They appear to be the result of cutting with a saw blade. Some manufacturers may indeed mold long lengths of material and then cut them to shorter lengths later. Other manufacturers send out tops with edges glossy and smooth on all edges.

The builder should check the delivery of pullman tops for exposed rough ends and have the field installer file and sand these rough ends as part of the installation.

Pullman Top Sink Clips

Imitation marble pullman tops for bathrooms have threaded screw inserts around the underside edges of the sink openings for underset sinks (Figure 23.31). The metal clips and screws that work with these inserts are supplied by the pullman top manufacturer and should be given to the builder when the tops are delivered and installed. The builder should make sure that he or she has these clips and screws before the start of the finish plumbing phase, when they will be needed.

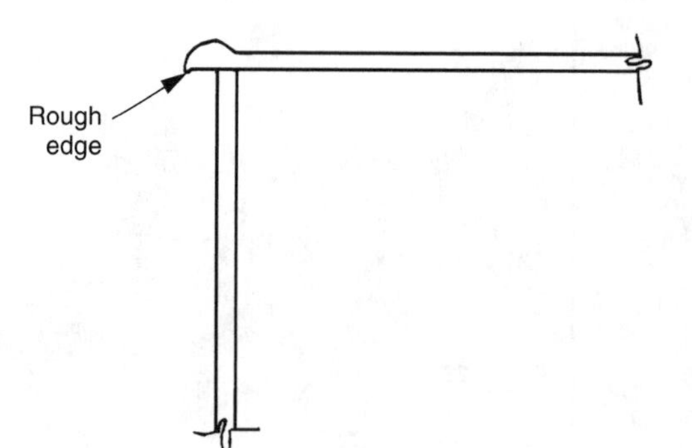

Figure 23.30

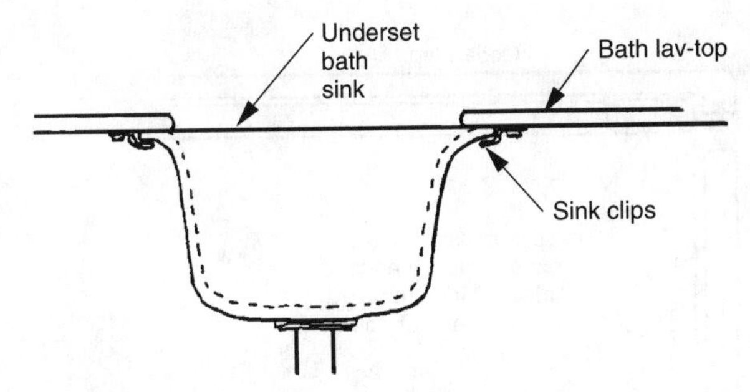

Figure 23.31

One-Piece Pullman Tops

Imitation marble pullman tops that angle around two or more walls should be one solid piece rather than several pieces joined together. The pullman top in Figure 23.32 turns two 45-degree corners with three pieces. It is difficult to get all of the edges on a top like this to join flush when pieced together. One edge or corner is usually high or low, and one or more of the joints may not fit tightly. Caulking is then needed to fill the resulting gaps.

A solid one-piece pullman top requires an accurate field measurement by the manufacturer and square and straight wall framing, but the better end-product is worth the added care involved.

Corian Tops in Kitchens

Illustrated in Figure 23.33 is a kitchen cabinet with a manufactured Corian top. In kitchens, Corian tops usually come with a lip that is turned down around the outside edge. The problem to watch for is that some styles of cabinets do not have a top rail above the cabinet doors. If this style of cabinet is mixed with a Corian top, the lip of the top may prevent the cabinet doors from opening.

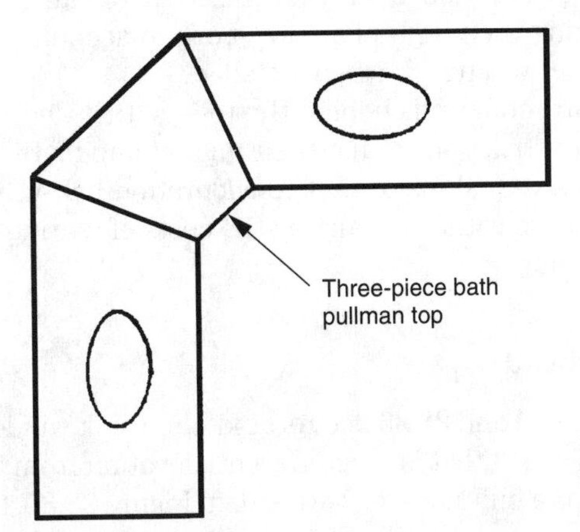

Figure 23.32

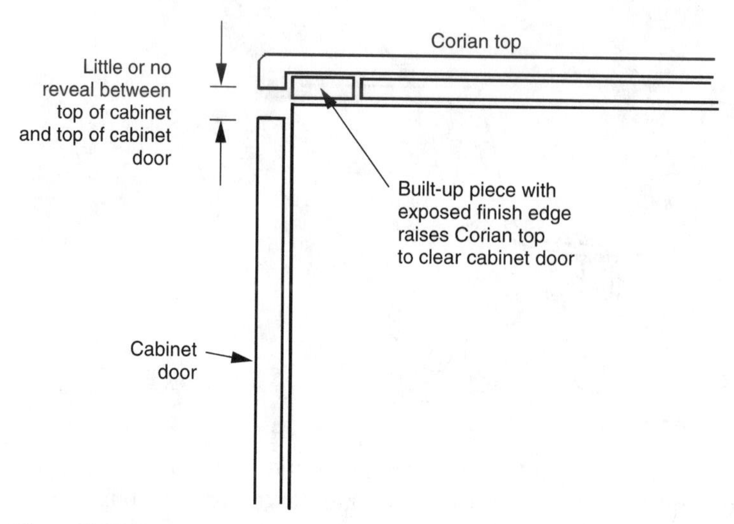

Figure 23.33

If this scenario happens, the top of the cabinet should be built-up with standard ¾-inch thick rough-top boards, with a finished material to match the cabinets at the front. In other words, have the cabinet company add a top rail.

Tiled Wall behind Toilet

If the standard dimension from a drywall-covered wall to the center of a toilet closet ring is 12 inches, when that wall is ceramic tiled, that dimension should be increased to 13½ inches in order to provide enough clearance for the toilet water tank. For example, the bathroom illustrated in Figure 23.34 has ceramic tile continuing behind the toilet from a bathtub or shower. In this example, the tile has been floated over a lath and cement mortar base. Because ceramic tile over a floated base projects past the surrounding drywall surfaces in a shower, the same thing also occurs where the wall is tiled behind the toilet. If the dimension from the framed wall to the toilet closet ring is not increased during the rough plumbing work to account for this increased projection, then the toilet will not fit when installed.

If you are going to tile the bathroom walls behind the toilets, that decision has to be made early enough during construction so that the rough plumbing to the toilets can be altered. This cannot be a casual marketing consideration half-way through the construction of the sales models without incurring the costs of tearing out floors and ceilings to modify the plumbing.

Tile at Windows with Bull-Nose Cornerbead

The problem shown in Figures 23.35 and 23.36 occurs at kitchen sink windows when ceramic tile is placed on a window sill that is at a different elevation from the top of the splash tile and the drywall has a bull-nose cornerbead. In Figure 23.35, the splash tile stops short of the curvature of the bull-nose drywall corner, leaving an awkward empty space between the splash tile and the window sill tile. Figure 23.36 shows a better method of solving this problem. The splash tile is continued around the bull-nose corner and up to the edge of the window frame using a small piece of tile placed at a 45-degree angle on the bull-nose curve. This is the same method used with a

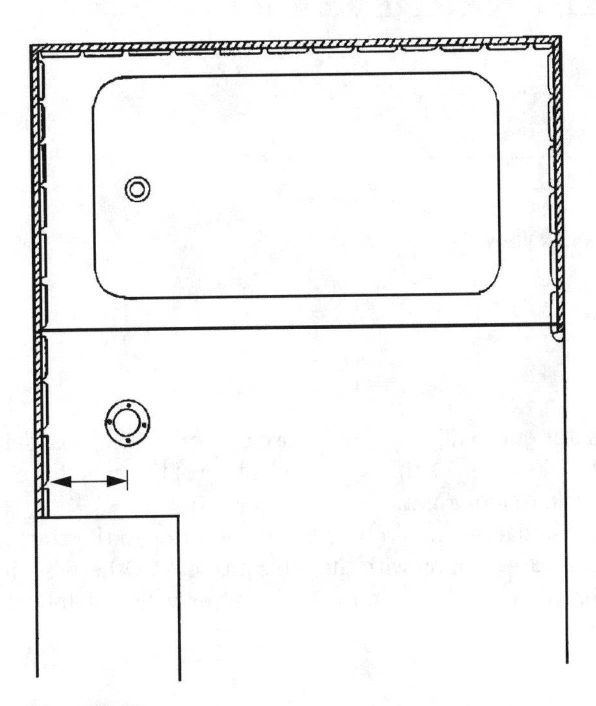

Figure 23.34

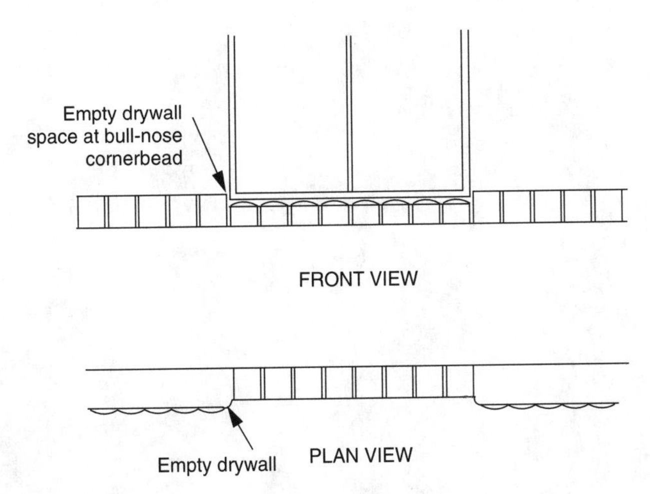

Figure 23.35

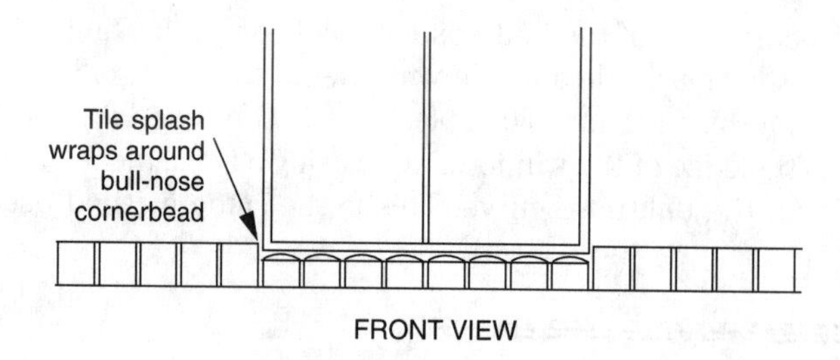

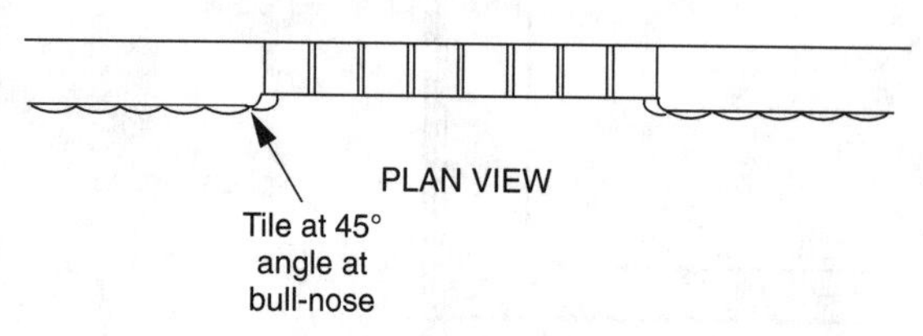

Figure 23.36

piece of baseboard as it is cut to fill-in the bull-nose corner at the floor. This method adds more work for the tile setter, as this 45-degree piece of tile must have both ends cut at a 22½-degree angle to look right.

The method of tile installation should be determined prior to the start of the tile work and discussed and agreed upon with the tile contractor. Otherwise, in production tract housing, the tile contractor and the tile setter will probably choose the quickest and easiest method.

Chapter

24

Insulation, Lath, and Stucco

Preliminary Insulation

Preliminary insulation is installed in certain areas of a house prior to insulating the entire house. It is required where the interior surfaces of exterior walls are covered over with either shear-panel plywood or a bathtub/shower enclosure (Figures 24.1 and 24.2). These areas would otherwise be covered up on both sides by exterior lath or wood siding, and therefore be made inaccessible for later insulation installation. By insulating these specific areas earlier, both sides of these walls can be covered over at their scheduled times.

The problem to avoid with preliminary insulation is that if it must be in place for several days or even weeks before the second side of the walls are covered by lath or siding, the wind may blow portions of the insulation out of the walls. It is therefore a good idea to specify in the insulation contract that all preliminary insulation be paper-backed and stapled in place.

Interior Wall Insulation

Figure 24.3 shows a second floor bedroom wall that has as its exterior surface the inside of the garage. Because this wall is inside the garage and will not get exterior stucco plastering or lathing paper, the insulation should be paper-backed and stapled in place.

Extra Bundle of Insulation Prior to Lathing

The builder should write into the insulation contract that an extra bundle of insulation be left at the job site for future repairs. The insulation can be stored for use prior to exterior lathing, should some small area of preliminary insulation be missed. The

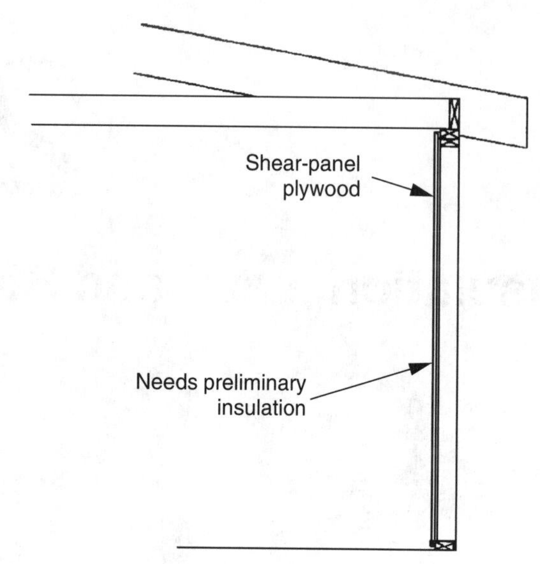

Figure 24.1

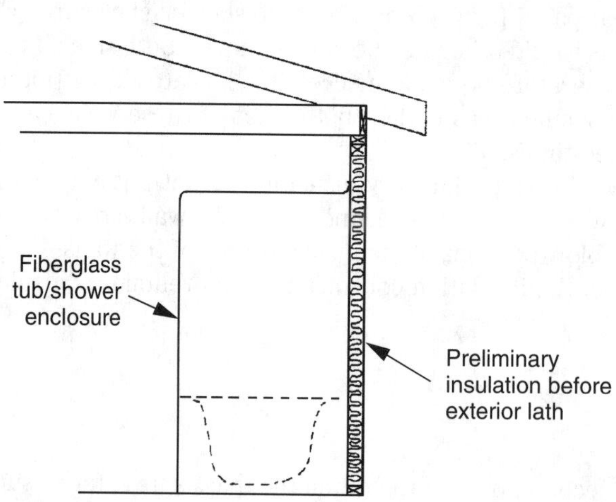

Figure 24.2

lather can then simply throw in a few pieces of insulation where needed and continue lathing without having to wait for the insulation subcontractor.

Balcony Handrail

Illustrated in Figure 24.4 is a second floor balcony with a metal handrail and wood cap between columns. If the columns are plastered, the metal handrail should be installed after the columns are lathed, but before the corner aid is installed. This allows the metal handrails to be cut, fit, and lag bolted to the columns. If not installed until after the columns have corner aid, the metal handrails will not fit around the column corners (Figure 24.5).

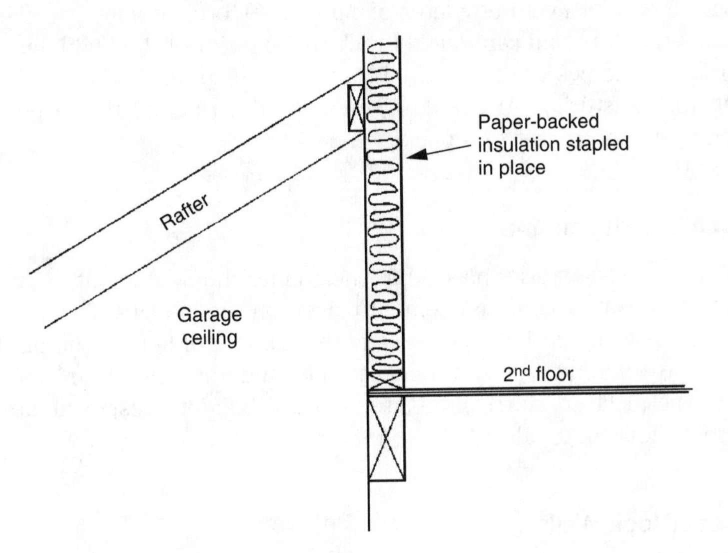

Figure 24.3

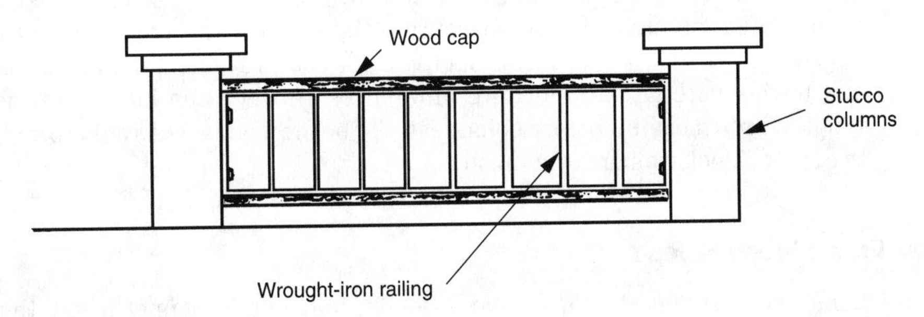

Figure 24.4

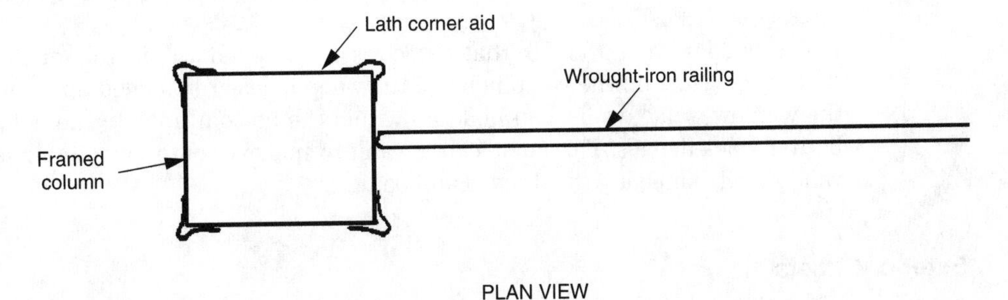

Figure 24.5

Spray Paint Broken Windows before Scaffold

Before stucco scaffolding is erected, the builder should spray paint all broken windows. By doing so, the builder knows exactly which windows were broken and which windows to charge the scaffold crew for after the stucco scaffolding is taken down. This helps the builder avoid any later disputes over broken windows. The builder and the plastering foreman can quickly walk the exterior of the buildings and agree that all broken windows are marked with spray paint prior to the start of lathing, scaffolding, and plastering. Any broken windows after that are the responsibility of the plastering subcontractor to replace.

Plaster Sand and Cement Placement

Before the start of the exterior plastering, the builder should find out where the plastering contractor wants sand and cement placed on the job site. This is contingent upon the length of the plastering hoses and the number of houses the plasterer can reach from one equipment set-up. Because the sand and cement are always delivered before the plasterers arrival on-site, it is the builder's responsibility to know where to spot these materials.

Brown and Color Coat Block Walls

Garden and perimeter property line walls that are built with masonry block and stucco plastered should be both stucco brown coated and color coated. If merely color coated directly over the masonry block surface, the grout cement joints between the blocks will bleed through the color coat, revealing the block wall. This is due to the material and moisture differences between the block and the cement grout. If brown coated before color coat, the brown coat acts as a uniform buffer between the block wall and the finish color coat.

Window Frame Weep Holes

Some metal frame windows have an open, horizontal channel below the window ledge with weep holes for water drainage (Figure 24.6). When water gets into the inside, active portion of the window frame bottom track, it simply drains out the weep holes (Figure 24.7).

The builder should check that stucco is not applied within this open channel, thereby clogging up the weep holes. If the weep holes are plugged up from draining the window frame, water can build up within the bottom track to the height of the bottom track flanges. This could cause water to spill over onto the interior side of the window sill, damaging the drywall and paint.

Exterior Outlets

One way to help prevent the accidental burying of exterior electrical outlet boxes during plastering is to pull the wires out of the boxes prior to the scratch coat.

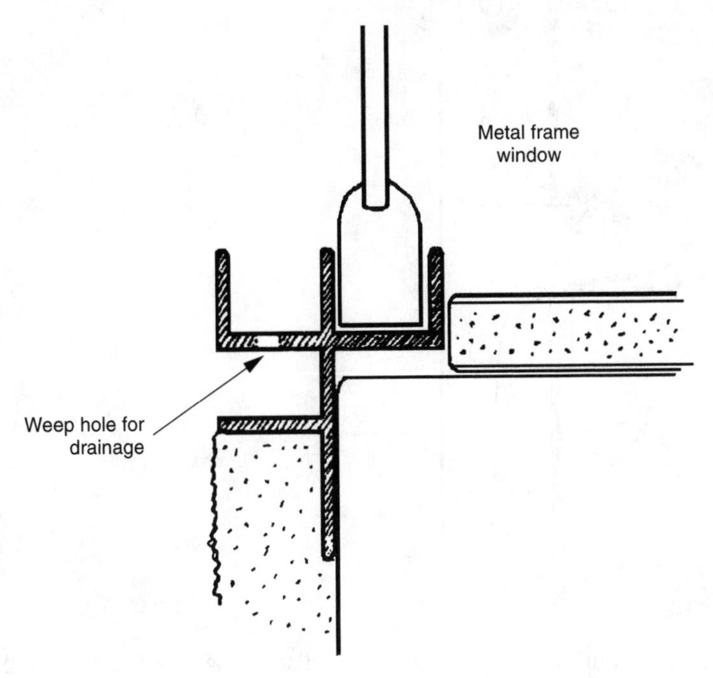

Figure 24.6

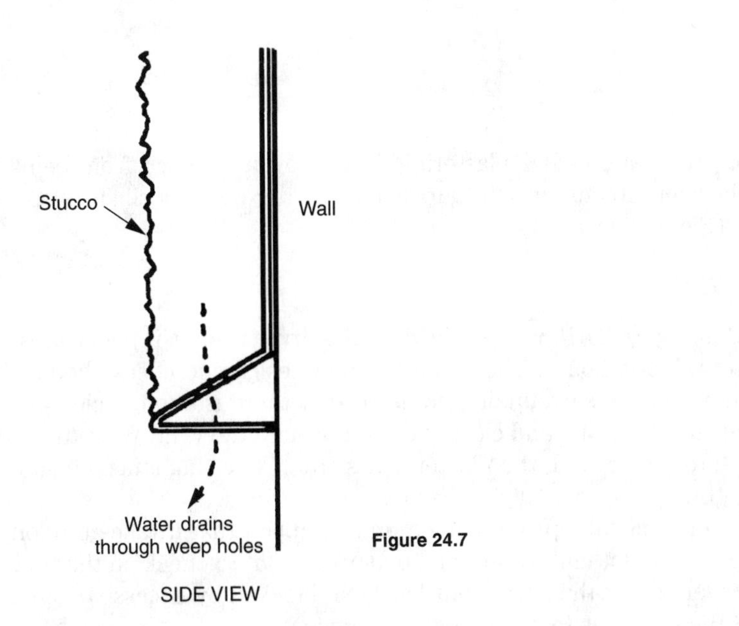

Figure 24.7

The electrician makes up the boxes by bending about 6 to 12 inches of extra wiring inside each outlet box. By pulling the wires straight out from the box, the copper wires are rigid enough to withstand the spray application of stucco and are easily seen by the person running the water hose to wash stucco off unwanted areas (Figure 24.8).

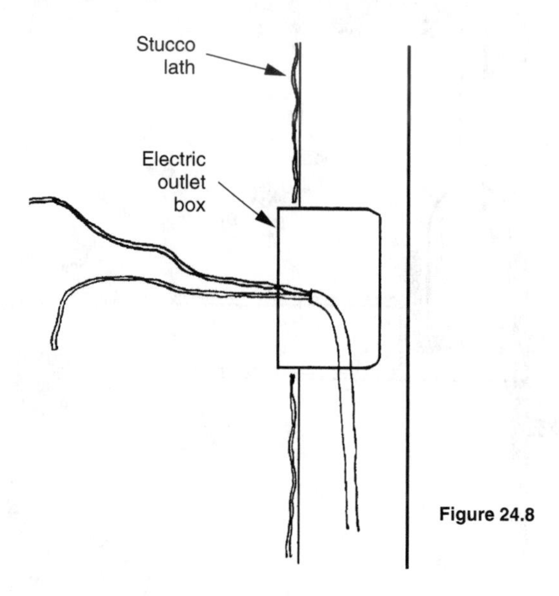

Figure 24.8

Pulling exterior outlet wires before stucco also helps check for any light fixture, receptacle, and entry doorbell outlet boxes that may have been covered over by the lather.

Cover Vents

Garage foundation vents and other screened vents into the garage should be covered with paper or plastic before plastering. This prevents stucco from being sprayed through the vent screens onto the garage interior drywall, mudsill, firedoor, and concrete slab (Figure 24.9).

No Stucco in Dryer Vent

Illustrated in Figure 24.10 is the side view of a dryer vent as it penetrates the exterior stucco wall of a house. The builder should design and check the height of the dryer vent above the surrounding grade so that there is enough clearance for the stucco plasterer to neatly and cleanly plaster around the vent. When the dryer vent is too close to the ground, the plasterer has trouble getting stucco below the vent without getting stucco actually in the vent.

Homebuyers will call up the builder, and complain that it takes too long for the dryer to dry their clothing. When the builder goes out to check on the problem, it is often discovered that the dryer vent has been blocked by excess stucco at the underside of the vent, due to not enough clearance.

Stucco Pickup after Grading

The final grading of lots around nearly completed houses is usually done with small sized tractors. Although these tractors can maneuver well around corners and within tight areas, occasionally the finish grader will back into a wall corner or bump into

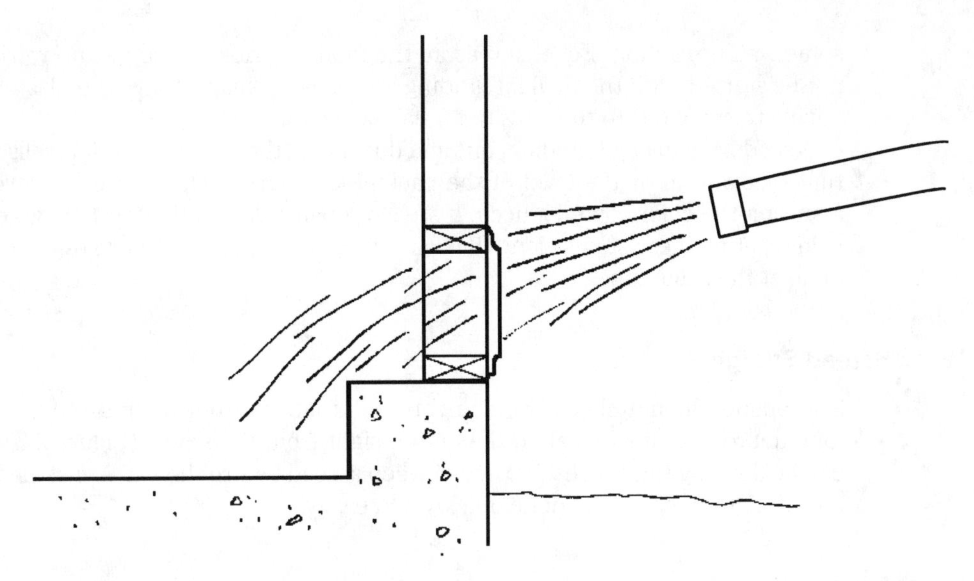

Figure 24.9

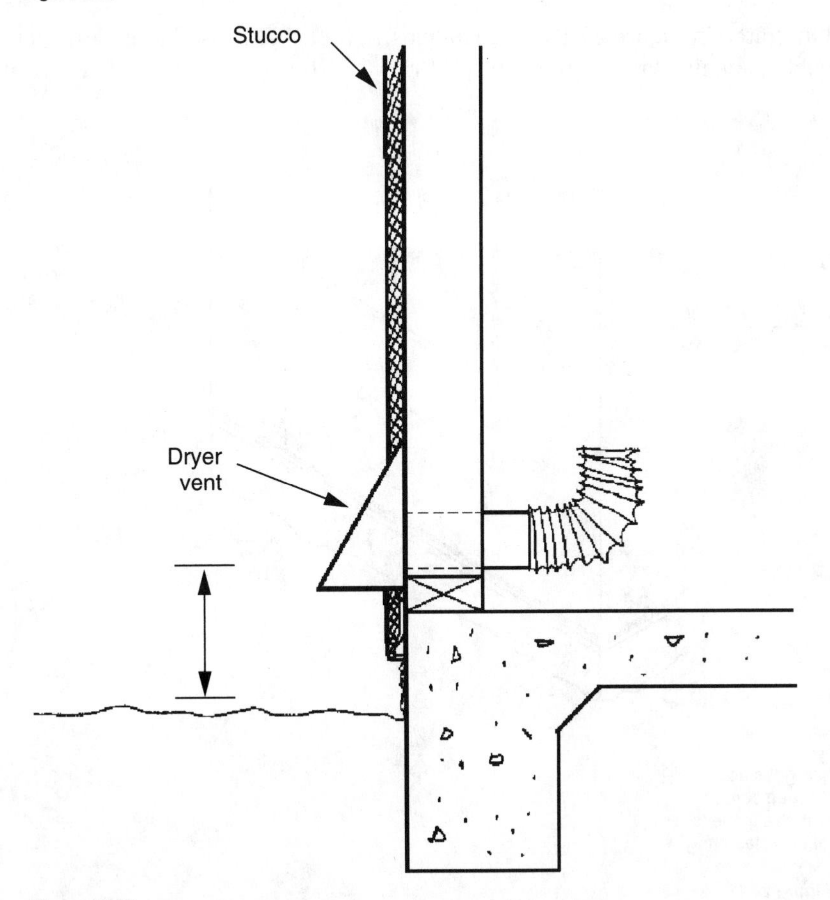

Figure 24.10

some wall projection. For this reason, the final exterior stucco pickup work should not be started until the final lot grading is finished. This will save the plasterer from having to come out to the job site for each mishap.

Also, during final lot grading, dirt and dust along the base of exterior walls is sometimes kicked up or the level of the backfill is lowered. This causes or reveals dirt stains on the color coated stucco. If stucco repair is left until after final grading, fog color coating these stains along the base of walls and general stucco repair can all be done at the same time.

Weep Screed Straight

The builder should check the straightness of lath bottom weep screed, especially long unbroken runs of wall in the line-of-sight from the street (Figure 24.11). This can be done by sighting each piece of weep screed before the lath is installed so that any straightening can be done quickly and easily.

Foam Eave Vents

Illustrated in Figure 24.12 is a foam plastic vent that is used as an attic vent at a roof gable peak in Spanish architecture. Prior to exterior lathing, the foam plastic vent is

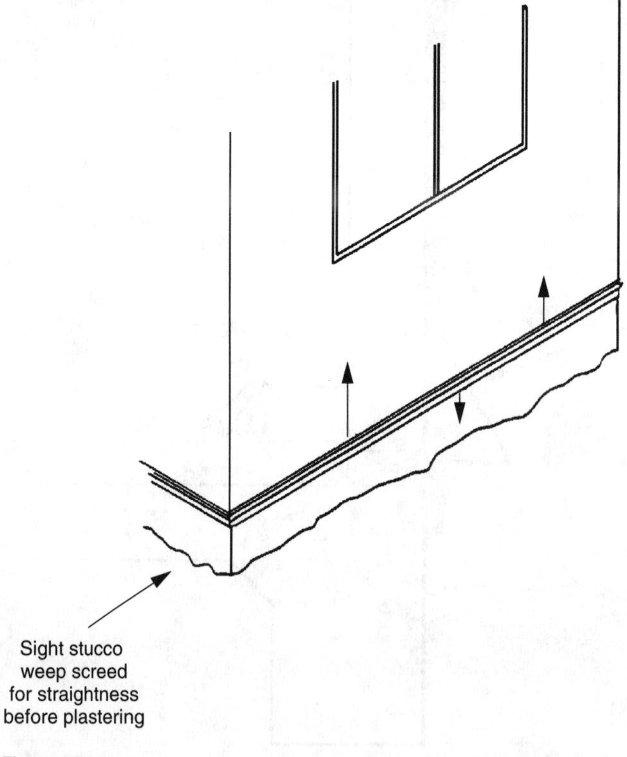

Figure 24.11

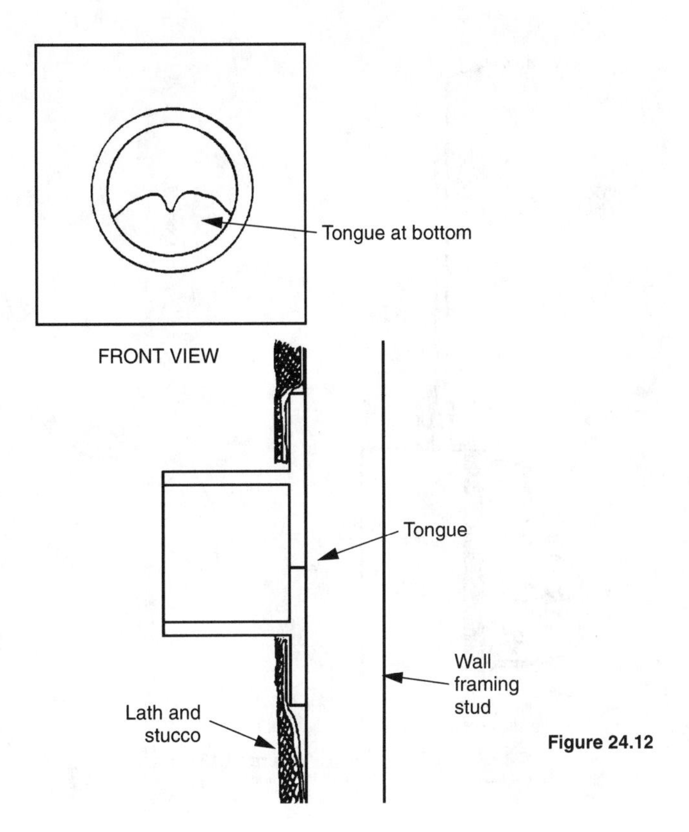

Figure 24.12

nailed to the wall framing below the roof gable peak. It is then lathed and plastered over, leaving only the round cylinder to resemble a clay pipe.

The problem with this type vent is that the tongue portion inside the vent is a water trap, used to prevent rain water from entering the building through it. The vent must therefore be installed with this tongue portion on the bottom.

Enough Water Pressure for Plastering

The builder should check before the start of exterior stucco plastering that the job site water source has enough water pressure for both the water hose to the mixer and the water hose for the spray person who is washing off the eaves, windows, and woodwork.

The time to discover that there is not enough water pressure from a particular hose-bib to get the plastering operation going is not when there are 12 to 15 people standing around waiting to work.

Protect Exterior Foam

Illustrated in Figure 24.13 is a side view of foam trim that is sometimes used to create architectural details around the exterior side of windows, doors, and soffits. This material is made of foam plastic, cemented to the brown coat of the stucco plaster, covered with a fabric mesh, then stuccoed with color coat along with the rest of the building.

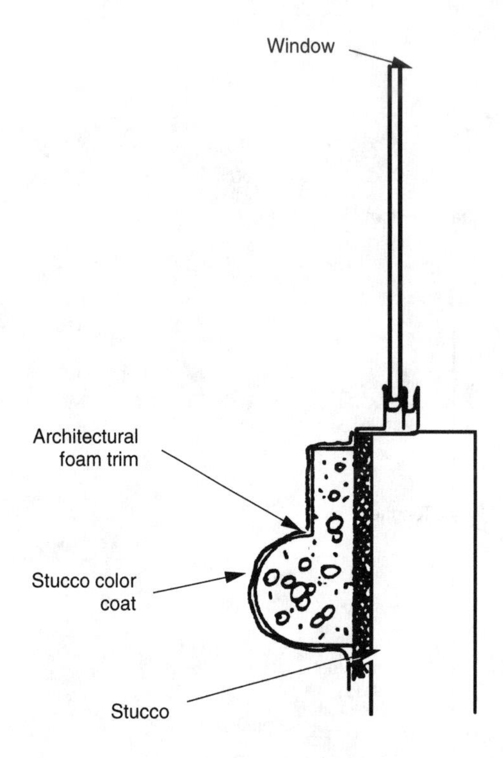

Figure 24.13

Because the foam only has a thin shell of stucco color coat, its surface can easily be dented or broken. The builder should check that window washers, painters, and window glass hack-out workers do not place ladders directly on this trim.

Clearance for Foam Trim

A common problem encountered with the use of contoured decorative foam around windows and doors is the lack of clearance when a door and window are close (Figures 24.14 and 24.15). In Figure 24.14, the foam plastic pieces were 8 inches wide, requiring a minimum of 18 inches or more between the window and the sliding glass door. If this minimum dimension is not considered on the plans, then some modification of the foam pieces or a substitution of narrower flat pieces used elsewhere, as between the windows and transoms, must also be used in the vertical space between the window and the slider.

Scrape Eaves

After the stucco brown coat, but before the paint undercoat, the plasterer should scrape off any remaining stucco from the overhang eaves that was missed by the water hose clean-up (Figure 24.16). This prevents the painter from painting over stucco that has adhered to the eave rafters or facia. Eventually, these chunks of stucco will chip and break off, leaving unpainted surfaces.

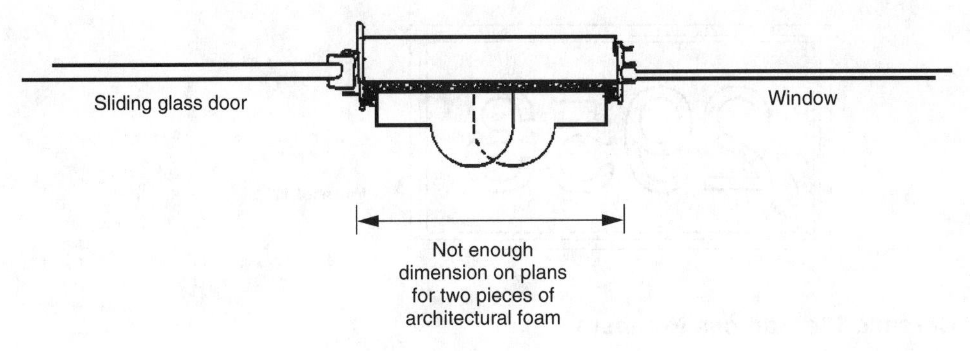

Figure 24.14

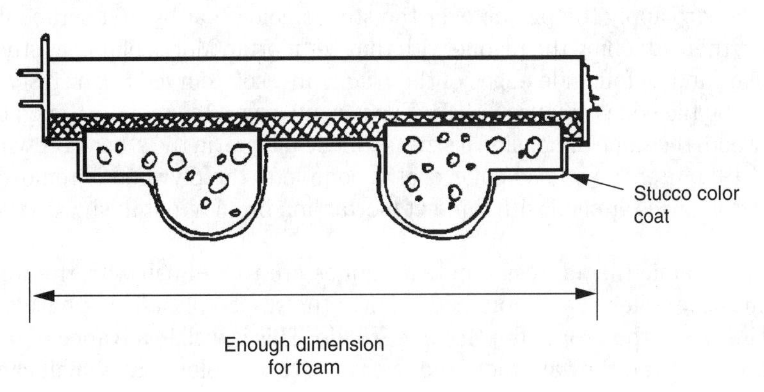

Figure 24.15

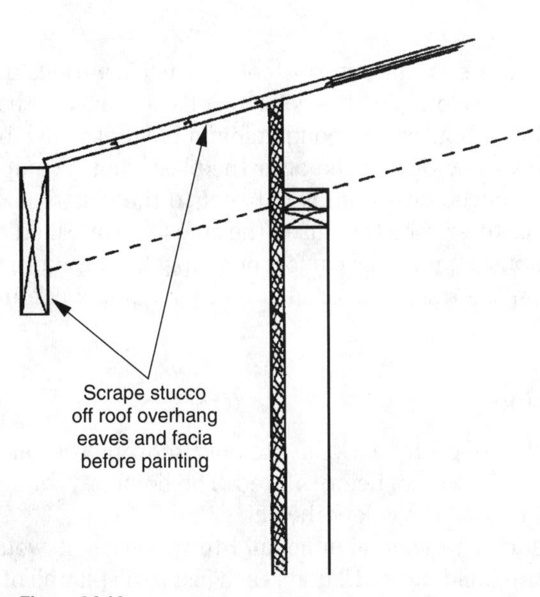

Figure 24.16

Figure 24.17

Ceramic Tile Address Numbers

Figure 24.17 illustrates a typical ceramic tile address number plaque. These plaques are about ⅜-inch thick. There are two ways to install tile address numbers over stucco. The first is to apply the plaque over the stucco color coat by first scraping a smooth surface, then attaching the plaque with thin-set mortar. More color coat stucco is then added around all four side edges of the plaque in a soft, curved Spanish style build-up to hide the tile edges (Figure 24.18). The second way is to embed a ½-inch thick piece of plywood, the same dimensional size as the tile plaque, in the stucco brown coat (Figure 24.19). After the stucco color coat is complete, the plywood is removed and the address plaque is applied with thin-set mortar and brought out flush with the wall surface.

If the ceramic tile address number plaques are to be flush with the wall surface, they must be ordered and obtained prior to the stucco plastering phase so that plywood inserts of the proper depth can be made. This is well in advance of the date the tiles are needed if they are mounted over the stucco color coat, which can be at the end of construction.

Check Stucco Color Coat Bags

As a routine check before the start of exterior color coat plastering, the builder should check the labeling or numbering on the color coat bags to ensure that the correct color is being used. Entire houses and condominium buildings have been color coated the wrong color, only to be discovered after the plastering is complete. This can happen when a color change addendum is not sent to the plasterer or the addendum is missed by the plasterer, who then used the stucco color called out on the original contract. In tract housing, mix-ups can occur simply from the plasterer mistaking the lot numbering on the tract map or incorrectly reading a confusing color scheme chart.

Condominium Building Stucco Colors

In tract housing, sales models of each floor plan are built showing the variety of exterior stucco, roof tile, and paint colors being offered. The homebuyers walk the individual model houses and choose the colors they like.

For attached housing, such as condominiums and townhouses, it would be impractical to build 4 or 5 entire buildings of 12 units each just to display all of the available exterior building colors. If each building is a different color scheme, however, the homebuyers are obviously entitled to see the exterior color schemes of future buildings that will be different from the sales model building.

One solution to this problem is to build full-scale mock-up walls outside the sales office showing each of the exterior color schemes. The 2×4 framed wall mock-ups do not have to be full height, but should include an entry door, a window, surrounding stucco or wood siding, wood trim and facia, and some roof tiles on top. This allows the homebuyers to see the color scheme of each future building type in full scale, rather than making a selection based on 4-inch × 4-inch sample squares of stucco colors, a single piece of roof tile, and paint color charts. This also prevents home-

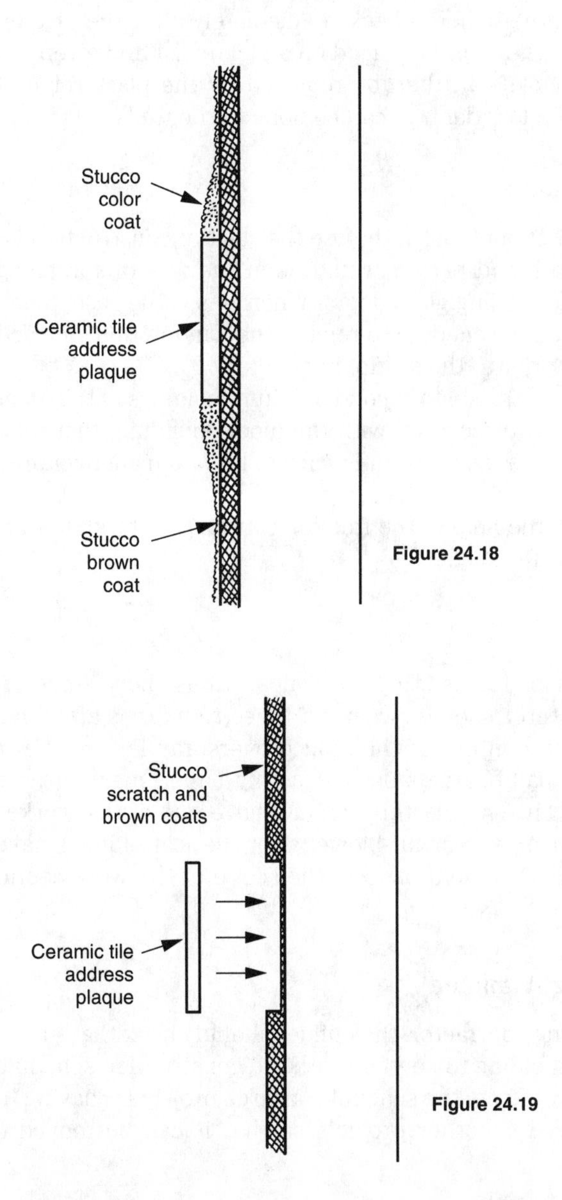

Figure 24.18

Figure 24.19

buyers from coming back later and saying they thought they were buying units in buildings that exactly matched the model building.

Leftover Planks inside Units

One thing the builder should not tolerate is the plastering stucco subcontractor leaving behind old scaffolding planks in the interiors of the units, especially on second and third floors. Nothing is worse than having to carry an old and splintered plank carefully down a tight stairway, without touching the freshly taped and painted drywall surfaces.

This should be an item the builder checks immediately after the plastering scaffolding is removed so that they can be placed on a pickup list and given to the plastering contractor. The planks can then be removed by the plasterer early in the drywall taping phase, when less damage can be done to the walls.

Sliding Glass Doors before Scaffold

Another thing the builder should look at before the start of construction is whether scaffolding for exterior lath and plaster would be in the way of spreading and installing the glass panels for sliding glass doors. When this is the case, a note should be placed on the construction schedule to remind the window subcontractor to get the glass panels installed prior to the scaffolding.

In one particular case, on three-story condominium buildings with balcony decks on the second floors, it was discovered with the model building, that 6×8 and 8×8 sliding glass door panels were too large and heavy to lift and maneuver around scaffolding that was already in place. On the second building, these panels were spread to the balcony decks with the help of the framing contractor's forklift, and installed prior to the start of the scaffolding.

Sweep Roofs after Stucco

On a typical two-story stucco house that has wall sections above some of the first floor roof areas, the plasterer should sweep off these roof areas after the plastering is complete. The sand content of the stucco overspray left on the roof after plastering acts like little ball bearings on the roofing paper, making the roof slippery and unsafe to walk on. The plasterer should have one of the workers sweep off each roof with a broom or with an air-blower after the scaffolding is taken down. This should be an item that is mentioned in the scope of the work section of the plastering contract.

Check that Enough Scaffolding Is Available

When selecting a plastering contractor, the builder should check that the contractor has access to enough scaffolding to keep up with the construction schedule. A typical problem that adversely affects the schedule on large projects is having to wait for scaffolding to become free on other projects so that it can be moved over and erected on your project. Problems that occur on another developer's project can hold up the removal of scaffolding, thus directly affecting your construction schedule. Through no fault of yours, your construction can come to a complete stop for several days because the plastering contractor did not have enough scaffolding to go around in the event of a holdup on another project.

Chapter

25

Roofing

Stocking Roof Tiles

Tiled roofs are loaded prior to interior drywall installation and exterior plastering. This allows any settling or deflection to occur due to the imposed weight of the roofing materials before these rigid wall coverings are applied.

The placement of roof tiles on the roof should leave at least a 6-foot clear working space next to walls (Figure 25.1). This allows room for scaffolding, dragging plastering hoses, and simply moving around the roof without tripping over stacks of roof tiles.

Broken Tiles

During the installation of concrete roof tiles, broken, cracked, or chipped tiles are discarded off the roof by the roofer. If the exteriors are covered with stucco lath and the sideyards are narrow, the roofers should only throw broken tiles off the back or front of each house.

If the roofers are allowed to throw broken tiles off the sides without looking and without care, some tiles may roll when they hit the ground and bump the sides of the houses, puncturing holes through the lath paper. This is especially true when the ground is dry and hard and the tiles are being thrown off a two-story roof. In this case, the tiles have a lot of velocity when they hit the ground.

When conditions of height and accessibility permit, it is a better practice to have the loaders and roofers place broken tiles on empty pallets on the roof. The pallets of broken tiles can then be lowered to the ground with the roofer's forklift as the loading or roofing progresses. This also makes for easy pickup by the cleanup company.

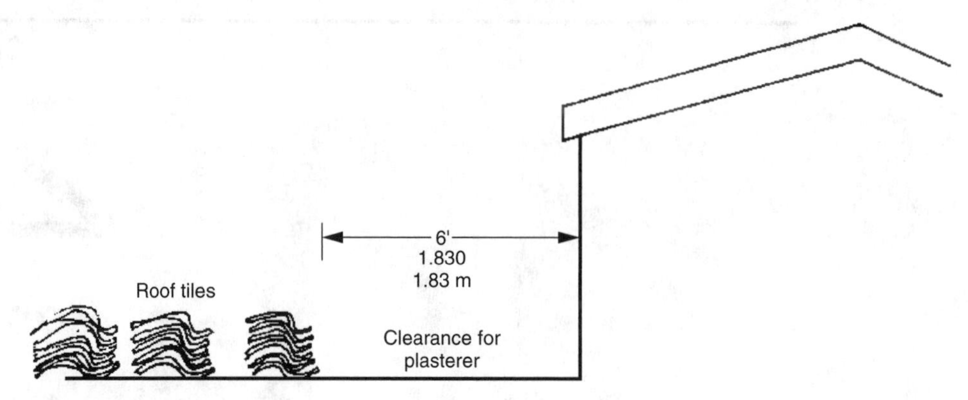

Figure 25.1

Roof Tile Throw-Off Locations

For multistory condominium and apartment buildings, it is sometimes difficult to find acceptable locations for the throwing of broken roof tiles off the roof when people are working at ground level. This issue should be addressed in the roofing contract.

Several problems are at issue here. First, it is both time-consuming and annoying for the roofers to have to carry broken tiles across the roof to a spot where they can be thrown safely off the roof. Ideally, the roofer would like to be able to toss any broken tile off the roof wherever he or she happens to be, without having to look down or shout a warning. Secondly, the safety of people working below and the lower portions of the buildings below must be considered. The builder does not want casually thrown roof tiles to roll once they hit the ground. They could hurt someone or strike the side of the building, causing stucco damage. A third consideration is that if the cleanup of roof tiles is not included in the roofing contract, then the regular cleanup contractor should not be expected to pick up broken roof tiles randomly spread all over the ground. This means that broken tiles must either be placed on empty pallets on the roof and then brought down by forklift or thrown off the roof into designated piles on the ground.

The problem is that some condominium and apartment buildings have tight courtyards, sides of the building inaccessible to forklifts, three-story heights, and numerous workmen on the ground so all of the above considerations cannot be satisfied. If the roofer can only be given three or four locations to throw the tiles to on the ground, requiring the roofers to carry the tiles over to these locations as part of the roof laying operation, then this condition should be considered and reflected in the roofing bid. Otherwise, the roofer will complain and request a time-and-material extra to cover the increase in labor time.

No Paint Overspray on Roofing Paper

When possible, the builder should try to schedule the paint undercoat of exterior roof facia board prior to the papering of the roof. This enables the painter to spray paint the facia without getting overspray on the roofing paper. Such paint overspray acts almost like small grains of sand or tiny ball bearings, making the perimeter edge of the roof next to the facia slippery and dangerous (Figure 25.2).

Roof Paper at Walls

Illustrated in Figure 25.3 is the transition from a sloped roof to the vertical side wall of a fireplace chimney. In this instance, Z-bar metal flashing is used with a 1×6 wood backing, which is typical for a flat to vertical condition.

The roofer must tuck the roofing paper tightly into the 90-degree corner. If casually and loosely rolled into the corner, the sheet metal flashing may break the paper, creating the potential for a roof leak.

Roof Jacks

Although it is the responsibility of the plumber and heating contractors to supply the roofer with metal roof jacks, the builder always ends up in the middle of disputes over roof jack shortages.

As a general rule, the builder should always try to control as many building materials on the job site as possible to ensure coordination. For this reason, all roof jacks should be delivered by the plumber and heating contractors to the builder, labeled per lot number and floor plan, and bundled with metal tie wire. The builder can then spread the roof jacks per lot as necessary, either by nailing them to a garage wall or by tacking them on the roof as they are needed. This allows the builder to be directly aware of and in control of any shortages before they are actually needed to complete a house and before it becomes a problem.

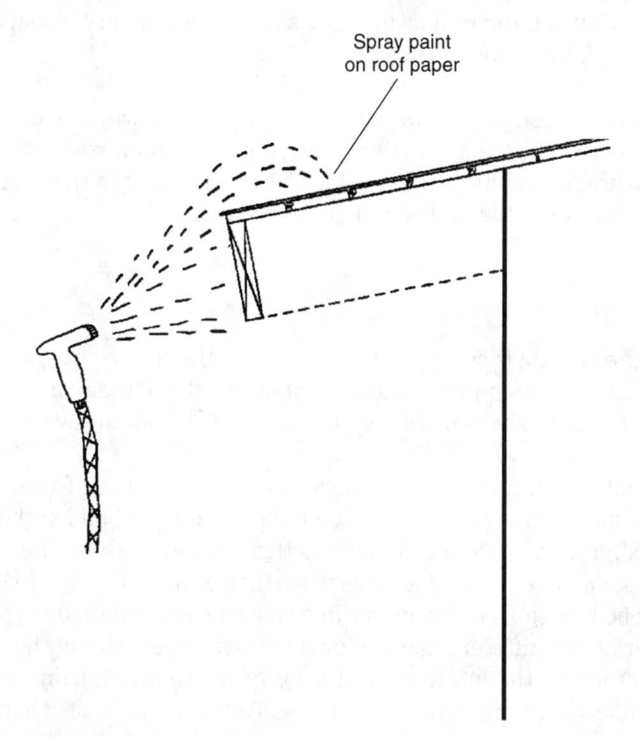

Figure 25.2

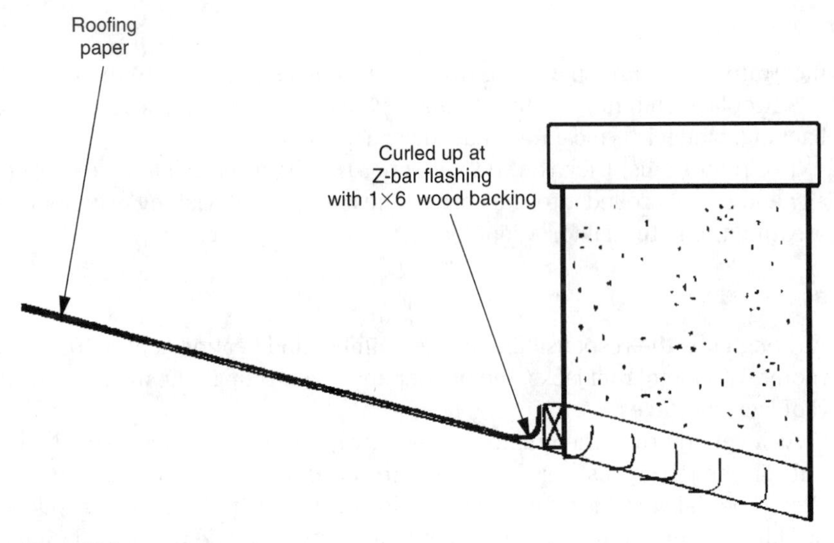

Figure 25.3

Tile Cutting Dust

For concrete or clay tile roofs, the tiles must be cut to fit into roof valleys, ridges, and walls. The cutting is done on the roof using a skil-saw with a masonry blade. This cutting creates dust that falls on top of the surrounding roof tiles.

After the end of each work day or when the roofer moves from one roof section to the next, the cutting dust should either be blown off the roof with an air blower or swept off with a broom. If the dust is left on the roof tiles and mixed with night or morning dew moisture, it becomes a paste film that is cemented to the roofing. This film changes the color of the tile and is difficult to wash off.

Roof Vent Paint

Sometime before the roofing is completed, the paint for the roof vents should be obtained. This paint is usually supplied by the painter, but the actual painting is done by the roofer because he or she is more adept at walking on roof tiles without breaking them.

The builder should be the middleman regarding the handling of the roof vent paint. The painter should give the paint to the builder, and the builder should pass the paint to the roofer as it is needed. This gives the builder an idea of how much of a particular color paint is being used and whether there will be enough to complete all the roofs without having to order more. Reordering can be difficult when there are sometimes four or five different paint colors that match the color of the roof tiles.

The builder can then be the informed arbitrator of the roof jack paint and not be caught in the middle of an argument over who gave how much of what color to whom, while roofs sit uncompleted.

Roof Mud Buckets

Near the end of construction, at the completion of the roofs, colored cement mortar is filled in at the ends of ridges, hips, and birds mouths for clay and concrete tile roofs. The builder should check that the roofers who are applying this mortar do not casually throw their empty buckets off the roof and onto the ground surrounding the buildings. Leftover colored mortar can splash out of the bouncing and rolling buckets and onto exterior walls, staining stucco, wood siding, and concrete walkways and driveways.

Mud Roof Tiles before Concrete Walkways

The builder should try to schedule and complete the mudding of clay or concrete roof tiles prior to the pouring of the concrete walkways and driveway. This eliminates the possibility that red or brown colored mortar will be accidentally spilled on top of freshly poured concrete, leaving stains that cannot be removed.

Chapter

26

Painting

Have Units Complete for Painter

The painter's main request when working on production housing is that the houses or apartments have all of the doors, wood trim, and cabinets in place before the interior painting starts. This is especially true now that oil base enamel paint has been replaced with water base enamel. The new water base enamels are more difficult to brush on doors and trim and have it come out as nice as the old oil base enamel. The painting of pickup items such as doors, which were not in place during the normal paint spraying phase, are therefore more difficult to paint by hand later because it takes longer and requires more work with the new enamel paints to get the same quality.

Units that are complete at the start of the interior painting allow the painter to finish his or her job more quickly. Every door, piece of casing, and strip of baseboard that can be included in each phase of the prep, undercoating, and enamel stages helps the unit get painted faster and more efficiently. When something is missing during the production painting because the finish carpenter was short of materials, this missing item is taken out of the production loop, and must then be prepped and painted individually, by hand. This requires the painter to set up for putty and caulking, then go back and set up for undercoat, then set up for sanding, then set up for enamel painting with a brush, all for one repair.

Tub Cleaning and Paint Masking

When walls, doors, wood trim, and cabinets are painted using an air-less spray-gun, the painter goes in prior to painting and covers and tapes everything off with plastic sheets. Because it usually takes the painter several hours and a lot of effort to mask a unit, the builder should check that no other activities have been scheduled that would undo this masking.

For example, tub cleaning generally occurs at roughly the same time period during construction as painting. Sometimes the builder mistakenly schedules both jobs simultaneously. This could never work, as the painter would have the bathtubs covered with plastic and would not want it removed so the tub could be cleaned.

Clean the Floors before Painting

When the walls and baseboard are covered with the same paint at the same time using a roller, the floors around the baseboard must be clean. If not, the sticky paint roller will pick up anything on the floor and roll it on the wall. It makes no sense to strain out chunks of dried paint and other materials from the paint before application, only to pick up dust, dirt, and sawdust from the floor (Figure 26.1).

Nail Holes in Particleboard

Nailing into particleboard, whether by hand or with a nail-gun, creates a small volcano-shaped mound around the nail hole (Figure 26.2). This does not otherwise occur when nailing into natural wood. Because particleboard is harder and more difficult to sand then natural wood, extra work is involved in preparing the nail holes prior to painting. The temptation for the painter is to use the same minimum care and attention for the particleboard nail holes as with nailing into soft pine door casing or baseboard. This is not a good idea because these small mounds are very noticeable if not sanded flush with the surface, especially when painted with enamel.

Sand Particleboard Edges

Particleboard shelving used in production housing is not manufactured to a uniform standard. Figure 26.3 illustrates typical particleboard shelving that has one of its edges specially finished so that it can be exposed. These finished edges can vary in quality from perfectly smooth to rough, pitted, and chipped.

When paint, especially enamel, is added to an already rough and chipped particleboard edge, it becomes sharp and jagged like a serrated knife. Because the painter

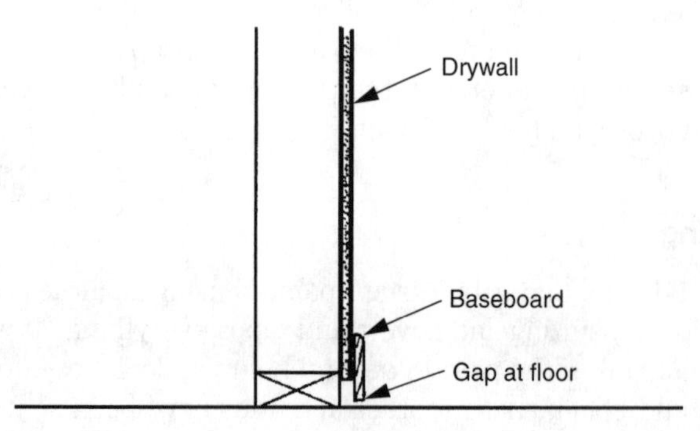

Figure 26.1

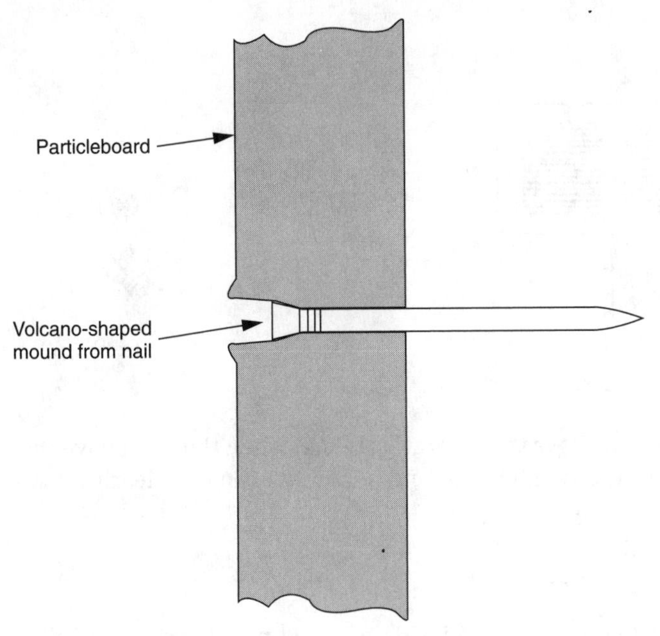

Figure 26.2

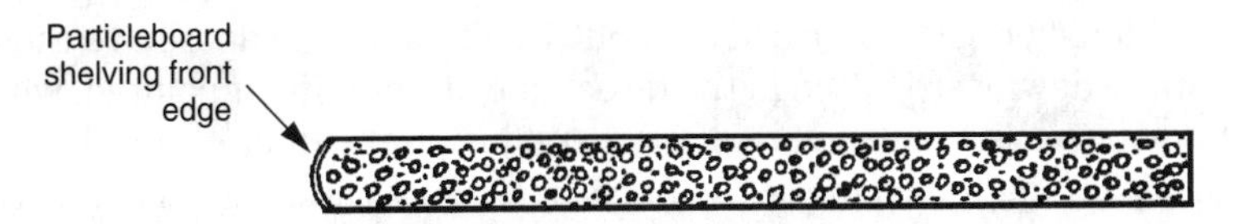

Figure 26.3

has no control over the quality of material brought on the job by other contractors, particleboard shelving with overly rough edges should be sanded smooth by the carpenters who installs the shelving. Because each shipment of particleboard delivered to the job site may differ, the builder must check to ensure that rough-edged shelving is sanded smooth before painting.

Caulk Registers

Air-conditioning wall registers should be installed before the caulk and putty paint preparation phase. The edges of the registers can then be caulked by the painters, along with the door casing, baseboard, and other trim. Caulking leaves the edges around the registers looking clean and finished (Figure 26.4).

Caulk Banjos

The nail holes and edges of cabinet banjos should be puttied, caulked, and painted. This is often missed by the painter because the banjo support cleat is hidden under

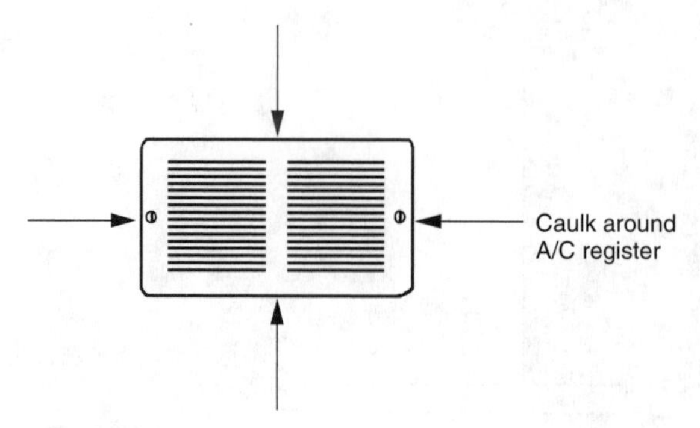

Figure 26.4

the lavatory top and above the toilet water tank. When the homebuyer cleans around the toilet, however, the unfinished and unpainted banjo is clearly apparent (Figure 26.5).

Black-Out A/C Ducts

It is a good practice to have the inside exposed metal surfaces of air-conditioning ducts painted black just after the drywall texture phase. This can be done with cans of spray paint by the painter or a laborer, or with the painter's spray rig.

Air-conditioning registers placed at about 7 feet above the floor have their flaps or blades slanted downward, often in the direct line of sight. If not painted, what is seen by the homebuyer is the unsightly bare metal inside the duct covered with drywall texture. By painting the inside of the duct black, only darkness is seen between the spaces of the register blades (Figure 26.6).

Put Back Drawers

It is common practice for the painter to remove the cabinet drawers for painting, staining, and lacquering. They are usually lined up in a nearby room, sprayed with paint, stain, or lacquer, and then left to dry. It is important for the painter to go back and reinstall the drawers after they are complete and dry. If left standing where they were painted, future workers could knock them over and damage them.

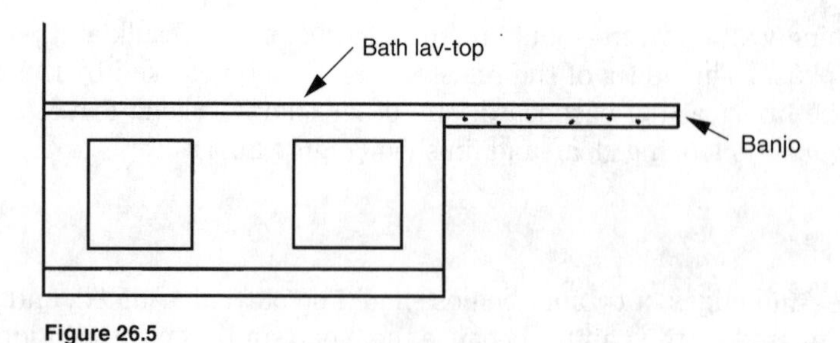

Figure 26.5

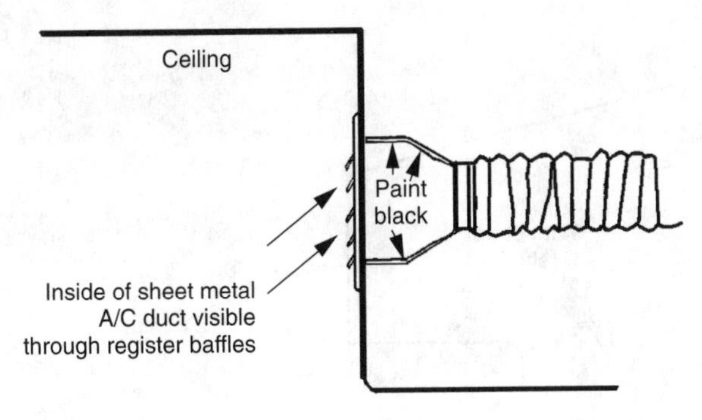

Figure 26.6

Stain Breadboards

The general rule in tract housing is not to install the kitchen breadboards until the houses are complete because they are often stolen. Breadboards should be installed just prior to the homebuyer occupancy, as part of the move-in kit of plumbing faucet aerators, shower heads, and so forth. When homebuyers have four or five different cabinet stain colors to choose from, however, the breadboards must be in place in each kitchen during the stain and lacquer phases. The breadboards can then be stained to match each kitchen, then be collected and stored until construction is complete. This is usually easier than having the painter stain all the breadboards at one time, with five different stain color options. It also provides a better color match.

Paint Pot Shelves

Drywalled pot shelves that are above head height should be painted, regardless of whether they are visible from a second floor hallway or staircase. Interior pot shelves are often used by the homebuyer for displaying all sorts of things. Unpainted surfaces that are missed by the builder will only result in a painting call-back after the homebuyer moves in and discovers them unfinished (Figure 26.7).

Paint Walls inside Cabinets

When kitchen and bath cabinets do not have wood veneer backs, the exposed drywall wall surfaces inside the cabinets should be painted by the painter (Figure 26.8). In kitchens and bathrooms, this paint should be enamel like the other surrounding walls. It is easy for both the painter and the builder to overlook painting under the cabinets because some physical effort has to go into checking this item.

In some cases, where the bathroom walls are painted before the cabinets are installed, all that is required is for these surfaces inside the cabinets to be touched up along with the other paint touchup work. This is usually necessary because of the smudges and dirt stains left on the walls from all of the plumbing work done underneath the sinks.

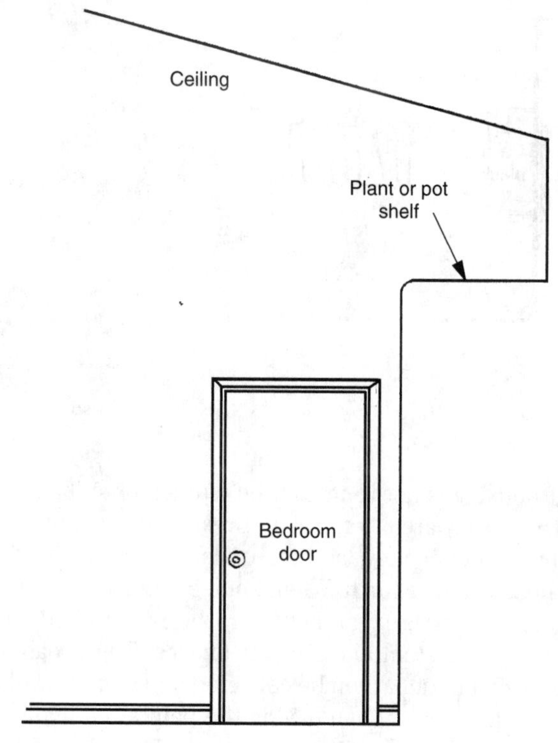

Figure 26.7

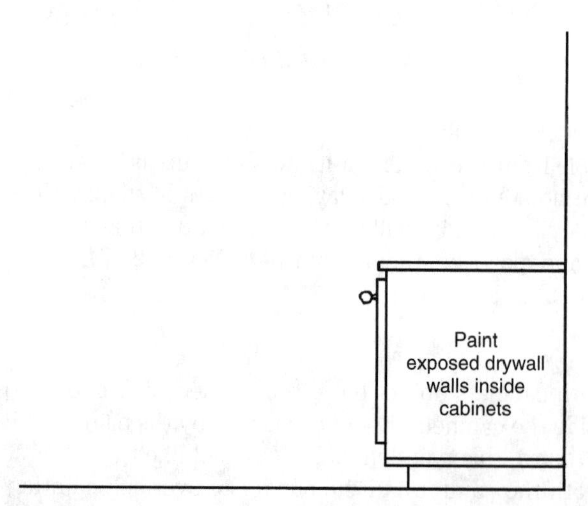

Figure 26.8

Black Window Frames

One of the more difficult construction problems the builder brings upon himself is to install anodized dark bronze colored window frames surrounded by drywall window sills. The joints between the dark frame and the white drywall provide an unwanted glaring point of evaluation of the construction. For these joints to come out straight and uniform, the drywall taping and follow up painting must be perfect. This is often too much to ask in tract housing.

The painter can come back just before the end of construction, after the walls are painted, and cut-in the window frames with dark bronze paint supplied by the window manufacturer. This is time-consuming though and will result in an extra to the painter if not previously agreed upon in the contract.

Caulk Medicine Cabinets

The front edge, and when reachable, the back edge of recessed medicine cabinets should be caulked where they meet the wall (Figure 26.9). This can be done by the painter or a job laborer with colored latex caulking to match the wall paint color when possible. Caulking greatly improves the joint appearance, and colored caulking to match the walls (usually only white, Navajo white, or bone white will match) removes the need to paint the caulking.

Nail Putty Color

When kitchen cabinets are stained, especially whitewash or light colored stains, the individual cabinet wood pieces often vary in shades of light and dark. A cabinet door can have rails, stiles, and panels all slightly different in color. The surrounding face frame pieces can also vary in degrees of light and dark. Even though the wood is the same, each piece receives the stain slightly different.

In production housing, the painter cannot custom color the nail hole filler putty to match variations in cabinet stain colors from rail to stile to panel to face frame. The

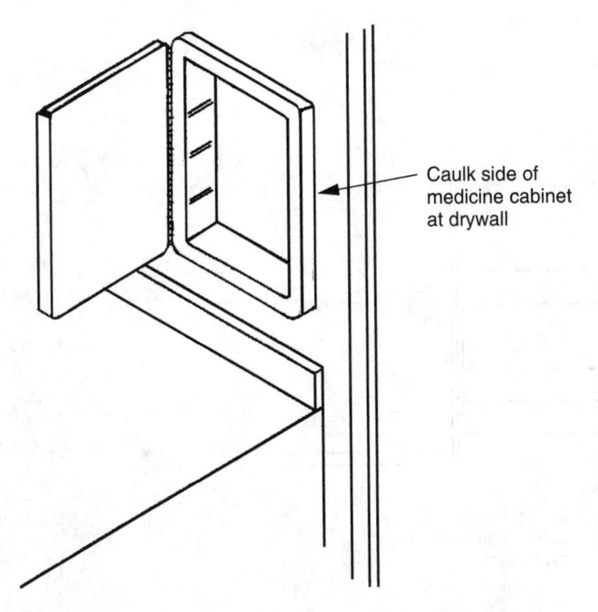

Figure 26.9

builder should specify in the homebuyer's manual or warranty that although the stain color selected by the homebuyer results in shade variations, the nail putty color will be one consistent shade. This will help prevent an unreasonable homebuyer from insisting that the nail putty colors exactly match the varying shades of stain colors on cabinets.

No Dust in Skirtboard Paint

The builder should check, prior to the start of interior painting, that the stair steps and landings are clean and free of dust (Figure 26.10). The builder should also check that the painter dusts off these stair steps and landings prior to painting. The problem to avoid is having the spray painting gun kick up dust when the painter is painting the stair skirtboard. This dust will then get mixed into the paint and end up on the skirtboard surface. The result is a rough, sandpaper-like surface.

If the proper precaution was not taken by the painter prior to painting the stair skirtboards, then the builder should have these surfaces scraped and sanded smooth again for repainting, during the final paint pickup phase prior to carpeting. The builder and painter should be aware of the problem, check for it at the right time, and get it corrected prior to the installation of carpet.

Paint Tops of Door Casing

Figure 26.11 shows the side view of interior door casing above the door header. This part of the casing sometimes doesn't get thoroughly painted because of the height and angle required for a spray-gun to reach it. The spray-gun nozzle is usually held below the level of the door header so that paint is sprayed horizontally onto the door casing. This often results in little to no paint hitting the top surface of the casing.

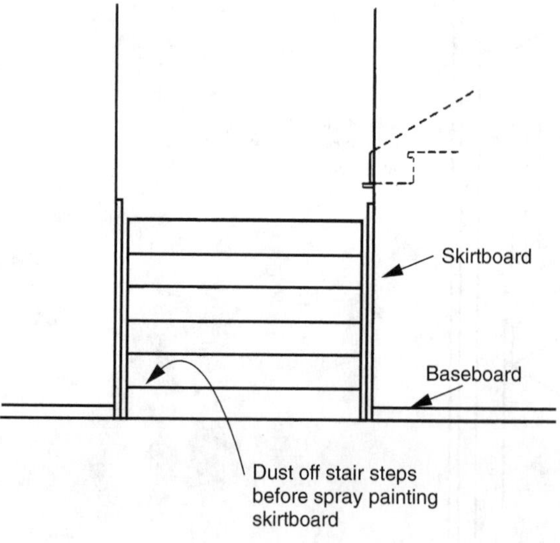

Figure 26.10

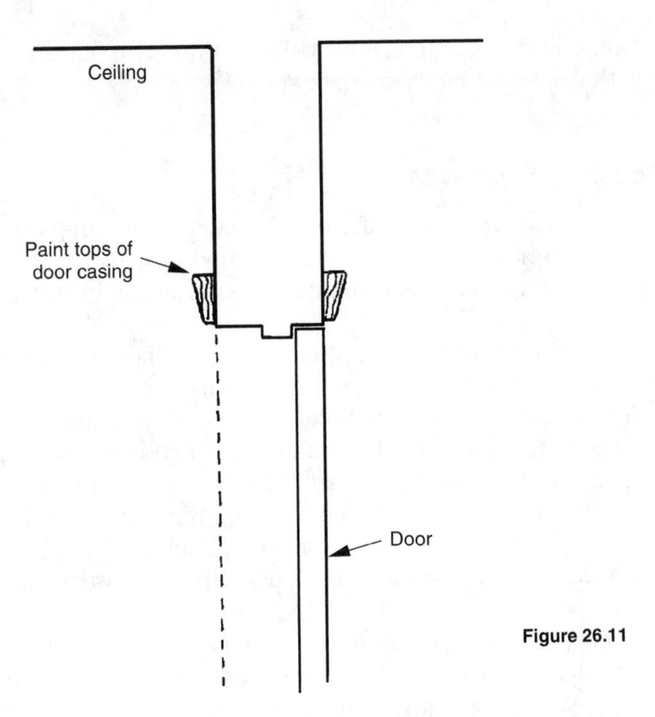

Figure 26.11

This is not a major problem because most people are looking up at the door header casing and don't see the top surface. However, the builder is paying for completely painted surfaces and all the homebuyer has to do to spot unpainted door casing tops is to get up on a ladder to hang pictures or clean off a high pot shelf.

Painting Baseshoe

The builder should ask the painter to raise the baseshoe off the floor or to spread it out on something at least a few feet above the ground prior to painting or staining so that dust does not get on it. For example, if the painter carelessly spreads out bundles of baseshoe directly on top of a subfloor, whatever dust remaining on the floor after it was swept, will be stirred up by the force of the air coming from the spraygun and end up on the baseshoe. This results in painted or stained baseshoe that has a rough, sandpaper-like surface—an unsightly comparison to the smooth, glossy look of properly painted baseshoe.

No Paint on Handrail Brackets

The painter should be required to either mask off stair handrail metal bracket supports or clean the paint from them during final paint touchup. This item should be included in the painting contract and discussed in the field before the start of painting. This issue should be enforced even to the point of having the painter remove the brackets from the handrail in order to remove any paint, and then reinstall them when clean. Few things look worse than polished or antique brass handrail brackets that are covered with paint overspray, drips, or smudges.

Painters to Clean Off or Touch up Fireplaces

Shown in Figure 26.12 is the front face of a prefabricated metal fireplace. Because fireplaces are installed during the rough framing period of construction, they are among only a few finish items, bathtubs being another, that are in place when the walls are painted.

The builder should write into the painting contract that the painter must either clean off or repaint the front metal surfaces of the fireplace fireboxes, which are usually a flat black color, during the final paint touchup work. This identifies the correction of paint overspray on fireplaces as a legitimate construction activity upfront and ensures that the painter will handle this as part of the normal paint pickup work. By indicating this painting on the contract, the debate over who caused the white specks on the black fireplace surfaces are no longer an issue. If cleaning or painting the fireplace front surfaces are in the bid portion of the painters work, then there is no further discussion.

Obviously, the whole problem can be avoided if the painter masks off the front of the fireplace prior to drywalling or painting. The drywall contractor almost always masks off the fireplaces before texturing. The painter sometimes masks off the fireplaces, but sometimes merely spray paints a few inches short of the opening, then picks up the unpainted areas with a roller or later during final paint touchup. Sometimes the texture or paint overspray is merely small particles that drift or blow in around the edges of the plastic masking.

One benefit of mentioning the paint touchup of fireplace fronts specifically in the painting contract is that it can then be scheduled as a separate activity, apart from the final paint pickup phase. Probably the best time to touchup the fireplace fronts is just after the ceramic tile or other decorative material has been installed around the fireplace openings, but before the installation of the fireplace glass doors. By painting it at this time, access to the black metal surfaces is easier and all other work to be done on the fireplace is complete.

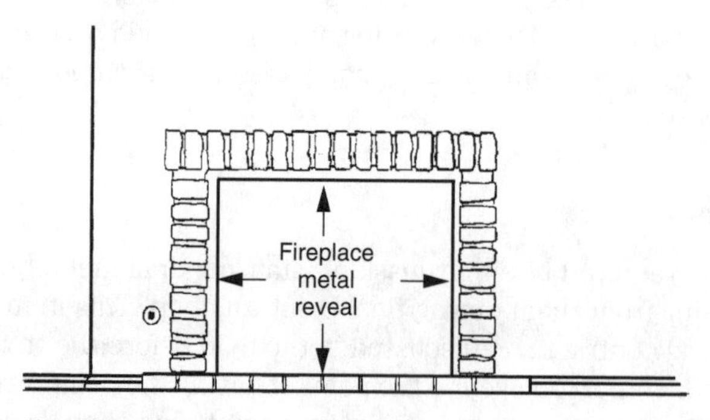

Figure 26.12

Painters Mask Off Floors around Cabinets and Doors

The painter should mask off the floors with paper around the cabinets and doors when staining and lacquering kitchen cabinets and undercoating and enameling interior doors. Although this is seldom done in production housing, this item should be added to the painting contract.

The reason is that if the painter is allowed to stain, lacquer, and paint at will, without protecting the floors, then areas of the subfloor around cabinets and interior doors are covered with paint in bands or swatches 6 to 12 inches wide. If the kitchen or bath floors get vinyl flooring, the vinyl adhesive will not stick to these stained, lacquered, or painted surfaces. The flooring installer is then forced to scrape off this paint from the subfloor as part of the flooring preparation. This is not only unfair to ask the vinyl flooring installer to clean up the painter's mess, but the mess is also very difficult to remove from the subfloors. Often the stain, lacquer, or paint will not easily scrape off the floors. The flooring installer must resort to using a disk sander, which often gets gummed up with the paint and lacquer. Furthermore, it is difficult to sand right up next to the cabinets without touching the cabinet sides and damaging their finish.

Another problem with allowing the painter to spray paint without masking off the floors, is that if there is any dust or sand particles on the floors at the time of spray painting, they will be cemented to the floor with the paint. If ¼-inch plywood underlayment is used underneath vinyl flooring, as is recommended elsewhere in this book, these sand particles remain underneath the plywood underlayment and crunch as you walk over the vinyl flooring. If not already cemented to the subflooring, these sand and dust particles would otherwise be swept off the floors by the vinyl installer as part of the floor prep.

Don't Paint Eaves White

From a construction standpoint, the underside eaves of roof overhangs should be painted any color but white. White paint magnifies all of the flaws in the tongue and groove starter board that is commonly used on roof overhangs. Knots, cracks, splits, and nail shiners show up more clearly with white paint.

Eaves can be attractively painted the same color, but several tones lighter, than the surrounding walls below. This gives two advantages. First, the transition from the stucco to the wood overhang is softened and disguised when the colors match rather than contrast. Second, any color other than white will help hide the woodwork flaws.

Paint Exterior Doors Soon

All surfaces of exterior doors should be primed and painted soon after they are hung. Even when exterior glue is used with lamination, the doors will delaminate if exposed for a period of time to moisture. Paint and stain seals the wood surfaces from moisture penetration. The builder should especially check to see that the painter is priming and painting the tops and bottoms of exterior doors.

The first coat of primer or stain should be applied within 24 hours of hanging the doors. The second and third coats should be applied no later than 5 days after the doors are hung. These are the general standards for the industry.

This time frame may not be convenient for the painter, who would like to undercoat and enamel the exterior doors along with the interior doors and woodwork. The 5-day time span between hanging the exterior doors and getting the rest of the woodwork ready for painting, may be too short. One solution is to have the finish carpenter hang exterior doors toward the middle or end of the finish carpentry work so that painting can begin soon afterwards. The other solution is to simply have the painter prime and paint the exterior doors independent of the rest of the woodwork. In either case, the builder should not accept waiting for sealing and protecting exterior wood doors merely to suit the painter.

Exterior Balcony Handrail

As a quality-control precaution, exterior balcony wood handrails should be sanded smooth before being painted or stained. When paint is added to rough-sawn and splintered surfaces, they become more jagged and sharp because the paint tends to make the surfaces brittle and hard. Even frayed surfaces around knots that have the soft hairy feeling of peach fuzz before being painted, become sharp and sandpaper-like after painting.

The mistake is thinking that because all other exterior wood plantons, facia board, walkway rails, and patio covers are rough-sawed or otherwise unsmooth lumber, the balcony handrail can also be rough and unsanded in keeping with the architectural scheme. Rough balcony handrails are quickly noticed and complained about by the homebuyer.

Galvanized Sheet Metal

Galvanized metal trim materials, such as mill-core and parapet wall cap, have a coat of protective oil applied during manufacturing to preserve the metal surfaces during storage. Galvanized sheet metal should be cleaned with soap and water, vinegar and water, or a commercial wash solution prior to painting. Special primers, called *fuse primers*, are made for galvanized metal. These primers bond to the surface and prevent later chipping and peeling. The builder should insist on the use of such primers, especially when metal surfaces are exposed to the weather. The roof parapet wall cap, balcony wall caps, and stucco expansion joints for an entire building can chip and peel within a short time if not properly prepared.

Painting Exterior Vents and Flashing

When a transparent stain is used on exterior wood facia, rafter tails, balcony deck handrail, and plantons, the builder should remember to have the painter come up with a matching, custom color of regular paint for the painting of exterior sheet metal vents and flashing, when they are next to these wood members. The transparent stain obviously cannot be used over the sheet metal.

Roof Jack Paint in One-Gallon Cans

The builder should specify in the painter's contract that the roof jack paint be given to the roofer in 1-gallon cans rather than 5-gallon buckets. This makes the handling of the paint container much easier up on the roof. It also makes the management of the paint easier and simpler, as the roofer does not have to find other smaller containers to pour paint into from a 5-gallon bucket.

Clean Off Decks for Sandblasters

When exterior wood facia, balcony deck rafters, corbels, and plantons are sandblasted in preparation for staining or painting, the builder should check that all balcony decks are free of drywall scraps, loose lumber, planks, and other debris. Because the sandblaster wears a protective hood while working and has a hard time seeing due to excess dust created when sandblasting, extra precaution should be taken to ensure that the sandblaster does not trip over something while working from a scaffold or from a second or third floor balcony deck.

Clean Off Roofs after Sandblasting

The roofs must either be swept clean with a broom or blown off with an air-blower following the sandblasting of roof facia board. The builder should include this issue in the sandblasting contract or purchase order and discuss it with the sandblasting crew prior to the start of the work.

Leftover sand particles from facia sandblasting on a sloping roof act as tiny ball-bearings, making the roof surface slippery and dangerous for succeeding tradesmen.

Roof facia and other exposed exterior woodwork is sandblasted prior to painting to achieve a particular architectural look.

Prime Wood Siding

Probably everyone has seen a housing project or an individual house that, after a few years, has its tongue and groove or shiplap wood siding badly cupping either inwardly or outwardly from the wall. This typically happens when only one side of the wood siding—the exterior side—is sealed with paint or stain. This occurs when the wood siding is installed raw and unsealed over wall framing that is covered with building paper. The siding is then spray painted with paint, stain, or lacquer on the exterior surface only. Because the one side that is painted or stained is protected from moisture penetration, and the raw unsealed side is not, half of the wood board will expand and contract with the change in moisture content and the sealed half will not. This is what creates the curvature or cupping in the siding board (Figure 26.13).

This problem can be eliminated or at least reduced by buying wood siding that is already factory primed and sealed, when the exterior surface of the siding is to be painted, or by priming the backsides of the siding boards with stain on the job site prior to installation, when the exterior surface of the siding is to be stained. However the builder wants to solve this potential problem, the idea should be to seal all of the surfaces of wood siding boards so that they have a chance of remaining straight.

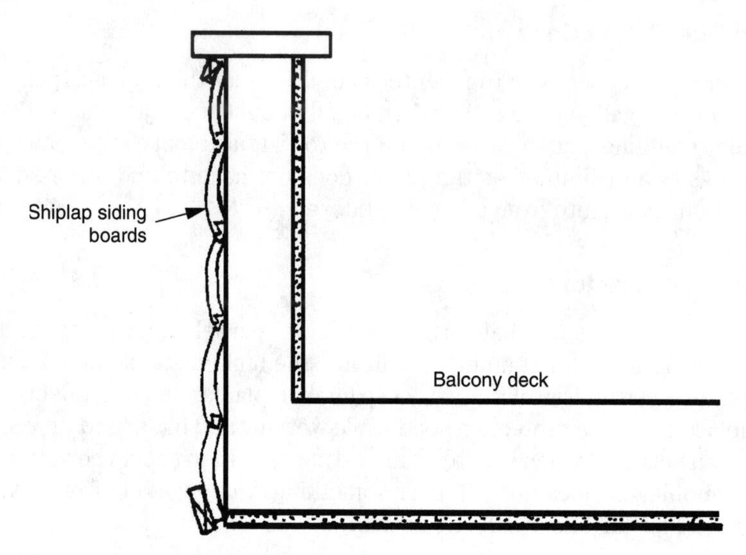

Figure 26.13

Chapter 27

Hardware

Select Easy-to-Install Doorknob Hardware

In production housing, where hundreds or even thousands of interior doorknobs must be installed, the builder should look at the ease of installation, along with the design and beauty of the hardware.

On one particular project, the developer chose an imported brand of hardware with an oval shaped solid brass doorknob that was both difficult and time-consuming to install. This hardware came with set screws that required two sizes of hex-head wrenches, plastic spacers, and parts that did not fit well together. To top it off, working space around the doorknob was tight.

Although this particular hardware was nice looking, the developer had no idea of the problems it caused. The finish carpentry contractor did not know what kind of hardware was being selected by the developer, and therefore bid the hardware at the standard rate. When the hardware installer got to the job site and discovered how difficult the job was going to be, he naturally asked the finish carpentry contractor for more money. This was worked out between the two parties, but the finish carpentry contractor had to eat the money that was lost in having to pay the installer more than was originally planned. Consequently, the hardware took longer to install, affecting the construction schedule.

From a customer service and unit prep standpoint, this hardware was very time-consuming to install. It seemed that every time the customer service person touched one of these doorknobs, it would cost 15 to 30 minutes. Whether the problem was that some parts did not fit together right, a set screw was missing, the backset was sticking, or something else, the repair always seemed to involve more than just a quick look.

For production housing, the builder or purchasing agent should examine closely the economics of the hardware installation, along with the marketing aspects of its appearance.

Repackage Hardware

Hardware is sometimes purchased directly by the housing developer or the general contractor and given to the finish carpenter to install as it is needed. This saves the developer the mark-up charge that the subcontractor would add to the hardware.

Hardware is usually delivered to the job site in two forms. Some hardware items, such as doorknobs, doorstops, peepholes, and house address numbers, come presorted in boxes labeled per individual house. Other items, such as towel bars, towel bar end posts, toilet paper holders and towel rings, are delivered in bulk form in separate boxes to the developer. During construction, each day that hardware is installed, the builder gives the finish carpenter some amount of the presorted boxes and adds what is needed for each house and floor plan from the bulk items.

In this situation, it is a good idea to open up, sort out, and repackage all the needed hardware items per house a few days or weeks before it is to be installed. This allows time to check that all the hardware is actually in the presorted boxes and time to add the bulk items without an impatient worker waiting for the hardware.

Getting the correct amount of hardware for each house saves hardware pickup and supervision time later. It also helps to have 10 or 15 boxes of hardware presorted, packaged, and labeled for 10 or 15 units so all the builder has to do is hand the boxes for one day's work to the hardware installer.

Door Stiles Wide Enough for Expensive Hardware

Illustrated in Figure 27.1 is the front view of a door that has the type of hardware where a rectangular square must be cut out of the door edge stile, then it is filled entirely by the doorknob hardware.

Because the builder is the only one who knows what type of hardware will be chosen for doors, he or she must tell the finish carpentry contractor what will be used. This allows the finish carpentry contractor to order a door type, which in this case, has a wide enough stile to handle the hardware.

The obvious mistake to avoid is not analyzing and coordinating the door selection with the hardware. For example, the builder does not want to discover after the doors are already hung that the cutout for the hardware is equal to or deeper than the width of the door stile.

Entry Door Hardware

The installation of entry door hardware should not be allowed to split or crack the edges of entry doors. This occurs when the rectangular portion of the hardware backset is forced into a door edge that has not been adequately chiseled out and recessed. The backset screws will also crack or split the entry door edge if holes are not predrilled for the screws (Figure 27.2). It is often the case that even the slightest crack or split in the entry door, once seen, renders the entire door defective in the mind of the homebuyer. Cracked doors that have gone unnoticed by the builder can be difficult and expensive to replace when the threshold, weatherstripping, and hardware are already installed.

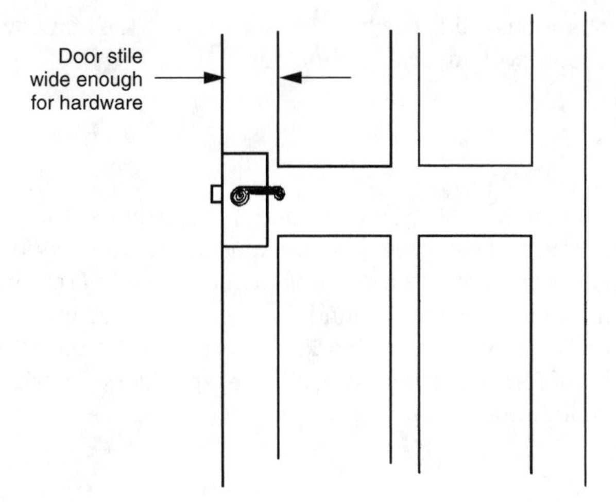

Figure 27.1

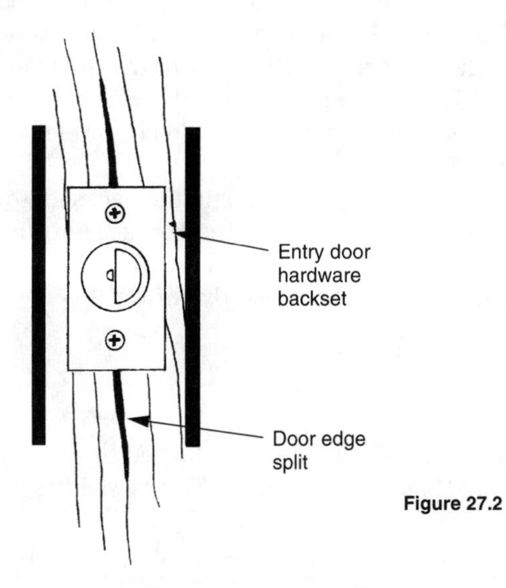

Figure 27.2

Thresholds after Hardware

The installation of thresholds and weatherstripping for exterior doors should be scheduled just after the keylock hardware is installed. This allows the threshold and weatherstripping to be pushed up snugly to the door while the lock backset latches and deadbolt throw-bolts are engaged. This results in a good weather seal, no slack or play between the door and the weatherstripping, and a solid feel to the door unit as a whole.

The problem to avoid is a weatherstripping installation that is so snug that it ruins the striker plate alignment. Doors should close easily against the weatherstripping, and latches and deadbolts should slide easily into place without the homeowner having to push against the doors with an arm or shoulder.

Lever Handle Doorknobs

For apartment or condominium lobby, laundry, and service doors, the doorknobs should be round rather than the lever type. Because most lever type doorknobs push down only so far and then stop, they often get broken in high traffic areas when people push downward on them too hard. A round doorknob requires gripping rather than pushing downward. To break a round doorknob by turning too hard takes a determined effort and a lot of force. Furthermore, a lever-type doorknob will wear out and break simply from unintentional overuse.

Backsets

Interior prefit or prehung doors for production housing come predrilled, bored, and mortised for finish hardware. Because doorknob backsets are manufactured with both round and square corners, the mortises on the sides of the doors are usually made with round corners (Figure 27.3). If a square-cornered backset comes with the doorknob hardware, the round corners of the mortise should be chiseled out square by the hardware installer (Figure 27.4).

When the mortises in the doors are round cornered and the backsets square, the temptation for the finish carpenter is to merely beat the backset into the mortise with a hammer without first chiseling out the corners (Figure 27.5). This results in a beat-up looking backset with all four corners bent slightly outward.

Swinging Pivot Doors

Hardware hinge clips for swinging doors should not be attached to wood wall framing that is covered by drywall only (Figures 27.6 and 27.7). Drywall has weak com-

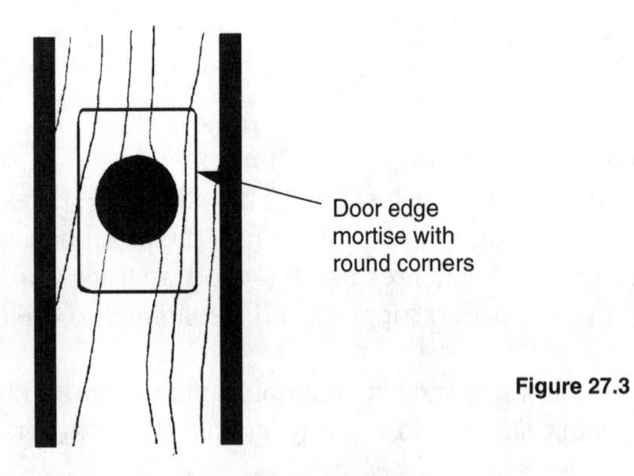

Figure 27.3

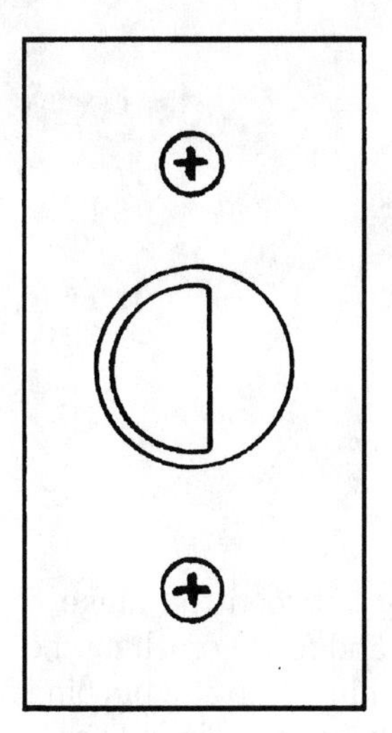

Figure 27.4

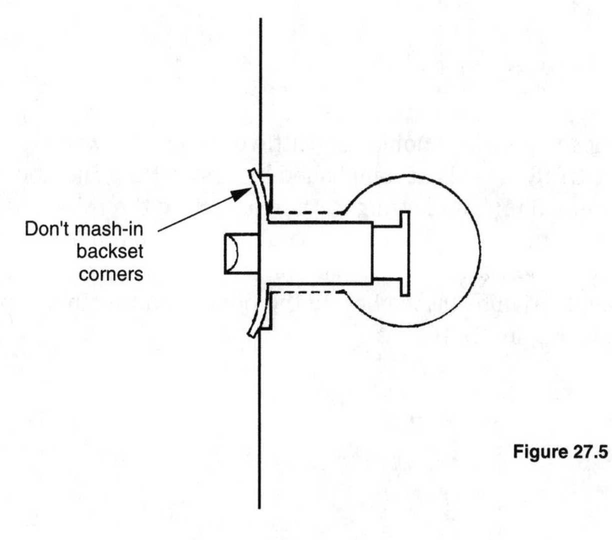

Figure 27.5

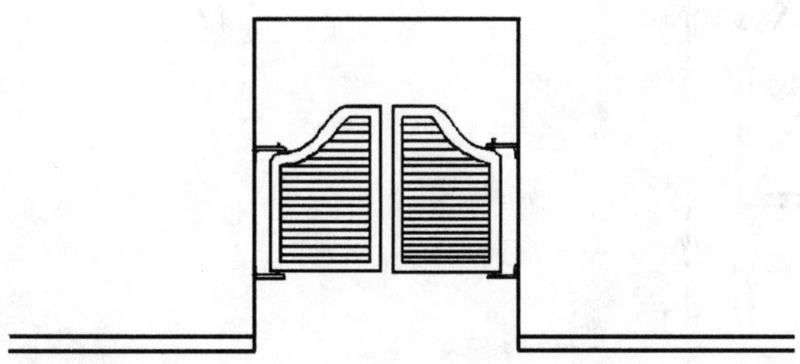

Figure 27.6

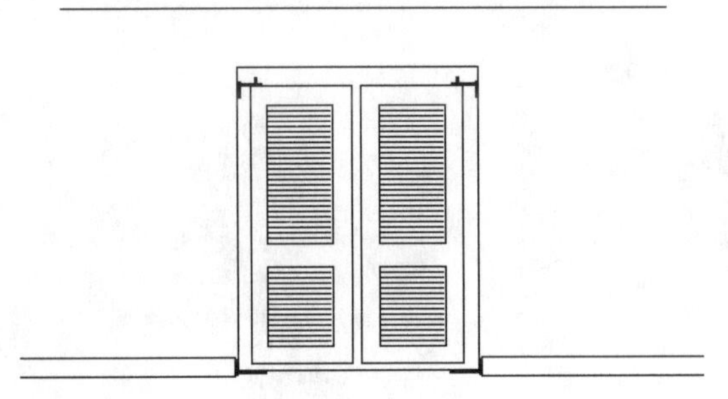

Figure 27.7

pressive strength and with the slightest binding between the hinge clip or metal bracket and the pivot pin, the clip will rock back and forth, crushing the drywall behind the clip (Figure 27.8). When this occurs, the hinge clip and pin bind further, the drywall paper breaks, and the crushed gypsum powder falls out from behind the hinge. Swinging pivot hinges of this type should be attached to wood jambs rather than drywall.

Pocket Door Latches

Pocket doors in the closed position should have little or no gap between the door and the doorjamb (Figure 27.9). This is accomplished by first setting the doorjamb side of the pocket door frame plumb and straight, then adjusting the roller hardware on the pocket door so that the door matches the jamb. Of equal importance is the installation of the pocket door latch. If any part of the latch projects past the door edge, it will strike the doorjamb surface before the door, resulting in a gap between the door and the jamb (Figure 27.10).

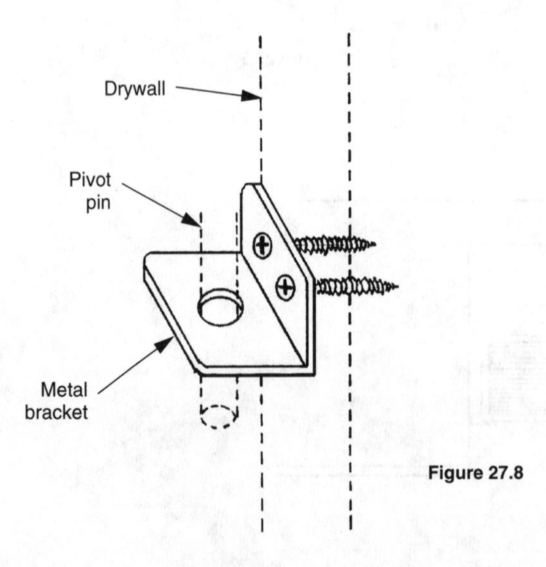

Figure 27.8

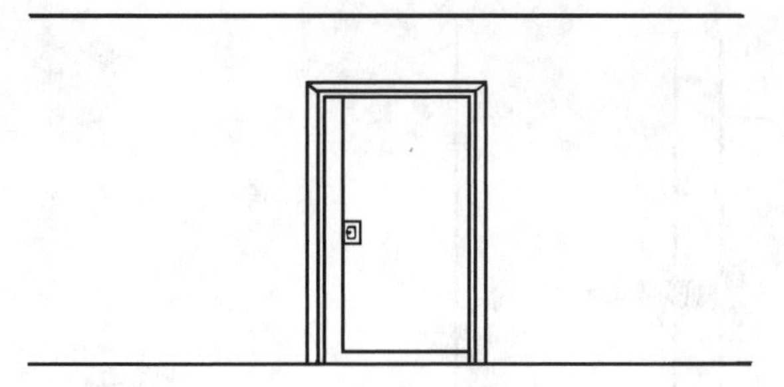

Figure 27.9

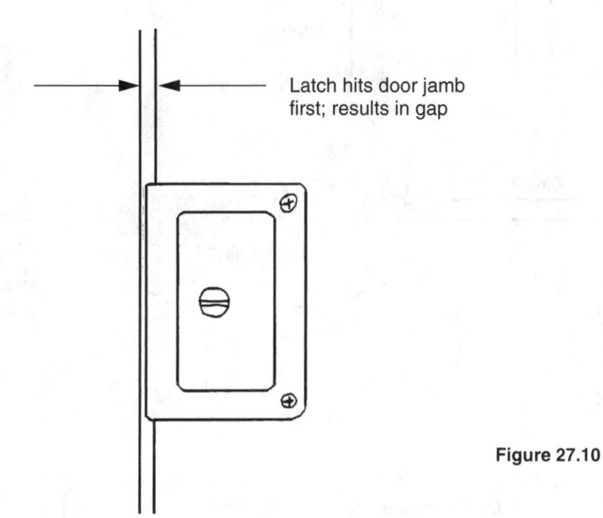

Figure 27.10

Toilet Paper Holders

Toilet paper hardware comes with attachment screws that are long enough to extend through drywall or plaster and into the wall framing studs or backing (Figure 27.11). If these same length screws are used to attach the paper holder to the side of the bathroom pullman, the screws will extend through the cabinet side and into the usable cabinet space below the sink (Figure 27.12). When the plans show the toilet paper holder mounted to the pullman cabinet, either shorter length screws must be ordered along with the paper holder hardware or the longer screws must be cut to a shorter length when the holders are installed.

Space between Toilet Paper Holders

The space between the two end brackets of a toilet paper holder should be the standard, uniform width of a roll of toilet paper (Figure 27.13). The hardware installer should use some sort of wood or paper template to ensure this standard spacing. Few things are worse than having a toilet paper holder end brackets installed too

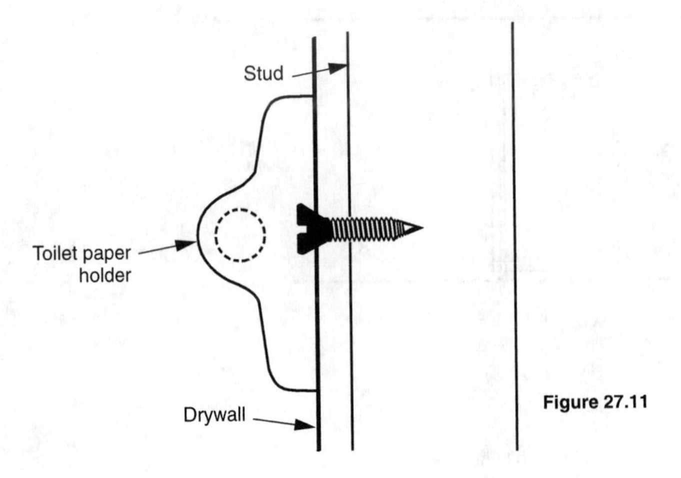

Figure 27.11

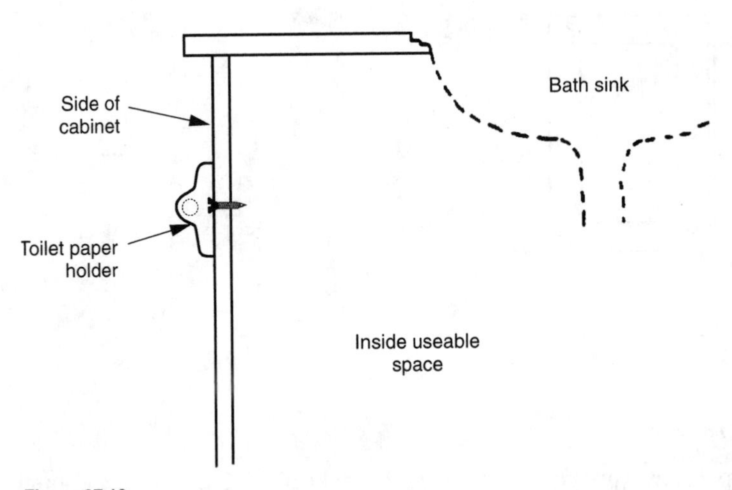

Figure 27.12

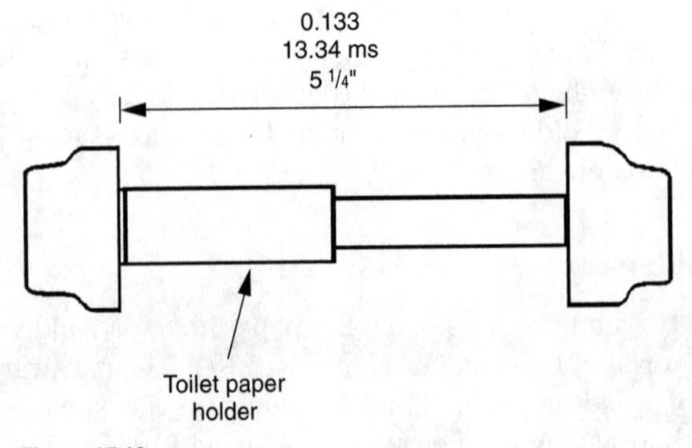

Figure 27.13

close together so the roll is unable to roll or having them set too far apart so that the roller bar and paper fall out.

Dummy-Knob Hardware

With hollow-core Masonite veneer closet doors, dummy-knob door handles should be attached using flanged metal studs rather than the wood screws that come with the dummy-knob hardware (Figure 27.14). The Masonite veneer is too thin to adequately hold the wood screws, and after a short time, the dummy-knob will work loose.

Check Length of Screws for Bi-Fold Knobs

A typical mistake in tract housing is to have louver bi-fold doors for the laundry closets that are 1⅜ inches thick, but with bi-fold door knob hardware that is prepackaged with 1⅛-inch long screws. The hardware installer or builder must then order or purchase the correct length screws separately, after discovering at the time of installation that the packaged screws are too short.

The builder should ask the hardware supplier to specify the correct length of screws to match the thickness of the bi-fold doors. This will prevent the builder from making an extra trip to the hardware store or lumber yard just for the screws. This is a very minor point, but it could be one less mistake to deal with (Figure 27.15).

Backing for Towel Bar Hardware

The builder should have the framer install wood backing in the walls for finish hardware to prevent the hardware installer from using molly-bolt fastener instead of screws at each mounting bracket. By having wood backing in the wall, molly-bolt fasteners cannot be used, and the hardware installer has no choice but to use the wood screws that come with each mounting bracket for towel bars, towel rings, and toilet paper holders.

The hardware installer uses molly-bolt fasteners when there is no wood in the wall at the spot where the hardware goes. Because the hardware installers must provide these molly-bolt fasteners themselves, they tend to be frugal when using them. For

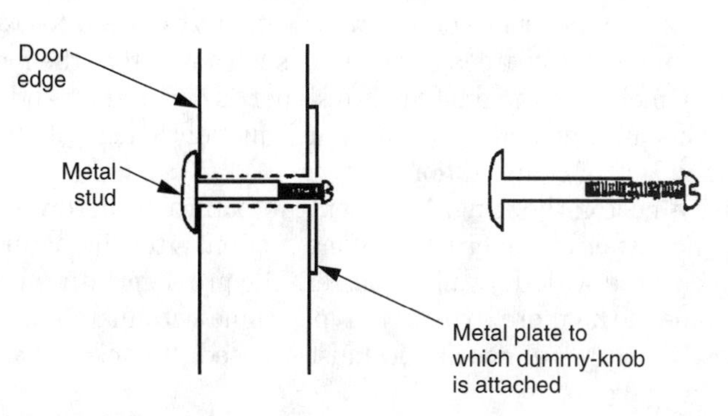

Figure 27.14

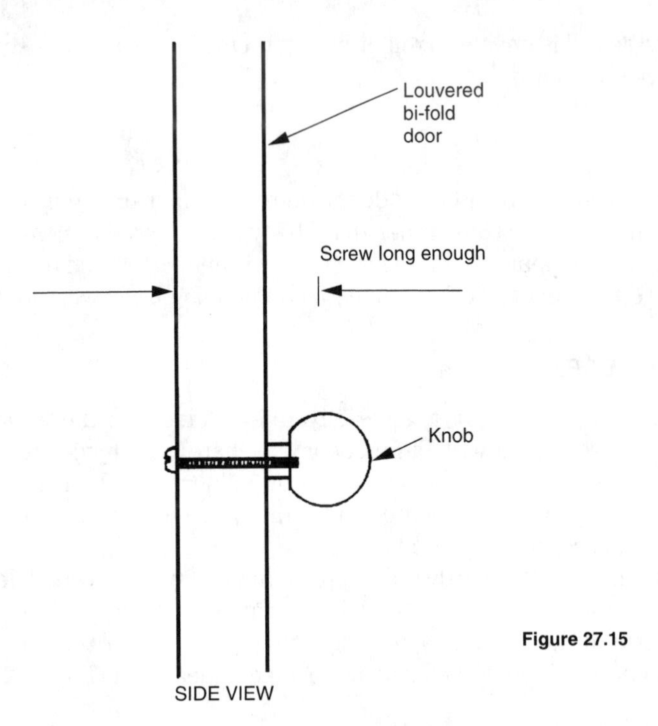

Figure 27.15

example, the installer has been known to use one molly-bolt fastener and one screw (to offset the cost of another molly-bolt fastener) when no wood backing is found.

The builder does not want a situation in which towel bars, towel rings, and toilet paper holders come loose after a short time because only one molly-bolt fastener was used per each bracket. By having backing installed in the walls for hardware, this problem is automatically prevented.

Custom Bath Hardware

In production housing, towel bars, hand-towel rings, and toilet paper holders are usually attached to bathroom walls using molly-bolt fasteners because they allow these fixtures to be placed anywhere on the walls without the need for wood backing. This type of bathroom hardware is designed in such a way that the means of attachment to the wall, including the molly-bolt fastener, is covered up and hidden by the fixture itself. This arrangement is ideal for production housing because of its simplicity, speed, and lack of complications.

In more expensive custom housing, however, the bathroom hardware is sometimes selected by the owner or buyer late in construction, after the bathroom walls have already been covered with drywall or plaster. The problem with this bathroom hardware is that some of the more expensive types come with matching decorative attachment screws, which are exposed and must be used. If backing has not been placed within the walls for bathroom hardware, the builder cannot substitute molly-bolt fasteners for the decorative screws. For example, some modern styles of bathroom hardware are made of clear, transparent plastic. Molly-bolt fasteners obviously cannot be used in place of the decorative screws because they would be visible behind the plastic.

The solution is to anticipate the locations for bathroom hardware, and place wood backing in the walls at those locations.

Peepholes

When drilling the hole for the peephole, it should be started from both sides of the door toward the center. A hole that is hastily drilled production-style from one side of the door only, will chip or split out the wood or veneer on the opposite side. The caution here is that the covering width of most peephole hardware is only slightly wider in diameter than the hole itself, and will not hide any splintered-out chips or frays (Figure 27.16).

Delete Peephole for Entry Doors with Glass

Illustrated in Figure 27.17 is the front view of an entry door. When entry doors come with glass lights at the top, as in this case, it does not make much sense to purchase

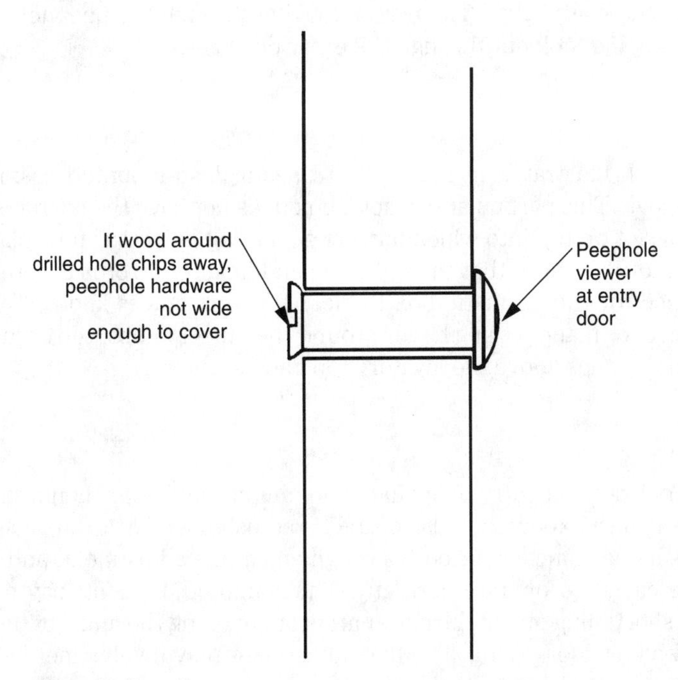

Figure 27.16

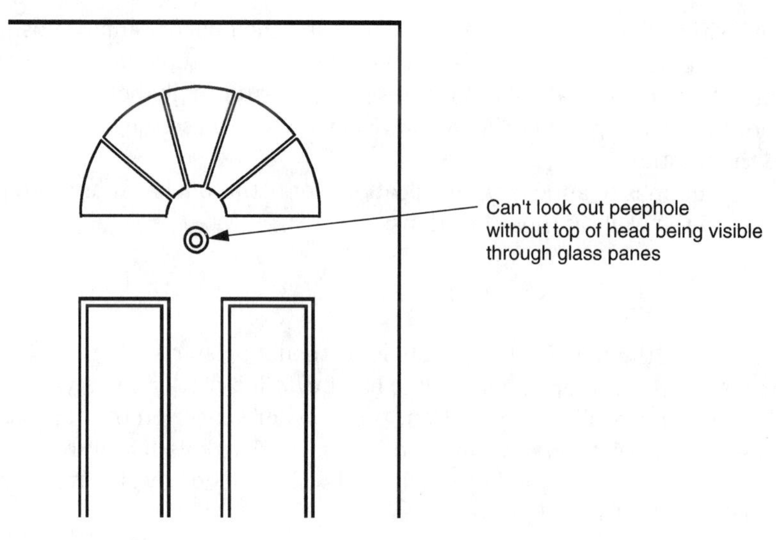

Figure 27.17

and install peepholes. The idea of a peephole is for the person inside the house to be able to look outside to see who is standing at the door without being seen. If a person were to look through the peephole in Figure 27.17, half of his or her head would be visible above the glass lights. The person looking through the peephole might as well straighten up and look out through the entry door glass.

Bath Cosmetic Box

The bathroom sink illustrated in Figure 27.18 has a flush-mounted cosmetic box with a mirror above. This particular cosmetic box does not have the mirror attached, but rather has a slot on top into which mirrors of varying heights can be placed.

The builder should be sure that the right size and height mirror is ordered, allowing enough space for the cosmetic box to clear the sink faucet knobs. To provide enough clearance for maneuvering hands around the faucet knobs, the cosmetic box should be several inches above the lavatory top back-splash.

Adjust Door Closers

Automatic door closers on entry doors in condominiums and apartments should not only be installed on the doors, but also correctly adjusted as part of the installation. Some door closers have means for both a rough and a fine adjustment, and both are needed for the closer to operate correctly. The rough adjustment may consist of lengthening or shortening one of the closer arms or changing the angle of the arm attached to the wall or door casing. The fine adjustment may involve merely tightening or loosening a special screw on the closer.

Trying to adjust the door closer with the fine adjustment screw alone may be quick and easy for the hardware installer, but the fine adjustment often does not have the range or strength to change the closing of the door from a hard slam to the desired slow and soft closing. To achieve this, the rough adjustment should be fixed as well.

Medicine Cabinet Joint

Shown in Figure 27.19 is the plan view of a medicine cabinet on a bathroom side wall. The joint between the medicine cabinet and the drywall cutout can be seen in the bathroom mirror. If the medicine cabinet is placed too close to the wall corner, it is difficult to get a finger around the medicine cabinet door to apply caulking to the large, unsightly joint. For this reason, the framed opening for the medicine cabinet should be placed at least two stud widths away from the wall framing corner.

Adjust Striker Plates

The striker plates for interior doorknobs have small metal tabs that can be bent outward to remove any slack between the door jamb stop and the door hardware (Figure 27.20). This removes the sloppy feeling of an interior door that can be rattled or jiggled back and forth even when the door is closed.

The builder should inform the hardware installer, and check during the final pickup walk, that the striker plates have been adjusted correctly. Each interior door should have a solid feel as it closes against the jamb stop, while the latch is heard to click into the striker plate hole.

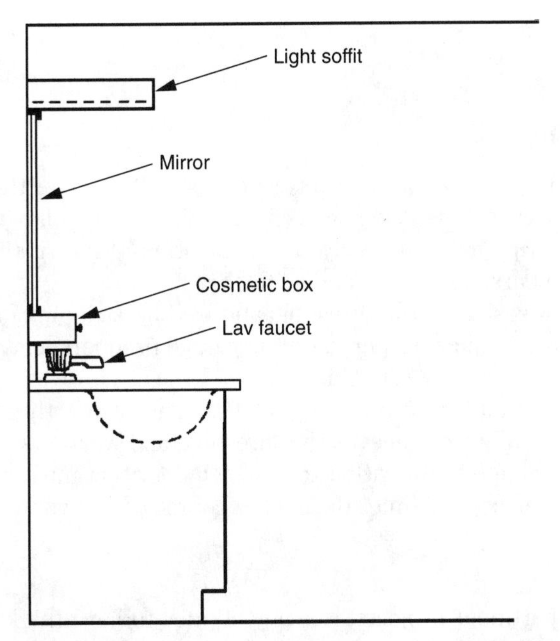

Figure 27.18

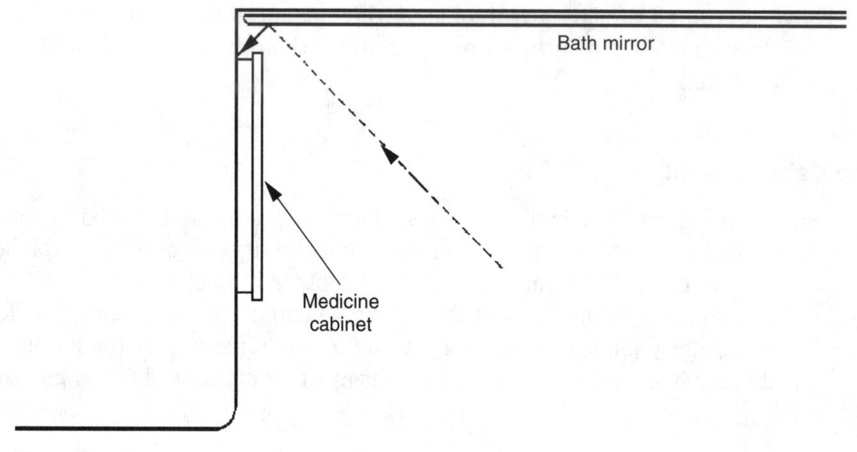

Figure 27.19

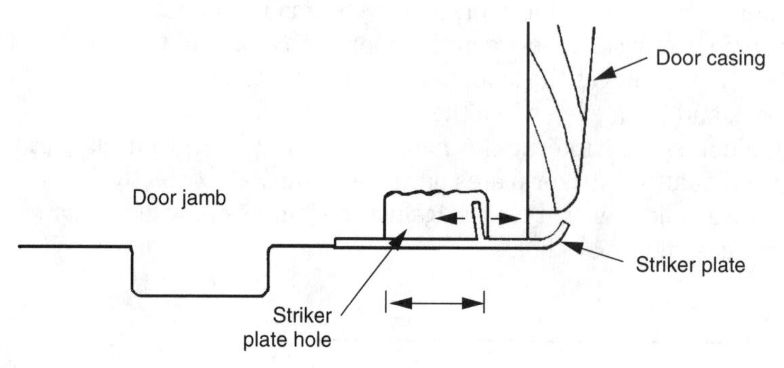

Figure 27.20

Spring Hinges and Weatherstripping

The normal construction sequence for door hardware installation is for the exterior entry and garage firedoors to have their locksets installed first, followed by the thresholds and weatherstripping, and finally the installation of interior hardware, such as doorknobs, towel bars, doorstops, and so forth.

The builder can make it easier for the threshold and weatherstripping installer by having the self-closing spring hinge, required on the garage firedoors, installed later along with the interior hardware rather than with the lockset hardware. This provides a garage firedoor without a spring hinge at the time of the threshold and weatherstripping installation, which relieves the threshold and weatherstripping installer from constantly having to fight with a self-closing door that must be wedged open and freed up for fitting several times during the course of the work.

Common Area Keys

For condominium and apartment projects, it is a good idea to have the keys cut in such a way that the builder's key opens not only the units but also the common area gate locks, the recreation building, the swimming pool gate, and the pool equipment room. This enables the builder to have access to every area of the project without having to use and keep track of several different keys. All of the above locks can be knocked out by the builder so that the keys given to the homebuyers will open only their units, the common area gates, and the recreation building.

Towel Bar and Shower Door

Shown in Figure 27.21 is a drawing in plan view of a bathroom shower door that cannot open past a towel bar. This situation occurred because the towel bar was installed by the finish carpentry hardware person prior to the installation of the shower door enclosure. If the towel bar is mistakenly placed too close to the shower, the problem is not perceived until the shower door is installed. In this case, the towel bar has to be moved over away from the shower door, the old drywall holes patched, and the area touched up with paint.

Size of Address Numbers

The builder should check with the city or county police and fire departments concerning any requirements on the size of house address numbers. For example, these departments may require a number size of at least 4 inches high in order to be easily identifiable. Not having the correct size numbers may cause the building to fail the police, fire, planning, or final occupancy inspection.

Different Key to Pool Room

In condominium and apartment projects with swimming pools, the key to the pool maintenance room should be different than the key that opens the gate into the pool area. This prevents condominium owners from being able to tamper with Jacuzzi or swimming pool temperature settings to suit themselves.

The builder must request that the door lock to the pool room be keyed differently than the locks to the common area gates. Keys to the pool room can then be given only to the pool maintenance person, the property management company, and the homeowners association or apartment manager.

Get Deposit for Builders Keys

A common problem that occurs on most construction job sites is tradesmen not returning builder's keys at the end of the workday. Builder's keys that are not returned

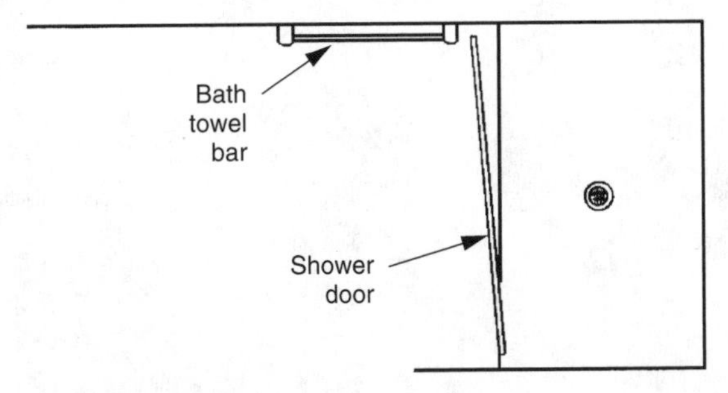

Figure 27.21

to the construction trailer at the end of the day or at the completion of the particular activity have less chance of ever being returned, especially if in the possession of tradesmen who make infrequent trips to that particular job site (e.g., a bath mirror installer, a shower door installer). The tradesmen either lose the key or forget from which job it came.

One way to prevent builder's keys from rapidly disappearing from the job site is to require some form of deposit from the tradesmen before handing out keys. For example, a 5 or 10 dollar bill, exchanged for a builder's key and pinned up on a clip board holding unused builder's keys and other deposits will ensure that builder's keys are returned after use.

Test Keys and Transmitters

After the homebuyer's loan arrangement is approved, keys to the house and garage door opener transmitters are given to the homebuyer. The builder should check beforehand that the keys are correct and open all of the locks. Likewise, the builder should make sure that the garage door opener transmitters have been coded and labeled correctly.

For example, if a block of houses or a 12-unit condominium building pass final inspection at the same time and are ready for occupancy, the builder should take the keys to these units and test every lock to ensure that they work. If the electricity is on in these units, the garage door opener transmitters can also be tested. This will prevent a major weekend debacle, should moving vans arrive at a new house that the homebuyers cannot open because the keys were labeled incorrectly at the hardware store and not tested in advance by the builder. This is a telephone call the builder does not want to get from the job site sales office on a Saturday or Sunday.

Chapter

28

Miscellaneous Finish

Mirrors and Medicine Cabinets

The bathroom pictured in Figure 28.1 has a full length mirror centered above the pullman and a medicine cabinet on each side wall. For a bathroom laid-out in this manner, the mirror should be ordered 6 to 8 inches shorter than the corner-to-corner length so that the medicine cabinets do not interfere with mirror installation. Another alternative is to schedule the mirror installation ahead of the medicine cabinets. The problem to avoid is not being able to install the corner-to-corner mirror because the two medicine cabinets are in the way.

If this problem occurs in one or two bathrooms only, then one of the medicine cabinets can simply be removed for the mirror installation. If this situation occurs on a large housing tract, the builder will not want to take out a large number of medicine cabinets that have already been installed.

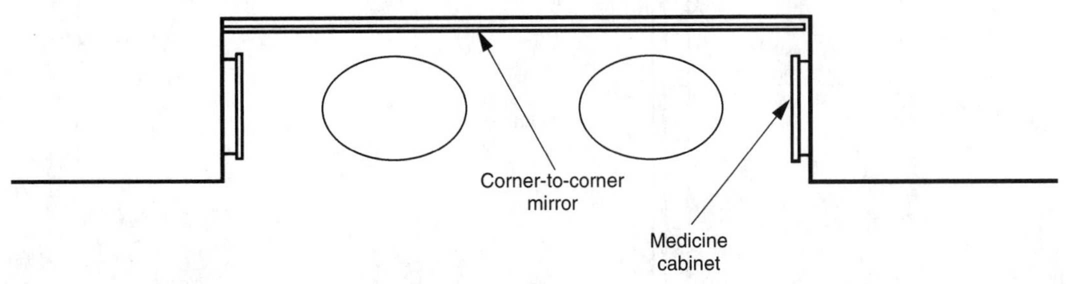

Figure 28.1

Mirror Bottom Track and Strip Light

When a bathroom vanity make-up area has vertical strip bar lighting on each side of the mirror, the mirror installer should cut the bottom mirror track to its exact length (Figure 28.2). If the track is cut a little too long, the homebuyer may not notice if the mirror were centered on the track and the excess did not exceed ¼ of an inch. With the addition of the side strip lights, however, this excess bottom track prevents the strip light from being installed tight to the mirror because it hits the bottom track before touching the mirror. A gap of drywall is therefore left exposed between the mirror edge and the strip light. Because the mirror is installed before the light fixtures, the mirror installer does not know that a problem has been created by the mirror bottom track being slightly too long.

Mirror Walls and Carpet

Figure 28.3 shows an exposed plastic spacer at the bottom of a wall mirror where the mirror meets a thin type of carpet, such as Berber. Because the mirror panels had to be raised up to fit tightly to the ceiling, the plastic spacer material was added to fill the gap at the bottom of the mirror. The thin carpet only added to the problem.

The builder should check the type of carpet selected by the homebuyer for units having mirror wall options, and let the mirror installers know that the fit of the mirror should be tight at the top and bottom when the type of carpet is a thin depth. Another solution is to install some type of mirror trim along the bottom of the mirror wall.

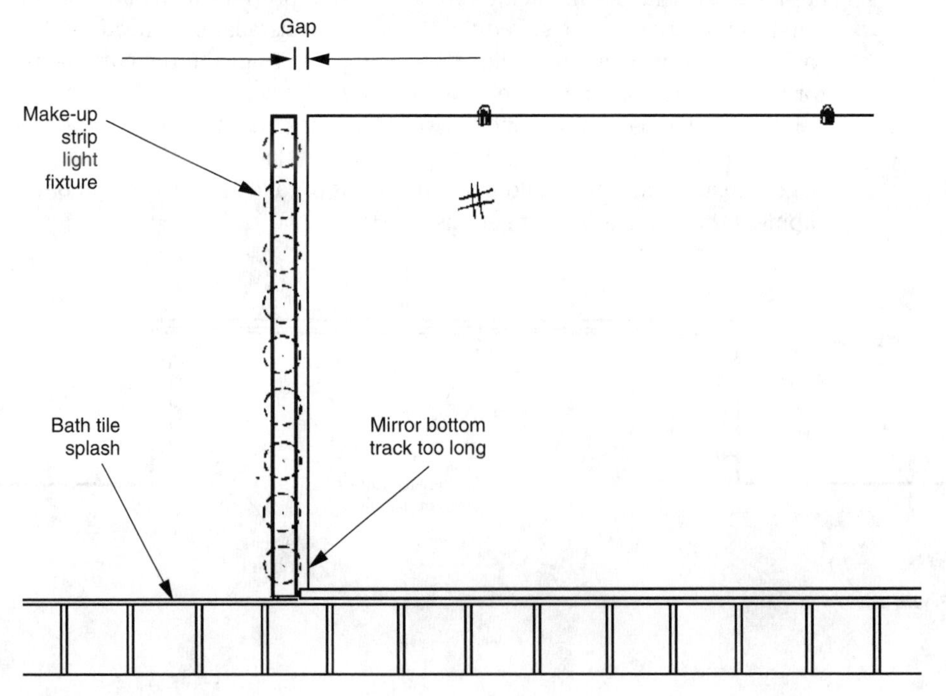

Figure 28.2

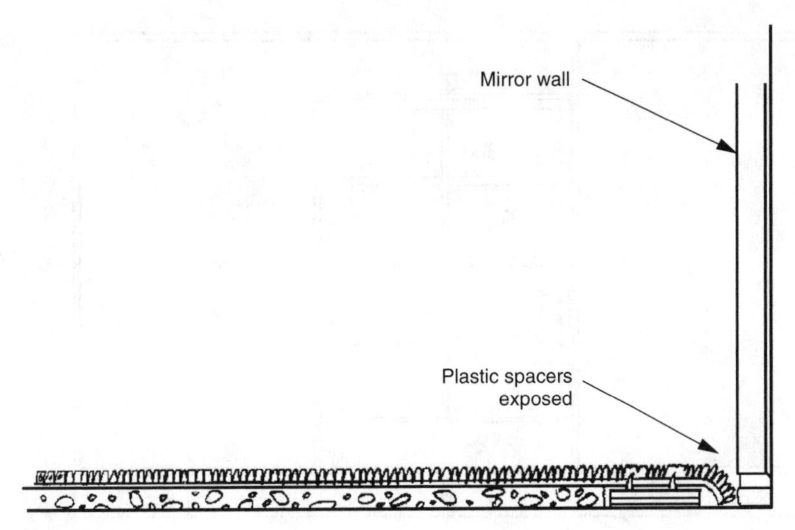

Figure 28.3

Long Mirrors

The builder should check beforehand whether long lengths of mirrors for bathroom walls will fit up stairways in one piece. If they do not, the time to find out is before they are delivered to the job site. If mirrors do not fit up the stairs, they can easily be cut into smaller pieces in the shop to match bathroom cabinet and tub dimensions.

Mirrors and Bull-Nose Cornerbead

The master bath window, pictured in Figure 28.4, is surrounded by the optional mirror design. Unfortunately, the dimension of the mirrors was measured to the edge of the window opening rather than to the edge of the bull-nose cornerbead curve. When installed, the mirror therefore extended beyond the flat portion of the wall, leaving an awkward gap at the edge of the mirror (Figure 28.5).

The correct method would have been to measure and install the mirrors one or two inches back from the window sill so that the mirror edges died before reaching the drywall bull-nose cornerbead curve. This would leave a drywall reveal around the window sills, but at least the gap would be eliminated.

Don't Install Chipped Mirrors

The builder should inform the mirror installers each time they come on the job site not to attempt to install mirrors that have tiny or barely noticeable chips out of corners or edges or that have any desilvering, however minor, on the backsides of the mirrors. The mirror installers should be informed that defective mirrors, no matter how minor the problem, should be left on the truck and not installed. This will save the installers the trouble of having to come back later to remove and replace the mirror. The installers should be made aware that each mirror will be looked at closely by

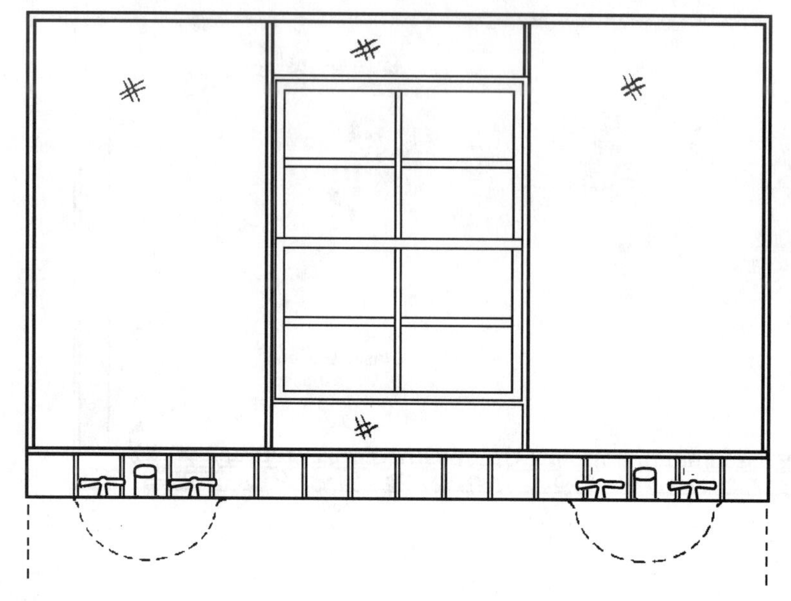

Figure 28.4

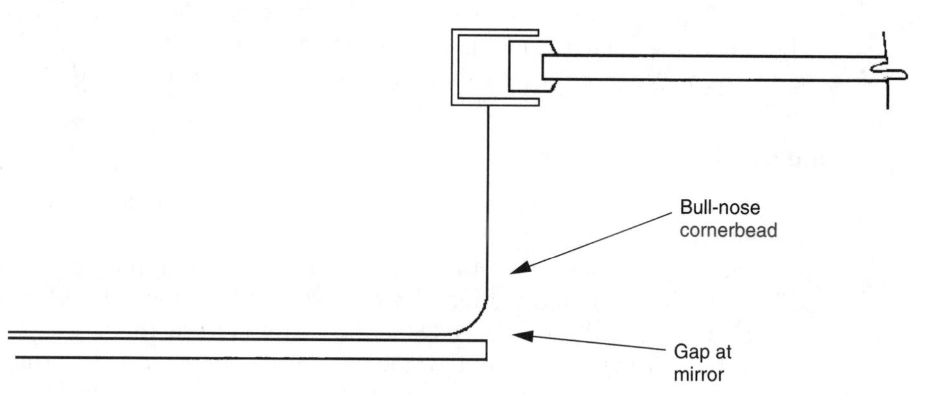

Figure 28.5

the builder, and therefore it is a waste of time and energy to install a mirror that will not pass inspection.

Schedule Mirrors before Medicine Cabinets

Bath wall mirrors should be installed before recessed medicine cabinets so that the mirror installers do not have to remove then replace the medicine cabinets. When the mirror installers are responsible for reinstalling the medicine cabinets, often times they are out of plumb and level. The hardware carpenter, who originally installed the medicine cabinets plumb and true, is then blamed for medicine cabinets that are crooked and out of plumb, thanks to the mirror installers. The best solution—install the mirror first.

Cheat Sheet for Oval Mirrors

Pedestal sinks are not usually installed in bathrooms that have hard surface flooring, such as ceramic or clay tiles, until after the flooring is installed. Because it is very difficult to disassemble the pedestal sink from the wall, the easiest method is to wait until the hard surface flooring is in, then install the sinks.

One problem associated with this is that in order to install the oval mirrors above the pedestal sinks, the contractor needs to know the height and centering dimensions with relation to the sink. If the sink is not in because the flooring needs to be installed, the builder should make up a cheat sheet drawings, such as the one shown in Figure 28.6, that gives the mirror installer the correct height and centering dimensions. This allows all of the oval mirrors in bathrooms containing pedestal sinks to be installed at the same time as all of the other bath mirrors, regardless of whether the sinks are in place.

This saves the mirror company from having to come out to install a mirror every time a builder finishes a bathroom and saves the builder from having to check whether mirrors were actually installed.

Drywall Finals

Drywall finals is the second and last general drywall repair. It comes after the finish trades are complete, but before the final interior paint touchup phase leading up to the installation of hard surface flooring and carpet. If the builder can get the drywall walls and ceilings close to perfect at this time, the only possible remaining drywall damage will be a few slight nicks or dents resulting from the flooring installation. These can be easily repaired by the builder's prep crew prior to the homebuyer walk-through.

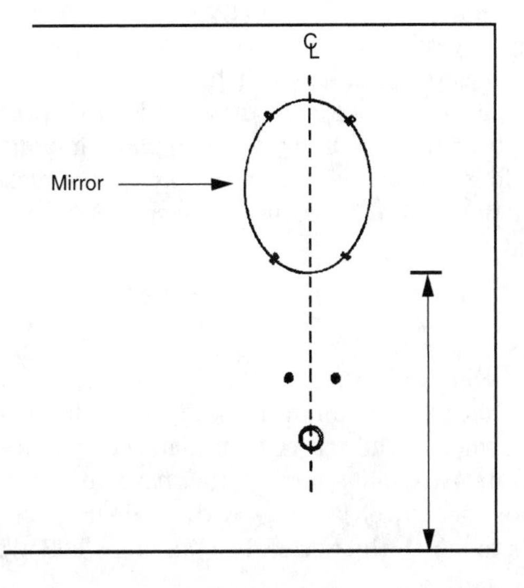

Figure 28.6

The best method for drywall finals is for the builder to go through each unit and circle with a pencil every nick, scratch, dent, and hole prior to the start of the drywall pickup work. This identifies everything the builder wants fixed at this final stage of the construction before the unit is shown to the homebuyer. By circling everything to be repaired, the builder takes the standard of quality judgment out of the hands of the drywall repairmen, and thereby removes any opportunity to strike a balance between speed and quality. Although this method is more work for the builder, and initially may seem tough on the drywall repairmen, getting walls and ceilings close to perfect at one time makes it easier on everyone in the long run. Obviously, dents and dings that are missed during the drywall finals, if left initially up to the drywall repairmen to find, results in call-back repairs for both the drywaller and the painter, not to mention supervision time from the builder.

Carpenters pencils work the best for circling drywall repair areas. To reach ceilings and high wall areas, a pencil can be taped to the end of a piece of door casing or baseboard.

Drop Light for Drywall Finals

One of the most important phases in late construction that affect the final quality of the unit is the final drywall pickup phase. When all of the interior finish electrical, plumbing, hardware, shower doors, mirrors, and closet doors are installed, nearly everything that has a chance to ding, dent, or scrape walls and ceilings during their spreading and installation are now in place. At this point, the drywall pickup crew comes in to fill and texture these dents and dings one last time before the flooring is installed and in preparation for the final paint touchup. If the builder can get every small nick and dent repaired during this final drywall pickup, the finish product will reflect a higher level of quality.

The problem during the drywall finals, however, is that it is still too early to get a final electrical inspection done so that the electricity can be turned on for better lighting. It is too easy for the drywall pickup crew to miss small nicks and dents in dark hallways, bathrooms, and closets that had no light.

One solution to this problem is to require the drywall pickup crew to use drop lights and extension cords during the final drywall pickup phase in order to provide their own lighting. If the builder follows the practice of circling the nicks and dents for repair prior to the start of the drywall finals, then he or she should also use a drop light when marking the walls.

Drywall and Stairways

During the letting of the subcontracts, both the builder and the drywall contractor should discuss the fact that the drywall in the stairways will no doubt take a beating from the other tradesmen as they carry their material up and down them. This is especially true for units with tight stairways that have curved walls, angled walls, pie-shaped stair steps instead of landings with 90-degree turns, and low ceilings that rise in sections to match the rise of the stair steps below. All of these factors make carrying materials, such as interior doors, skirtboard, baseboard, closet doors, bath mirrors, and shower doors, up the stairways difficult without hitting the surrounding walls and ceilings.

Both the builder and the drywall contractor should acknowledge beforehand that there will be more drywall prepaint and final repairs in stairways like these than is normal. If all parties recognize this at the start, then later disputes over what is the reasonable and acceptable level of work required to bring the stairways up to the standards of the rest of the house can be avoided. In other words, the builder or job site superintendent will not be approached by the drywall contractor to approve extras for the increased work required in the stairways.

Color Texture for Drywall Finals

Drywall finals refer to the second phase of drywall repairs. This occurs after the boxout and sweep following the finish trades, but before final paint touchup and flooring installation.

The drywall repairman should add a slight amount of color to the drywall patching and texturing of each repair. This is usually done by adding a small amount of yellow tint latex coloring to the drywall mud as it is mixed for patching and texturing. This color additive is available at most paint stores. The color yellow is easy to see on walls and ceilings, but at the same time is light enough to be covered with one coat of paint.

Coloring the drywall repairs enables the painter to quickly and easily find the last remaining areas to be patched on the walls and ceilings. These areas can then be painted prior to the flooring installation, reducing the extent of final paint touchup.

Some interior paint colors, such as Navajo white, are almost identical to the color of drywall mud. If the drywall repairman during drywall finals does not add coloring to the repairs when colors are so close, it is often difficult for the painter to find all of the large and small patches in dark closets, bathrooms, and hallways. This results in more final paint touchup in the units after the flooring is in or more unpainted areas noticed by the homebuyer when the lights are turned on after move-in.

Drywall Perfect under Wallpaper

In the sales models, the builder should pay special attention to the quality of the drywall prepaints and finals in the rooms that will be wallpapered. Because wallpaper goes on as part of construction in sales models, prior to any last-minute ding and dent repair, the drywall surfaces must be in perfect condition at the time of wallpapering. If the walls are not perfectly smooth and the paper does not require a blank stock prepapering, any defects in the drywall will telegraph through the wallpaper.

In other words, the close inspection of the drywall in sales models must take place earlier than is normal in the production units because once the wallpaper is hung, there are no more opportunities to make minor repairs of small dings, dents, and scrapes on the drywall surfaces that were missed during the drywall finals.

Wrought-Iron Handrail between Two Fixed Points

Illustrated in Figure 28.7 is a wrought-iron handrail at a balcony deck. The handrail is sandwiched between a stuccoed wall and a stuccoed pony wall. For the wrought-iron handrail to fit snugly against both stucco surfaces, and therefore look good, it has to be equal in dimension to the width of the opening. But this makes it difficult to slide into place, especially with stuccoed surfaces which are seldom uniformly flat. Look at any building project that has this wrought-iron handrail design and you will find skid marks along the stucco where the handrail had to be forced into place.

There are two solutions to this problem. The first is to not design a rigid member, such as a wrought-iron handrail, between two fixed points, whenever possible. The second is to anticipate the damaging of exterior stucco around metal handrail that must be forced into place and to cover this item in the stucco contract so that it does not become an extra for the plastering contractor.

Prepaint Wrought-Iron in Shop

The builder should consider having wrought-iron handrail panels shop painted before delivering because there might not be enough space to paint them after installation (Figure 28.8). For example, Figures 28.9 and 28.10 show wrought-iron handrail panels that were placed a few inches above a concrete entry porch and a few inches away from a stucco column. In both examples, painting is made difficult by the tight spaces around the panels after installation. Not only is it difficult for the painter to obtain complete and thorough paint coverage in these areas, but it is also difficult for the

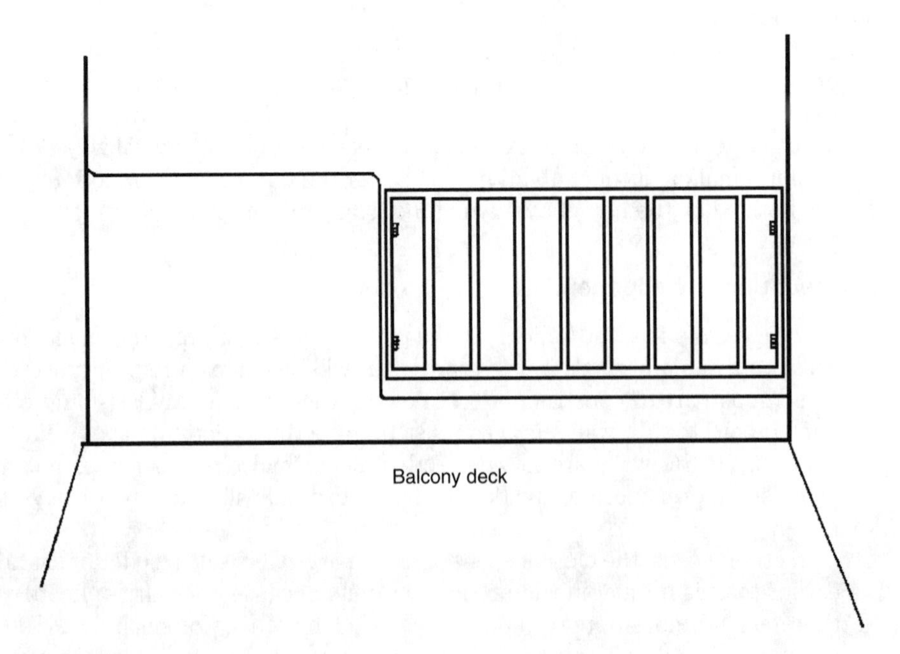

Figure 28.7

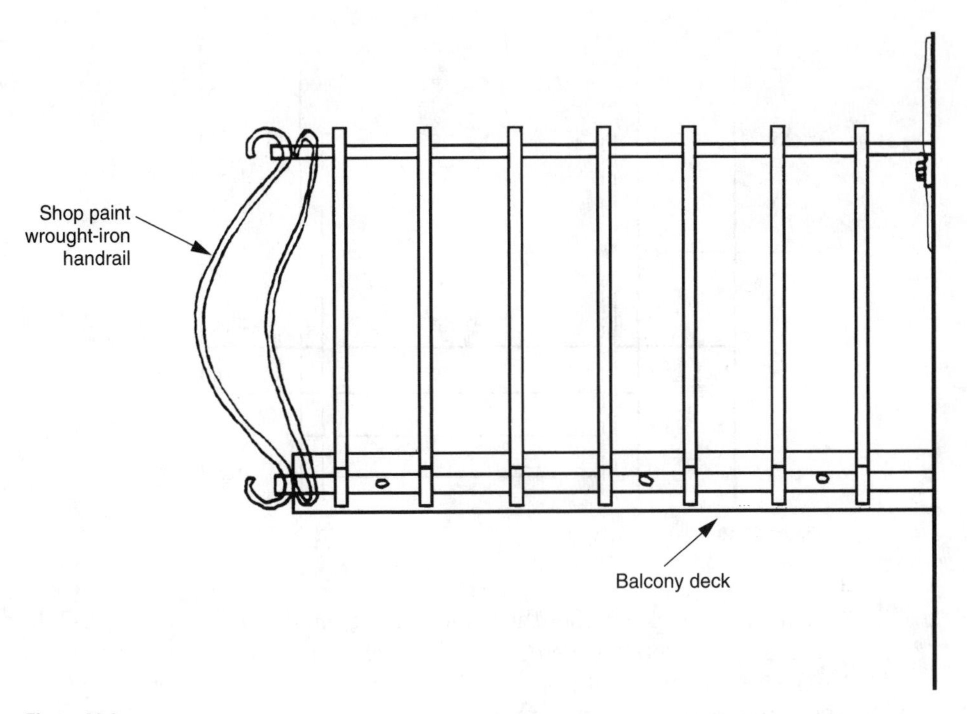

Figure 28.8

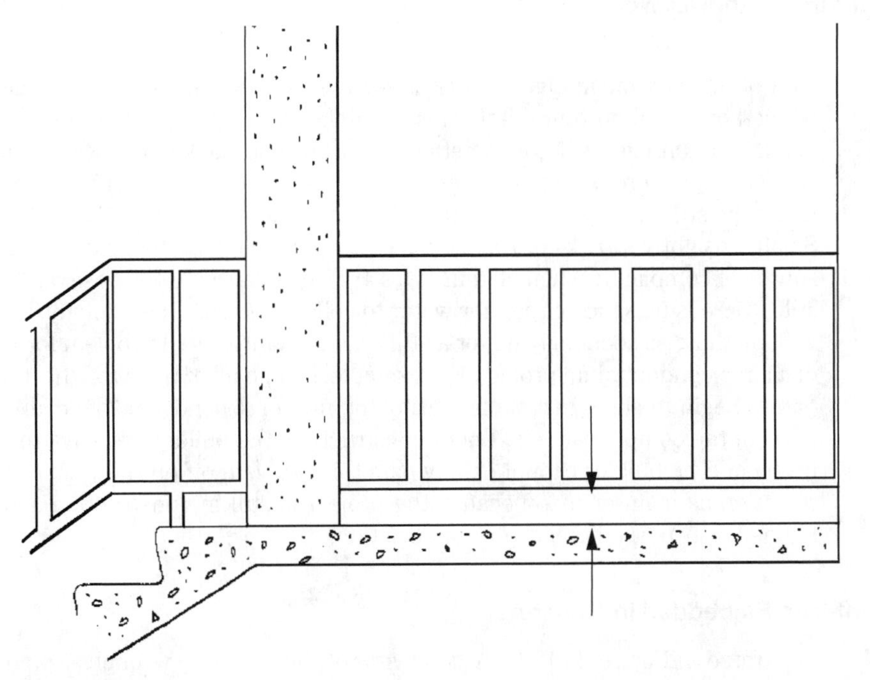

Figure 28.9

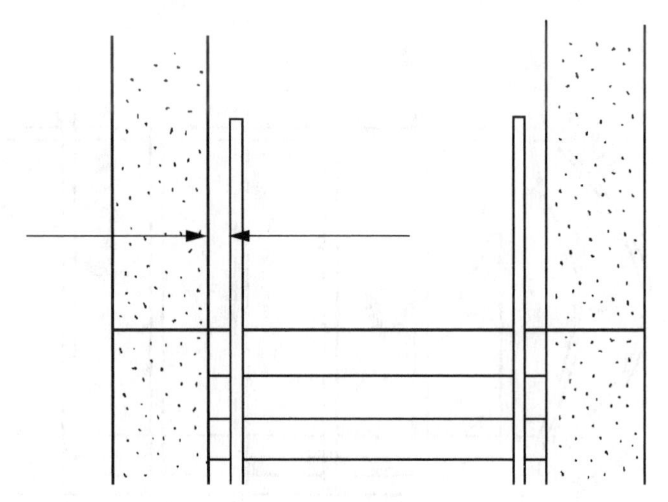

Figure 28.10

builder to check whether the painting has been thoroughly completed. If these areas are not completely covered by paint, the shop primer coat of paint will eventually wear off, the wrought-iron will begin to rust, and rust stains will begin to show up on the concrete at the base of the panels.

Wrought-Iron Handrail Welding

Illustrated in Figure 28.11 is the side view of a wrought-iron handrail. The downward view of the handrail in Figure 28.12 shows the welding joint at the vertical stile piece to the horizontal bottom rail. In this example, the stile is welded in place with two beads, one on each side of the stile. The front and back sides of the stile are not welded, and therefore open. The problem with this is that rain water gets underneath the stile through the two unwelded sides and rusts out the bottom of the stile. Small, unsightly streaks of rust stains later form below each vertical stile and are a source of complaint. When all four sides are welded around the base of each vertical stile, there is no space or gap for water to collect and rust the wrought-iron.

Something as seemingly minor as this can turn into a real problem for the builder of a large condominium project. For example, should all of the wrought-iron handrail panels begin to show rust stains during the first or second year of occupancy while parts of the complex are still under construction, the builder will have to rectify the problem. The builder cannot simply pass this problem off onto the homeowners association as maintenance because the project is still in construction and not fully handed over to the homeowners association.

Wrought-Iron Embedded in Concrete

Illustrated in Figure 28.13 is the side view of the base of a wrought-iron handrail as it enters a cored hole in a concrete porch. The handrail panel is plumbed and leveled, then pour stone is added in the cored hole around the wrought-iron base to secure the post in place. If the pour stone is rodded-off even with the concrete slab, it will shrink slightly upon drying, leaving a small crater around the base of the post. Landscape irrigation water or rainwater will collect in this small crater and rust out the base of the post if the painter did not get paint all the way down to the concrete.

A better way to finish the pour stone around the post is to shape it into a slight cone, sloping it downward away from the post (Figure 28.14). This prevents water from collecting and causing damage.

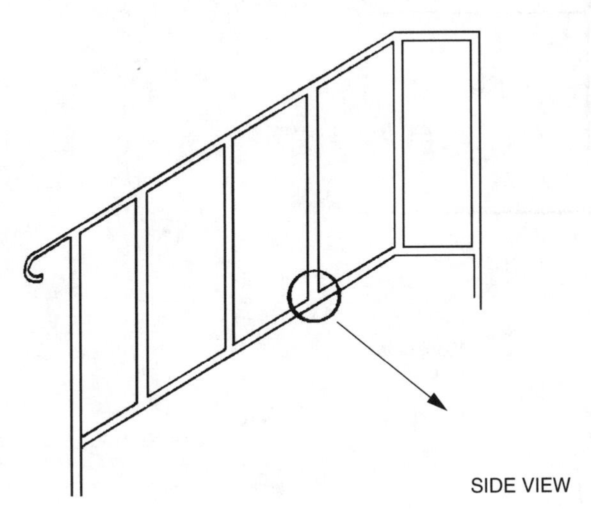

Figure 28.11

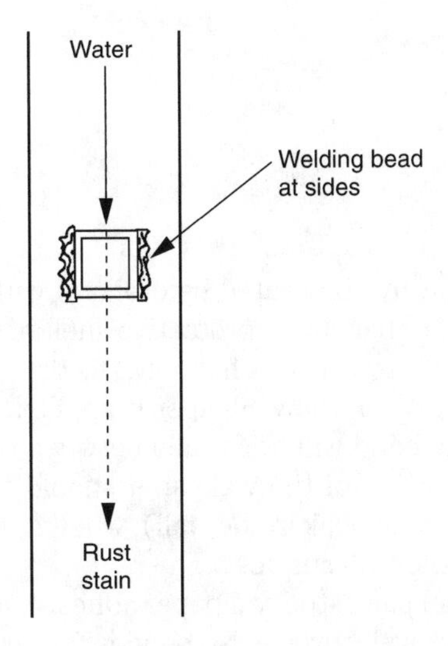

Figure 28.12

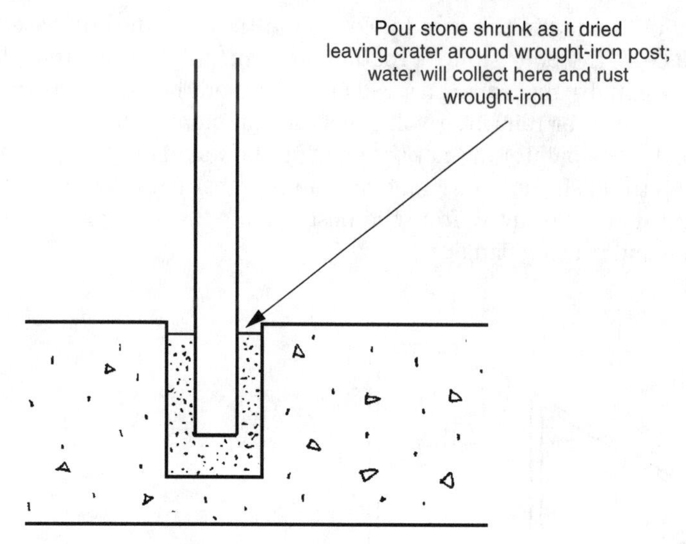

Figure 28.13

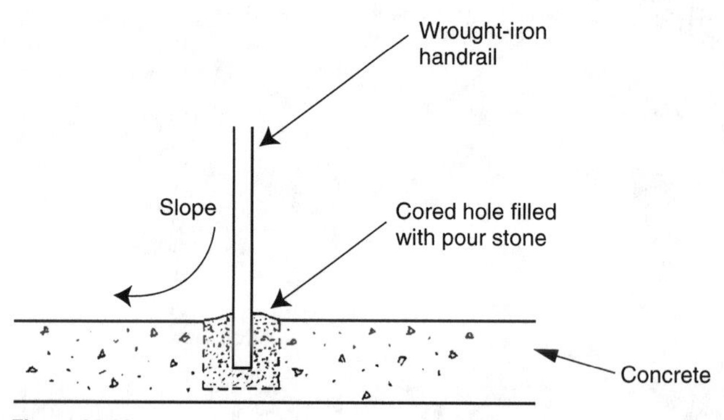

Figure 28.14

Wallpaper Sizing

Walls that are to be wallpapered should always be coated beforehand with wall sizing or enamel paint. This creates a moisture barrier or protective shell between the drywall and the wallpaper adhesive. If painted with latex flat only, the drywall retains the ability to *breathe* or soak in moisture. When the wallpaper is applied, the adhesive will soak into the drywall and a strong bond will be formed between the wallpaper and the drywall surface paper. If a portion of the wallpaper should have to be removed later for whatever reason (a plumbing leak in the wall), when pulled, it will take the drywall paper with it, damaging the wall surface.

If the wall is coated with sizing or enamel paint, the wallpaper adhesive adheres to the sizing or paint without reaching the drywall surface. No bond is formed with the drywall paper and the wallpaper can be easily peeled off with no damage to the wall.

Wallpaper Adhesive

Wallpaper adhesives vary with the type of wallpaper being installed. Some are easily removed from surrounding door casing and other wood trim with a wet sponge long after it has dried. Other adhesives are difficult to remove when dry, and therefore, must be thoroughly cleaned off woodwork and other surfaces during wallpaper installation. The builder should check to make sure the wallpaper hanger is washing the wallpaper adhesive off all surrounding surfaces.

Balcony Spindle Spacing

Every building code has a maximum allowable space between handrail spindles on exterior balcony decks. This prevents small children from falling through the spaces between the vertical spindles. This maximum space used to be 6 inches in the U.B.C., but now is 4 inches.

Often times the architect specifies a spindle spacing on the plans that uses the nominal size of the lumber rather than the actual lumber dimensions. For example, 2-inch × 2-inch spindles, spaced 6 inches on center, will result in gaps between the spindles of 4½ to 5 inches because rough-sawed wood can vary from ⅜ to ¾ of an inch under nominal size. This will not pass the 4-inch requirement during final building inspection. For a 4-inch spacing, the spindles should be set apart with a 4-inch block spacer instead of using an on-center dimension that results in spacing under or over the desired 4 inches (Figure 28.15).

Thresholds

Thresholds should be installed beneath exterior doors in such a way that the doors touch them, but do not hinder door opening and closing. In tract housing, the threshold installer often cuts the bottoms of exterior doors as part of the threshold installation. The goal should be to cut the doors so that the threshold and weatherstripping provide a weather seal, yet do not rub or bind-up the door operation.

A threshold and weatherstrip installation that is too snug a fit around exterior doors may seem like good craftsmanship at the time, but it may lead to problems. Two or three months later the builder may get a customer service complaint letter

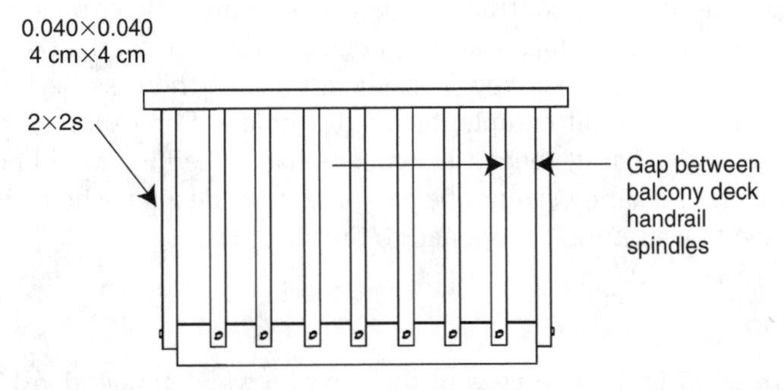

Figure 28.15

stating that the entry door rubs on the threshold or that it is difficult to turn the key to the deadbolt lock because the door fits too tightly against the weatherstripping.

Scratched Glass

Because windows are often surrounded by exterior wall surfaces that are sprayed with stucco, there is a great possibility that the glass will get scratched. This can be a major and costly problem in housing construction.

What makes this problem difficult to monitor and prevent is that three phases of construction must be properly performed to prevent window scratches. The glass must be coated thoroughly with a protective substance or covered by plastic sheet masking. If covered by a temporary coating, the plasterer must thoroughly wash down the glass with water after stucco spraying. And the window washer must use sharp razor blades and care during the final cleaning of the windows.

The most important of the three may be the final window washing. No more than two windows should be scraped with one industrial quality razor blade. A putty-knife should not be used for scraping and cleaning the window glass, even if the blade is frequently sharpened. A putty-knife blade, even when sharp, is not thin enough to get underneath the coating of residual stucco that is left on the surface of the glass. The putty-knife will push rather than lift this material across the window, which scratches the glass.

Garage Door Side Jambs

The garage door vertical side jambs are sometimes nailed into the foundation mudsill at about the same location as the lag bolt at the bottom of the garage door springs. Therefore, the lag bolt, which attaches the bottom of the garage door spring to the garage jamb, should have its hole predrilled. This will help prevent the garage door jamb from splitting from too many nails and a lag bolt in the same area (Figure 28.16).

Garage Lockups

During the middle to later stages of a tract housing project, after the garage doors are hung, garages become available for the storage of various subcontractors materials.

The builder should make sure he or she receives a key or the combination to each garage that is locked-up for subcontractors materials. Some subcontractors have the attitude that because it is their materials to safeguard, they have a right to the use of a garage with their own exclusive key. This should not be the case because it may cause problems in the overall completion of the house. For example, when the builder cannot get into a garage because someone has locked it and did not leave a key, pickup work on that garage cannot be completed. Tradesmen who need to work in the garage have to wait until it is opened to finish.

Garage Door Tension Rod

Shown in Figure 28.17 is the end view of the top of a wood garage door. The steel tension rod at the top of the garage door can be adjusted by tightening or loosening the nuts at each end of the threaded rod. This enables the top of the door to bow or straighten to match the garage door header. By adjusting this rod, the top of the garage door can fit snugly against the garage door opening. The builder should quality-control check this tension rod adjustment to match the garage door header as part of the garage door installation.

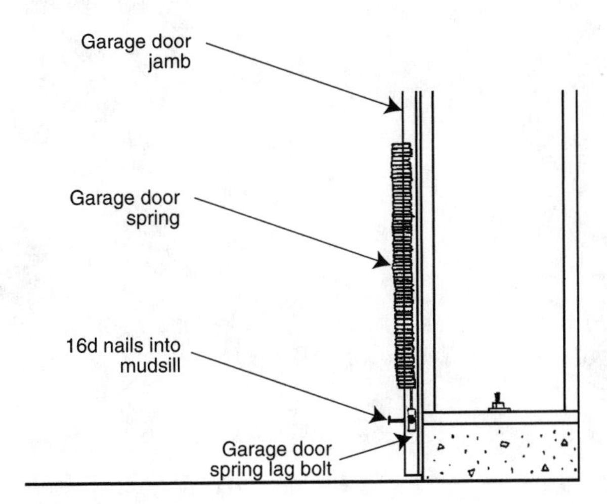

Figure 28.16

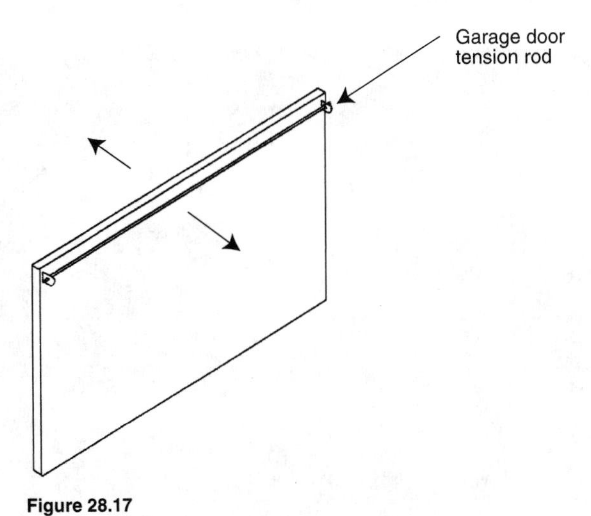

Figure 28.17

Chapter

29

Flooring

Screw Plywood Subfloors before Carpet

One of the more annoying problems the customer service person is faced with solving are floor squeaks discovered after the homebuyer has moved in. This usually involves moving furniture, pulling back carpet, fixing the floor squeak, restretching and tucking in the carpet, and putting the furniture back in place.

One way to reduce the number of potential floor squeaks is to include in the framing contract the screwing off of the plywood subfloors prior to finish flooring. This should be scheduled just after the box-out and sweep following the finish trades completion. At this point the floors are clean and free of debris. Special screw guns with rapid loading rolls of screws and long barrels that allow operation without having to bend over enabling the entire subfloor sections to be screwed off quickly and easily.

Subfloor Corrections

Some flooring companies have a quality-control person who goes through the units prior to the start of the flooring installation and marks the subfloors for needed corrections. These corrections can include: high spots, due to beams that are upwardly bowed; low spots, due to beams or joists that bow downward; soft areas in the plywood, caused by something being dropped on the subfloor during construction; and small holes in the plywood, also caused by something being dropped. The quality-control person typically marks the areas that need repair with red or black crayon keel or with upside-down cans of spray paint. A list of the areas is also sometimes made and given to the builder. The quality-control check can also be done by the builder.

The problem is that the soon-to-be new homebuyers, near the end of the construction, get anxious to move into their new homes and consequently come out to the job site on weekends to see how close the construction is to being completed. If subfloor corrections are written all over the floors, the homebuyers will make a mental or written note of these floor humps, low spots, and soft areas, and check that they were all corrected during the walk-through.

Sometimes a subfloor correction can be softened or toned down by simply grinding or sanding down a high, uneven plywood joint, rather than cutting out the subfloor and planing down the upwardly bowed beam. After padding and carpet, the floor hump is barely noticeable if you did not know about it prior to flooring. The point is not to tip-off new homebuyers about marginal subfloor problems that can be repaired to most peoples satisfaction without treating each one as a radical, major repair. This cannot be done if every minor subfloor problem is circled and labeled with red spray paint for everyone to see. The new homebuyer interprets the circled and labeled area as a definite problem that needs to be fixed rather than a subfloor area to be looked at by the builder and possibly written off as a nonproblem depending upon the type of flooring going in.

A better method is to have the flooring quality-control person make a written list that describes the problem and its location. If the builder cannot find the problem from the written list, it is probably not that major to begin with, and certainly not worth telegraphing to every homebuyer who comes to the job site.

Grind Down High Spots for Wood Flooring

The builder should make sure that any finish wood flooring installation also includes the grinding down of plywood subfloor high spots before the flooring is laid. This should be part of the normal floor preparation prior to installing wood plank or parquet finish flooring.

Grinding before installing wood flooring accomplishes two things: minor differences in the subfloor can be corrected and major humps in the floor can be identified and fixed. To fix the humps, the builder has to remove the plywood and shave the bowed joist or beam. It is much better to grind the minor high spots and identify and fix the major problems than to install the floor, only to face these difficulties later.

Flooring and Toilets

The builder should work out a policy ahead of time concerning conflicts between the setting of toilets in bathrooms and the installation of hard surface flooring. When homebuyer selections are made with enough time prior to the completion of construction, then hard surface flooring, such as wood plank, ceramic tile, or clay paver tile, can be installed in bathrooms within the normal sequence of construction. In this case, the toilets are not set by the plumber during the finish plumbing work, but are held back to be installed after the flooring is down. This eliminates the flooring people from having to remove a toilet to install their flooring. It also eliminates the plumber from having to reinstall a toilet after it has been removed by the flooring installers.

In inventory units, however, that have not been sold at the completion of construction, thus flooring selections have not been made as yet, the building inspector will usually require that toilets be in place for the final building inspection. This is where the flooring versus toilets issue needs to be resolved upfront.

The best policy is to have the flooring people remove the toilets that are in the way during flooring installation, then have the plumber come back and reset the toilets when the flooring is complete. Because the plumber is going to be responsible for the warranty work on leaky toilets, it is better to have him or her reset the toilets rather than have the responsibility for an incorrectly installed toilet shared between the plumber and the flooring contractor.

This must be spelled out in both the flooring and plumbing contracts and some provision must be made for paying the plumber for coming back out a second time to reinstall a toilet that has been removed for hard surface flooring.

Floor Tile Layout in Bathrooms

Illustrated in Figure 29.1 is the plan view of ceramic flooring tiles installed in a small bathroom. In this example, the tile installer started the installation with full field tiles at the back wall of the bathroom, behind the toilet. This resulted in a cut row of tile at the bathroom doorway. In this particular case, this meant that the dark gray clay color of the inside of the tile was exposed because a thin Berber carpet, which was

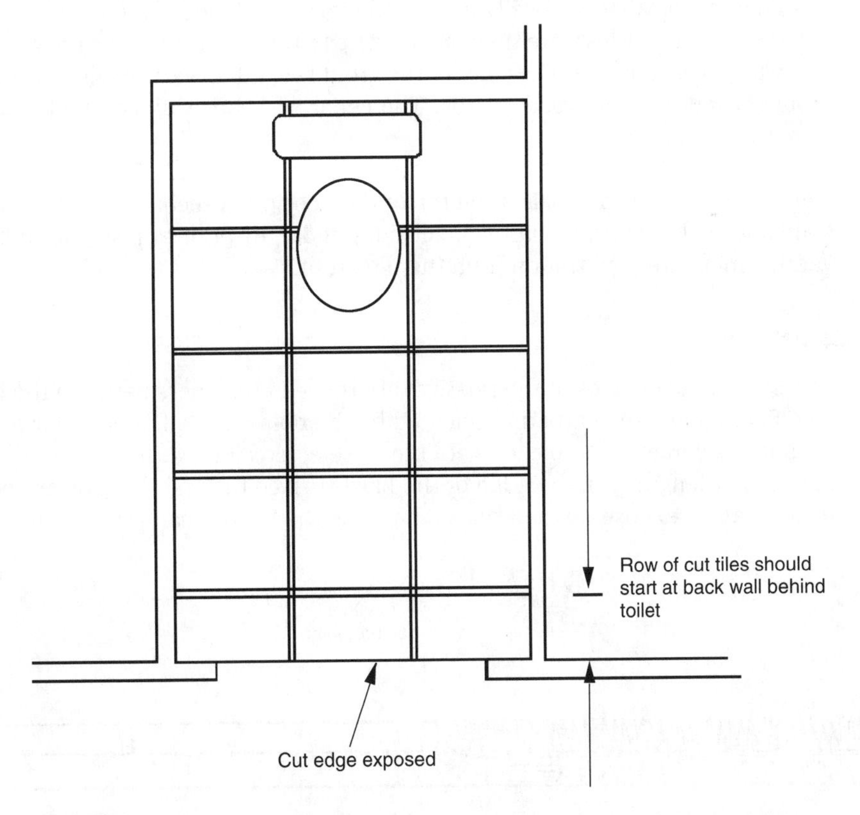

Figure 29.1

at a slightly lower elevation than the floor tile, meet the tile at the doorway. If the cut row of tiles was placed at the back wall of the bathroom instead, a full field tile would have ended up at the bathroom doorway, resulting in the colored and finished factory tile edge being exposed at the doorway.

The builder should go over the layout of floor tiles with the flooring subcontractor and installers prior to the start of the hard surface flooring in an effort to resolve such potential problems.

Floor Tile Edges

The outside edges of ceramic and clay floor tiles should be filled with mortar or grout where the edges of the tiles meet carpeting. If an empty space or gap is left underneath these tile edges that meet carpeting in a walking area, the edges can be easily cracked by foot pressure (Figure 29.2).

Don't Grout at Wood Siding

Exposed raw wood siding boards should not come in contact with wet concrete or ceramic floor tile grout. The moisture will seep up the wood and stain the bottom of the siding a gray or bleached color. In one instance, cedar tongue and groove siding boards were installed vertically in several elevator lobbies of a condominium building. Ceramic floor tiles were then installed up to the siding, with the tile grout in contact with the woodwork. The color of the grout seeped upward as the moisture of the grout soaked into the wood boards. The cedar was thus stained one to two inches from the bottom.

This mistake could have been avoided by installing the tile flooring first, then installing the cedar siding boards on top of the flooring. A second solution would be to stain and seal the siding boards on all sides and ends prior to installation, thus preventing moisture penetration from the tile grout.

Tile and Carpet Joints

The builder should look at the relative difference in thickness between the hard surface flooring and the carpeting selected by the homebuyer. If a hard surface material, such as ceramic, marble, or slate tile, is selected along with a type of carpet that has a short length nap, the edge of the hard surface tile and the mortar bed underneath may be exposed. When this is the case, the builder can then discuss ahead of

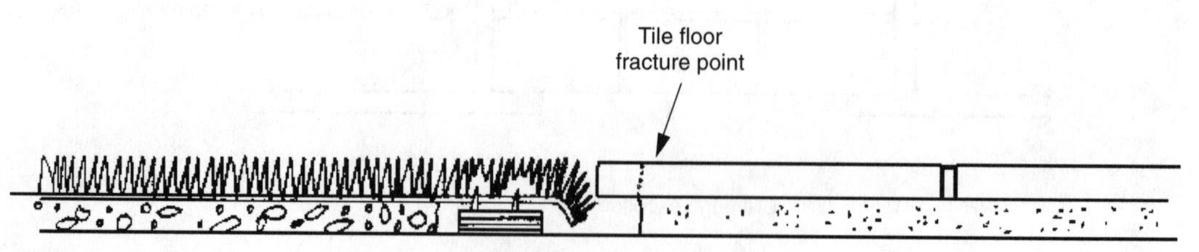

Figure 29.2

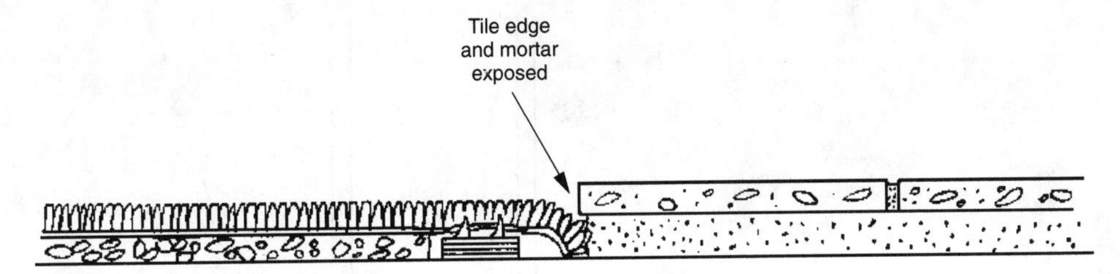

Figure 29.3

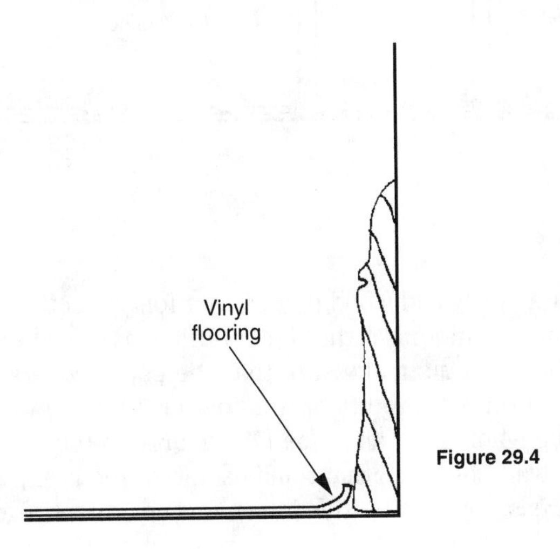

Figure 29.4

time with the flooring company the possibility of using tiles with finished edges at the hard surface to carpet joint and coloring the mortar at this location to match the hard surface material (Figure 29.3).

Trim Vinyl at Baseboard

The builder should check that the vinyl flooring is trimmed correctly along the baseboard and cabinets. The edges of the vinyl should not turn or bend up against the baseboard or cabinets (Figure 29.4). If the vinyl is not trimmed correctly, the baseshoe cannot be placed tightly on top of the floor. The baseshoe installer must then take out his utility knife and trim the vinyl before the baseshoe can go down.

Ramp between Tile and Marble

Illustrated in Figure 29.5 is a floor level view of a doorway between an entry and a family room. The marble tiles in the entry have been laid over a cement floated base and are, therefore, higher than the ceramic tile laid in the family room. When this combination occurs, the ceramic tile in the doorway should ramp at an angle upward toward the marble. In order to do this, the flooring must be scheduled with marble first and tile second.

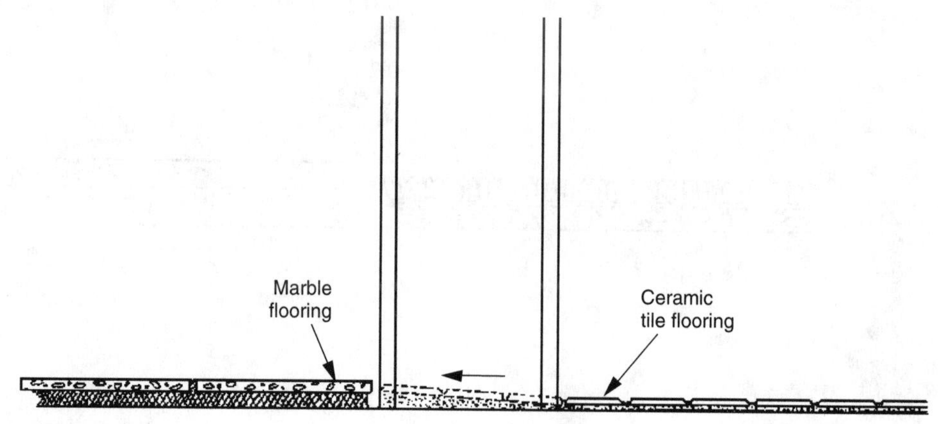

Figure 29.5

Paper over Wood Flooring

When paper is used to cover a newly laid wood floor, either for protection during the remainder of construction or as runners so that foot traffic stays on the paper and not on the flooring, the builder should make sure that the paper covers the entire floor. If the paper is rolled out on the wood floor in 2-foot or 3-foot wide runners or if it is spread out to cover the whole floor and taped down around the perimeter but is short a few inches at each wall, the wood that remains uncovered is exposed to the sunlight, while the wood areas covered by the paper are not. After a few weeks, when the paper is removed, the wood floor will be different colors wherever the non-papered areas and the papered areas meet.

To avoid having to possibly rip out and replace a wood floor, the builder should completely cover the floor from baseboard to baseboard, taping it in place around all edges with adhesive tape that will not leave glue behind on the floor when the tape is pulled up.

Caulk Vinyl around Doors

When vinyl flooring is laid in kitchens and bathrooms, the gaps between the vinyl edges and the wall baseboard are covered with wood baseshoe trim. This baseshoe cannot be installed around door jambs and door casing. These areas should therefore be caulked by the vinyl flooring installer (Figure 29.6).

Hard Surface Flooring and Appliances

There are three ways to handle hard surface flooring underneath kitchen appliances. The first way is to install flooring under the appliance, all the way back to the rear wall. This enables the appliance to be placed without having to adjust the front and back legs for leveling. The second way is to extend the flooring only far enough into the opening to end up underneath the front portion of the appliance. With this method, the appliance legs must be adjusted to level the appliance to the different

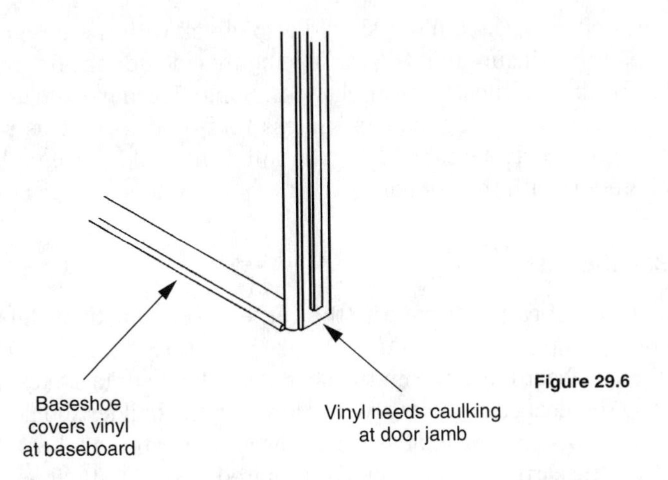

Figure 29.6

flooring elevations. The third way is to install all of the kitchen appliances, and then simply install the flooring up to the appliances.

For hard surface flooring materials, such as wood, ceramic tile, clay paver tiles, slate, or marble tiles, the first method is the one that should be used for several reasons. The thicknesses of hard surface flooring materials can range from ½ inch to 1½ inches or more, depending on the thickness of the material itself and the thickness of any required underlayment material. If the second method of installation is used, with the flooring only going part way under the appliance, the difference in adjustment between the front and rear legs of the appliance may not be enough to level the appliance. If the third method is used, which is to install the flooring up to the appliance, the dishwasher may be wedged into place between the flooring and the kitchen countertop, preventing the dishwasher from being able to be pulled out for repairs. With methods two or three, there is the chance that the raw subfloor may be exposed at the sides or front of the appliance. The first method is the best alternative because it provides a continuous level surface for installing the appliances without having to adjust the appliance legs and without the possibility of the subfloor showing.

A few words of caution, however, are needed involving hard surface flooring in kitchens. In methods one and two, the appliances cannot be in place when the flooring is laid because the flooring goes under the appliances and the appliances would be in the way. In unsold inventory units, where the buyers will have an option to select different types of flooring when the units are sold later, the appliances should be left in their boxes and placed in the living room or other room adjacent to the kitchen with the units obviously kept locked. When the units are sold and the kitchen flooring installed, the plumbers come back and install the appliances and the move-in kit of faucet aerators, shower heads, and so forth. The builder should inform all of the workers that are in the unit after the flooring installation that the unboxed appliances are not to be dragged over the new hard surface flooring. Tile, marble, and especially wood flooring can be scratched in this way. These workers include the carpet installers, the finish cleanup crew, and the plumbers.

If method one is chosen, the builder should also check with the design center or the flooring contractor to insure that the yardage figures include the flooring that extends all the way back underneath the appliances. Some flooring estimators will calculate only to the front of the appliances, unless told otherwise. This results in a flooring material shortage at the time of installation if the builder wants the flooring all the way back underneath the appliances.

Carpet Tack-Strip and Baseboard

During the installation of carpet tack-strip, the carpet-layer sometimes hits the baseboard with his or her hammer, taking out a chunk of wood (Figure 29.7). One or two such divets in the baseboard falls within the acceptable limits of mistakes the painter can spackle during the final paint pickup work. However, the builder should check that if the carpet-layers are leaving behind baseboard divets in greater numbers than one or two per house, then the carpet-layers should be required to spackle their own mistakes.

If the carpet-layers do repair their own baseboard divets, more than one application of spackle is often required because the first coat will shrink upon drying, leaving a flat or slightly hollow spot at the repair.

Tack-Strip under Cabinets

The tack-strip should always be placed at the very edge of carpet, even when the edge of the carpet is beneath the toe space of a cabinet. Figure 29.8 illustrates a mistake involving the placement of carpet tack-strip in front of a bathroom lavatory cabinet. Carpet-layers have a special tool for installing tack-strip in hard-to-reach areas. This tool should have been used in this instance.

If the builder allows the tack-strip to be installed outside and in front of the toe space (much easier and quicker for the carpet-layers), the sharp tack-strip nails end up directly beneath the location where bare feet will be standing. Unless the carpet is thick, the tack-strip nails have enough length to reach bare feet when the carpet

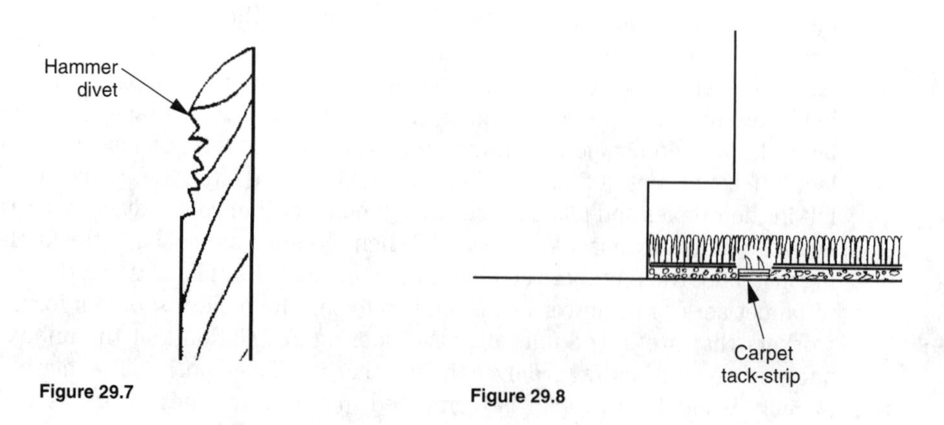

Figure 29.7

Figure 29.8

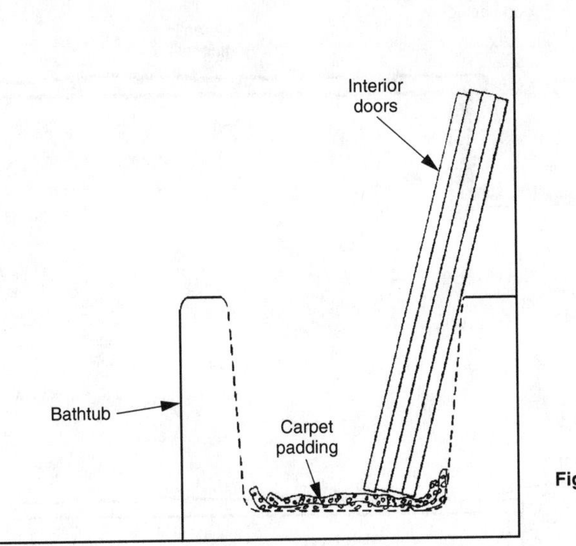

Figure 29.9

nap is pushed down with body weight. The correct location for tack-strip is under cabinets, as shown in Figure 29.8.

Doors Placed in Bathtubs during Carpet-Laying

In two-story houses and condominiums that are at least a floor away from the garage, the carpet installers are always looking for a place to put interior doors so that they are out of the way during the carpet-laying without having to carry them to and from the garage. The solution for them is to stack the interior doors within a nearby bathtub.

If the builder allows the carpet-layers to stack the doors in the tubs, the bottom of the bathtubs should at least be protected by placing the doors on top of carpet padding. This will prevent the doors from scratching the bottom surface of the bathtubs (Figure 29.9).

Wood Nosing at Landings

Illustrated in Figure 29.10 is the top floor or landing of a flight of stairs that has tile as the floor covering. In this example, some type of wood or other flooring material should be used at the stair tread as nosing to give the carpet something to tuck into. Wood nosing makes a good transition from carpet to tile. Tile is usually too brittle and may break if used as the nosing, especially if projected out past the stair riser.

Another solution is to continue the hard surface tile flooring down the top stair riser face (Figure 29.11). The carpet on the top stair tread can then tuck into the vertical tile on the riser.

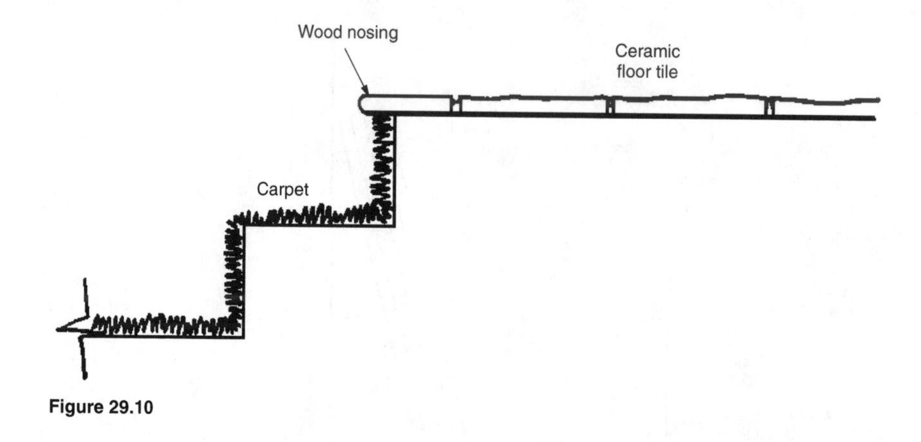

Figure 29.10

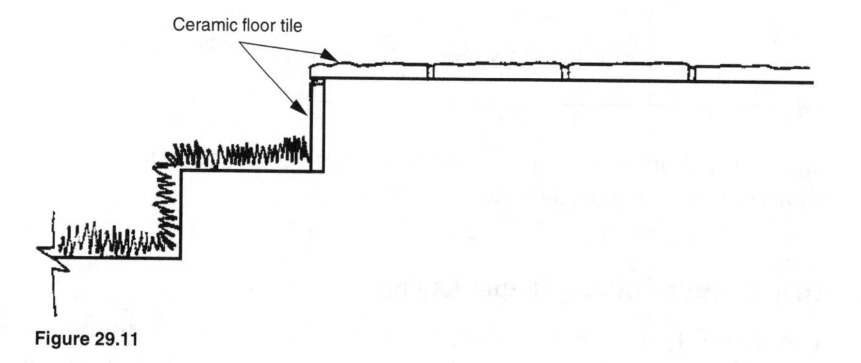

Figure 29.11

Metal Trim at Stair Nosing

Illustrated in Figure 29.12 is the side view of a plywood subfloor and a carpeted stairway below. In this example, the subfloor—the kitchen floor—is covered with vinyl. The problem is what to do with the plywood stair tread nosing, which must project out 1 inch to match the rest of the stair treads. On a project in which many of the homebuyers select wood flooring, ceramic tile, marble tile, or clay paver tile, this particular condition of vinyl meeting carpet at a stair tread can be easily overlooked. However, eventually some homebuyer will choose vinyl flooring in the kitchen. Then the vinyl installer will lay the vinyl in the kitchen, the carpet installer will lay the carpet throughout the house including the stairs, and no one will have considered this one exposed plywood nosing.

To rectify this situation, a finished metal trim piece (also shown in Figure 29.12) can be added to fit around the stair nosing and cover the top of the step, the edge, and the bottom. The builder should check that the vinyl installer brings this metal trim to the job site at the time of the vinyl installation.

Ramp Berber Carpet to Tile

Figure 29.13 illustrates the difference in thickness between Berber carpet and a hard surface flooring, such as ceramic tile or marble, that was installed over cement board

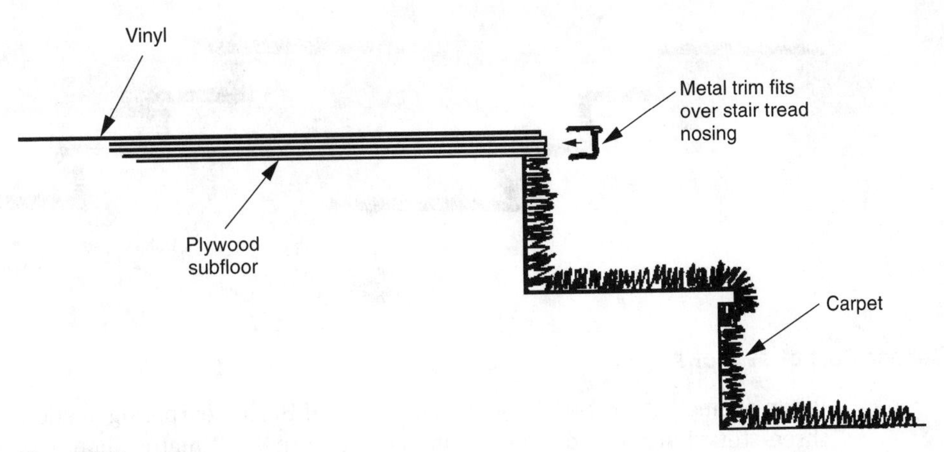

Figure 29.12

underlayment, on top of plywood subflooring. When there is a transition at entries, kitchens, and bathrooms between Berber carpet (or any other thin knap carpet) and hard surface flooring, the builder should require the carpet-layers to install a sloping ramp using redwood roofing shingles. This softens the elevation difference between the two flooring materials and prevents an almost certain call-back and repair later when the homebuyer complains about this height differential.

Carpet at Stairs

The two most common methods of carpeting stair steps are shown in Figure 29.14. The waterfall carpeting technique allows the carpet to fall over the tread, extending in one piece down the riser, usually at an inward angle. With the upholstering technique, the carpet tucks around and underneath a stair tread nosing, with a separate piece of carpet cut and attached to the vertically straight stair riser.

Upholstered is the better looking of the two methods; however, the installation usually costs a little more for the extra work involved. Waterfall is typically used in apartments or less expensive houses. When deciding which technique to use, there are a few issues to consider. First, if the upholstered technique is the look desired, the stair steps must be built with nosings (1 inch standard) at the front of each tread and vertically straight risers. If the waterfall technique is the method to be used, then the stairs should be built with no nosings and angled risers. A second consideration is the recognition that two methods are available for carpeting stair steps and that the builder should choose one method before the start of construction. A third consideration is for the builder to check that the carpet installers are actually doing an upholstered application at the stairs when that is the method purchased and contracted.

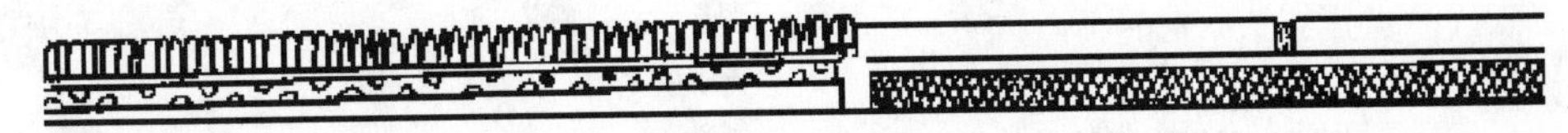

Figure 29.13

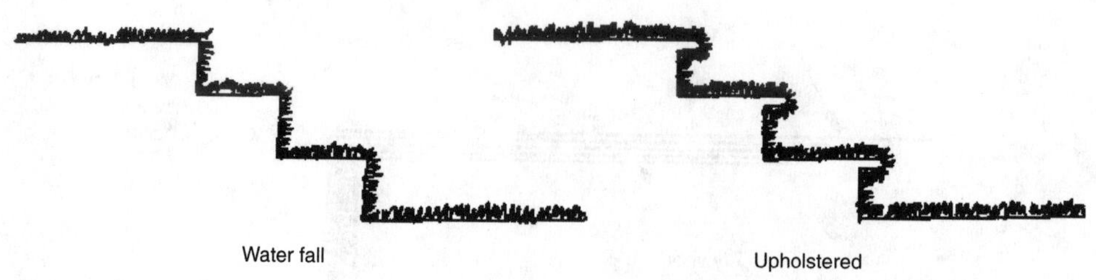

Figure 29.14

Berber Carpet at Stairs

When estimating the number of square yards of Berber carpeting needed for two- or three-story houses and condominiums, the carpet estimator should consider the grain pattern of Berber carpet when figuring stair coverage. Because the grain of the carpet runs lengthwise with the roll, a 4-foot wide piece for the 4-foot wide stairs must be cut parallel with the grain and lengthwise with the 12-foot wide roll, otherwise the carpet will bird's mouth at the stair nosing (Figure 29.15). If the carpet is cut and installed with the grain running lengthwise down the stairs, the carpet will not bird's mouth at the stair tread nosing (Figure 29.16).

The carpet estimator cannot therefore simply add up the square footage of the various rooms in the house, and then throw in the square footage of the stairs be-

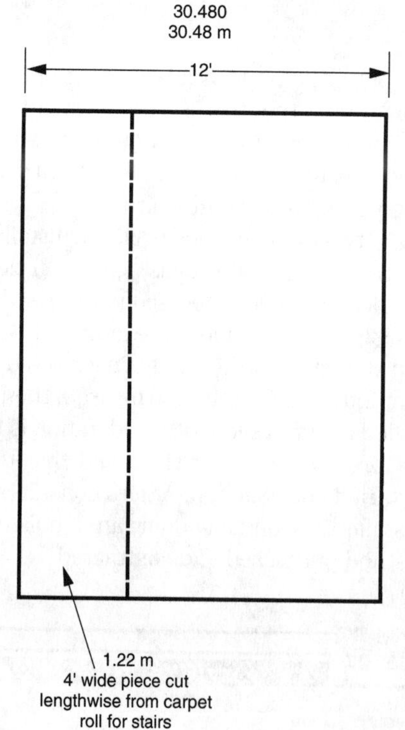

Figure 29.15

cause this method does not consider the carpet grain pattern. After the carpet is correctly cut out of the roll and installed by the carpet-layer, the total yardage needed to complete the house will come up short. Scraps and leftover pieces of Berber cannot be figured into the yardage estimate like other types of carpeting, unless the grain patterns of the scraps and leftovers happen to work out with the width and length of the stairways.

Sculptured Carpet at Stair Steps

On one particular project, the new homebuyer saw in a decorating magazine a carpet which had V-grooves shaved out of the carpet in an 18-inch square diamond pattern. This buyer had the new carpet in his living room cut slightly larger than needed, taken to a specialist who cut the carpet into squares, shaved the edges of the carpet on all four sides of each square, and then reassembled the squares with a new carpet backing to form a sculptured carpet to match the example in the magazine.

Because this procedure was new to everyone—the flooring contractor, the builder, and the homebuyer—it was overlooked which became apparent after the sculptured carpet was installed. The carpet was beautiful when laid flat on the floor, but did not look good when bent around two stair steps at a landing, as shown in Figure 29.17. The 180-degree bend around the stair step nosing exposed the carpet backing through the shorter knap at the shaved V-grooves that expanded and opened up at each step. This was not anticipated by anyone beforehand. In hindsight, some transition or border of standard carpet should have been installed at the stair steps, with the sculptured carpet ending short of the stairway.

Patterned Carpet and Seams

Shown in Figure 29.18 is the plan view of two adjacent bedrooms, with doorways that are at 45-degree angles with each room. If a woven, patterned Berber type carpet is selected by the homebuyer, the builder should check with the flooring contractor that he or she estimated enough extra yardage into these bedrooms so that the patterns can be lined up and joined at the doorway seams.

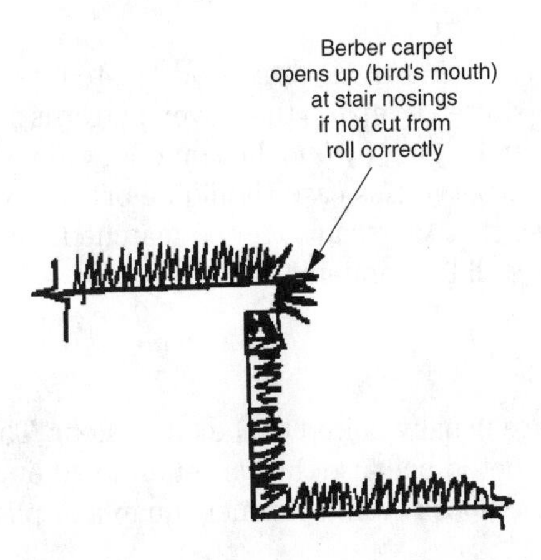

Figure 29.16

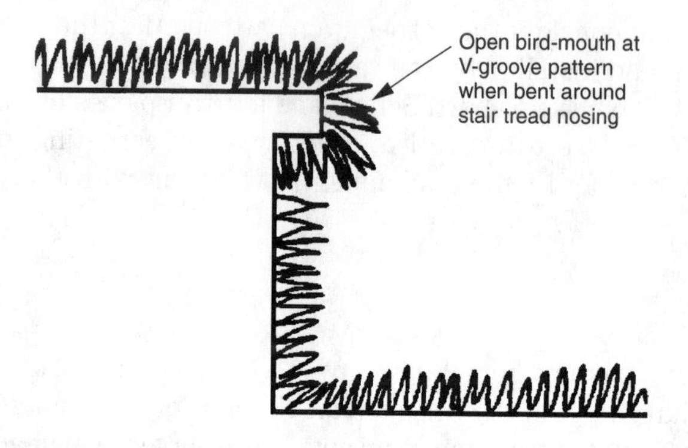

Figure 29.17

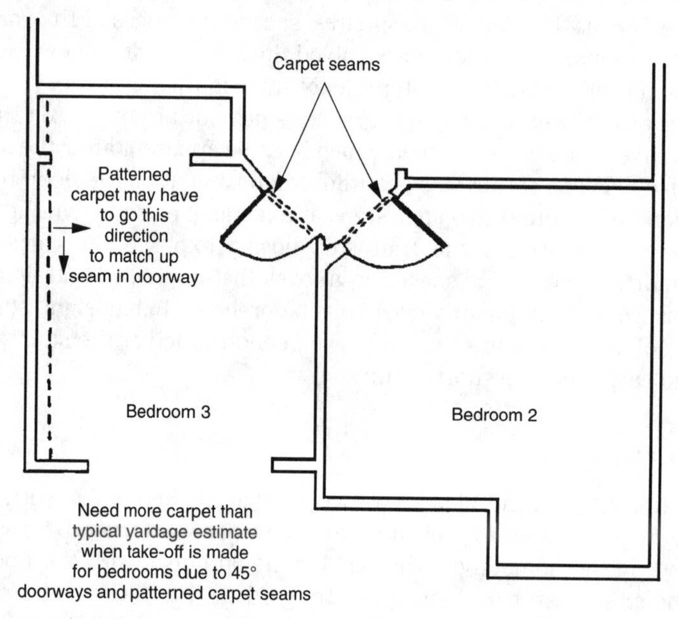

Figure 29.18

If the carpet yardage is estimated at the normal yardage adequate for other types of carpet, when the Berber carpet is shifted to match the woven patterns at the doorway to hallway seams, the carpet may be long by 6 inches on one wall and short by 3 inches on the opposite wall. The carpet in this case should be ordered with a foot extra width at each bedroom wall so that the seams can be matched and still have enough carpet left at each bedroom wall for trimming.

Metal Circular Stairs

Prefabricated metal circular stairs are usually bolted to the lower floor. This is not a problem for lower floors that are carpeted because the carpet and padding will hide the metal flanges and bolts. When the floor is wood parquet squares or planks, however, the wood flooring is too thin to cover the flanges and bolts (Figure 29.19). In this case, the builder should try to devise some way of recessing the lower floor subfloor at the circular stairs so that the flanges and bolts end up below the floor level. The metal staircase must then be built with a first step that is slightly higher than the other steps so that when the staircase is completed and the flooring is installed, the rise for each step will be the same.

By lowering the bottom flanges and bolts of the metal staircase into the floor, the wood flooring on top can be cut out and fit around the staircase stile post instead (Figure 29.20). The attachment flange plate and bolts are then hidden from view underneath the wood flooring.

Check Toilets for Cracks

Toilets should be checked after they are initially installed by the plumbers, and then checked again after carpet tack-strip installation. Occasionally, a carpet tack-strip installer, because of the lack of and awkwardness of the working space around a toilet, will miss the tack-strip nail with the hammer and hit the base of the toilet instead. If a cracked toilet is not discovered until after the carpet is installed in the bathroom, the carpet-layer will probably say that the toilet was already cracked before he or she arrived. When the plumber is questioned about the toilet, he or she will naturally say that it was not installed cracked.

To prevent this problem, the builder should insist that a special nailing bar, called a *stair tool*, be used for tack-strip installation around the base of the toilet. The builder should also check the toilets after both trades are finished.

Baseboard Touchup

Final drywall and paint pickup are completed before the installation of carpet. When laying carpeting, however, the baseboard and lower wall sections are marked and scuffed up. The carpet edge trimming tool often leaves a pencil-like line on the baseboard. To fix this problem, the builder should include in the painting contract an additional minimum of 2 hours touchup on the baseboard after carpet installation. These 2 hours should be designated for baseboard touchup, as any other final paint touchup should be completed as part of the final paint pickup phase before carpet laying.

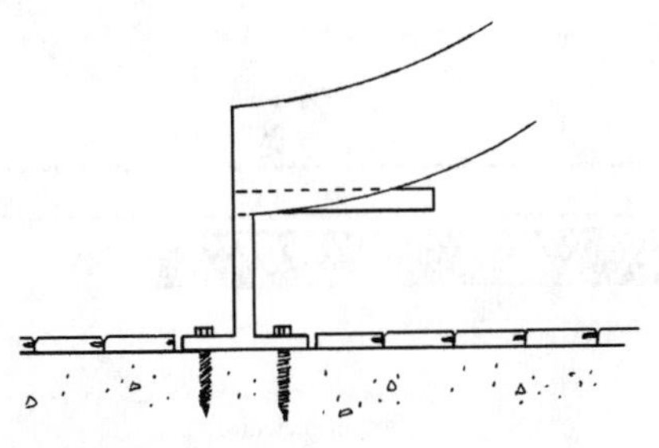

Figure 29.19

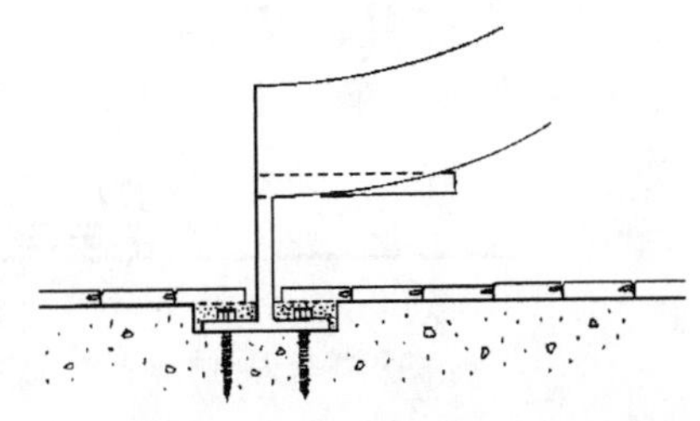

Figure 29.20

Don't Scribe Cut Vinyl around Toilets

A practice that should be avoided in production housing is the scribe cutting of bathroom vinyl flooring around toilets. This is often done in an attempt to avoid the cost of removing and resetting toilets. The correct way to install vinyl flooring in a bathroom is to remove the toilet, install the vinyl flooring with a hole cut out for the toilet closet ring, and then have the plumber come back and reinstall the toilet. This eliminates the chance of the vinyl edges peeling back and curling up from the toilet base later. It also eliminates the vinyl seam that must be made behind the toilet and sloppy cutting of the vinyl around the toilets. Doing it the right way by pulling the toilets makes the vinyl flooring installation faster and cleaner.

Asking the vinyl installer to cut around toilets is another example of trying to cut costs in areas of construction that should not even be considered in the first place. Part of installing vinyl flooring in bathrooms is the cost of pulling and reinstalling of toilets. Asking the vinyl installer to cut out around toilets in order to save $30 or $50 per toilet should not be an option.

Check Composition Board at Tile

Illustrated in Figure 29.21 is the side view of cement composition board. It is typically used as underlayment under ceramic tile flooring or a tile or marble fireplace hearth and on top of plywood subflooring. The builder should check that the hard surface tile flooring installers do not leave this underlayment board projecting slightly beyond the tile flooring so that it interferes with the carpeting installation. This occurs when the tile flooring installers original guesstimate of the dimensions for the underlayment board ends up slightly larger than the tile installation, after all the tiles are placed and grouted. If the underlayment extends beyond the flooring a ¼ inch or ½ inch, the hard surface flooring installer should chip this excess amount off rather than ignore the problem and leave it for the next tradesman to fix.

This can be a real problem for the builder when the hard surface flooring installers and carpet-layers are two different subcontractors. The hard surface installers will

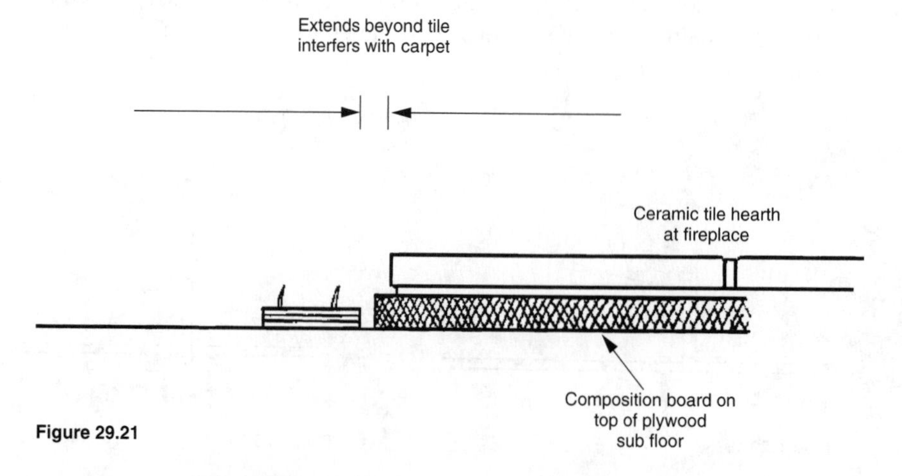

Figure 29.21

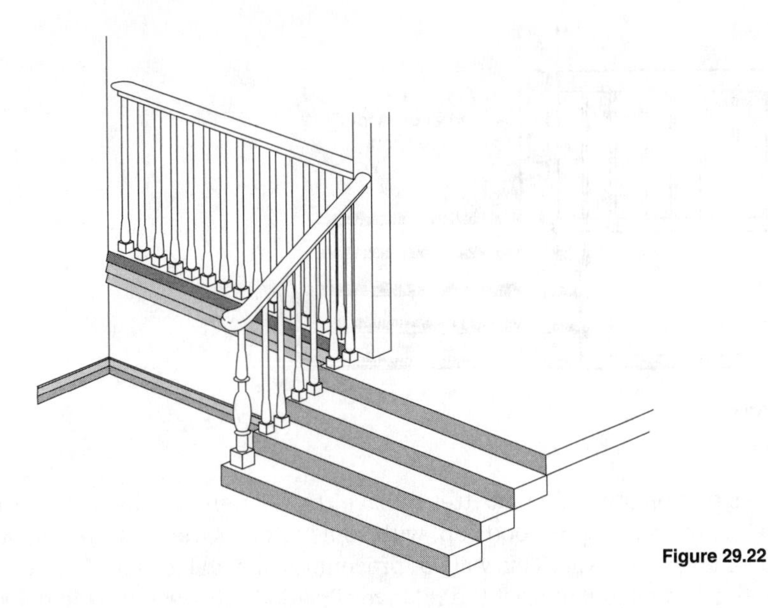

Figure 29.22

not look out for the interests of the carpet-layers, because they work for different companies.

Handrail Spindles and Carpet

Figure 29.22 shows a wood handrail with wood spindles that are nailed directly to the stair tread plywood. This design requires the carpet on these stairs to be cut and seamed around each individual handrail spindle. Not only does this design not look good, but it is difficult and time-consuming to cut and seam the carpet around possibly dozens of wood spindles, and have the seams come out unnoticeable as they should.

A better method is to have the handrail spindles rest upon a wood cap border. This not only looks better, but provides the straightedge of a wood board into which the carpet can be tucked rather than tucking it into a number of spindles (Figure 29.23).

Floor Tile at Handrail Cap

The entry tile floor, pictured in Figure 29.24, ended up slightly higher than the bottom wood cap for the wood handrail partition at a sunken living room. In this particular example, not only did the tile grout crack as expected at the joint with the wood cap, but for some reason, the wall framing directly underneath the handrail partition either settled slightly, the wood cap dried out and shrunk, or the entry floor raised slightly up, resulting in the edge of the tile grout being higher than the wood cap. When looking at the entry tile floor from the sunken living room, you could see the edge of gray-colored tile grout sticking up an ⅛ of an inch above the wood cap. The attempt to have the tile floor die into the wood cap at the same level did not work.

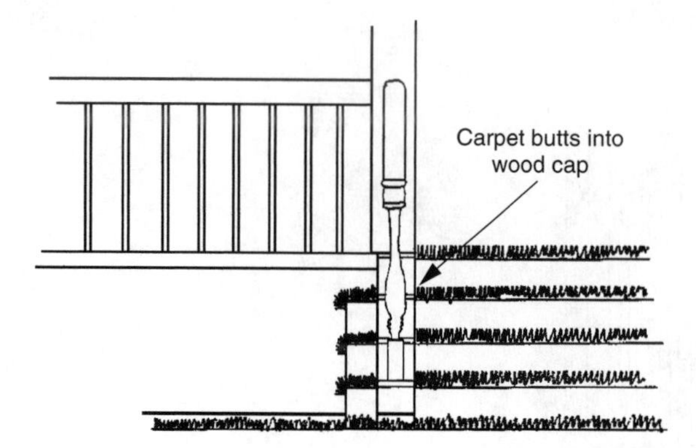

Figure 29.23

In this situation, it would be better to have the floor and the handrail designed so that the tile butts into the wood cap, with some leftover wood cap reveal above the finish floor level of the tile. This will not prevent the inevitable crack between the tile grout and the wood, but at least the relative elevations between the tile floor and the handrail floor cap do not have to be perfectly even, as in this case where the builder wanted a flat transition from tile to wood cap.

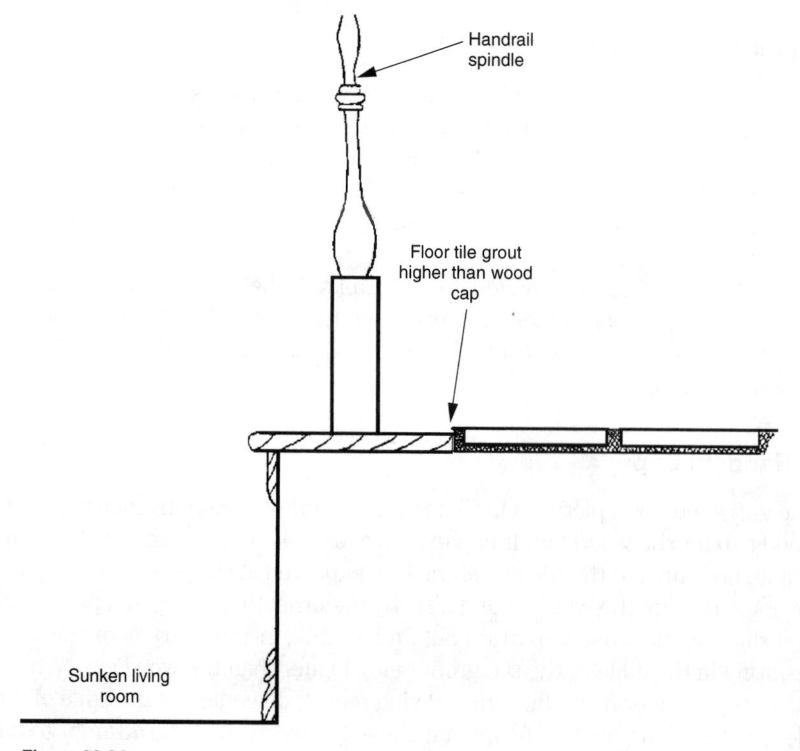

Figure 29.24

Index

Illustrations are in **boldface.**

D

E

F

G